Surface Design: Applications in Bioscience and Nanotechnology

Edited by
Renate Förch, Holger Schönherr, and A. Tobias A. Jenkins

Related Titles

H.-J. Butt, K. Graf, M. Kappl

Physics and Chemistry of Interfaces

ISBN: 978-3-527-40413-1

C. S. S. R. Kumar (Ed.)

Nanotechnologies for the Life Sciences

10 Volume Set

ISBN: 978-3-527-31301-3

K. Wandelt (Ed.)

Surface and Interface Science

Volume 5: Nanostructures and Quantum Phenomena

ISBN: 978-3-527-40689-0

Volume 6: Solid/Liquid and Biologiacal Interfaces

ISBN: 978-3-527-40499-5

Volume 7: Applications of Surface Science

ISBN: 978-3-527-40500-8

D. S. Goodsell

Bionanotechnology

Lessons from Nature

ISBN: 978-0-471-41719-4

C.M. Niemeyer, C. A. Mirkin (Eds.)

Nanobiotechnology

Concepts, Applications and Perspectives

ISBN: 978-3-527-30658-9

Surface Design: Applications in Bioscience and Nanotechnology

Edited by
Renate Förch, Holger Schönherr, and A. Tobias A. Jenkins

WILEY-VCH

WILEY-VCH Verlag GmbH & Co. KGaA

The Editors

Dr. Renate Förch
Max Planck Institute for Polymer Research
Mainz, Germany

Prof. Dr. Holger Schönherr
University of Siegen
Department of Physical Chemistry
Siegen, Germany

Dr. A. Tobias A. Jenkins
University of Bath
Department of Chemistry
Bath, United Kingdom

Library of Congress Card No.: applied for

British Library Cataloguing-in-Publication Data
A catalogue record for this book is available from the British Library.

Bibliographic information published by the Deutsche Nationalbibliothek
The Deutsche Nationalbibliothek lists this publication in the Deutsche Nationalbibliografie; detailed bibliographic data are available on the Internet at http://dnb.d-nb.de.

Printed in the Federal Republic of Germany
Printed on acid-free paper

Typesetting Kühn & Weyh, Satz und Medien, Freiburg
Printing Strauss GmbH, Mörlenbach
Bookbinding Litges & Dopf Buchbinderei GmbH, Heppenheim

ISBN: 978-3-527-40789-7

Foreword

Thomas S. Kuhn's theory of scientific revolution became a landmark in the intellectual history of the twentieth century. He wrote in his 1962 book '*The Structure of Scientific Revolutions*' that science was not a steady, cumulative acquisition of facts and knowledge but rather 'a series of peaceful interludes punctuated by intellectually violent revolutions.' During those revolutions, he wrote, 'one conceptual world view is replaced by another' and this takes place when, within the hereto existing paradigms, scientists cannot account for new findings any longer. Such paradigm shifts, as it has been suggested, are currently underway in polymer science. On the foundations of polymer science new fields like macromolecular nanotechnology, and biomacromolecular science are emerging and the rapidly changing intellectual circumstances and technical possibilities are rewriting the science of soft matter. This volume is a typical product reflecting this changing scientific landscape and paradigm shift.

In 1998, facilitated by a joint Ph.D. student, we decided that we should have joint 2–3-day long group work meetings to take place, alternately, in Mainz at the Max Planck Institute for Polymer Research (MPI), and in Enschede, in the Netherlands, at the University of Twente, recently in the MESA$^+$ Institute for Nanotechnology. To date, another eleven meetings followed this first one, more or less on an annual basis. With the appointment of Toby Jenkins at the University of Bath the originally bipolar venue became a triangle and the Mainz–Twente–Bath workshop series, or MTB Seminars as we call them today, was born. The lectures are usually offered by graduate students and postdoctoral scientists, and we occasionally combine these with guest lectures offered by more senior colleagues who just happened to be around visiting, or were encouraged to join from other groups. These 'guest' groups initially included the departments of David Reinhoudt and Jan Feijen from Twente. Beyond the reach of bureaucrats (we have never received any funding to finance the costs, which were always covered by our regular research grants, ensuring intellectual independence) a tremendous, dynamic and intellectually highly challenging discussion seminar series was born, which combined the more fundamental surface–biointerface interests of the Knoll group at MPI with the 'midstream' platform work of the Vancso group backed up by engineering in MESA$^+$ and the biologically tinted work of the Jenkins group in Bath.

Surface Design: Applications in Bioscience and Nanotechnology
Edited by Renate Förch, Holger Schönherr, and A. Tobias A. Jenkins

ISBN: 978-3-527-40789-7

As this seminar series, in a way, paralleled the paradigm shift in polymer science, we all thought that it would be useful to share some major topics in the form of a book with interested readers.

Once asked what would be the most fitting name for the MTB series, Toby Jenkins' immediate reply was: 'Designer Surfaces', which certainly influenced the title of this monograph. In this volume the reader finds a collection of chapters in which we give a representative account of the topics covered during MTB meetings in their most recent 'editions'. We believe that these chapters should be useful for all those readers, who are interested in receiving a (somewhat biased) cross-section of the field of 'surface design and engineering' primarily for biomaterial and life-science applications. We complemented the book with tutorial reviews focusing on selected fundamentals that – we all believe – would be useful to follow the more specialized chapters and may help graduate students and newcomers to this field. The tutorial section is followed by four blocks of chapters. First, designer concepts of functional, engineered polymer surfaces and interfaces as platforms are described; and on this basis, via selected representative examples, their applications in sensing are treated. The next block deals with nanoparticles and containers, which can be conceptually considered as efforts aimed at engineering soft matter towards the third dimension, from surfaces via interfaces, towards a full three-dimensional (3D) control. Current interest focuses on the design, fabrication and applications of full 3D systems and it is a cliché to say that the development of this area is still in its early stages. All these topics would be unimaginable without sophisticated tools like scanning probe microscopy and plasmonic devices; hence the last block of chapters is devoted to analysis techniques. The book is then completed with a glossary of surface analytical tools.

We can ask ourselves what might we expect in the future? Looking forward, the possibilities are tremendous. We expect scientific progress in areas like solving the protein-folding problem, providing designer principles for artificial–biological interphases with controlled protein fouling, scaffolds for tissue engineering, and concepts for assembling working machines powered and controlled by molecular motors. We expect to make progress towards the full control of three-dimensional functional designer structures. These should exhibit a carefully assembled structural architecture to serve functions in devices like ultrafast molecular computing, photonics, and sensing; in (nano)medicine and biomaterials, foodstuff, and energy production. Of course, this list is by no means complete. We will look at Nature for inspiration, and for guiding principles as we proceed along this road, satisfying our scientific curiosity and providing things for use.

It is a great pleasure to close this foreword with some acknowledgements. Of course, our contributions to the MTB seminars and to this book have been primarily facilitating, encouraging, providing the logistics, and perhaps here and there steering the discussions. We have always strongly emphasized the importance of interactions among graduate students and young scientists, and never begrudged any support, financial or other, for mutual visits and joint experiments. After all, success and progress depends on those who do the work, that is, on our graduate students and postdocs. Results of these collaborations are described in countless joint papers. New friendships were born, and old ones became stronger during the social parts of the MTB seminars, which took place in boats on the River Avon in Bath, during bowling and barbecues in Twente, in British pubs and bars, in 'Bierhallen' in Mainz and in castles along the Rhein river. This has been a great time and we look forward to the continuation. Due to changes in personal circumstances venues and participating parties may change but there is one thing for sure that will not change, that is our enthusiasm and support for this seminar series.

The book the reader holds in his/her hands would not have been possible without the tireless and great contributions of the three editors, Renate Förch from Mainz, Holger Schönherr from Twente, and Toby Jenkins from Bath. Renate, Holger, Toby, thank you very much.

Wolfgang Knoll
G. Julius Vancso

Contents

Surface Design: Applications in Bioscience and Nanotechnology
Edited by Renate Förch, Holger Schönherr, and A. Tobias A. Jenkins

ISBN: 978-3-527-40789-7

List of Contributors

Ahmed I. Abou-Kandil
Max-Planck-Institute for Polymer Research
Ackermannweg 10
55128 Mainz
Germany

Marta Álvarez
Max-Planck-Institute for Polymer Research
Ackermannweg 10
55128 Mainz
Germany

Heinrich F. Arlinghaus
Westfälische Wilhelms-Universität Münster
Physikalisches Institut
Wilhelm-Klemm-Straße 10
48149 Münster
Germany

Katja Bender
University Hospital Cologne
Institute of Legal Medicine
Melatengürtel 60–62
50823 Cologne
Germany

Edmondo Benetti
University of Twente
MESA+ Institute for Nanotechnology
Department of Materials Science and Technology of Polymers
P.O. Box 217
7500 AE Enschede
The Netherlands

Ilona Bredebusch
Westfälische Wilhelms-Universität
Medizinische Klinik und Poliklinik B
Gastroenterologische Molekulare Zellbiologie
Domagkstraße 3A
48149 Münster
Germany

Petra J. Cameron
University of Bath
Department of Chemistry
Bath, BA2 7AY
United Kingdom

Li-Qiang Chu
Max-Planck-Institute for Polymer Research
Ackermannweg 10
55128 Mainz
Germany

Surface Design: Applications in Bioscience and Nanotechnology
Edited by Renate Förch, Holger Schönherr, and A. Tobias A. Jenkins

ISBN: 978-3-527-40789-7

Wen-Fei Dong
Max-Planck-Institute of Colloids
and Interfaces
14476 Golm/Potsdam
Germany

Jakub Dostálek
Austrian Research Centers – ARC
Nano-System-Technologies
Donau-City-Straße 1
1220 Wien
Austria

Hatice Duran
Max-Planck-Institute for Polymer
Research
Ackermannweg 10
55128 Mainz
Germany

Nina V. Dziomkina
University of Twente
MESA+ Institute for Nanotechnology
Department of Materials Science and
Technology of Polymers
P.O. Box 217
7500 AE Enschede
The Netherlands

Arthur Echler
Max-Planck-Institute for Polymer
Research
Ackermannweg 10
55128 Mainz
Germany

Anika Embrechts
University of Twente
MESA+ Institute for Nanotechnology
Department of Materials Science and
Technology of Polymers
P.O. Box 217
7500 AE Enschede
The Netherlands

Chuan Liang Feng
BioMaDe Institute
Nijenborgh 4
9747 AG Groningen
The Netherlands

Renate Förch
Max-Planck-Institute for Polymer
Research
Ackermannweg 10
55128 Mainz
Germany

Marina I. Giannotti
University of Barcelona / CIBER–BBN
Institut de Bioenginyeria de Catalunya
Edificio Hélix PCB
C/ Baldiri Reixac, 15–21
08028 Barcelona
Spain

Ullrich Gunst
Westfälische Wilhelms-Universität
Münster
Physikalisches Institut
Wilhelm-Klemm-Straße 10
48149 Münster
Germany

Ming-Yong Han
University of Twente
MESA+ Institute for Nanotechnology
Department of Materials Science and
Technology of Polymers
P.O. Box 217
7500 AE Enschede
The Netherlands

Mark A. Hempenius
University of Twente
MESA+ Institute for Nanotechnology
Department of Materials Science and
Technology of Polymers
P.O. Box 217
7500 AE Enschede
The Netherlands

Chun Jen Huang
Max-Planck-Institute for Polymer Research
Ackermannweg 10
55128 Mainz
Germany

Dominik Jańczewski
Institute of Materials Research and Engineering
A*STAR (Agency for Science, Technology and Research)
3 Research Link
Singapore 117602

A. Tobias A. Jenkins
University of Bath
Department of Chemistry
Bath, BA2 7AY
United Kingdom

Ulrich Jonas
FORTH / IESL
Voutes Str.
P.O. Box 1527
71110 Heraklion
Greece

Martin Jübner
University of Cologne
Institute for Legal Medicine
Melatengürtel 60–62
50823 Cologne
Germany

Amal Kasry
Max-Planck-Institute for Polymer Research
Ackermannweg 10
55128 Mainz
Germany

Wolfgang Knoll
Max-Planck-Institute for Polymer Research
Ackermannweg 10
55128 Mainz
Germany

E. Stefan Kooij
University of Twente
$MESA^+$ Institute for Nanotechnology
Department of Materials Science and Technology of Polymers
P.O. Box 217
7500 AE Enschede
The Netherlands

Ingo Köper
Max-Planck-Institute for Polymer Research
Ackermannweg 10
55128 Mainz
Germany

K.H. Aaron Lau
Max-Planck-Institute for Polymer Research
Ackermannweg 10
55128 Mainz
Germany

Xiaosong Li
Max-Planck-Institute for Polymer Research
Ackermannweg 10
55128 Mainz
Germany

Yujie Ma
University of Twente
$MESA^+$ Institute for Nanotechnology
Department of Materials Science and Technology of Polymers
P.O. Box 217
7500 AE Enschede
The Netherlands

Helmuth Möhwald
Max Planck Institute of Colloids and Interfaces
14476 Golm/Potsdam
Germany

Melba Navarro
Institute for Bioengineering of Catalonia (IBEC)
c/Baldiri Reixac 13
08028 Barcelona
Spain

Swapna Pradhan-Kadam
Max-Planck-Institute for Polymer Research
Ackermannweg 10
55128 Mainz
Germany

Monique Roerdink
University of Twente
MESA$^+$ Institute for Nanotechnology
Department of Materials Science and Technology of Polymers
P.O. Box 217
7500 AE Enschede
The Netherlands

Christina E. Rommel
Westfälische Wilhelms-Universität
Medizinische Klinik und Poliklinik B
Gastroenterologische Molekulare Zellbiologie
Domagkstraße 3A
48149 Münster
Germany

Jürgen Schnekenburger
Westfälische Wilhelms-Universität
Medizinische Klinik und Poliklinik B
Gastroenterologische Molekulare Zellbiologie
Domagkstraße 3A
48149 Münster
Germany

Andreas Scholten
University of Cologne
Institute for Legal Medicine
Melatengürtel 60–62
50823 Cologne
Germany

Holger Schönherr
University of Siegen
Department of Physical Chemistry
Adolf-Reichwein-Straße 2
57076 Siegen
Germany

Weiqing Shi
University of Twente
MESA$^+$ Institute for Nanotechnology
Department of Materials Science and Technology of Polymers
P.O. Box 217
7500 AE Enschede
The Netherlands

Jing Song
University of Twente
MESA$^+$ Institute for Nanotechnology
Department of Materials Science and Technology of Polymers
P.O. Box 217
7500 AE Enschede
The Netherlands

Oya Tagit
University of Twente
MESA$^+$ Institute for Nanotechnology
Department of Materials Science and Technology of Polymers
P.O. Box 217
7500 AE Enschede
The Netherlands

Matthias Tamm
Max-Planck-Institute for Polymer Research
Ackermannweg 10
55128 Mainz
Germany

Ewa Tocha
Dow Olefinverbund GmbH
P.O. Box 1163
06258 Schkopau
Germany

Nikodem Tomczak
Institute of Materials Research and Engineering
A*STAR (Agency for Science, Technology and Research)
3 Research Link
Singapore 117602

G. Julius Vancso
University of Twente
MESA$^+$ Institute for Nanotechnology
Department of Materials Science and Technology of Polymers
P.O. Box 217
7500 AE Enschede
The Netherlands

Inga Vockenroth
Max-Planck-Institute for Polymer Research
Ackermannweg 10
55128 Mainz
Germany

Nicolas Vogel
Max-Planck-Institute for Polymer Research
Ackermannweg 10
55128 Mainz
Germany

Thomas L. Williams
University of Sussex
School of Life Sciences
Department of Biochemistry
Brighton, BN1 9QG
United Kingdom

Basit Yameen
Max-Planck-Institute for Polymer Research
Ackermannweg 10
55128 Mainz
Germany

Szczepan Zapotoczny
Jagiellonian University
Faculty of Chemistry
Ingardena 3
30-060 Cracow
Poland

Shan Zou
University of Twente
MESA$^+$ Institute for Nanotechnology
Department of Materials Science and Technology of Polymers
P.O. Box 217
7500 AE Enschede
The Netherlands

1
Tutorial Reviews

Surface Design: Applications in Bioscience and Nanotechnology
Edited by Renate Förch, Holger Schönherr, and A. Tobias A. Jenkins

ISBN: 978-3-527-40789-7

1.1
Coupling Chemistries for the Modification and Functionalization of Surfaces to Create Advanced Biointerfaces

Holger Schönherr

1.1.1
Introduction

Biocompatibility, biointegration and functionality, among other properties, of natural and artificial materials are determined to a large extent by the materials' surface properties [1]. Full control over the surface properties represents a crucial issue for the applicability of biomaterials and materials that are interfaced with biological media or specimens [2]. In addition to physical attributes, including surface roughness and elastic modulus, these properties are imparted by the presence and arrangement of biologically active molecules. Some of these molecules are responsible for specific interactions or signaling, while others suppress unwanted nonspecific interactions of biomolecules present in the corresponding media with the surface.

Implants may serve as an illustrative example, for which surface roughness and chemical composition have been demonstrated to determine the success of a desired process, for instance the integration of the implant with the newly grown bone (osteoinduction and osseointegration of implant material) [3]. Likewise, in biosensors, the interaction of analytes in solution with receptors immobilized on a sensor chip surface can depend crucially on the orientation of the receptor, its surroundings on a molecular length scale and its separation from the surface (Chapter 1.4). Again, the performance and overall function is intimately related to surface chemistry. Since many materials, in general, do not possess suitable or optimized surface structure and functionality, physical and chemical modification and functionalization protocols are applied to improve the surface properties. In this tutorial review chapter, we will focus on the aspect of *surface chemistry* and its role in fabricating advanced biointerfaces.

As alluded to above, advanced biointerfaces are met in many applications, most notably in implants, but also in biosensors and assays that aim at unraveling the interaction of cells with surfaces. In these areas not only the 'appropriate' surface chemistry (e.g. the exposure of particular proteins in an oriented manner) is important, but also the homogeneity on various length scales and the spatial organization in 2D and 3D. One prominent example is the interaction of cells with

Surface Design: Applications in Bioscience and Nanotechnology
Edited by Renate Förch, Holger Schönherr, and A. Tobias A. Jenkins

ISBN: 978-3-527-40789-7

patterned surfaces that expose a cell-adhesive protein and a passivating function in patterns with different sizes. As shown in studies by Whitesides, Ingber and coworkers [4], the size of the pattern may control cell growth, proliferation and apoptosis. More recent work has highlighted the importance of protein clusters and characteristic length scales for cell–surface interactions [5]. Here, control over surface chemical composition down to the level of individual polypeptide molecules and spacings of < 50 nm is required, as will be discussed in more detail in Chapter 3.4. Similar challenges are met in the area of single (bio)molecule work [6], where the immobilization of individual molecules at preselected locations and the study of their properties and interactions using single molecule optics or scanning probe microscopy approaches sets stringent requirements on surface design (Figure 1).

Apart from obvious challenges for surface analytics in the quantitative analysis of surface coverages and lateral heterogeneities on the length scales mentioned [7], this level of sophistication mandates a closer look at the coupling chemistries that are being utilized. One area of particular attention are the peculiarities and limitations of surface coupling chemistries used for bioconjugation. Patterning and related lithographic techniques will not be treated as these topics are central issues of Chapters 2.2, 2.4, 3.2 and 3.4 of this book.

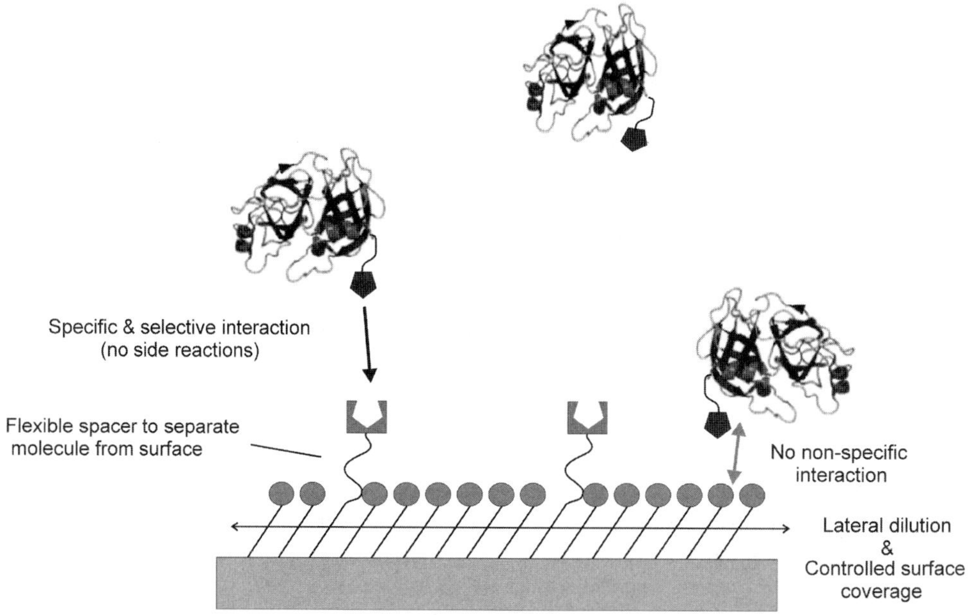

Figure 1 Schematic that illustrates some of the concepts and requirements for surface chemistry and the parameters that must be controlled. The example sketches the immobilization of a protein via one particular functional group that is selectively recognized by a receptor moiety immobilized in a mixed monolayer on a solid support.

For this tutorial review it is out of scope to treat this broad topic exhaustively in terms of the different classes of chemical reactions and approaches employed. Rather than being encyclopedic in character, the chapter highlights the most prominent classes of surface reactions and will provide the reader with selected illustrative examples. The topic will be developed starting with an overview on the nature of the surfaces and substrates that need to be functionalized to obtain advanced biointerfaces, followed by a brief introduction of frequently applied thin-layer approaches to mediate the attachment of biologically relevant molecules. Subsequently, the focus is shifted to the general peculiarities and differences between surface and solution chemistry. The second main part comprises a discussion of the bioconjugation approaches and is structured according to covalent and noncovalent routes. Finally, means to immobilize proteins and other biomolecules in an oriented manner will receive attention.

1.1.2 Surfaces and Self-Assembled Monolayers

Advanced biointerfaces are obtained by reacting surfaces of flat substrates or complex 3D structures with biologically active and other functional molecules. This approach, as also highlighted in Figure 10(c), is widely followed since the time-consuming and sometimes tedious syntheses of biomolecules of interest with pending side chains equipped with reactive moieties for surface immobilization can be circumvented. Before addressing the underlying surface-chemical approaches and some prominent classes of reaction, the nature and structure of the surfaces must be considered.

In the past decades, appropriate functionalization protocols have been designed for various classes of materials, including metals, oxides and polymers. These methods rely on strong interactions of suitable (macro)molecules with the underlying surface [8]. In most approaches a precursor layer is deposited that serves at least two purposes: It passivates the underlying substrate (barrier function) and provides attachment sites for further chemical reactions and bioconjugation. Classic examples are the modification of metal and metal oxide surfaces with self-assembled monolayers (SAMs) by assembly from solution or the gas phase (Figure 2) [9–12] or thin polymeric films, deposited, e.g., by spin- or dip-coating (Figure 3) [13, 14] or by plasma polymerization (Chapter 1.4) [15]. Some materials, for instance polymers, can be directly activated in the gas phase using plasma treatments or by wet chemistry for subsequent chemical modification (Chapter 2.4).

The direct deposition of biologically relevant molecules (and immobilization via nonspecific physisorption) on the mentioned surfaces is only successful in certain cases. While DNA and certain robust proteins do not denature and loose functionality upon adsorption, many biological molecules do. In addition, the stringent requirement for certain applications sketched out earlier require enhanced control over, e.g. protein position, coverage and in particular orientation. This control can only be realized in parts using simple physisorption-based approaches. For passivating coatings, e.g. to prevent the nonspecific adsorption of proteins from serum

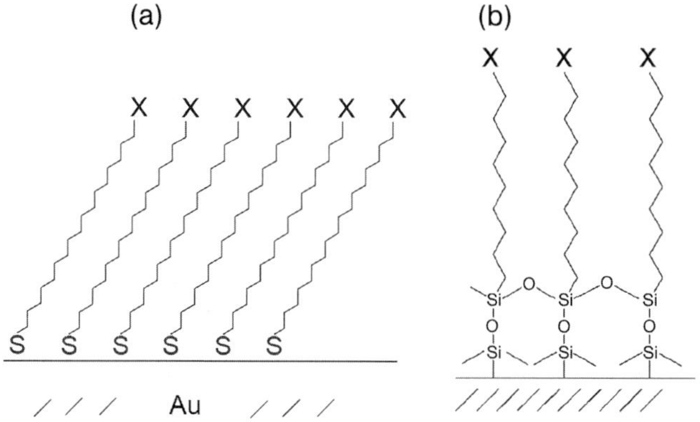

Figure 2 (a) Schematic structure of self-assembled monolayer of end-functionalized alkane thiols and disulfides on gold. The range of functional groups X that can be introduced is almost unlimited due to the rare crossreactivity of the functional groups typically employed and the thiol or disulfide chain end [9, 11]. (b) Schematic of silane-based SAM. SAMs on gold or silicon may suffer from degradation under oxidative conditions (ozone or oxygen and UV light) and condition of extreme pH, respectively. In addition, SAMs may possess intrinsic molecular scale defects, such as pinholes.

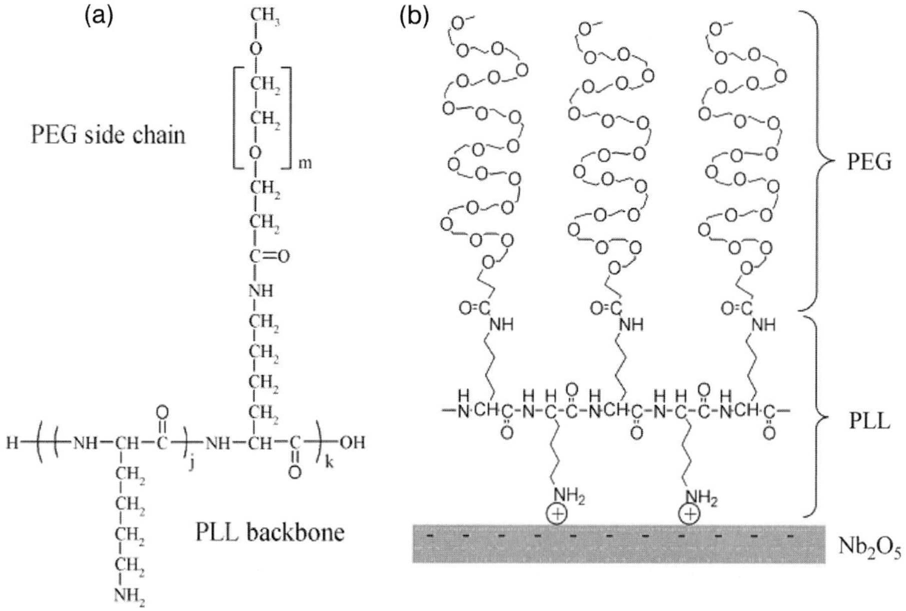

Figure 3 (a) Molecular structure of the PLL-g-PEG copolymer (shown in its uncharged form). (b) Idealized scheme of the interfacial structure of a monolayer of PLL-g-PEG adsorbed on a metal-oxide substrate (Nb_2O_5) via electrostatic interactions between the negatively charged metal oxide surface and positively charged amino-terminated PLL side chains (at neutral pH). (Reproduced with permission from reference [17]; S. Pasche, S. M. De Paul, J. Voros, N. D. Spencer, M. Textor, Langmuir 2003, 19, 9216, Copyright 2003 by American Chemical Society).

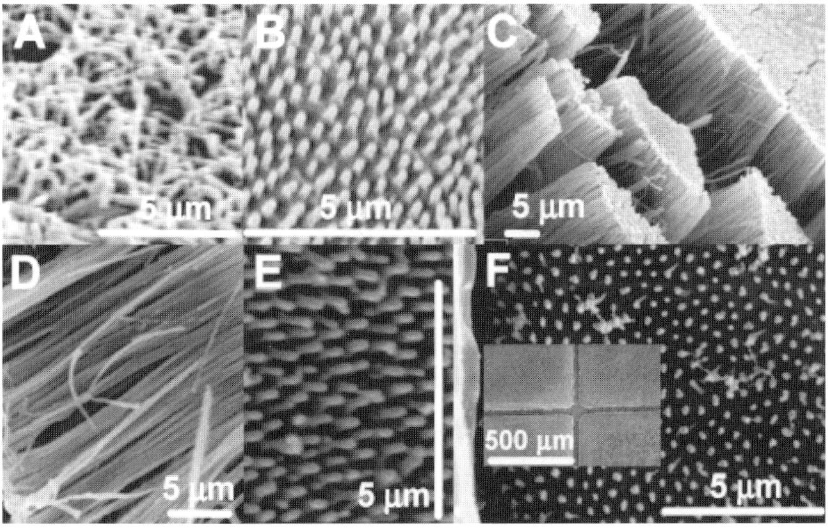

Figure 4 SEM images of PMMA nanopillars of varying aspect ratio, *R*. (A) Top view of collapsed nanopillars (60-µm tall and 175 nm diameter, $R = 343$). (B) 1-µm tall and 175-nm diameter pillars, $R = 5.7$. (C) and (D) $R = 343$. (D) Close-up view of pillars in (C). (E) $R = 5.7$ pillars surface modified to yield carboxylic acid groups. (F) Top view of 5-µm tall and 150-nm diameter ($R = 33.3$) nanopillars integrated into a 50-µm wide, 50-µm deep, and 5-cm long PMMA fluidic channel; the inset image is of the microchannel cross-tee containing free-standing nanopillars. (Reproduced with permission from reference [22]; G. Chen, R. L. McCarley, S. A. Soper, C. Situma, J. G. Bolivar, Chem. Mater. 2007, 19, 3855, Copyright 2007 by American Chemical Society).

or body fluids, multisite nonspecific adsorption of poly-L-lysine-poly(ethylene glycol) conjugates (PLL-PEG) have been successfully implemented (Figure 3) [16, 17]. The implementation of more sophisticated approaches often relies on surface chemical reactions starting from a mediating layer (*vide infra*). These also include surface-initiated polymerizations (Chapter 2.2) [18].

The choice of a mediating layer, such as a SAM or a polymeric coating, is guided by considerations regarding the properties of the substrate and the requirements of the terminal functional groups in terms of reactivity. To link a given exposed functional group with an active molecule the concept of *heterobifuctional spacers* has been introduced with success (*vide infra*). The molecules are designed to provide selective reactions at each chain terminus for (i) the substrate functional group and (ii) the attachment site of the biologically active molecule of interest.

The application of the synthetic schemes is not in limited to flat substrates. To increase coverage per unit surface area and to release the constraints and limitations of ordered SAMs, dendrimers and other quasi-3D structures, including thin polymer films have been explored [19, 20] Likewise, true 3D structures were shown to be efficiently functionalized with a procedure originally developed for flat substrates [21]. These structures include topographically structured supports

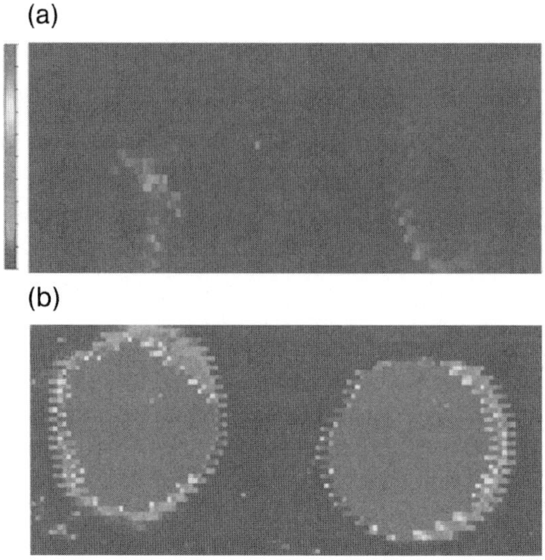

Figure 5 13.2 mm × 3.06 mm fluorescence images of two-element oligonucleotide arrays on (a) planar PMMA and (b) nanopillared PMMA ($R = 343$) surfaces. Scale is 0 to 10 000 counts. Pixel size is 101.6 μm. (Reproduced with permission from reference [22]; G. Chen, R. L. McCarley, S. A. Soper, C. Situma, J. G. Bolivar, Chem. Mater. 2007, 19, 3855, Copyright 2007 by American Chemical Society).

and 3D specimens obtained by etching or machining silicon or glass, photoresist materials, polymers, etc.

Free-standing nanopillars of poly(methyl methacrylate) (PMMA) [22], obtained by photopolymerization within micrometer-tall pores of anodic aluminum oxide, were reported to be stable under conditions of light-directed and solution-phase surface modification (Figure 4) [23]. These nanopillars provide enhanced functionality due to the increased surface area, as demonstrated, e.g., in oligonucleotide hybridization experiments and proteolytic reactions (Figure 5).

The stability of the linkage between the supporting substrate and the immobilized molecules is a clear requirement in some applications, while others rely on the reversibility of the attachment, in particular for biosensors that need to be regenerated. Dextran hydrogels are a prime example of high loading capacity coatings used in surface plasmon resonance biosensing (see Chapter 1.4) [19].

1.1.3 Reactions at Surfaces

Surface reactions differ in many aspects from reactions carried out in bulk solution. In a review article by Chechik, Crooks and Stirling [24], the effects of the ordered environment of monolayers were grouped into solvent effects (e.g. altered surface $pK_{1/2}$ values), steric effects and electronic/anchimeric effects. In particular, solvent and anchimeric effects do not necessarily possess analogues in solution. Among the prominent additional differences are increased susceptibility to (trace) contaminations (clean surfaces are a prerequisite), limited mobility of immobilized functional groups and spacers, hindered access of reactants to reactive sites in organized assemblies, crowded transition states, the presence of surface forces, and problems related to reaction by-products.

The last issue is easy to comprehend as reaction by-products cannot be separated from the desired product when the unwanted side product is attached to a surface-bound species. One example of a side reaction is discussed in Chapter 3.2, where the photochemical reaction of the nitroveratryloxycarbonyl (NVoc) protection group may yield an imine instead of the desired free amine. A second prominent side reaction is the hydrolysis of active esters, such as the NHS- or p-nitrophenyl esters, in the aqueous media used for reactions of biologically relevant molecules.

The photodeprotection of an NVoc protected amine group proceeds via decarboxylation to yield an aldehyde, the free amine and CO_2. The aldehyde photoproduct formed may react with the amine group to give an imine before the aldehyde diffuses away [25]. If the amine is immobilized on the surface, the imine side product would be surface bound as well and could potentially affect the properties of the surface studied. In a solution reaction the work-up of the reaction mixture, comprising the separation of the imine by-product and purification of the amine product using, e.g., chromatography, would be a standard procedure. For the surface reaction, however, this is impossible. Hence surface reactions must be, by definition, highly selective and side reactions must be absent or suppressed in order to obtain the required control over surface properties.

For a surface reaction on a solid support the number of reactive surface-bound functional groups is intrinsically limited. An upper limit of available sites can be estimated from the area requirement per molecule in SAMs on gold, which is of the order of ~ 20–25 $Å^2$/molecule [9]. These values correspond to coverage of $5\text{–}4 \times 10^{14}$ molecules/cm^2. Hence, the surface would be fully covered by an equivalent of ~ $7 \times 8 \times 10^{-8}$ g of fully upright-standing molecules of a molar mass of 100 Da. This simplified treatment suggests that nanograms or less of contaminants in a system may already lead to severe problems, if these contaminants interact strongly with the surface. Thus, a very small number of molecules may contaminate such a surface through physisorption. If the reactants cannot displace the contaminant molecules, the areas covered by these species remains unreacted and a heterogeneously functionalized surface would result.

Compared to solution, the immobilized molecules, functional groups or spacers possess limited mobility because the attachment causes a reduction in the degrees of freedom (translation is largely impossible). Thus, reactive groups immobilized on surfaces do not freely diffuse like in solution and hence may collide less frequently with reactant molecules. Therefore, their reactivity is affected and lower overall yields may result. For the solution-based reactants, both the activating species such as 1-ethyl-3-(3-dimethylaminopropyl)carbodiimide hydrochloride (EDC) and *N*-hydroxysuccinimide (NHS) on the one hand, and the targeted biomolecule, e.g. DNA, a protein or an antibody, on the other hand, may suffer from hindered access to the respective reactive sites in organized assemblies (e.g. SAMs). This may lead to more crowded transition states and unfavorable transition-state entropies. Consequently, reaction rates are decreased and yields may be lower.

This effect can be illustrated for the immobilization of the protein fragment BT Toxin on a solid support. As discussed above, a SAM that exposes reactive functional groups can be employed. As will be outlined below in more detail, proteins can be covalently coupled via pending lysine residues to a surface using NHS chemistry. In the following example, a reactive NHS-ester terminated SAM was prepared following two alternative routes that should be nominally identical (Figure 6):

(a) via the chemical activation of the carboxylic acid groups of a –COOH terminated SAM assembled from the commercially available 16-mercaptohexadecanoic acid on Au(111) using EDC/NHS chemistry; or
(b) by assembly of the corresponding disulfide 16,16′-dithiobis(*N*-hydroxysuccinimidylhexadecanoate) (N.B.: Alkane thiols and their corresponding disulfides were shown to yield practically identical SAMs).

After the reaction with BT toxin in buffered medium the samples were carefully rinsed with pure buffer and Mill-Q water followed by drying in a stream of nitrogen. The subsequent AFM analysis (Figure 7), in which the covalently attached proteins can be clearly resolved, showed a marked difference in surface coverage. While the sample prepared by route (b) was densely covered with on average 530 molecules/μm^2, the chemically activated surface (route (a)) showed a five-fold reduced coverage of merely 100 molecules/μm^2. This example demonstrates that the activation of the carboxylic-acid-terminated SAM proceeded with rather limited yield, likely due to steric crowding at the carbonyl centers (for intermediate steps of the reaction mechanism, see Figure 11 below). Even though the reaction proceeds to a much higher extent for the NHS-ester disulfide, close inspection of the AFM image in Figure 7(b) suggests that there are unoccupied protein-free areas, which were either blocked by adsorbed contaminations or are sites where the side reaction of NHS-esters in aqueous medium (hydrolysis to the carboxylic acid) has already set in.

These drawbacks of reduced reactivity can in part be avoided by an appropriate choice of the substrate. As shown in studies on the reactivity of ester groups in hydrolysis reactions [26, 27] the constraints due to steric crowding present in highly organized monolayer architectures can be in part circumvented by using macromolecular systems or disordered SAMs composed of short-chain adsorbates (Figure 8) [28] or assembled as binary mixtures of adsorbates with different chain lengths [26].

In this example, the reactivity of NHS-ester SAMs in the base-catalyzed hydrolysis in aqueous NaOH was investigated with contact angle, FTIR spectroscopy and XPS measurements [29]. A pronounced chain-length dependence was found indicating that reduced reactivity (smaller rate constants and deviations from the

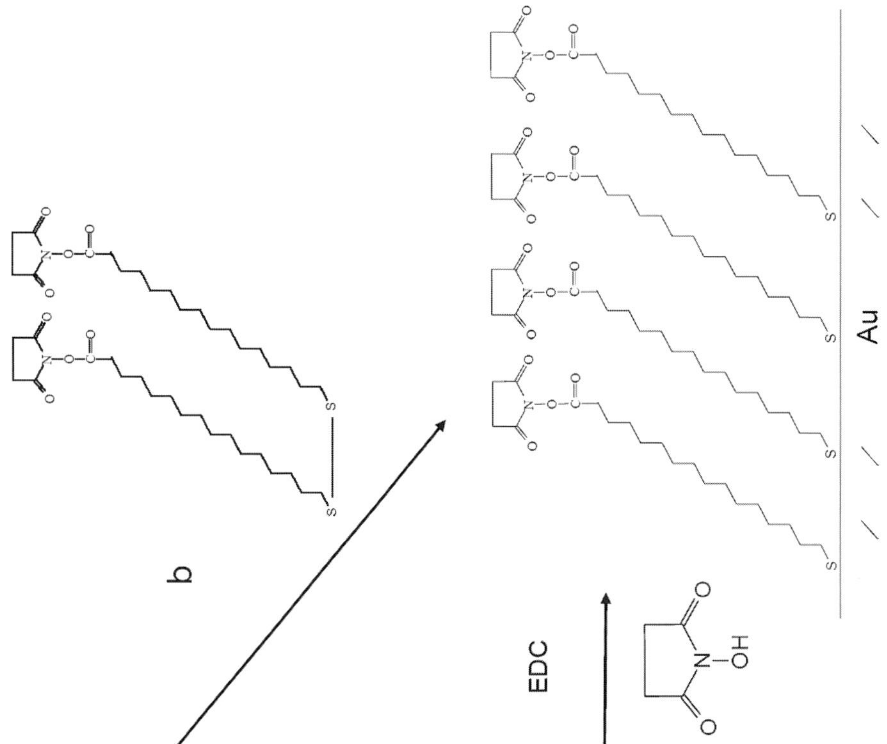
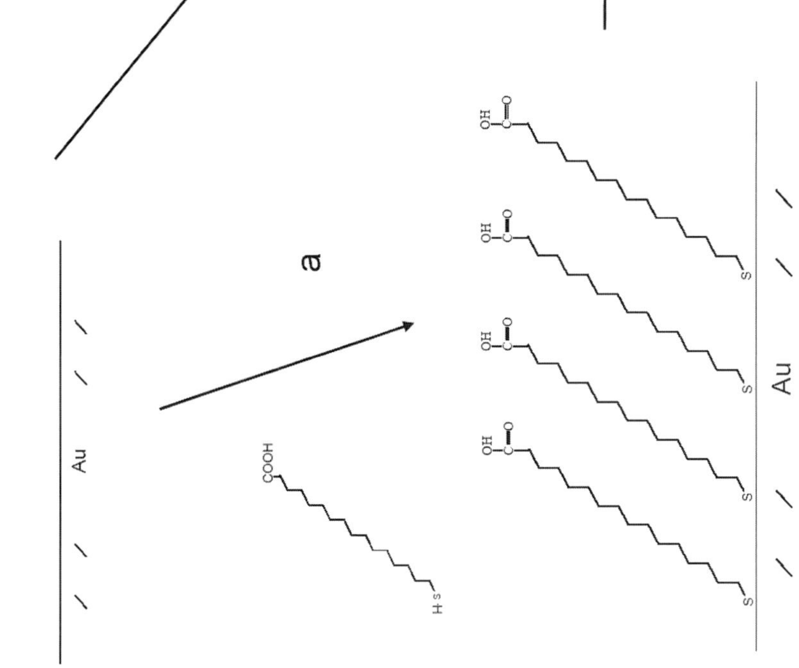

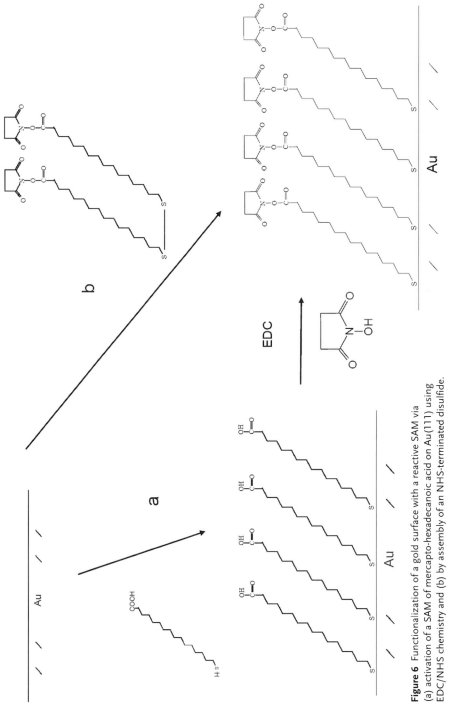

Figure 6 Functionalization of a gold surface with a reactive SAM via (a) activation of a SAM of mercapto-hexadecanoic acid on Au(111) using EDC/NHS chemistry and (b) by assembly of an NHS-terminated disulfide.

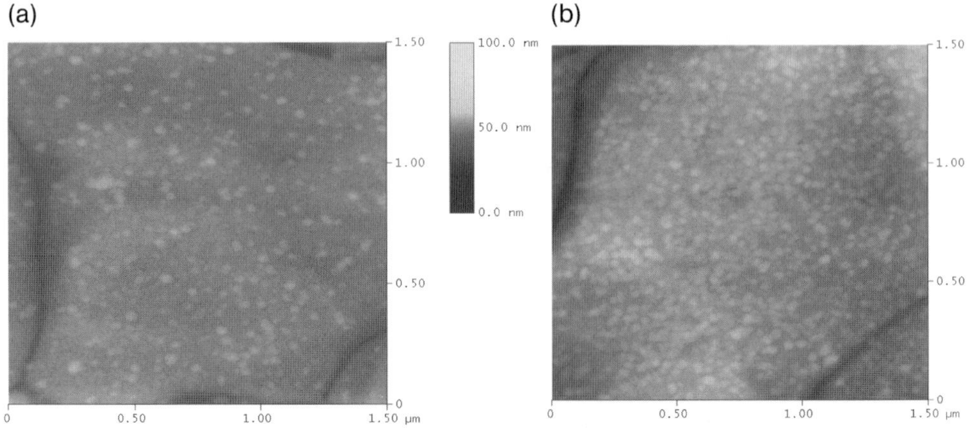

Figure 7 Tapping-mode AFM height images of BT-toxin covalently coupled to (a) a NHS-ester SAM prepared via route (a) in Figure 6 and (b) to a SAM of a NHS ester disulfide (route (b) in Figure 6) (A. T. A. Jenkins, S. Liu, H. Schönherr, unpublished data).

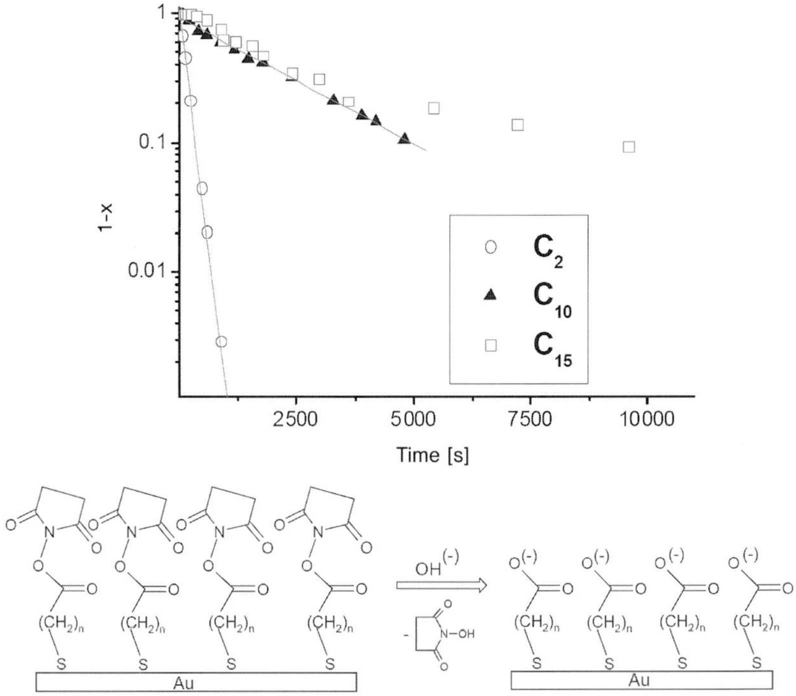

Figure 8 Fractional surface coverage of NHS ester-terminated monolayers (determined by contact angle and FTIR spectroscopy measurements) vs. reaction time for the hydrolysis in 0.01 M NaOH ($T = 30$ °C). C_n denotes the number n of methylene groups in the alkane chain of the disulfides.

expected exponential kinetics for long-chain SAMs) are related to higher conformational order of the SAM.

The presence of surface forces represents another peculiarity for which there is no analogue in solution-based chemical reactions. Since the attachment reactions occur close to solid surfaces in liquid media, long-range intermolecular interactions between the reacting molecules and the surface molecules or atoms cannot be *a priori* neglected. Examples comprise electrostatic interactions and van der Waals forces. Also, the effect of nearby functional groups must be considered in some cases [30].

The stability of a binary complex of self-complementary 2-ureido-4[1H]-pyrimidinone(UPy) in SAMs on gold was found to be different from solution. While the half-life of such a complex estimated based on the reported equilibrium complexation constant in $CHCl_3$ of $10^7\ M^{-1}$ is 170 ms, the complexes formed at SAMs of 2-ureido-4[1H]-pyrimidinone-hydroxyalkane disulfide adsorbates were found to be stable in pure $CHCl_3$ for more than 24 h (Figure 9) [31]. Secondary interactions between the external trifluoromethyl- or ferrocenyl-derivatized pyrimidinone guests and the underlying SAM were deemed responsible for this quasi-irreversibility of the complexation reaction in neat $CHCl_3$ at ambient temperatures. If the surface-immobilized UPy moieties are bound to an oligomeric PEG spacer, the complexation constants are similar to those reported for solution [32].

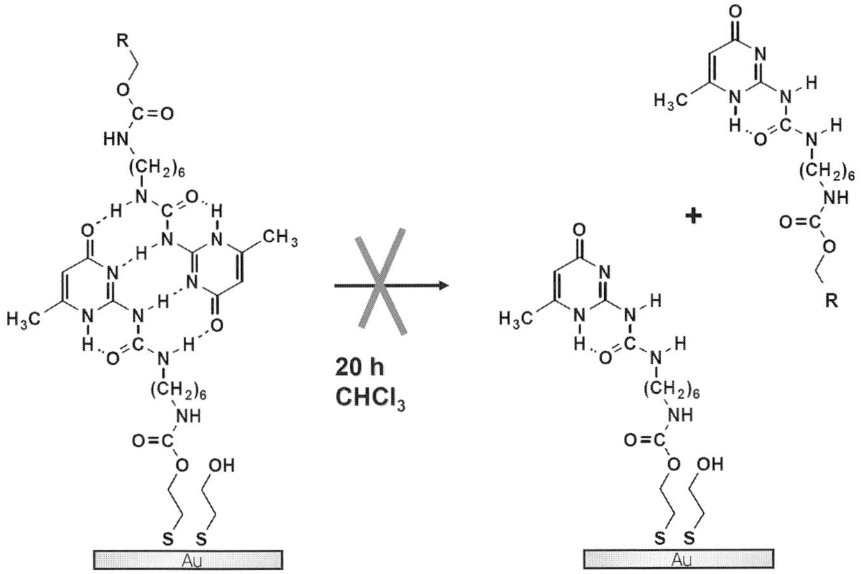

Figure 9 Schematic of binary complex of two self-complementary 2-ureido-4[1H]-pyrimidinone moieties held together by four hydrogen bonds. Despite the dynamic character of the reaction in solution, the interaction at the solid/liquid interface is affected by secondary interactions.

1.1.4 Coupling Chemistries

We can broadly differentiate different methods to immobilize biomolecules based on the nature of the attachment. In the following paragraphs selected examples of covalent and noncovalent coupling reactions are discussed. With increased sophistication and refinement of the approaches both manners can be employed to immobilize for instance proteins in defined orientation. In view of the functionality of the modified surfaces, this represents an important goal.

1.1.4.1 Covalent Attachment Reactions

Among the covalent coupling reactions we differentiate substitution and addition reactions. The obvious difference between these two classes of reactions is the presence of a leaving group in the former case. While a good leaving group may increase the reactivity, such as in the case of reactive (active) esters vs. conventional esters, the transition state may be more complex since the bulky leaving group contributes to the steric hindrance.

Substitution Reactions

A well-established and classic reaction in aqueous medium is the substitution employing active esters, such as the already briefly introduced NHS ester (Figure 10a) [33–35]. Under mild conditions (bio)molecules with primary amino groups are efficiently coupled. By optimizing the relevant parameters pH, concentration, ionic strength, and reaction time, the optimized conditions need to be established for each type of protein. Since many proteins contain a number of lysine residues, often located on the outer periphery of the proteins, this reaction has received attention. However, it has been noted that the abundance of lysines results often

Figure 10 (a) Reaction scheme for forming amide bonds with a SAM of 11-mercaptoundecanoic acid (MUA) on a gold surface. In the first step, the MUA carboxylic acid groups are reacted with the carbodiimide, EDC, and then N-hydroxysulfosuccinimide (NHSS) to form the NHSS ester. Subsequent reaction of this activated intermediate with an aqueous solution of either ammonia (NH_3) or poly(L-lysine) (PL) results in the attachment of these amines to the surface by formation of an amide bond. (Reproduced with permission from reference [36]; B. L. Frey, R. M. Corn, Anal. Chem. 1996, 68, 3187, Copyright 1996 by American Chemical Society).
(b) Structure of adsorbates (top) and illustration of the differences between the common intermediate method and the method involving the synthesis of alkanethiol-based ligands (bottom). In the common intermediate method, a SAM bearing carboxylic acid groups is formed by the cochemisorption of alkanethiols 1 and 2. This SAM, after activation with NHS and EDC, serves as a common intermediate for the attachment of different ligands by amide-bond formation. In the more commonly used mixed SAMs method, an alkanethiol-based ligand (e.g., 5) is synthesized and characterized and then used to form mixed SAMs. (Reproduced with permission from reference [34]; J. Lahiri, L. Isaacs, J. Tien, G. M. Whitesides, Anal. Chem. 1999, 71, 777, Copyright 1999 by American Chemical Society).

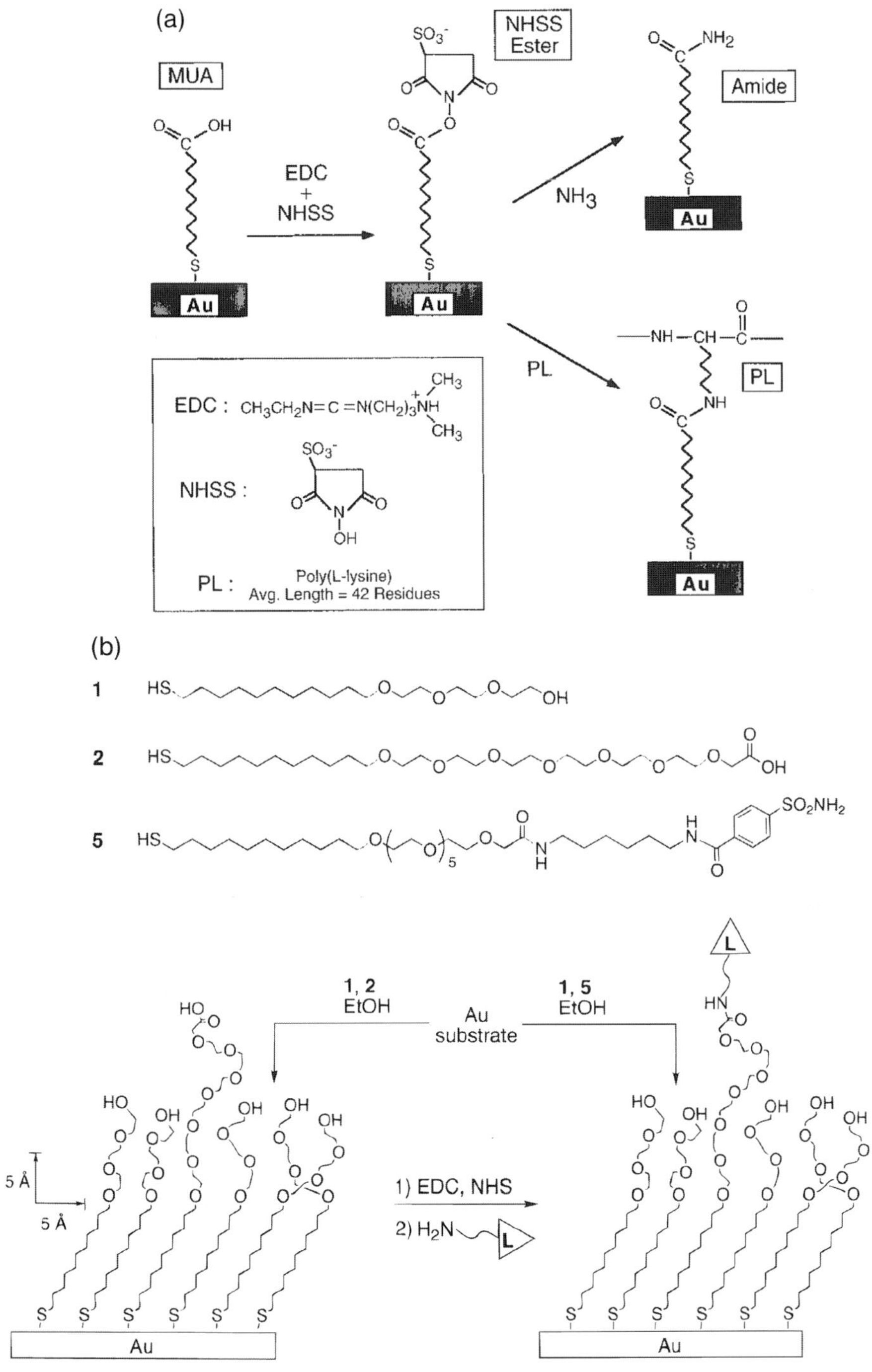

Figure 10 (legend see p. 14)

in multipoint attachment, thereby increasing the heterogeneity and restricting the conformational flexibility of the immobilized protein. The resulting amide linkages is relatively stable. In addition to proteins, polypeptides, amino end-functionalized DNA or PEG oligomers (Figure 10b) can be efficiently immobilized [20].

The yield of the activation step has been reported to be limited also in case of the sulfonated NHS esters [36]. Frey and Corn reported a reaction yield of 50% that was assumed to be limited by the steric packing of the NHSS ester intermediate. However, after three cycles of carboxylic acid activation to the NHSS ester and reaction with NH_3, nearly all of the carboxylic-acid-terminated molecules (~80%) were converted to amides.

NHS-ester-based coupling reactions have also been applied on polymer film surfaces [37]. Among others, DNA, proteins and PEG have been immobilized in a controlled manner. A related strategy involves the activation of surface-bound carboxylic acid groups with trifluoroacetic acid anhydride to form *in situ* an anhydride SAM [38]. This reactive monolayer can be subsequently modified in aprotic media with molecules that carry a terminal primary amino group.

The limited yield of the activation reaction can be attributed to the sterically demanding transition states and intermediate reaction products (Figure 11).

If the pH of the buffer is well controlled, the hydrolysis of the NHS ester is only a negligible side reaction (compare Figure 8). However, this side reaction limits the shelf life of commercially available DNA array support slides that employ

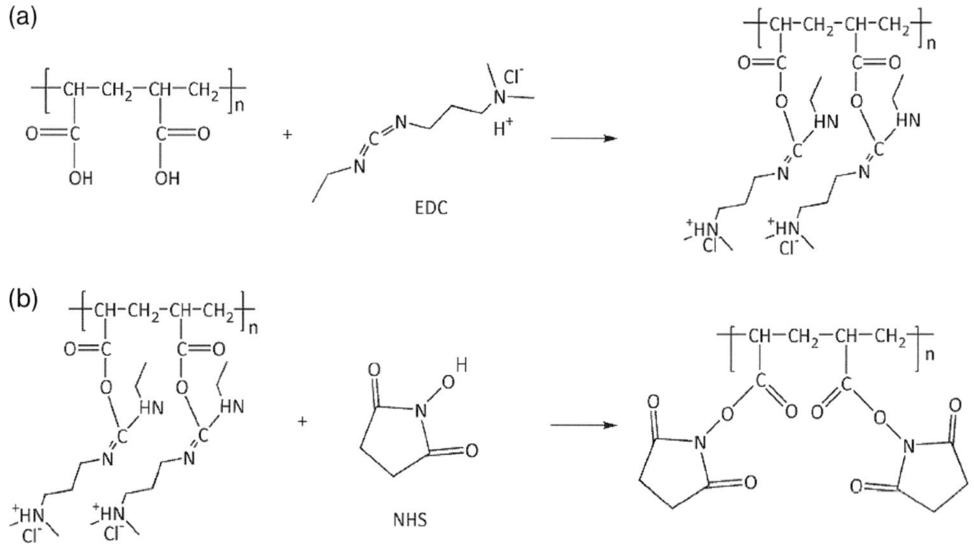

Figure 11 Activation of carboxylic acid groups in poly(acrylic acid) in a polymer-analogous reaction with (a) EDC and (b) NSH. In organized assemblies the bulky intermediate products can be problematic and may lead to lower yields compared to solution.

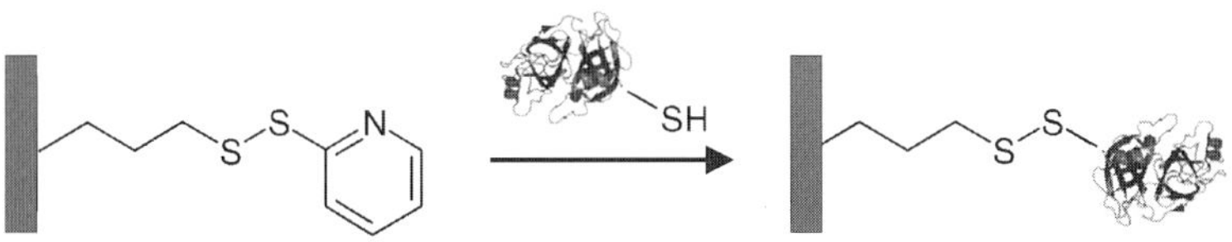

Figure 12 A disulfide exchange reaction can be utilized to immobilize proteins via the thiol group of a pending cysteine residue.

NHS chemistry [7]. In such films the inefficient surface-activation reaction is circumvented by using NHS-ester-equipped polymers [20].

A second valuable reaction is the substitution reaction between a pyridinyl disulfide and the terminal thiol functionality of a biomolecule or protein [39]. Sulfhydryl-terminated DNA is readily available and many proteins possess pending cysteine groups. The resulting disulfide is of moderate stability, which can be exploited in a regeneration of the surface or a reductive cleavage using dithiothreitol (DTT).

Addition Reactions

In contrast to the substitution reactions introduced in the previous section, addition reaction possess the advantage that there is, in the absence of side reactions, *a priori* no reaction product other than the desired conjugate.

An early established reaction, for instance for DNA immobilization, is the Schiff base formation (Figure 13) between a surface-immobilized aldehyde and a primary amine on the biomolecule [40]. The imin reaction product is formed either as *cis* or *trans* isomer and is in dynamic equilibrium with the free reactants. This equilibrium can be exploited for reversible and reuseable biochips. However, if robust attachment is sought, a reduction of the imine with borane chemistry is typically carried out to yield a more stable secondary amine.

Further frequently employed addition reactions include the Michael addition (Figure 14) [36, 41]. In this coupling reaction a terminal thiol of, e.g., a cysteine residue is added to a surface-bound maleimide or a vinylsulfone. The maleimide

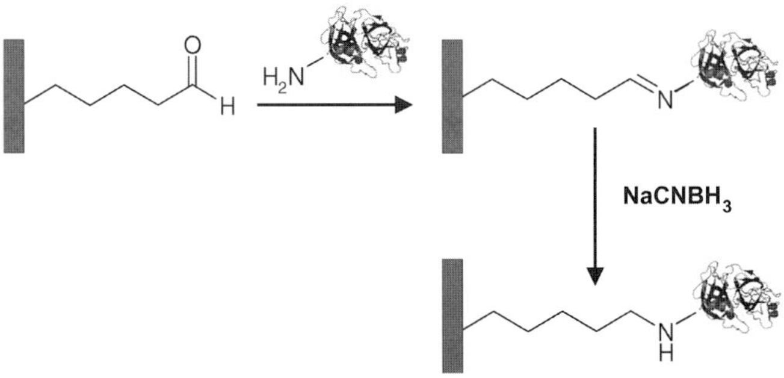

Figure 13 Schiff base formation and subsequent reductive treatment using sodium cyanoborohydrid.

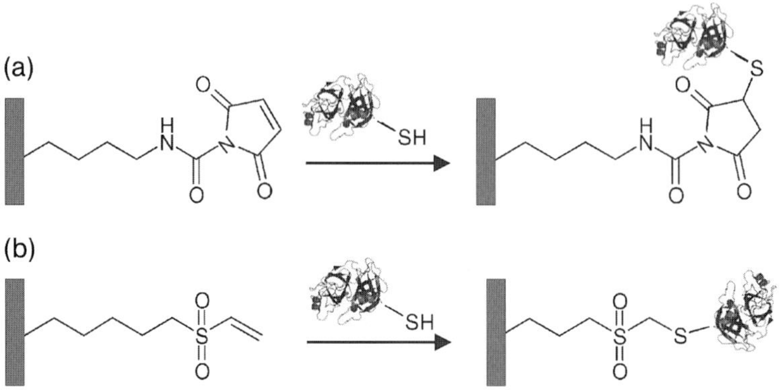

Figure 14 Michael addition reactions for bioconjugation.

has been reported to exhibit some crossreactivity with amines at pH values above the optimum pH range of 6.5–7.5.

More advanced reactions ensure a stereospecific attachment of appropriately derivatized biomolecules. In a Diels–Alder $4\pi + 2\pi$ cycloaddition, as introduced by Mrksich and coworkers, pentadienyl-derivatized molecules have been immobilized on SAMs that expose maleimide moieties [42]. An additional feature of this versatile reaction is the possibility to trigger the conversion by electrochemistry (i.e. on demand) via *in-situ* generation of the dienophil, i.e. quinone (Figure 15). This reaction has been exploited to immobilize the cell-adhesive peptide sequence RGD (Arg-Gly-Asp) [43] to SAM surfaces for cell–surface interaction studies [44].

A unique class of reactions has been developed in the last decade based on the long-known 1,3-Dipolar cycloaddition (Huisgen reaction). Using a copper catalyst the reaction between an azide and a terminal acetylene proceeds with essentially quantitative yield under mildest conditions to yield a stable 1,2,3-triazole. The tolerance of this reaction against other functional groups is remarkable. As a prime example of the so-called click chemistry, the 1,3-Dipolar cycloaddition has been pioneered by Sharpless and others and has found in the meantime ample application in the field of bioconjugation and surface modification [45, 46].

The reactions sketched above can be performed on SAMs, on pending side groups of a polymer that has been spin-coated on a solid substrate or on complex 3D structures [47]. One prerequisite is the presence or introduction of the surface-

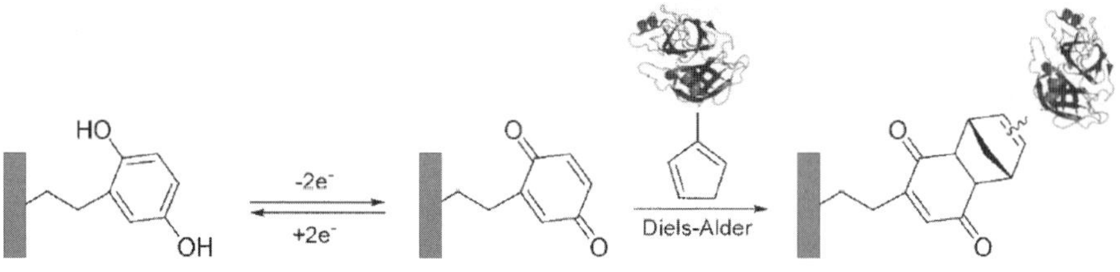

Figure 15 Diels–Alder $4\pi + 2\pi$ cycloaddition.

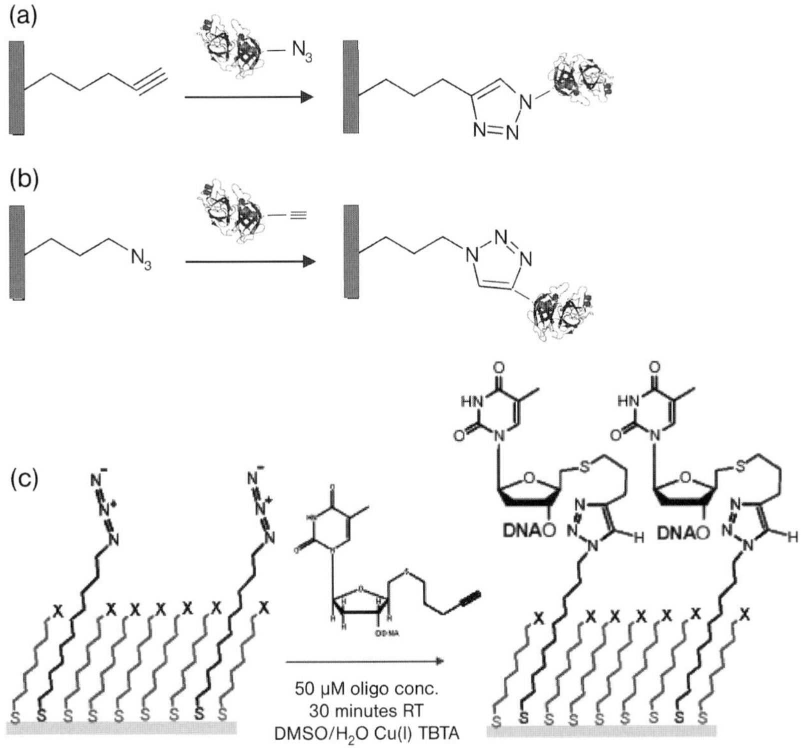

Figure 16 1,3-Dipolar cycloaddition of (a) acetylene or (b) surface-bound azide. (c) Surface modification of a mixed SAM on gold by chemoselective coupling of acetylene-bearing oligonucleotides. (Reproduced with permission from reference [46]; N. K. Devaraj, G. P. Miller, W. Ebina, B. Kakaradov, J. P. Collman, E. T. Kool, C. E. D. Chidsey. J. A. Chem. Soc. 2005, 127, 8600, Copyright 2005 by American Chemical Society).

immobilized functional group that is subsequently exploited in the attachment reaction. Unless the SAMs, etc., are equipped with the functional groups, the modification processes comprise multiple steps and hence the overall yields may be compromised. To reduce the number of reaction steps, so-called heterobifunctional spacers have been introduced with success. These spacers provide, in addition to the spatial separation of the surface and the active site, orthogonal functionalities that can be exploited in the two required subsequent reaction steps. Examples are isocyanate–maleimide or NHS-ester–maleimide molecules.

Figure 17 shows a sequence of reaction steps of a multistep reaction reported by Corn and coworkers [48]. Here, substitution reactions as well as addition reactions are employed together with heterobifunctional spacers to immobilize a maleimide or thiol-terminated biomolecule.

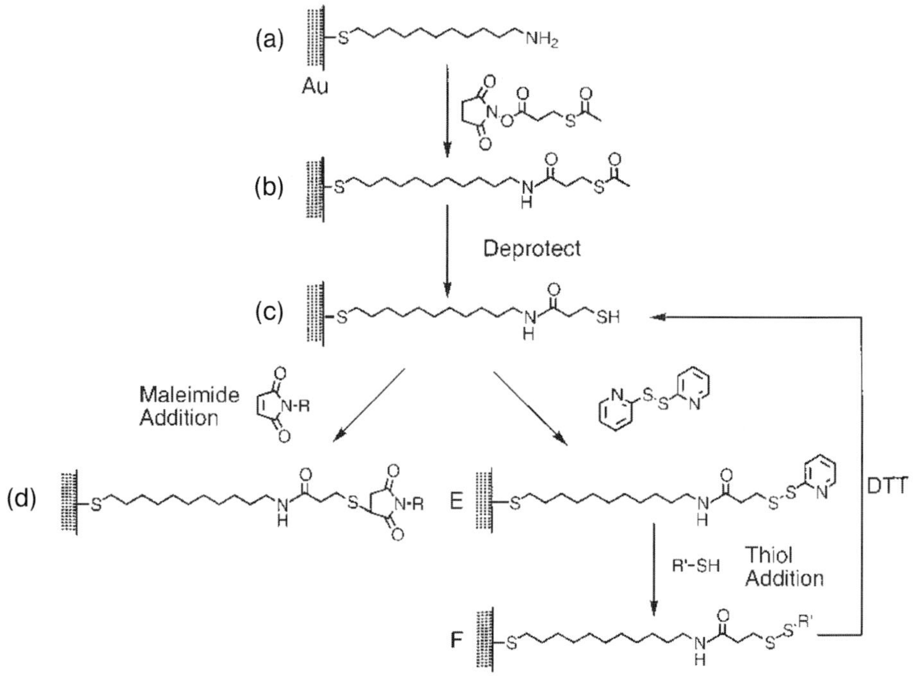

Figure 17 Multistep functionalization of an amino-terminated SAM employing heterobifunctional spacers and protection groups to generate and further functionalize a sulfhydryl-terminated adsorbate. (Reproduced with permission from reference [48]; E. A. Smith, M. J. Wanat, Y. F. Cheng, S. V. P. Barreira, A. G. Frutos, R. M. Corn, Langmuir 2001, 17, 2502, Copyright 2001 by American Chemical Society).

1.1.4.2
Noncovalent Attachment Reactions

Noncovalent bioconjugation reactions are frequently used as alternatives to the covalent immobilization schemes discussed above. In particular, the selectivity of the surface reactions can be enhanced. Two of the most prominent examples will be introduced here because of their relevance and widespread use.

The system comprising the proteins streptavidin or avidin, on the one hand, and biotin (vitamin H) on the other hand, is of special interest because it has one of the largest free energies of association yet observed for noncovalent binding of a protein and small ligand in aqueous media (Figure 18a). The interaction between biotin and the streptavidin is characterized by an extraordinarily high binding constant of ~ 10^{15} M^{-1} (avidin and biotin possess a similarly high binding constant of ~ 10^{13} M^{-1}) [49]. In practice, this interaction is so strong that the bond can be regarded as permanent. The complexes are also very stable over a wide range of temperature and pH. In addition, the reaction is very specific and because of the position of two pairs of biotin binding sites on opposite faces of the protein, it can thus serve as a molecular platform for the construction of designed interfacial architectures. These architectures are built up by immobilizing biotin, either using a polymeric coupling layer (Figure 18a) or surface-chemical modifications of SAMs (Figure 18b). The latter example also shows a beautiful extension of this concept to light-induced protection groups [50]. Upon irradiation, biotin is deprotected and hence available for binding (strep)avidin.

Since biotin is available with various functional groups, many alternative pathways are possible [49]. Pioneering work by Knoll and coworkers has shown that the optimum surface coverages in SAMs are on the order of 0.1; hence a matrix thiol is being coadsorbed to provide sufficient lateral dilution and to avoid the peculiar effects of surface reactions in highly organized assemblies [51, 52].

Many of the reactions introduced above require a defined surface chemistry or biomolecules with defined pending functional groups. Other than in synthesized DNA, where the chain-end function can be determined by synthetic design, lysine and cysteine residues of naturally occurring, isolated proteins are rarely unique. Thus, the covalent attachment does not provide control over the correct orientation of the immobilized protein because of lack of regiospecificity. Another drawback not yet mentioned refers to the blocking of a reactive site of a protein as a result of the immobilization procedure, thus reducing the overall activity of the protein.

To address this shortcoming metal-ion chelator systems have been further developed and refined. A most prominent approach is the use of the *N*-nitrilotriacetic acid (NTA)/His_6-tag chelator system that has been invented primarily as a universal approach for the isolation and purification of gene products in a single step [53, 54]. The NTA is typically immobilized on a surface and captures, in the presence of nickel ions, histidine-tagged biomolecules by exploiting the high affinity of histidine for Ni^{2+} ions. Thereby, the His-tagged protein is bound to the surface [55]. Since the His_6-tag is small and flexible the function of the protein is preserved. Furthermore, by adding imidazole or EDTA the complexation can be reversed, thereby giving access to reusable biochip surfaces. In Figure 19(a) and (b) a typical chelator thiol is shown, which can be employed in a diluted SAM with the matrix thiol shown [56]. Using the PLL-PEG adsorbate system similar chelate complexes can be formed (Figure 19c) [57]. More recently, protein binding by multivalent chelator surfaces was shown to be a strategy to overcome the known problems with low affinity and fast dissociation of the mono-NTA–His_6-tag interaction. As demonstrated by Tampe and coworkers, immobilized histidine-tagged proteins were uniformly oriented and retain their function (Figure 19e) [58].

Other approaches to control the orientation of, e.g., antibodies compromise the use of antibody fragments. Jenkins and coworkers have shown, for example, the effect of orientation in the detection of human chorionic gonadotrophin (hCG) by surface plasmon field-enhanced fluorescence spectroscopy (SPFS) using a sandwich assay with antibodies (Chapter 1.4) [59]. They observed an order of magnitude improvement in the limit of detection of an oriented Fab-hCG fragment compared to randomly immobilized entire hCG.

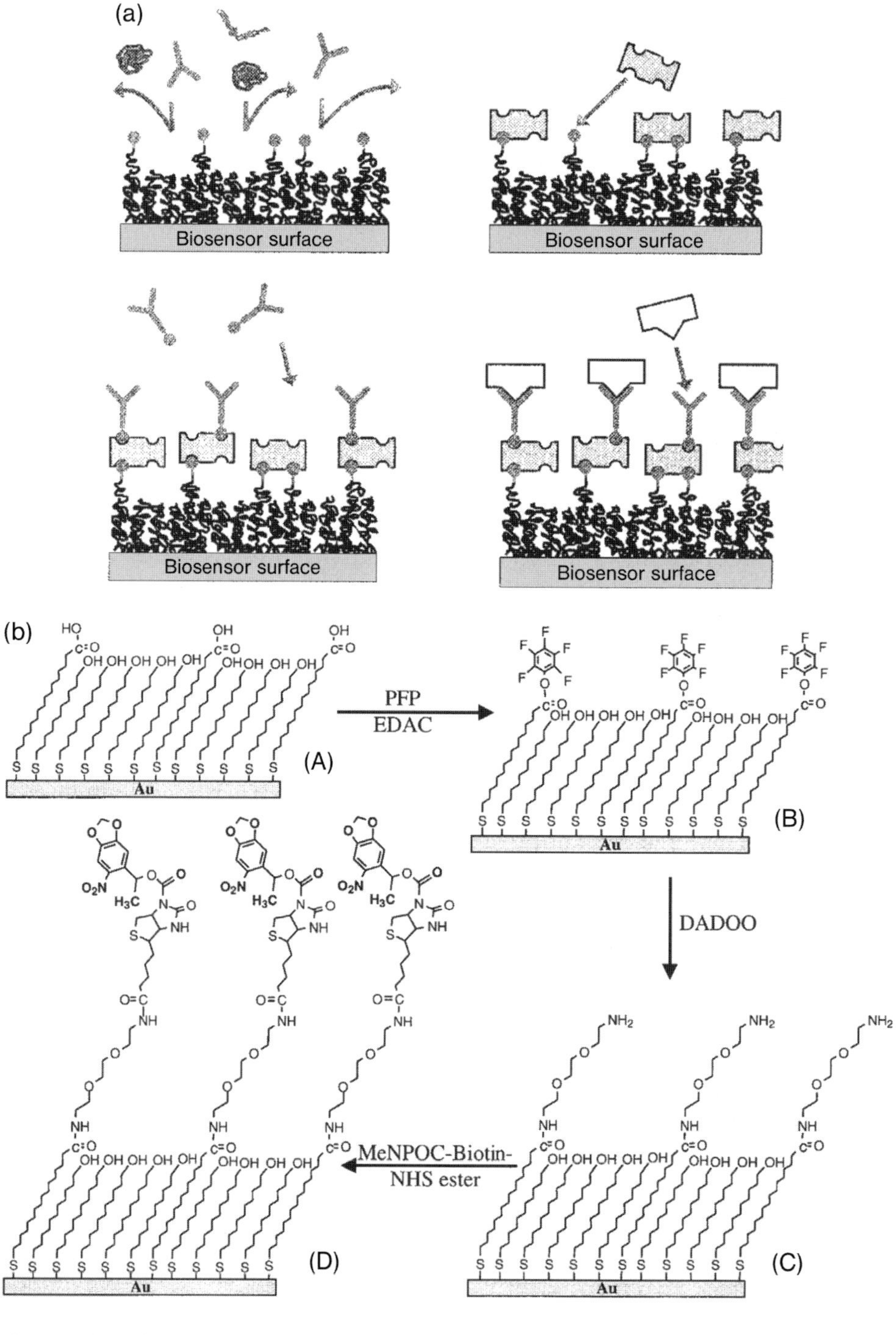

Figure 18 (legend see p. 23)

Finally, oriented antibody attachment can be mediated by protein A or protein G. Protein A binds to mammalian immunoglobulins, especially IgG. More specifically, protein A has the ability to bind to the Fc portion of IgG. If Protein A is immobilized on a surface its Fc receptors become points of attachment sites to which antibodies can be immobilized [60, 61] In a first step of the surface modification, the protein A is coupled to the surface using an approach that lacks the regiospecificity, e.g. by NHS ester chemistry. The interaction of IgG with the protein, however, is regiospecific, thereby providing a surface with correctly oriented IgG antibodies.

The reactions and examples introduced in this tutorial review chapter show, as mentioned in the introduction, only a narrow cross-section of relevant chemistries for coupling chemistries utilized for the modification and functionalization of surfaces to fabricate advanced biointerfaces. Many reactions that are successfully applied in the area of bioconjugation can be carried out with good results on the surface [62]. However, in particular on flat samples, such as typical biosensor chips, the peculiarities of surface chemical reactions may lead to unexpected behavior [24]. Therefore, the analysis of the surface prepared by a number of complementary advanced surface analytical techniques is a necessity [7, 63]. In particular, for label-free biosensing, heterogeneous responses of the sensor surface have an impact on the signal. Here, surface plasmon resonance (SPR) imaging has helped to unravel inhomogeneous coverages of ligands [64]. It is anticipated that advances and breakthroughs in the analytical instrumentation (e.g. by time-of-flight secondary ion mass spectrometry [65]) will render the imaging of surface chemical composition at smaller length scales than those accessible by imaging SPR possible and thereby help to clarify the role and impact of heterogeneities at biointerfaces.

Figure 18 (a) Schematic of biointerface fabrication using multisite adsorption of poly-L-lysine-poly(ethylene glycol) conjugates by dip-coating. The charged lysine residues adhere via electrostatic interactions on the oxidic biosensor surface, the PEG brushes prevent nonspecific adsorption of proteins, while the incorporated terminated biotin can be exploited to immobilize tetrafunctional streptavidin. The remaining binding sites of streptavidin serve as attachment sites for biotin-modified antibodies. (Reproduced with permission from reference [16]; N.-P. Huang, J. Voros, S. M. De Paul, M. Textor, N. D. Spencer, Langmuir 2002, 18, 220, Copyright 2002 by American Chemical Society). (b) Route to biotin-terminated SAMs: (A) First, a gold substrate is functionalized with a binary mixture of a hydroxyl-terminated and a carboxylic acid-terminated thiol to provide a reactive SAM on gold (Figure 1a). The carboxylic acid end groups enable subsequent derivatization of the SAM, while the hydroxyl groups resist nonspecific protein adsorption. The carboxylic acid end groups in the mixed SAM are converted to a reactive pentafluorophenyl ester (Figure 1b, B), and coupled to a bifunctional amine-terminated oligo(ethylene glycol) linker (Figure 1b, C). The terminal amine groups in the oligo(ethylene glycol)-derivatized SAM are conjugated with a caged biotin NHS-ester (Figure 1b, D). Finally, the caged biotin is deprotected by UV light through a photomask and subsequent incubation with streptavidin results in selective binding of streptavidin to the irradiated regions. (Reproduced with permission from reference [50]; Z. Yang, W. Frey, T. Oliver, A. Chilkoti, Langmuir 2000, 16, 1751, Copyright 2000 by American Chemical Society).

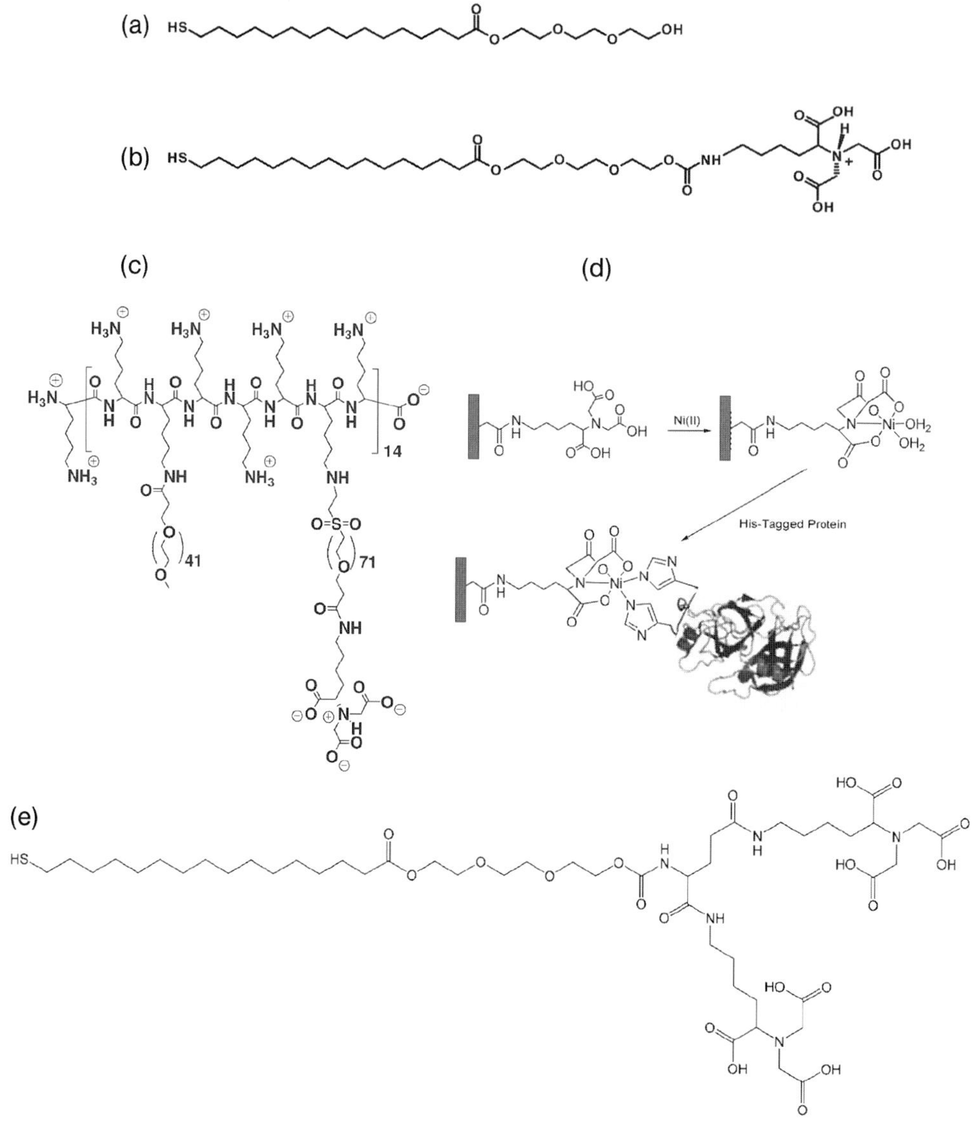

Figure 19 The thiols shown in panels (a) and (b) have been employed in mixed SAMs on gold to bind His-tagged proteins in the correct orientation. Similar protein capture has been demonstrated using the modified PLL-PEG polymer shown in panel (c). The details of the chelate formation are shown for a protein with 2 pending His tags in panel (d). (e) Structure of multivalent chelator thiol.

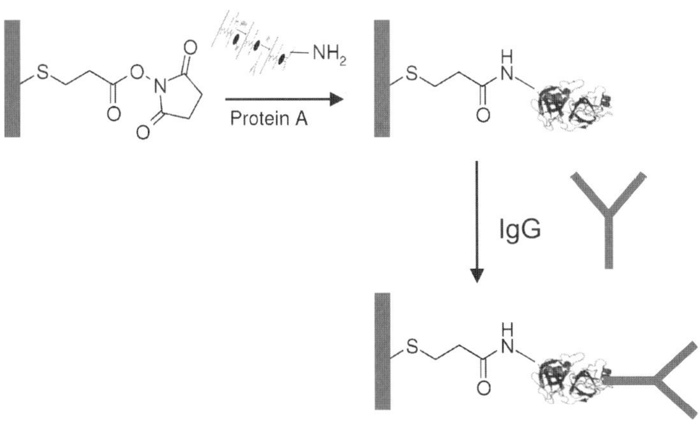

Figure 20 Oriented immobilization of IgG via covalently captured protein A.

Acknowledgment

The author is greatly indebted to the following people for fruitful and enlightening discussions, collaboration and guidance: Prof. Dr. G. Julius Vancso, Dr. Mark A. Hempenius, Prof. Dr. David N. Reinhoudt, Prof. Dr. Jurriaan Huskens (all University of Twente), Dr. Barbara Dordi (Sabic Innovative Plastics), Dr. Chuanliang Feng (BioMade), Prof. Dr. Steven D. Evans (University of Leeds), Dr. Victor Chechik (The University of York), Dr. Toby Jenkins (University of Bath), Prof. Dr. Wolfgang Knoll, Prof. Dr. Hans-Jürgen Butt, Dr. Renate Förch (all Max-Planck-Institute for Polymer Research, Mainz), Prof. Dr. Curtis W. Frank (Stanford University) and Prof. Dr. Helmut Ringsdorf (University of Mainz).

References

1 Kasemo, B. (2002) *Surface Science*, **500**, 656.
2 Castner, D.G. and Ratner, B.D. (2002) *Surface Science*, **500**, 28.
3 Habibovic, P., Yuan, H., van der Valk, C.M., Meijer, G., van Blitterswijk, C.A. and de Groot, K. (2005) *Biomaterials*, **26**, 3565.
4 Chen, C.S., Mrksich, M., Huang, S., Whitesides, G.M. and Ingber, D.E. (1997) *Science*, **276**, 1425.
5 Arnold, M., Cavalcanti-Adam, E.A., Glass, R., Blummel, J., Eck, W., Kantlehner, M., Kessler, H. and Spatz, J.P. (2004) *ChemPhysChem*, **5**, 383.
6 Visnapuu, M.-L., Duzdevich, D. and Greene, E.C. (2008) *Molecular BioSystems*, **4**, 394.
7 Gong, P. and Grainger, D.W. (2004) *Surface Science*, **570**, 67; Gong, P., Harbers, G.M. and Grainger, D.W. (2006) *Analytical Chemistry*, **78**, 2342.
8 Schönherr, H., Degenhart, G.H., Dordi, B., Feng, C.L., Rozkiewicz, D.I., Shovsky, A. and Vancso, G.J. (2006) *Advances in Polymer Science*, **200**, 169.
9 Love, J.C., Estroff, L.A., Kriebel, J.K., Nuzzo, R.G. and Whitesides, G.M. (2005) *Chemical Reviews*, **105**, 1103.

10 Netzer, L. and Sagiv, J. (1983) *Journal of the American Chemical Society*, **105**, 674.
11 Ulman, A. (1996) *Chemical Reviews*, **96**, 1533.
12 Sieval, A.B., Vleeming, V., Zuilhof, H. and Sudholter, E.J.R. (1999) *Langmuir*, **15**, 8288.
13 Yan, M.D., Cai, S.X., Wybourne, M.N. and Kenna, J.F.W. (1994) *Bioconjugate Chemistry*, **5**, 151.
14 Kenausis, G.L., Voros, J., Elbert, D.L., Huang, N.P., Hofer, R., Ruiz-Taylor, L., Textor, M., Hubbell, J.A. and Spencer, N.D. (2000) *The Journal of Physical Chemistry B*, **104**, 3298; Huang, N.P., Michel, R., Voros, J., Textor, M., Hofer, R., Rossi, A., Elbert, D.L., Hubbell, J.A. and Spencer, N.D. (2001) *Langmuir*, **17**, 489.
15 Förch, R., Zhang, Z.H. and Knoll, W. (2005) *Plasma Processes and Polymers*, **2**, 351.
16 Huang, N.-P., Voros, J., De Paul, S.M., Textor, M. and Spencer, N.D. (2002) *Langmuir*, **18**, 220.
17 Pasche, S., De Paul, S.M., Voros, J., Spencer, N.D. and Textor, M. (2003) *Langmuir*, **19**, 9216.
18 Senaratne, W., Andruzzi, L. and Ober, C.K. (2005) *Biomacromolecules*, **6**, 2427; (b) Edmondson, S., Osborne, V.L. and Huck, W.T.S. (2004) *Chemical Society Reviews*, **33**, 14.
19 Löføs, S. and Johnsson, B.J. (1990) *Chemical Society – Chemical Communications*, **21**, 1526; Xu, F., Persson, B., Löføs, S. and Knoll, W. (2006) *Langmuir*, **22**, 3352.
20 Feng, C.L., Zhang, Z., Förch, R., Knoll, W., Vancso, G.J. and Schönherr, H. (2005) *Biomacromolecules*, **6**, 3243.
21 Dusseiller, M.R., Schlaepfer, D., Koch, M., Kroschewski, R. and Textor, M. (2005) *Biomaterials*, **26**, 5917.
22 Chen, G., McCarley, R.L., Soper, S.A., Situma, C. and Bolivar, J.G. (2007) *Chemistry of Materials*, **19**, 3855.
23 McCarley, R.L., Vaidya, B., Wei, S., Smith, A.F., Patel, A.B., Feng, J., Murphy, M.C. and Soper, S.A. (2005) *Journal of the American Chemical Society*, **127**, 842.
24 Chechik, V., Crooks, R.M. and Stirling, C.J.M. (2000) *Advanced Materials*, **12**, 1161.
25 Cameron, J.F. and Frechet, J.M.J. (1991) *Journal of the American Chemical Society*, **113**, 4303.
26 Schönherr, H., Chechik, V., Stirling, C.J.M. and Vancso, G.J. (2000) *Journal of the American Chemical Society*, **122**, 3679.
27 Dordi, B., Pickering, J.P., Schönherr, H. and Vancso, G.J. (2004) *Surface Science*, **570**, 57.
28 Dordi, B., Schönherr, H. and Vancso, G.J. (2003) *Langmuir*, **19**, 5780.
29 Schönherr, H., Feng, C.L. and Shovsky, A. (2003) *Langmuir*, **19**, 10843.
30 Chechik, V. and Stirling, C.J.M. (1997) *Langmuir*, **13**, 6354.
31 Zou, S., Zhang, Z., Förch, R., Knoll, W., Schönherr, H. and Vancso, G.J. (2003) *Langmuir*, **19**, 8618.
32 Zou, S., Schönherr, H. and Vancso, G.J. (2005) *Journal of the American Chemical Society*, **127**, 11230; Embrechts, A., Schönherr, H. and Vancso, G.J. (2008) *The Journal of Physical Chemistry. B*, **112**, 7359.
33 Staros, J.V., Wright, R.W. and Swingle, D.M. (1986) *Analytical Biochemistry*, **156**, 220.
34 Lahiri, J., Isaacs, L., Tien, J. and Whitesides, G.M. (1999) *Analytical Chemistry*, **71**, 777.
35 Yang, M., Teeuwen, R.L.M., Giesbers, M., Baggerman, J., Arafat, A., de Wolf, F.A., van Hest, J.C.M. and Zuilhof, H. (2008) *Langmuir*, **24**, 7931; Wojtyk, J.T.C., Morin, K.A., Boukherroub, R. and Wayner, D.D.M. (2002) *Langmuir*, **18**, 6081.Duhachek, S.D., Kenseth, J.R., Casale, G.P., Small, G.J., Porter, M.D. and Jankowiak, R. (2000) *Analytical Chemistry*, **72**, 3709; Wagner, P., Hegner, M., Kernen, P., Zaugg, F. and Semenza, G. (1996) *Biophysical Journal*, **70**, 2052.
36 Frey, B.L. and Corn, R.M. (1996) *Analytical Chemistry*, **68**, 3187.
37 Feng, C.L., Embrechts, A., Bredebusch, I., Bouma, A., Schnekenburger, J., García-Parajó, M., Domschke, W., Vancso, G.J. and Schönherr, H. (2007) *European Polymer Journal*, **43**, 2177.

38 Chapman, R.G., Ostuni, E., Yan, L. and Whitesides, G.M. (2000) *Langmuir*, **16**, 6927.

39 Corey, D.R., Munoz-Medellin, D. and Huang, A. (1995) *Bioconjugate Chemistry*, **6**, 93; Kunath, K., Merdan, T., Hegener, O., Haberlein, H. and Kissel, T.J. (2003) *Gene Medicine*, **5**, 588.

40 Peelen, D. and Smith, L.M. (2005) *Langmuir*, **21**, 266; D'Souza, S.F. and Godbole, S.S. (2002) *Journal of Biochemical and Biophysical Methods*, **52**, 59; Choi, H.J., Kimb, N.H., Chung, B.H. and Seong, G.H. (2005) *Analytical Biochemistry*, **347**, 60; Betancor, L., Lopez-Gallego, F., Hidalgo, A., Alonso-Morales, N., Mateo, C., Fernandez-Lafuente, R. and Guisan, J.M. (2006) *Enzyme and Microbial Technology*, **39**, 877; MacBeath, G. and Schreiber, S.L. (2000) *Science*, **289**, 1760.

41 Sid Masri, M. and Friedman, M. (1988) *Journal of Protein Chemistry*, **7**, 49.

42 Houseman, B.T., Huh, J.H., Kron, S.J. and Mrksich, M. (2002) *Nature Biotechnology*, **20**, 270; de Araffljo, A.D., Palomo, J.M., Cramer, J., Kohn, M., Schroder, M., Wacker, R., Niemeyer, C., Alexandrov, K. and Waldmann, H. (2006) *Angewandte Chemie – International Edition*, **45**, 296.

43 Ruoslahti, E. (1996) *Annual Review of Cell and Developmental Biology*, **12**, 697.

44 Yousaf, M.N., Houseman, B.T. and Mrksich, M. (2001) *Angewandte Chemie – International Edition*, **40**, 1093.

45 Duckworth, B.P., Xu, J., Taton, T.A., Guo, A. and Distefano, M.D. (2006) *Bioconjugate Chemistry*, **17**, 967; Collman, J.P., Devaraj, N.K., Eberspacher, T.P.A. and Chidsey, C.E.D. (2006) *Langmuir*, **22**, 2457; Lummerstorfer, T. and Hoffmann, H. (2004) *The Journal of Physical Chemistry B*, **108**, 3963.

46 Devaraj, N.K., Miller, G.P., Ebina, W., Kakaradov, B., Collman, J.P., Kool, E.T. and Chidsey, C.E.D. (2005) *Journal of the American Chemical Society*, **127**, 8600.

47 Goddard, J.M. and Hotchkiss, J.H. (2007) *Progress in Polymer Science*, **32**, 698.

48 Smith, E.A., Wanat, M.J., Cheng, Y.F., Barreira, S.V.P., Frutos, A.G. and Corn, R.M. (2001) *Langmuir*, **17**, 2502.

49 Wilchek, M. and Bayer, E. (1990) *Avidin-Biotin Technology Methods in Enzymology*, Vol. 184, Academic Press, San Diego, CA.

50 Yang, Z., Frey, W., Oliver, T. and Chilkoti, A. (2000) *Langmuir*, **16**, 1751.

51 Spinke, J., Liley, M., Schmitt, F.J., Guder, H.J., Angermaier, L. and Knoll, W. (1993) *Journal of Chemical Physics*, **99**, 7012.

52 Knoll, W., Park, H., Sinner, E.-K., Yao, D. and Yu, F. (2004) *Surface Science*, **570**, 30.

53 Arnold, F.H. (1992) *Metal-Affinity Protein Separations*, Academic Press, San Diego, CA.

54 Hochuli, E. (1990) *Genetic Engineering*, **12**, 87; Hochuli, E., Bannwarth, W., Dçbeli, H., Gentz, R. and Stuber, D. (1988) *Bio-Techniques*, **6**, 1321.

55 Nieba, L., Nieba-Axmann, S.E., Persson, A., Hamalaine, M., Edebratt, F., Hansson, A., Lidholm, J., Magnusson, K., Karlsson, A.F. and Pluckthun, A. (1997) *Analytical Biochemistry*, **252**, 217.

56 Gamsjaeger, R., Wimmer, B., Kahr, H., Tinazli, A., Picuric, S., Lata, S., Tampe, R., Maulet, Y., Gruber, H.J., Hinterdorfer, P. and Romanin, C. (2004) *Langmuir*, **20**, 5885.

57 Zhen, G., Falconnet, D., Kuennemann, E., Vörös, J., Spencer, N.D., Textor, M. and Zürcher, S. (2006) *Advanced Functional Materials*, **16**, 243.

58 Tinazli, A., Tang, J., Valiokas, R., Picuric, S., Lata, S., Piehler, J., Liedberg, B. and Tampe, R. (2005) *Chemistry – A European Journal*, **11**, 5249.

59 Vareiro, M.L.M., Liu, J., Knoll, W., Zak, K., Williams, D. and Jenkins, A.T.A. (2005) *Analytical Chemistry*, **77**, 2426.

60 Turkova, J. (1999) *Journal of Chromatography B*, **722**, 11.

61 Dutra, R.F., Castro, C.M.H.B. and Azevedo, C.R. (2000) *Biosensors and Bioelectronics*, **15**, 511.

62 Li, X.-M., Huskens, J. and Reinhoudt, D.N. (2004) *Journal of Materials Chemistry*, **14**, 2954; Rusmini, F., Zhong, Z. and Feijen, J. (2007) *Biomacromolecules*, **8**, 1775.
63 Grainger, D.W. and Castner, D.G. (2008) *Advanced Materials*, **20**, 867.
64 Schasfoort, R.B.M. and Tudos, A.J. (eds) (2008) *Handbook of Surface Plasmon Resonance*, RSC Publishing.
65 Takahashi, H., Emoto, K., Dubey, M., Castner, D.G. and Grainger, D.W. (2008) *Advanced Functional Materials*, **18**, 2079.

1.2
Tutorial Review: Surface Plasmon Resonance-Based Biosensors

Jakub Dostálek, Chun Jen Huang, and Wolfgang Knoll

1.2.1
Introduction

Over the last two decades, rapid and sensitive detection of chemical and biological species have become increasingly important in various areas such as medical diagnostics (e.g. detection of disease markers and drug development), food control (e.g. detection of food borne pathogens and toxins) or environmental monitoring (e.g. detection of pollutants). The urgent need for novel analytical tools in these areas has stimulated research in biosensors. Among various biosensor technologies [1–3], those based on surface plasmon resonance (SPR) are rapidly gaining popularity. SPR biosensors allow for sensitive, label-free and real-time detection of molecular binding events through the monitoring of refractive-index changes induced by the capture of analyte of interest on their surface. Since their invention at the end of the last century [4], SPR biosensors have become an established technology for the interaction analysis of macromolecules in pharmaceutical and life-sciences research and they have made great strides to expand to the other application areas.

The aim of this chapter is to provide a brief and practical guide to SPR biosensors. Further, we describe the fundamentals of surface plasmon resonance optics (Sections 1.2.2 and 1.2.3), the implementation of SPR biosensors (Section 1.2.4), sensitivity (Section 1.2.5), evaluation of the sensor output (Section 2.2.6) and current state-of-the-art applications of SPR biosensor technology (Section 2.2.7).

1.2.2
Surface Plasmons

Surface plasmons (SPs) are optical waves that originate from coupled collective oscillations of the electron plasma and the associated electromagnetic field on metallic surfaces [5], see Figure 1. SPs propagate along a planar interface between a semi-infinite metal and a dielectric with the (complex) propagation constant β:

Surface Design: Applications in Bioscience and Nanotechnology
Edited by Renate Förch, Holger Schönherr, and A. Tobias A. Jenkins

ISBN: 978-3-527-40789-7

$$\beta = \frac{2\pi}{\lambda}\sqrt{\frac{n_m^2 n_d^2}{n_m^2 + n_d^2}} = \frac{\omega}{c}\sqrt{\frac{n_m^2 n_d^2}{n_m^2 + n_d^2}} \tag{1}$$

where λ is the wavelength, ω is the angular frequency, c is the speed of light in vacuum, n_d is the refractive index of the dielectric and n_m is the (complex) refractive index of the metal. SPs can be excited on a surface of a metal at wavelengths below its plasma frequency ω_p. For the excitation of surface plasmons in the visible and near-infrared part of the spectrum, mostly noble metals such as gold, silver or aluminum are used. The dispersion relation of SPs propagating on a surface of these metals is shown in Figure 2(a).

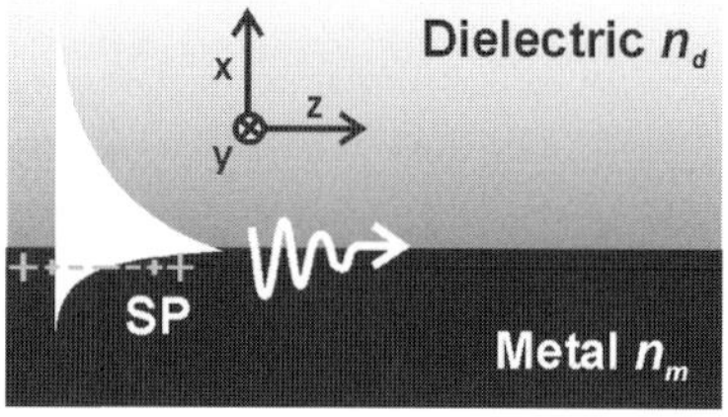

Figure 1 Surface plasmon propagating along a metal/dielectric interface.

The electromagnetic field of SP is transverse magnetically (TM) polarized. By using the Cartesian coordinates shown in Figure 1, this field can be described by the following nonzero components: magnetic intensity parallel to the interface H_y, electric intensity parallel to the interface E_z and electric intensity perpendicular to the interface E_x. As seen in Figure 2(b), the SP field decays exponentially from the metal/dielectric interface with the penetration depths $L^\mathrm{m}{}_\mathrm{pen}$ and $L^\mathrm{d}{}_\mathrm{pen}$ into the metal and dielectric, respectively. The penetration depth is defined as the distance perpendicular to the surface at which the field amplitude decreases by a factor 1/e ($L^j{}_\mathrm{pen} = \mathrm{Re}\{\beta^2 - k_0{}^2 n_j{}^2\}^{-1/2}$ where j is m or d for the metal and dielectric, respectively, and Re{} is a real part of a complex number). The energy of SP dissipates while it propagates along the metal surface due to the losses within the metal film. This damping can be quantified by the propagation length L_pro as the distance along the metallic surface at which the intensity of the SP mode drops to 1/e ($L_\mathrm{pro} = (2\mathrm{Im}\{\beta\})^{-1}$ where Im{} is the imaginary part of a complex number).

In order to illustrate typical characteristics of SPs, let us assume a wavelength in the red part of the spectrum $\lambda = 0.633$ μm and the interface between gold with the refractive index of $n_\mathrm{m} = 0.1 + 3.5\mathrm{i}$ and a dielectric with the refractive index of $n_\mathrm{d} = 1.33$. For these parameters, the SP exhibits the propagation length and penetration depths summarized in Table 1.

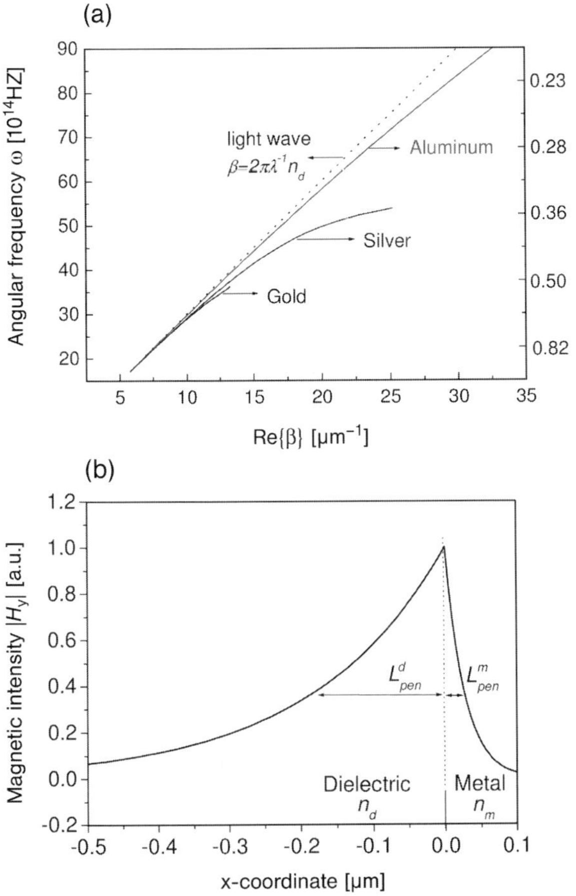

Figure 2 (a) Dispersion relation of the SP at the interface between air ($n_d = 1$) and gold, silver and aluminum. (b) Distribution of the magnetic intensity field H_y of the SP propagating along the interface of gold ($n_m = 0.1 + 3.5$) and a dielectric ($n_d = 1.33$) at a wavelength of $\lambda = 0.633$ μm.

Table 1 Characteristics of surface plasmon propagating along a gold surface.

Characteristics of surface plasmon			
Wavelength	$\lambda = 0.633$ μm	Penetration depth in gold	$L^m{}_{pen} = 0.027$ μm
Gold refractive index	$n_m = 0.1 + 3.5i$	Penetration depth in dielectric	$L^d{}_{pen} = 0.183$ μm
Dielectric refractive index	$n_d = 1.33$	Propagation length	$L_{pro} = 7.2$ μm

1.2.3
Optical Excitation of Surface Plasmons

A light wave can efficiently couple to a surface plasmon when the component of its wave vector that is parallel to the interface matches the real part of the surface plasmon propagation constant Re{β}. However, as shown in Figure 2(a) the wave vector of a light wave in the dielectric $2\pi\lambda^{-1}n_d$ is always lower than Re{β}. Therefore, direct excitation of SPs on a planar metal/dielectric interface is not possible and a coupler needs to be used to enhance the wave vector of the light wave. The most often used SP couplers utilize the attenuated total reflection method (ATR) and a high refractive index prism (prism coupler). Another possibility is to employ the diffraction on periodic modulated metallic surfaces (grating coupler). Schematics of the prism coupler and grating coupler are shown in Figure 3.

In the ATR prism coupler with Kretschmann geometry, a light beam is launched into a glass prism (refractive index n_p) with a thin metal film (refractive index n_m) and a lower refractive index dielectric (refractive index $n_d < n_p$) on its base (see Figure 3a). The light beam is total internal reflected at the prism base under an angle θ and penetrates into the metal film via its evanescent field. For thin metal films, the evanescent field reaches the outer interface between the metal and the lower refractive index dielectric at which a surface plasmon propagates with the propagation constant (1). As the prism refractive index n_p is larger than that of the dielectric n_d, the component of the light beam wave vector that is parallel to the surface $2\pi\lambda^{-1}n_p\sin(\theta)$ can be matched to that of the SP:

$$\frac{2\pi}{\lambda}n_p \sin(\theta) = \mathrm{Re}\{\beta\} \qquad (2)$$

If the condition (2) is fulfilled, the coupling between the light beam and SPs can be established. As seen in Figure 4(a), the excitation of SP is observed as a resonant dip in the reflectivity spectrum. In addition, this figure shows that the coupling strength to SP can be controlled by changing the thickness of the metal film d_m. For instance, for the wavelength of 0.633 μm, a prism with the refractive

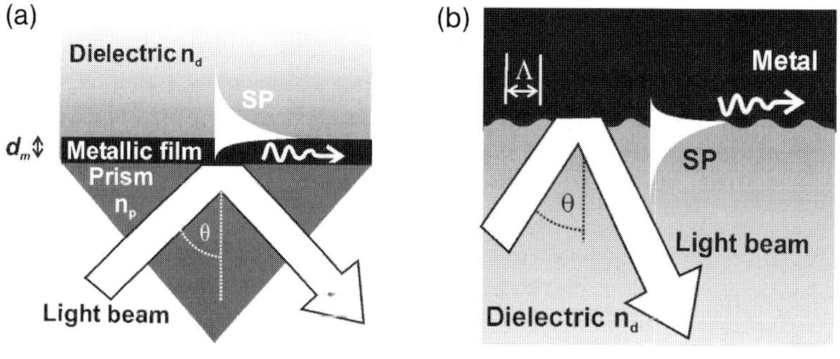

Figure 3 Optical excitation of surface plasmons by means of (a) prism coupler and (b) grating coupler.

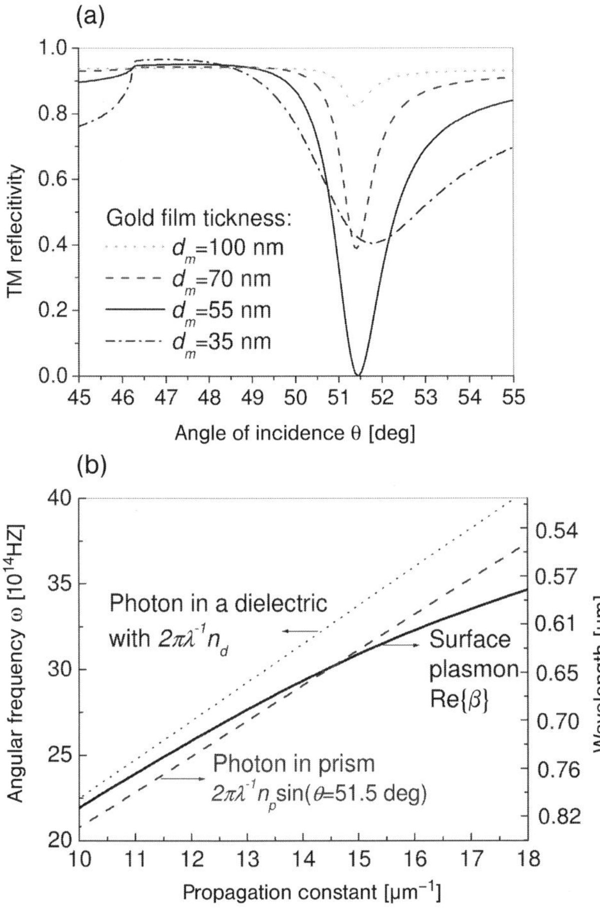

Figure 4 (a) Angular reflectivity spectra for the excitation of SPs at the wavelength of $\lambda = 0.633$ μm on an interface between gold ($n_m = 0.1+3.5i$) and a dielectric ($n_d = 1.33$) using a prism coupler ($n_p = 1.845$) and a gold film with the thickness d_m between 35 and 100 nm. (b) The comparison of the dispersion relation of a SP and photons in the prism and the dielectric for the same parameters.

index of $n_p = 1.845$, a gold film with the refractive index of $n_m = 0.1+3.5i$ and a dielectric with $n_d = 1.33$, the optimum thickness that enables the full coupling to SPs is of $d_m = 55$ nm. Figure 4(b) shows the dispersion relation of a light wave in the dielectric and in the prism compared to that of surface plasmons for the same materials. It reveals that the dispersion relation of the light wave in the prism crosses that of SP for the angle of incidence $\theta = 51.5$ deg and a wavelength of $\lambda = 0.633$ μm, which are identical to those for which the condition (2) is fulfilled and the minimum of SPR dip in Figure 4(a) occurs.

1.2.4
Implementation of SPR Biosensors

Surface plasmon resonance (SPR) biosensors are devices that incorporate two key components: (a) optical setup for the excitation and interrogation of *surface plasmons* and (b) *biomolecular recognition elements* (BRE). The biomolecular recognition elements are typically immobilized on a metallic sensor surface and they can specifically capture the analyte molecules from analyzed liquid samples, see Figure 5. The binding of analyte molecules induces a change in the refractive index in the vicinity to the metallic surface δn_d, which alters the propagation constant of surface plasmons $\delta\beta$ and consequently changes the coupling conditions to surface plasmons. Therefore, molecular binding events can be detected by measuring variations of the characteristics of a light beam that is resonantly coupled to surface plasmons, e.g. as a shift in the SPR reflectivity dip.

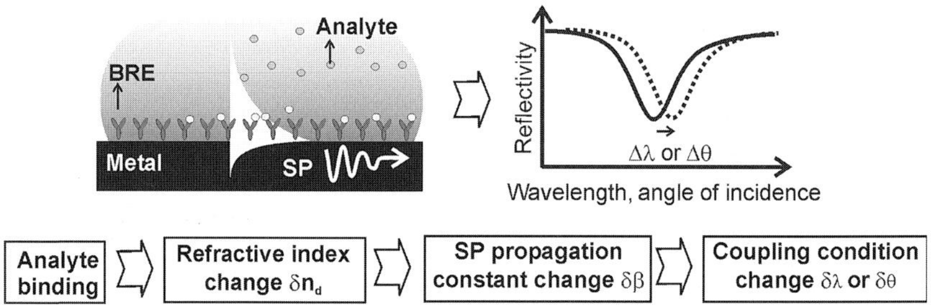

Figure 5 Principle of operation of SPR biosensors utilizing the detection of binding-induced refractive-index variations.

SPR Optical Platforms

Since the invention of SPR biosensors in the beginning of the 1980s [4], prism couplers utilizing the Kretschmann geometry have been mostly employed for the excitation of surface plasmons. In these devices, a sensor chip consisting of a glass slide with a thin metal film and attached biomolecular recognition elements is optically matched to a prism, see Figure 6(a). Besides the prism coupler, grating couplers were used in the design of SPR biosensors, see Figure 6(b). The periodic relief surface of a metallic grating enables the diffraction coupling of a light wave to surface plasmons and its first implementation in an SPR biosensor was demonstrated at the end of the 1980s [6]. In contrast to prism couplers, grating coupler-based SPR biosensors do not require an optical matching of a sensor chip. Furthermore, the sensor chip can be prepared from plastics [7] using mass-production-compatible technologies, such as hot embossing or injection molding. Moreover, optical waveguides [8] and optical fibers [9] have been investigated for the design of miniaturized SPR biosensors.

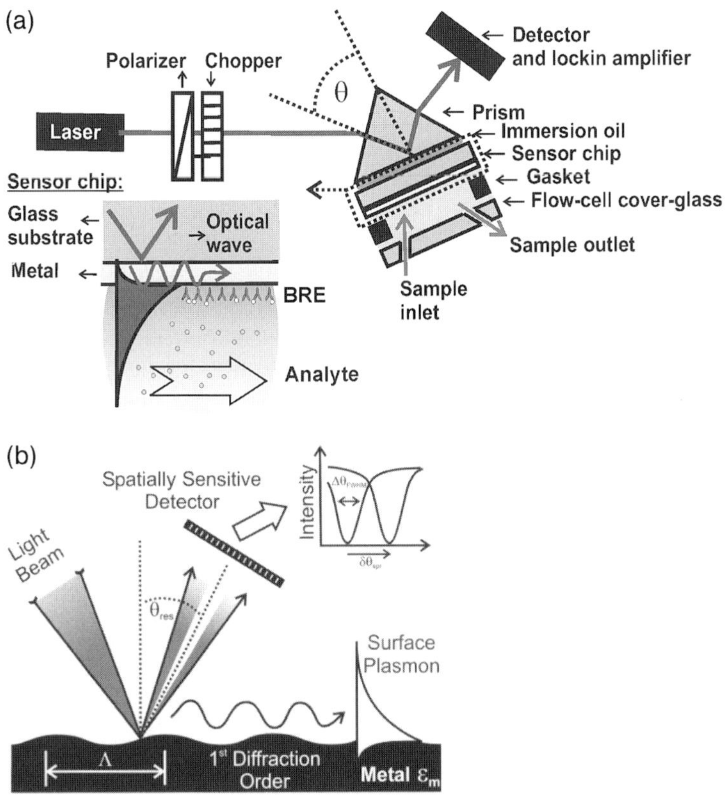

Figure 6 SPR optical platforms supporting SPR biosensors based on (a) prism coupler and angular modulation of SPR and (b) grating coupler and angular modulation of SPR. (Reprinted with permission from [10]).

Modulation of SPR

For the detection of binding-induced changes in the surface plasmons propagation constant $\delta\beta$, variations in the characteristics of a light wave coupled to surface plasmons are measured. These variations can occur in the angular as well as wavelength spectrum of a light wave at the output of a SPR coupler. In the sensors utilizing angular or wavelength modulation of SPR, the angular or wavelength reflectivity spectrum is measured and changes in the spectral position of the SPR dip minimum is determined by using a metric such as centroid, polynomial or Lorentzian curve fitting [11–13].

In devices based on angular modulation of SPR [14], a monochromatic parallel light beam is launched into a SPR coupler that is mounted on a motorized rotation stage and the angular dependence of light-wave intensity is detected using a photodiode, see Figure 6(a). In addition, SPR instruments with angular modulation of SPR that avoid using moving mechanical parts have been designed. In these instruments, a convergent light beam is launched into a coupler and the angular distribution of intensity of the light beam is detected by a spatially sensitive sensor such as a charge coupled device (CCD), see Figure 6(b). Another approach based on an acousto-optical deflector for fast scanning of angular SPR reflectivity was reported recently [15]. In the wavelength modulation of SPR [16], a parallel polychromatic light beam is coupled to surface plasmons and its wavelength spectrum is detected by a spectrometer [17]. In addition, the application of an acousto-optical filter for the wavelength scanning was demonstrated [18].

The change in the SP propagation constant $\delta\beta$ can also be measured using the intensity modulation of SPR. In this method, both angle of incidence and the wavelength are set fixed in the region where the slope of the SPR reflectivity dip is maximum. The shift of the SPR dip is detected from variations in the intensity of the reflected light beam. The intensity modulation has found its application in SPR measurements with spatial interrogation referred as to SPR imaging or SPR microscopy [19].

Sensors based on the intensity modulation of SPR can typically detect changes around 10^{-5} refractive index units (RIU) [20]. The analysis of the whole spectrum by using the angular and wavelength modulation of SPR allows for more accurate measurements [21]. Typically, instruments relying on this approach are capable of detecting refractive-index variations as low as 10^{-7} RIU. Currently, the best SPR sensors have reached the resolution of 10^{-8} RIU [15, 22].

Biomolecular Recognition Elements

Antibodies and oligonucleotides are commonly used for the specific capture of target molecules at the surface of SPR biosensors. These biomolecules exhibit uniquely specific binding to a wide range of analytes and recently their derivatives such as single-chain antibody fragments (scFvs) or peptide nucleic acid (PNA) have been developed, see Figure 7. Single-chain Fv (scFv) antibody fragments have a molecular mass of about 25 kDa and they are the smallest antibody entities comprising the antigen-binding site [23, 24]. Owing to their small size, they can be immobilized on the sensor surface with a higher density than whole antibodies. Phage and ribosome display techniques [25] allow for the *in-vitro* generation of high-affinity scFv molecules against virtually any molecular targets. For DNA sensing, PNAs composed of identical bases as DNA and of an uncharged backbone can be used for the analyte recognition. PNA can bind to DNA molecules with higher strength than the complementary DNA due to the lower electrostatic repulsion. Furthermore, other biomolecular recognition elements including carbohydrate molecules [26], artificial small nucleic acid ligands referred to as aptamers [27] and whole bacterial phages [28] were shown to be suitable for applications in SPR biosensors.

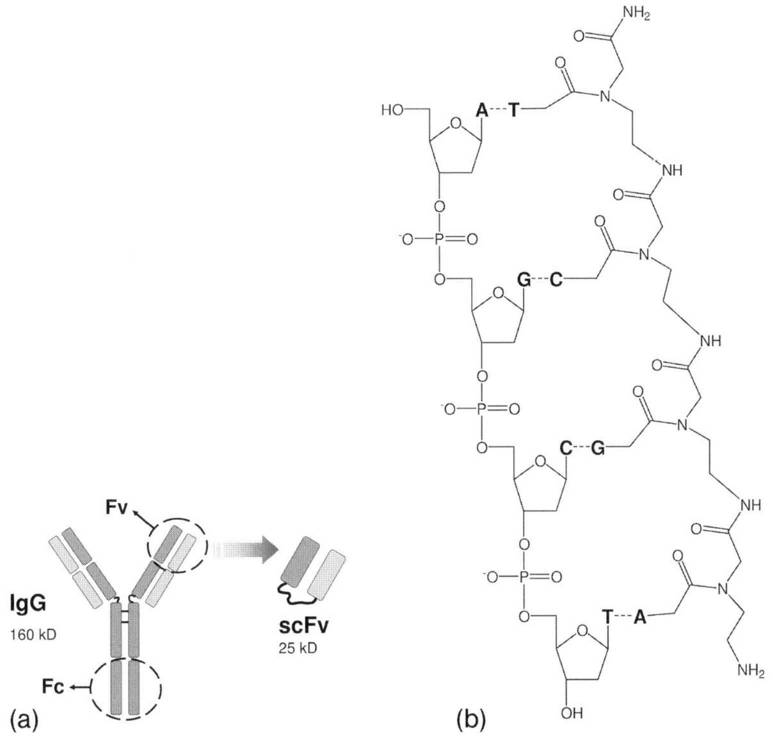

Figure 7 (a) Schematic of IgG molecule and its single-chain fragment scFv; (b) the specific interaction of DNA (left) and PNA (right) molecules.

Immobilization of Biomolecular Recognition Elements

In SPR biosensors, stable linkage between biomolecular recognition elements and the sensor surface need to be established. Moreover, the immobilized biomolecules have to retain their biological activity and the surface architecture should exhibit nonfouling properties minimizing the nonspecific interactions with analyzed samples. Within this section, most common immobilization methods used for the design of SPR biosensors are summarized. Let us note that a more general overview of various techniques developed for the immobilization of biomolecules can be found in Chapter 1.5 entitled 'Modern Biological Sensors'.

In SPR biosensors, gold films are preferably used for the excitation of surface plasmons due to their stability. Physical adsorption enables the direct immobilization of biomolecules on a gold surface through ionic, hydrophobic and polar interactions [29]. This method allows for simple coupling, however, the binding strength is typically weak and the orientation and homogeneity of immobilized molecules cannot be well controlled. In addition, conformation changes induced by the close proximity to the gold surface often reduce the biological activity of immobilized molecules.

More stable linking to metallic surfaces can be achieved by the covalent coupling. Most commonly, a self-assembled monolayer (SAM) of alkane thiol-derivatives is immobilized on a metallic surface to provide desired functional groups, see Figure 8. For instance, SAM of carboxylic-acid-terminated thiols can be activated by an active ester chemistry (e.g. by using N-1-Ethyl-3-[3-dimethylaminopropyl] carbodiimide hydrochloride [EDC] and hydroxysulfosuccinimide [NHS]) and covalently linked with the desired molecules via their primary amine groups [30]. Other reactive groups, such as thiol, or aldehyde, were also exploited for covalent linking of different molecules [31].

For the affinity coupling of biomolecules, the interaction between streptavidin (or its derivatives) and biotin is often employed. Recently, biotin-terminated alkane thiols are available for the self-assembly on metallic surfaces [32]. To such surface, various biotin-labeled molecules can be attached by using streptavidin as a linker, see Figure 9(a). Immunoglobulin G (IgG) antibodies are important biomolecular recognition elements as they are commercially available for wide range of analytes. As Fc fragment of IgG antibodies can affinity bind to proteins A and G, IgG molecules can be immobilized on a sensor surface that was modified with these proteins. This method allows for efficient immunoreactions at the surface as the coupling through Fc fragment provides the proper orientation of IgG binding sites towards the analyte [34]. Currently, new affinity coupling methods based on the interaction between histidine-tagged and chelated metal ions have been introduced to SPR biosensors. This method takes advantage of proteins modified by genetic engineering with affinity tags that can be used for site-specific immobilization on SPR biosensor chips [35, 36].

In order to immobilize biomolecular recognition elements with a high surface density, three-dimensional binding matrices were developed. The most commonly used binding matrix is based on grafted carboxymethyl dextran (CMD), see Figure 9(b). CMD-based matrix offers a high binding capacity, good nonfouling

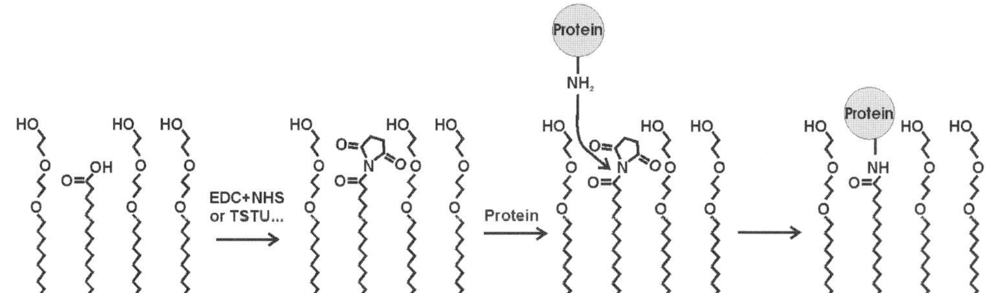

Figure 8 A mixed thiol SAM on a gold surface with PEG nonfouling background and carboxyl groups used for the amine coupling of a protein.

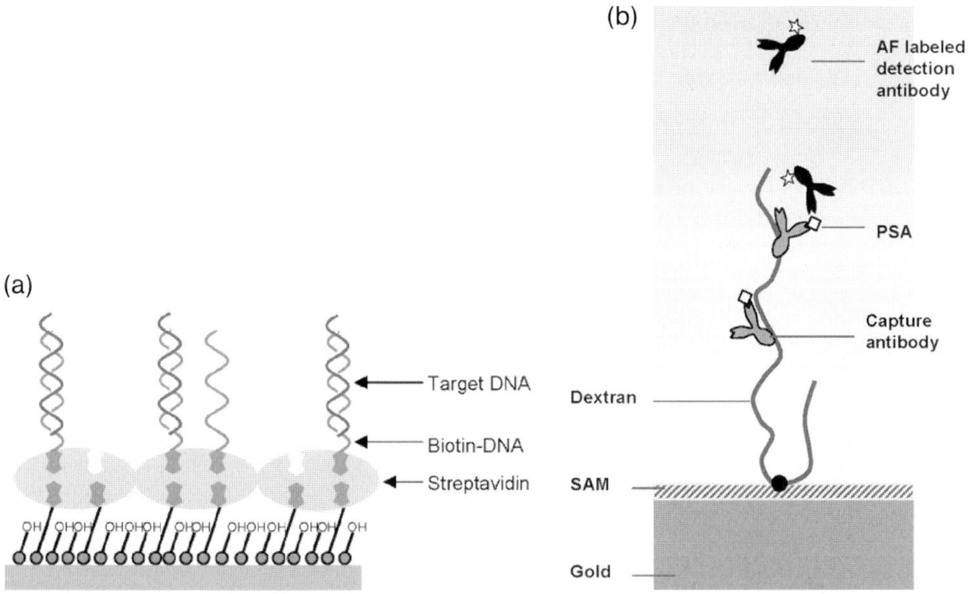

Figure 9 Biomolecular recognition elements attached at a gold surface in the form of (a) single layer attached using a thiol linker and (b) into a three-dimensional dextran matrix. (Reprinted with permission from [32, 33]).

properties and the biological activity of coupled biomolecules is retained to a high degree due to the coupling to its flexible chains [37].

For the design of surfaces with nonfouling properties, mixed thiol SAMs are commonly used. Poly(ethylene glycol) (PEG)-terminated thiol is one of the most efficient protein-repelling and biocompatible materials. PEG has been applied in many bio-related fields, including tissue engineering, drug delivery and biosensors [38]. In addition, other candidates for design of nonfouling surfaces have been investigated including agarose, mannitol, polyacrylamide, pluronics, and bovine serum albumin [39].

1.2.5 Sensitivity of a SPR Biosensor to Biomolecular Binding

SPR biosensors are optical devices that measure refractive-index changes induced by the binding of target molecules to the sensor surface. Further, let us investigate the sensitivity of SPR biosensors that utilize the angular modulation of SPR and a prism coupler depicted in Figure 6(a). The output of an angular spectroscopy-based SPR biosensor is the change in the resonant angle of incidence $\delta\theta_{res}$, which corresponds to a shift of a SPR dip in the angular reflectivity spectrum, see Figure 10. For SPR biosensors based on the intensity modulation of SPR, the measured

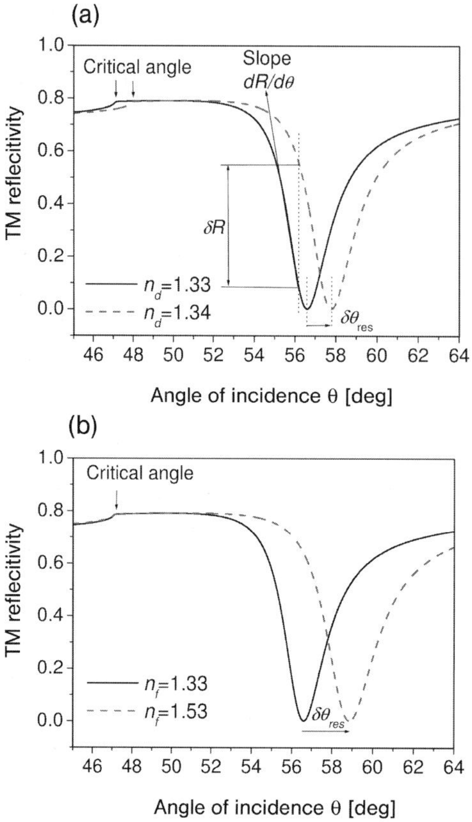

Figure 10 Angular reflectivity spectrum for (a) the bulk refractive-index change ($n_d = 1.33$ and 1.34) and (b) for a surface refractive-index change (a thin film with the thickness of 10 nm and refractive index $n_f = 1.33$ and 1.53). The coupling to surface plasmons at the wavelength of $\lambda = 0.633$ μm by a 90-degree prism with $n_p = 1.845$ and a 55-nm thick gold film with $n_m = 0.1{+}3.5i$. As depicted in Figure 6(a), the angle of incidence θ outside the prism was assumed.

change in the intensity of reflected light δR can be converted to the shift in the resonant angle of incidence $\delta\theta_{res}$ as:

$$\delta\theta_{res} = \delta R/(dR/d\theta) \tag{3}$$

where $dR/d\theta$ is the slope of the SPR reflectivity curve, see Figure 10(a).

Refractive-index variations due to the binding of target molecules can occur within the whole evanescent field of SP (bulk refractive-index change) or within close proximity to the metallic surface (surface refractive-index change). For small changes in the bulk refractive index δn_d, the resonant angle shifts linearly with the refractive-index change:

$$\delta\theta_{res} = S_b \delta n_d \tag{4}$$

where the factor S_b is referred as to the *bulk refractive-index sensitivity*. The sensitivity S_b depends on the wavelength and on optical parameters of the involved materials [40]. If the refractive-index change occurs on the metallic surface in a thin layer with a thickness comparable to or lower than the SP penetration depth L^d_{pen}, the sensitivity is different. For a layer with a refractive index n_f and thickness d that is much lower than the SP penetration depth $d << L^d_{pen}$, a change in its optical thickness $\delta n_f d$ induces a shift $\delta\theta_{res}$:

$$\delta\theta_{res} = S_s \delta n_f d \tag{5}$$

where the factor S_s is referred to as the *surface refractive-index sensitivity*. The surface refractive-index sensitivity can be obtained as $S_s = 2S_b/L_p$ from perturbation theory [41] or numerically from Fresnel reflectivity theory.

From the magnitude of the binding induced change $\delta\theta_{res}$, the surface concentration of captured molecules can be obtained as:

$$\gamma = \frac{\delta\theta_{res}}{S}\frac{\partial c}{\partial n_f} \tag{6}$$

where S is equal to the surface refractive-index sensitivity S_s if the analyte binds directly to the surface (e.g. binding of medium size protein to a monolayer of receptors). For the binding that occurs within the whole evanescent field of SP, the coefficient S is equal to $S_b\, d^{-1}$ (e.g. ligands immobilized within a three-dimensional binding matrix with the thickness $d > L^d_{pen}$). The coefficient $\partial c/\partial n_f$ describes the refractive index increase with the concentration of analyte molecules. For most proteins, the coefficient $\partial c/\partial n_f$ lies in the range 0.14–0.2 μL mg^{-1}, see Liedberg *et al.* [16].

Further, we numerically calculated the bulk and surface refractive-index sensitivities for the wavelength of $\lambda = 0.633$ μm and a prism coupler consisting of a 90 degree prism with the refractive index of $n_p = 1.845$, a 55-nm thick gold film with $n_m = 0.1+3.5i$, and an aqueous medium with $n_d = 1.333$ on the top. As shown in Table 2, the bulk and surface refractive-index sensitivities are equal to $S_b = 118$ deg and $S_s = 1.28$ deg nm^{-1}, respectively. The surface refractive-index sensitivity obtained analytically from bulk refractive index is of 1.17 deg nm^{-1}.

Table 2 Sensitivity of an angular modulation of SPR-based sensor with a 90 degree prism and a wavelength of $\lambda = 0.633$ μm.

SPR prism coupler		**Sensitivity of RI changes and molecular binding**	
Prism refractive index	$n_p = 1.845$	Bulk RI sensitivity	$S_b = 118$ deg
Gold film thickness	$d_m = 55$ nm	Surface RI sensitivity	$S_s = 1.28$ deg nm^{-1}
Gold film refractive index	$n_m = 0.1+3.5i$	Protein induced RI	$\partial c/\partial n_f = 0.14–0.2$ μL mg^{-1}
Sample refractive index	$n_d = 1.33$	Monolayer surface conc.	$\gamma = 3–5\, \delta\theta_{res}$ ng mL^{-1} deg^{-1}

1.2.6
Evaluation of Binding Affinity Constants

SPR biosensors can monitor refractive-index changes on the sensor surface in real time, which allows for the observation of kinetics of binding events. This feature made SPR biosensors an attractive tool for biomolecular interaction analysis (BIA) as it enables the measurement of binding rate constants [42] and thermodynamic properties [43] of investigated reactions. Further, we briefly discuss the analysis of the SPR biosensor output and the design of BIA experiment. Let us note that more detail information on application of SPR biosensors for biomolecular interaction analysis can be found in numerous reviews and books [42–44].

Let us assume a reaction between molecules A and molecules B that fulfils the following conditions: (i) molecules B are identical and can bind only with one molecule A, (ii) molecules A do not react with other molecules A and (iii) each individual binding event of a molecule A to a molecule B is independent on the occupancy of other molecules B. The dynamic equilibrium of the reaction can be given as:

$$A + B \underset{k_d}{\overset{k_a}{\rightleftarrows}} AB \tag{7}$$

where k_a and k_d are the association and dissociation constants, respectively. In a SPR biosensor, molecules B are immobilized on a surface and molecules A are contained in a sample that is flowed through a flow cell on its top, see Figure 11(a). In a typical experiment, the sensor output $\delta\theta_{res}$ is measured in time in two phases: (a) association phase when a sample containing molecules A is flowed through the flow cell and (b) dissociation when a solution without molecules A is flowed. As seen in Figure 11(b), during the association phase a gradual increase in the sensor output is observed due to the capture of molecules A on the surface. During the dissociation phase a decrease in the sensor response is measured due to the release of molecules A from the surface. From measured kinetics $\delta\theta_{res}(t)$, the association constant k_a and dissociation constant k_d constants of the reaction can be determined by fitting with an appropriate model.

In general, the reaction kinetics within a flow cell depends on association and dissociation constants k_a and k_d as well as on the spatial coordinates, flow (convection) and diffusion of molecules. The reaction kinetics on the surface can be solved numerically [45] or analytically. An analytical solution can be found when the sample flow is laminar and the time needed for the diffusion of an analyte across the flow cell (with the height h) is much larger than that for which the molecules are dragged through the flow cell (with the length L) due to convection (flow). Then, the concentrations of molecules A, B and of the complex AB can be averaged over the spatial coordinates obtaining the following pseudo-first-order differential equation:

$$\frac{d\gamma}{dt} = k_{on}\alpha(\beta - \gamma) - k_{off}\gamma \tag{8}$$

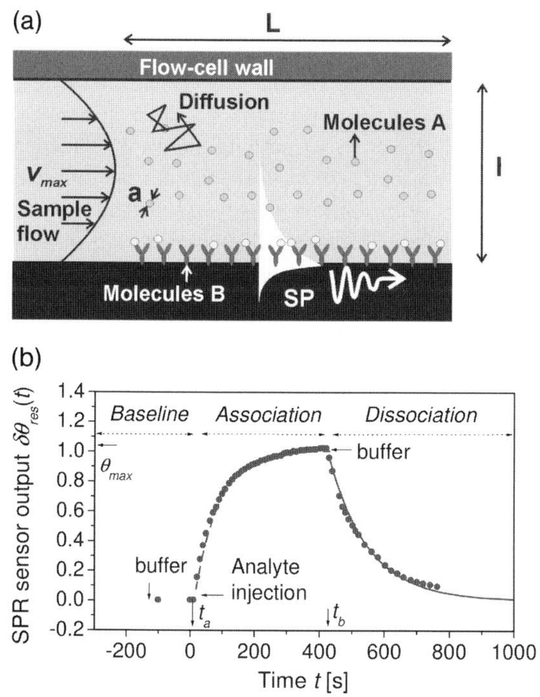

Figure 11 (a) Scheme of a flow cell attached to a SPR biosensor surface. (b) Typical time evolution of a SPR sensor output upon the adsorption (association phase) and desorption (dissociation phase) of molecules A contained in a sample to molecules B anchored to the surface (circles and line show the measured and fitted data, respectively).

where γ is the concentration of captured molecules A on the surface, α is the concentration of molecules A in the sample, β is the concentration molecules B on the surface, and t is time. In this equation, the term $k_{on}\alpha(\beta - \gamma)$ describes the association and the term $k_{off}\gamma$ the dissociation of the complex AB. The k_{on} and k_{off} are effective association and dissociation constants defined by:

$$k_{on} = \frac{k_a}{1 + k_a[\beta - \gamma(t)]/k_m} \tag{9}$$

$$k_{off} = \frac{k_d}{1 + k_a[\beta - \gamma(t)]/k_m} \tag{10}$$

Equations (8) to (10) take to account three processes that control the reaction kinetics: the capture of molecules A by molecules B (coefficient k_a), the release of molecules A from molecules B (coefficient k_d) and the diffusion of molecules A to and from the surface (coefficient k_m). The mass-transport coefficient k_m can be obtained directly by solving the partial differential equations for the flow and diffusion in a flow cell [45] or by using a two-compartment model [44] as:

$$k_m = \xi\left(\frac{v_{\max}D^2}{hL}\right)^{1/3} \tag{11}$$

where D is the diffusion constant, v_{max} is the flow velocity in the middle of the flow cell and ξ is a constant close to one ($\xi = 1.378$ was derived by Edwards *et al.* [45] and 0.98 was published by Sjolander *et al.* [46]). For spherical molecules with the diameter a, a solution with a viscosity of η, the absolute temperature T and for the Boltzmann constant k_B the diffusion constant D can be expressed via the Stokes formula as:

$$D \approx \frac{k_B T}{6\pi a \eta} \tag{12}$$

In order to quantify the effect of the diffusion to the reaction kinetics, let us introduce a Damköhler number defined as follows:

$$Da = k_a\beta\left(\frac{hL}{v_{\max}D^2}\right)^{1/3} \tag{13}$$

For large Damköhler number Da >> 1, the mass transport of molecules A to the surface is much slower than the association rate $k_m << k_a\beta$ and the reaction is referred as to *mass-transport controlled*. In this regime, the effective association and dissociation constants are $k_{on} \approx k_m\beta^{-1}$ and $k_{off} \approx k_m k_d (k_a\beta)^{-1}$, respectively, for $\gamma << \beta$. For the analysis of biomolecular interactions it is desirable to minimize the effect of the mass transport. This can be achieved by lowering the surface coverage β, increasing the flow velocity v_{max} or by decreasing the flow-cell depth h. Through these parameters, the Damköhler number can be reduced to Da << 1, for which the mass transport is much faster than the reaction association rate $k_m >> k_a\beta$. Under this condition, the reaction is referred as to *interaction controlled* and the effective binding constants approaches the association and dissociation constants $k_{on} \approx k_a$ and $k_{off} \approx k_d$.

In the interaction-controlled regime, k_a and k_d can be determined by fitting the measured sensor output $\delta\theta_{res}(t)$ with the model described above. Further, let us assume that at the time t_a, a sample with the (molar) concentration α is injected into the sensor and flowed through the flow cell until the time t_b (association phase; $t_a < t < t_b$). Afterwards, a blank solution with $\alpha = 0$ is flowed through the flow cell (dissociation phase; $t > t_b$). In addition, let us suppose that no molecules A are captured on the sensor surface prior to the analysis (i.e. $\gamma = 0$ for $t < t_a$).

Solution of the kinetics described by Equations (8) to (10) can be found separately for the association and dissociation phases. For the dissociation phase, it can be shown that the sensor output exponentially decays in time as:

$$\delta\theta_{res}(t) = S\frac{\partial n_f}{\partial c}M_w\gamma(t_b)e^{-k_d(t-t_b)} = \delta\theta_{res}(t_b)e^{-k_d(t-t_b)} \quad (14)$$

where the term $S\times(\partial n_f/\partial c)$ is the sensitivity factor described in the Section 1.2.5 **(AQ4)**, M_W is the molecular weight of molecules A, $\gamma(t_b)$ is the (molar) concentration of captured molecules A at the time t_b and $\delta\theta_{res}(t_b)$ is the sensor output at the time t_b. For the association phase, the time evolution of the sensor output is:

$$\delta\theta_{res}(t) = S\frac{\partial n}{\partial c}M_w\frac{k_a\alpha\beta}{k_a\alpha+\beta}\left(1-e^{-(k_a\alpha+k_d)(t-t_a)}\right)$$
$$= \delta\theta_{res}^{eq}\left(1-e^{-(k_a\alpha+k_d)(t-t_a)}\right) \quad (15)$$

where the constant $\delta\theta_{res}{}^{eq}$ is the sensor output for the equilibrium concentration $\gamma_{eq} = k_a\alpha\beta(k_a\alpha + k_d)$ for which the association rate $k_a\alpha(\beta-\gamma)$ and dissociation rate k_{off} γ are equal.

Functions (14) and (15) can be fitted to the measured kinetics $\delta\theta_{res}(t)$ to determine the affinity constants k_a and k_d. Firstly, the dissociation constant k_d is obtained by fitting the kinetic $\delta\theta_{res}(t)$ for the dissociation phase ($t > t_b$) with the function (14). Afterwards, the kinetic for the association phase ($t_a < t < t_b$) is fitted with the equation (15) into which k_d is plugged and k_a is determined. Let us note that all parameters $\delta\theta_{res}(t_b)$, $\delta\theta_{res}{}^{eq}$, k_a and k_d can be set as variable parameters during the fitting by, e.g., the method of least squares. In order to illustrate the previous theory, we included parameters describing the typical flow, diffusion and binding conditions for common protein–protein interactions in Table 3.

Table 3 Parameters describing the typical flow, diffusion and binding conditions in SPR biosensors; diffusion coefficient *D* was calculated for water at the temperature 25 °C and a molecule A with the diameter a = 10 nm; immunoglobulin molecules (IgG molecular weight of 160 kDa) were assumed to be immobilized on the surface with the surface concentration β = 1–10 ng mm^{-2}.

Biomolecular interaction analysis			
Flow cell dimensions	h = 0.1 mm, L = 5 mm, w = 5 mm	Association and dissociation constants	$k_a = 10^3$–10^7 mol^{-1} L s^{-1}, $k_d = 10^{-4}$–10^{-1} s^{-1}
Flow rate	θ = 0.1 mL min^{-1}	Diffusion coefficient	$D = 2 \times 10^{-5}$ mm^2 s^{-1}
Flow velocity	$v_{max} = 1.5\theta/hw$ = 5 mm s^{-1}	Diffusion rate	$k_m \sim 3 \times 10^{-3}$ mm s^{-1}
Surface coverage	$\beta \sim 6$–60×10^{-15} mol mm^{-2}	Damköhler number	DA ~ 10^{-3}–10^2

1.2.7
Applications of SPR Biosensors

Since numerous commercial SPR biosensor instruments are available [47–50], this technology has become routinely used in modern laboratories. In general, SPR biosensors found their applications in two main areas: for biomolecular interaction analysis (BIA) and for detection of chemical and biological species.

Biomolecular Interaction Analysis

SPR biosensors provide a generic tool for the analysis of various biomolecular interactions such as those between proteins and peptides [51], protein–DNA interactions [20], and DNA hybridization [52]. In conjuction with the development

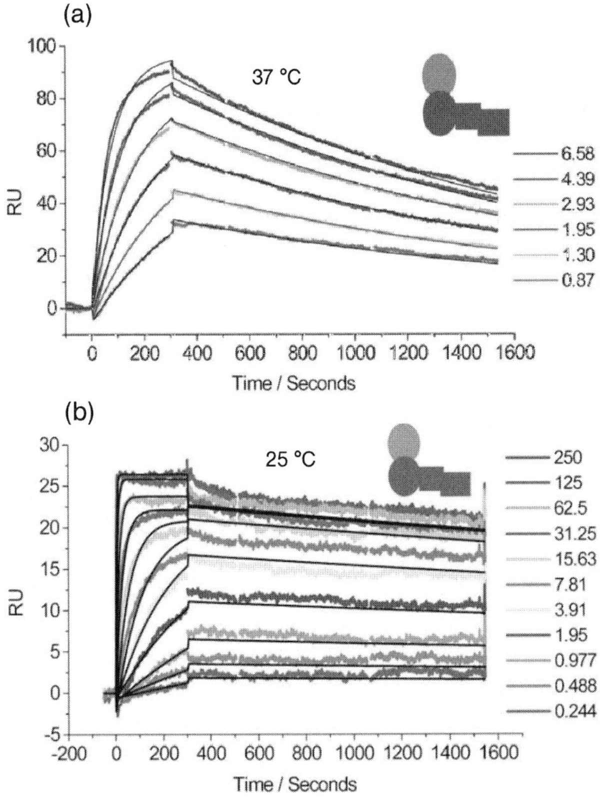

Figure 12 Binding kinetics of IκBα ankyrin repeat domain to NF-κB(p50/p65) for different concetrations (shown on the right-hand side in nM) and the temperatures of (a) 37 °C and (b) 25 °C fitted with a model. (Reprinted with permission from [56]).

of high-throughput SPR biosensor devices (mostly utilizing on SPR imaging/microscopy [26, 53]), applications in areas such as proteomic analysis, drug discovery and pathway elucidation are pushed forward [30, 54, 55]. An example of the analysis of a protein–protein interaction is given in Figure 12. In this work, Berquist *et al.* [56] investigated the binding kinetics of IκBα ankyrin repeat domain to NF-κB(p50/p65) at different concentrations and temperatures. By applying the binding kinetic model the stability, affinity constants and termodynamic properties of this reaction were determined.

Detection of Chemical and Biological Species

In general, the implementation of SPR biosensor technology for detection of chemical and biological species presents a complex task owing to the diversity of sample matrices (liquid or solid samples), type of deployment (portable versus permanently installed sensor system) and detection environment (field, mobile lab, industrial plant, etc.). SPR biosensors for detection of a wide range of analytes related to important areas such as medical diagnostics (e.g. disease markers such as prostate specific hormone [57]), environment monitoring (e.g. pesticides [58, 59]) and food control (e.g. toxins such as domoic acid [60] or pathogens [61]) were covered in numerous extensive reviews and books [54, 62–68].

Over recent years, various compact SPR optical platforms suitable for portable sensor devices were developed. These include SPR biosensors utilizing angular modulation of SPR developed by Elkind *et al.* [69], Kawazumi *et al.* [70] and Thirstrup *et al.* [71] Currently, an interesting approach based on a diffraction grating coupler and a wavelength modulation for miniaturized spectroscopic surface plasmon resonance (SPR) sensors was reported [72]. Integration of a miniature SPR biosensor platform into a portable system was demonstrated by Chinowsky *et al.* [12].

SPR biosensors can be used for the detection of target analytes using various assays. The direct detection format is typically used for medium-size and large analytes (> 10 kDa) that produce sufficiently large refractive-index changes [17, 73]. For detection of smaller analytes, mostly sandwich assay [58], competitive assay [74] or inhibition assay [59] are used. In a sandwich immunoassay, analyte is flowed over the sensor surface and target molecules are captured by attached antibodies. Subsequently, a solution with second antibodies is introduced to the sensor and its binding to the captured analyte is detected, see Figure 13(a). In a competitive assay, free analyte and the analyte conjugated to a larger molecule compete for the same recognition site at the sensor surface, see Figure 13(b). For the competitive assay, the sensor response is inversely proportional to the concentration of the free analyte. In an inhibition immunoassay, a derivative of the target analyte is immobilized to the sensor surface. The analyzed sample is preincubated with an antibody against the target analyte and the binding of the unreacted antibody molecules to the analyte derivative is measured, see Figure 13(b). The inhibition assay-based sensors exhibit a sensor response that is inversely proportional to the concentration of the analyte in the sample solution.

After the capture of analyte molecules to the surface, typically a regeneration step is applied. The regeneration can be performed by changing pH [60], using detergents [75] or enzymes [59, 68] by which analyte bound to the biorecognition element is released, leaving the sensor available for subsequent measurements, see Figure 14.

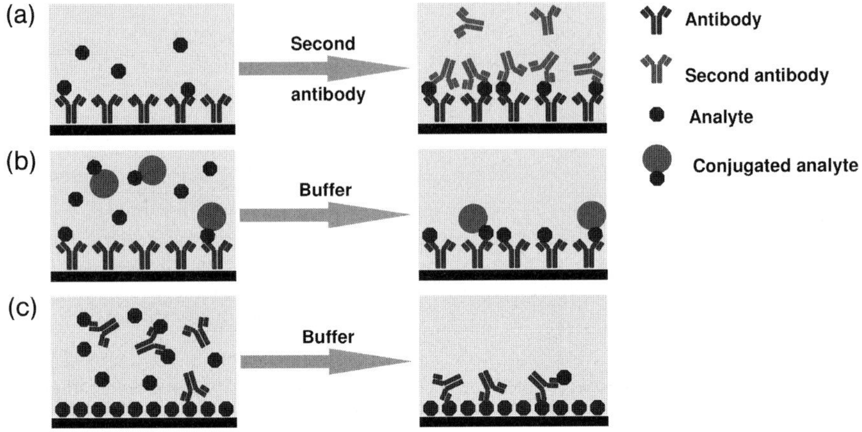

Figure 13 Schematic images of three detection formats in the SPR biosensors. (a) sandwich assay; (b) competitive assay; (c) inhibition assay.

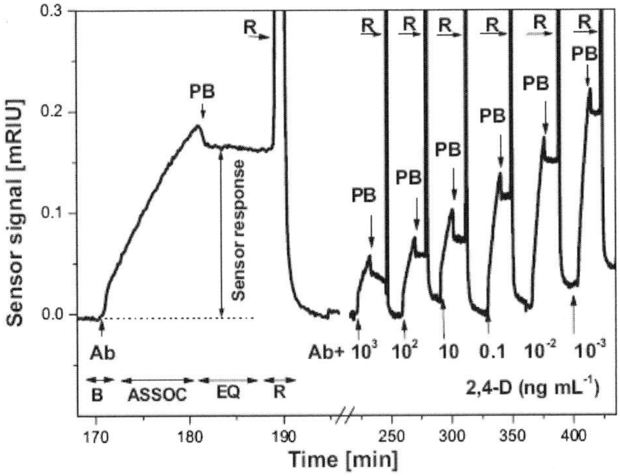

Figure 14 Example of a sensorgram measured with a SPR biosensor utilizing inhibition immunoassay-based detection of herbicide 2,4-dichlorophenoxyacetic acid (2,4-D). The measurement consisted of a series of cycles in which a sample with 2,4-D was (concentration indicated in the graph) injected followed by the regeneration. (Reprinted with permission from [76]).

1.2.8 Summary

Biosensors based on surface plasmon resonance (SPR) have become an established technology in materials and life-science research. Recently, applications in other important areas ranging from medical diagnostics, environmental monitoring to food control are pushing forward. The highly multidisciplinary research in surface plasmon resonance (SPR) biosensors combines optics, surface engineering and biology. Within this chapter, we describe fundamentals of SPR optics, design of key components and recent state-of-the-art applications of SPR biosensors.

Acknowledgments

This publication was financially supported by the European Commission in the Communities 6th Framework Programme, Project TRACEBACK (FOOD-CT-036300), and coordinated by Tecnoalimenti. It reflects the author's views and the Community is not liable for any use that may be made of the information contained in this publication.

References

1 Gonzalez-Martinez, M.A., Puchades, R. and Maquieira, A. (2007) Optical immunosensors for environmental monitoring: How far have we come. *Analytical and Bioanalytical Chemistry*, **387**, 205–218.

2 Rasooly, A. and Herold, K.E. (2006) Biosensors for the analysis of food- and waterborne pathogens and their toxins. *Journal of Aoac International*, **89**, 873–883.

3 Vo-Dinh, T. and Cullum, B. (2000) Biosensors and biochips: advances in biological and medical diagnostics. *Fresenius Journal of Analytical Chemistry*, **366**, 540–551.

4 Liedberg, B., Nylander, C. and Lundstrom, I. (1983) Surface-Plasmon Resonance for Gas-Detection and Biosensing. *Sensors and Actuators*, **4**, 299–304.

5 Rather, H. (1983) *Surface Plasmons on Smooth and Rough Surfaces and on Gratings*, Springer Verlag, Berlin.

6 Cullen, D.C., Brown, R.G.W. and Lowe, C.R. (1987) Detection of Immuno-Complex Formation Via Surface-Plasmon Resonance on Gold-Coated Diffraction Gratings. *Biosensors*, **3**, 211–225.

7 Lawrence, C.R., Geddes, N.J., Furlong, D.N. and Sambles, J.R. (1996) Surface plasmon resonance studies of immunoreactions utilizing disposable diffraction gratings. *Biosensors & Bioelectronics*, **11**, 389–400.

8 Dostalek, J., Ctyroky, J., Homola, J., Brynda, E., Skalsky, M., Nekvindova, P., Spirkova, J., Skvor, J. and Schrofel, J. (2001) Surface plasmon resonance biosensor based on integrated optical waveguide. *Sensors and Actuators B – Chemical*, **76**, 8–12.

9 Slavik, R., Homola, J. and Brynda, E. (2002) A miniature fiber optic surface plasmon resonance sensor for fast detection of staphylococcal enterotoxin B. *Biosensors & Bioelectronics*, **17**, 591–595.

10 Dostalek, J. and Homola, J. (2008) Surface plasmon resonance sensor based on an array of diffraction gratings for

highly-parallelized observation of biomolecular interactions. *Sensors and Actuators B: Chemical*, **129**, 303–310.

11 Nenninger, G.G., Piliarik, M. and Homola, J. (2002) Data analysis for optical sensors based on spectroscopy of surface plasmons. *Measurement Science & Technology*, **13**, 2038–2046.

12 Chinowsky, T.M., Jung, L.S. and Yee, S.S. (1999) Optimal linear data analysis for surface plasmon resonance biosensors. *Sensors and Actuators B – Chemical*, **54**, 89–97.

13 Tobiska, P. and Homola, J. (2005) Advanced data processing for SPR biosensors. *Sensors and Actuators B – Chemical*, **107**, 162–169.

14 Matsubara, K., Kawata, S. and Minami, S. (1988) Optical Chemical Sensor Based on Surface-Plasmon Measurement. *Applied Optics*, **27**, 1160–1163.

15 VanWiggeren, G.D., Bynum, M.A., Ertel, J.P., Jefferson, S., Robotti, K.A., Thrush, E.P., Baney, D.A. and Killeen, K.P. (2007) A novel optical method providing for high-sensitivity and high-throughput biomolecular interaction analysis. *Sensors and Actuators B – Chemical*, **127**, 341–349.

16 Liedberg, B., Lundstrom, I. and Stenberg, E. (1993) Principles of Biosensing with an Extended Coupling Matrix and Surface-Plasmon Resonance. *Sensors and Actuators B – Chemical*, **11**, 63–72.

17 Homola, J., Dostalek, J., Chen, S.F., Rasooly, A., Jiang, S.Y. and Yee, S.S. (2002) Spectral surface plasmon resonance biosensor for detection of staphylococcal enterotoxin B in milk. *International Journal of Food Microbiology*, **75**, 61–69.

18 Jory, M.J., Bradberry, G.W., Cann, P.S. and Sambles, J.R. (1995) A Surface-Plasmon-Based Optical Sensor Using Acoustooptics. *Measurement Science & Technology*, **6**, 1193–1200.

19 Rothenhausler, B. and Knoll, W. (1988) Surface-Plasmon Microscopy. *Nature*, **332**, 615–617.

20 Shumaker-Parry, J.S. and Campbell, C.T. (2004) Quantitative methods for spatially resolved adsorption/desorption measurements in real time by surface plasmon resonance microscopy. *Analytical Chemistry*, **76**, 907–917.

21 Karlsson, R. and Stahlberg, R. (1995) Surface-Plasmon Resonance Detection and Multispot Sensing for Direct Monitoring of Interactions Involving Low-Molecular-Weight Analytes and for Determination of Low Affinities. *Analytical Biochemistry*, **228**, 274–280.

22 Slavik, R. and Homola, J. (2007) Ultrahigh resolution long range surface plasmon-based sensor. *Sensors and Actuators B-Chemical*, **123**, 10–12.

23 Yao, D.F., Yu, F., Kim, J.Y., Scholz, J., Nielsen, P.E., Sinner, E.K. and Knoll, W. (2004) Surface plasmon field-enhanced fluorescence spectroscopy in PCR product analysis by peptide nucleic acid probes. *Nucleic Acids Research*, **32**.

24 Kikuchi, Y., Uno, S., Nanami, M., Yoshimura, Y., Iida, S., Fukushima, N. and Tsuchiya, M. (2005) Determination of concentration and binding affinity of antibody fragments by use of surface plasmon resonance. *Journal of Bioscience and Bioengineering*, **100**, 311–317.

25 Worn, A. and Pluckthun, A. (2001) Stability engineering of antibody single-chain Fv fragments. *Journal of Molecular Biology*, **305**, 989–1010.

26 Smith, E.A., Thomas, W.D., Kiessling, L.L. and Corn, R.M. (2003) Surface plasmon resonance imaging studies of protein-carbohydrate interactions. *Journal of the American Chemical Society*, **125**, 6140–6148.

27 Win, M.N., Klein, J.S. and Smolke, C.D. (2006) Codeine-binding RNA aptamers and rapid determination of their binding constants using a direct coupling surface plasmon resonance assay. *Nucleic Acids Research*, **34**, 5670–5682.

28 Balasubramanian, S., Sorokulova, I.B., Vodyanoy, V.J. and Simonian, A.L. (2007) Lytic phage as a specific and selective probe for detection of Staphylococcus aureus – A surface plasmon resonance spectroscopic study. *Biosensors & Bioelectronics*, **22**, 948–955.

29 Rusmini, F., Zhong, Z.Y. and Feijen, J. (2007) Protein immobilization strategies for protein biochips. *Biomacromolecules*, **8**, 1775–1789.

30 Navratilova, I., Papalia, G.A., Rich, R.L., Bedinger, D., Brophy, S., Condon, B., Deng, T., Emerick, A.W., Guan, H.W., Hayden, T., Heutmekers, T., Hoorelbeke, B., McCroskey, M.C., Murphy, M.M., Nakagawa, T., Parmeggiani, F., Qin, X.C., Rebe, S., Tomasevic, N., Tsang, T., Waddell, M.B., Zhang, F.F., Leavitt, S. and Myszka, D.G. (2007) Thermodynamic benchmark study using Biacore technology. *Analytical Biochemistry*, **364**, 67–77.

31 Slavik, R., Homola, J. and Vaisocherova, H. (2006) Advanced biosensing using simultaneous excitation of short and long range surface plasmons. *Measurement Science & Technology*, **17**, 932–938.

32 Su, X.D., Wu, Y.J., Robelek, R. and Knoll, W. (2005) Surface plasmon resonance spectroscopy and quartz crystal microbalance study of streptavidin film structure effects on biotinylated DNA assembly and target DNA hybridization. *Langmuir*, **21**, 348–353.

33 Knoll, W., Park, H., Sinner, E.K., Yao, D.F. and Yu, F. (2004) Supramolecular interfacial architectures for optical biosensing with surface plasmons. *Surface Science*, **570**, 30–42.

34 Oh, B.K., Lee, W., Chun, B.S., Bae, Y.M., Lee, W.H. and Choi, J.W. (2005) The fabrication of protein chip based on surface plasmon resonance for detection of pathogens. *Biosensors & Bioelectronics*, **20**, 1847–1850.

35 Tinazli, A., Tang, J.L., Valiokas, R., Picuric, S., Lata, S., Piehler, J., Liedberg, B. and Tampe, R. (2005) High-affinity chelator thiols for switchable and oriented immobilization of histidine-tagged proteins: A generic platform for protein chip technologies. *Chemistry – a European Journal*, **11**, 5249–5259.

36 Lata, S., Gavutis, M., Tampe, R. and Piehler, J. (2006) Specific and stable fluorescence labeling of histidine-tagged proteins for dissecting multi-protein complex formation. *Journal of the American Chemical Society*, **128**, 2365–2372.

37 Xu, F., Persson, B., Lofas, S. and Knoll, W. (2006) Surface plasmon optical studies of carboxymethyl dextran brushes versus networks. *Langmuir*, **22**, 3352–3357.

38 Milton, H.J. (1992) *Poly (Ethylene Glycol) Chemistry: Biotechnical and Biomedical Applications*, Kluwer Academic Publishers.

39 Nelson, C.M., Raghavan, S., Tan, J.L. and Chen, C.S. (2003) Degradation of micropatterned surfaces by cell-dependent and -independent processes. *Langmuir*, **19**, 1493–1499.

40 Homola, J., Koudela, I. and Yee, S.S. (1999) Surface plasmon resonance sensors based on diffraction gratings and prism couplers: sensitivity comparison. *Sensors and Actuators B – Chemical*, **54**, 16–24.

41 Adam, P., Dostalek, J. and Homola, J. (2006) Multiple surface plasmon spectroscopy for study of biomolecular systems. *Sensors and Actuators B – Chemical*, **113**, 774–781.

42 Fagerstam, L.G., Frostellkarlsson, A., Karlsson, R., Persson, B. and Ronnberg, I. (1992) Biospecific Interaction Analysis Using Surface-Plasmon Resonance Detection Applied to Kinetic, Binding-Site and Concentration Analysis. *Journal of Chromatography*, **597**, 397–410.

43 Day, Y.S.N., Baird, C.L., Rich, R.L. and Myszka, D.G. (2002) Direct comparison of binding equilibrium, thermodynamic, and rate constants determined by surface- and solution-based biophysical methods. *Protein Science*, **11**, 1017–1025.

44 Myszka, D.G., He, X., Dembo, M., Morton, T.A. and Goldstein, B. (1998) Extending the range of rate constants available from BIACORE: Interpreting mass transport-influenced binding data. *Biophysical Journal*, **75**, 583–594.

45 Edwards, D.A., Goldstein, B. and Cohen, D.S. (1999) Transport effects on surface-volume biological reactions. *Journal of Mathematical Biology*, **39**, 533–561.

46 Sjolander, S. and Urbaniczky, C. (1991) Integrated Fluid Handling-System for Biomolecular Interaction Analysis. *Analytical Chemistry*, **63**, 2338–2345.

47 http://www.biacore.com.

48 http://www.ibis-spr.nl/.

49 http://www.plexera.com.

50 http://www.sensia.es.

51 Shliom, O., Huang, M., Sachais, B., Kuo, A., Weisel, J.W., Nagaswami, C., Nassar, T., Bdeir, K., Hiss, E., Gawlak, S., Harris, S., Mazar, A. and Higazi, A.A. (2000) Novel interactions between urokinase and its receptor. *Journal of Biological Chemistry*, **275**, 24304–24312.

52 Feriotto, G., Corradini, R., Sforza, S., Bianchi, N., Mischiati, C., Marchelli, R. and Gambari, R. (2001) Peptide nucleic acids and biosensor technology for real-time detection of the cystic fibrosis W1282X mutation by surface plasmon resonance. *Laboratory Investigation*, **81**, 1415–1427.

53 Wolf, L.K., Gao, Y. and Georgiadis, R.M. (2007) Kinetic discrimination of sequence-specific DNA – Drug binding measured by surface plasmon resonance imaging and comparison to solution-phase measurements. *Journal of the American Chemical Society*, **129**, 10503–10511.

54 Karlsson, R. (2004) SPR for molecular interaction analysis: a review of emerging application areas. *Journal of Molecular Recognition*, **17**, 151–161.

55 Boozer, C., Kim, G., Cong, S.X., Guan, H.W. and Londergan, T. (2006) Looking towards label-free biomolecular interaction analysis in a high-throughput format: a review of new surface plasmon resonance technologies. *Current Opinion in Biotechnology*, **17**, 400–405.

56 Bergqvist, S., Croy, C.H., Kjaergaard, M., Huxford, T., Ghosh, G. and Komives, E.A. (2006) Thermodynamics reveal that helix four in the NLS of NF-kappa B p65 anchors I kappa B alpha, forming a very stable complex. *Journal of Molecular Biology*, **360**, 421–434.

57 Yu, F., Persson, B., Lofas, S. and Knoll, W. (2004) Surface plasmon fluorescence immunoassay of free prostate-specific antigen in human plasma at the femtomolar level. *Analytical Chemistry*, **76**, 6765–6770.

58 Minunni, M. and Mascini, M. (1993) Detection of pesticide in drinking-water using real-time biospecific interaction analysis (BIA). *Analytical Letters*, **26**, 1441–1460.

59 Mouvet, C., Harris, R., Maciag, C., Luff, B., Wilkinson, J., Piehler, J., Brecht, A., Gauglitz, G., Abuknesha, R. and Ismail, G. (1997) Determination of simazine in water samples by waveguide surface plasmon resonance. *Analytica Chimica Acta*, **338**, 109–117.

60 Yu, Q., Chen, S., Taylor, A., Homola, J., Hock, B. and Jiang, S. (2005) Detection of low-molecular-weight domoic acid using surface plasmon resonance sensor. *Sensors and Actuators B*, **107**, 193–201.

61 Koubova, V., Brynda, E., Karasova, L., Skvor, J., Homola, J., Dostalek, J., Tobiska, P. and Rosicky, J. (2001) Detection of foodborne pathogens using surface plasmon resonance biosensors. *Sensors and Actuators B – Chemical*, **74**, 100–105.

62 Homola, J., Yee, S.S. and Gauglitz, G. (1999) Surface plasmon resonance sensors: review. *Sensors and Actuators B – Chemical*, **54**, 3–15.

63 Homola, J. (2008) Surface plasmon resonance sensors for detection of chemical and biological species. *Chemical Reviews*, **108**, 462–493.

64 Dostalek, J., Ladd, J., Jiang, S.Y. and Homola, J. (2006) in *Surface Plasmon Resonance Based Sensors* (ed. J. Homola), Springer, pp. 3–45.

65 Dostalek, J. and Homola, J. (2006) in *Surface Plasmon Resonance Based Sensors* (ed. J. Homola), Springer, pp. 3–45.

66 Ladd, J., Tayllor, A. and Jiang, S. (2006) in *Surface Plasmon Resonance Based Sensors* (ed. J. Homola), Springer, pp. 3–45.

67 Phillips, K.S. and Cheng, Q. (2007) Recent advances in surface plasmon resonance based techniques for bioanalysis. *Analytical and Bioanalytical Chemistry*, **387**, 1831–1840.

68 Shankaran, D.R., Gobi, K.V.A. and Miura, N. (2007) Recent advancements in surface plasmon resonance immunosensors for detection of small molecules of biomedical, food and environmental interest. *Sensors and Actuators B – Chemical*, **121**, 158–177.

69 Elkind, J.L. Stimpson, D.I., S. A.A., B. D.U. and Melendez, J.L. (1999) Integrated analytical sensors: the use of the

TISPR-1 as a biosensor. *Sensors and Actuators B*, **54**, 182–190.

70 Kawazumi, H., Gobi, K., Ogino, K., Maeda, H. and Miura, N. (2005) Compact surface plasmon resonance (SPR) immunosensor using multichannel for simultaneous detection of small molecule compounds. *Sensors and Actuators B*, **108**, 791–796.

71 Thirstrup, C., Zong, W., Borre, M., Neff, H., Pedersen, H. and Holzhueter, G. (2004) *Sensors and Actuators B*, **100**.

72 Telezhnikova, O. and Homola, J. (2006) New approach to spectroscopy of surface plasmons. *Optics Letters*, **31**, 3339–3341.

73 Krenn, J.R., Ditlbacher, H., Schider, G., Hohenau, A., Leitner, A. and Aussenegg, F.R. (2003) Surface plasmon micro- and nano-optics. *Journal of Microscopy – Oxford*, **209**, 167–172.

74 Shimomura, M., Nomura, Y., Zhang, W., Sakino, M., Lee, K., Ikebukuro, K. and Karube, I. (2001) Simple and rapid detection method using surface plasmon resonance for dioxins, polychlorinated biphenylx and atrazine. *Analytica Chimia Acta*, **434**, 223–230.

75 Lotierzo, M., Henry, O., Piletsky, S., Tothill, I., Cullen, D., Kania, M., Hock, B. and Turner, A. (2004) Surface plasmon resonance sensor for domoic acid based on grafted imprinted polymer. *Biosensors and Bioelectronics*, **20**, 145–152.

76 Dostalek, J., Pribyl, J., Homola, J. and Skladal, P. (2007) Multichannel SPR biosensor for detection of endocrine-disrupting compounds. *Analytical and Bioanalytical Chemistry*, **389**, 1841–1847.

1.3
Tutorial Review: Surface Modification and Adhesion

Renate Förch

1.3.1
Introduction

Adhesion processes are probably amongst the most important and most common surface-related phenomena. The term *adhesion* refers to the *attractive forces* that can be observed between *unlike molecules* or unlike materials. On the contrary, we speak of *cohesion* when we observe attractive forces between *like molecules*. The different types of adhesive forces that can be observed range from valence forces all the way to mechanical interactions and occur in all areas of nature and technology. The term adhesion covers different types of mechanisms ranging from bioadhesion processes involving the interaction of specific adhesion molecules that allow for the binding between, for example, cells to surfaces all the way to composites and heavy construction materials. Adhesive forces are observed anywhere from subzero temperatures to a few hundred degrees centigrade (e.g. aerospace). As a result of this extreme range of conditions under which interfacial adhesion takes place, many theories have developed to describe the phenomena. The types of mechanisms used to describe a particular situation are generally based on the chemical/biochemical nature of the adhesive/adherend combination. Thus, the strength of the adhesive forces between two materials depends on a number of different factors based on the morphology, the chemical structure and the wettability of the materials to be bonded.

A number of different types of adhesion mechanisms have been proposed to explain why one material sticks to another. The most common of these are briefly described below:

- *Mechanical adhesion* is said to occur if the two materials interlock mechanically either because of intrinsic roughness of the surfaces or because of the interlocking of chains or fibers. Examples include materials used in dental restoration and the interlocking nature of Velcro and other textile adhesive forms.

- *Electrostatic adhesion* occurs for materials that may exchange electrons to form a difference in the electrical charge at the joint, in its structure resembling a capacitor. Such a structure may thus create attractive electrostatic forces and

Surface Design: Applications in Bioscience and Nanotechnology
Edited by Renate Förch, Holger Schönherr, and A. Tobias A. Jenkins

ISBN: 978-3-527-40789-7

hence a measurable adhesion between the materials. In these cases, electrostatic theory describes the formation of an electrical double layer at the interfaces and consequent Coulombic attraction largely accounts for adhesion and the resistance to separation. An example of this type of adhesion is the stickiness of a polymer film on a layer of paper or another polymer film.

- *Diffusive adhesion* is said to occur if the materials at the joint blend by diffusion. Adhesion in this case is considered to be a 3-dimensional volume process, rather than a 2-dimensional surface process. This would, for example, occur if both materials are mobile and soluble in each other. It is particularly effective with polymer chains when one end of the molecule diffuses into the other material. Diffusive adhesion is also the mechanism that takes place during sintering and welding. When ceramic or metallic powders are pressed together and heated, the atoms at the interface diffuse from one material into the other, forming very strong joints.
- *Dispersive adhesion or adsorption* occurs if sufficiently intimate contact is achieved at the interface, such that the materials will adhere because of interatomic and intermolecular forces such as van der Waals forces. This is probably the most widely applied theory on interfacial adhesion. In this case, the molecules involved are polar with respect to average charge density and have regions of net positive and net negative charge. This polarity may be a permanent property of the molecule or a transient effect as a result of the random movement of electrons within the molecule.

  Wetting is the initial physical process occurring during dispersive adhesion. It is a 'first step' in many adhesion processes, being of tremendous importance in thin-film growth and in many life-science situations. The polarized molecule will be trapped in a weakly adsorbed state, also known as *physical adsorption* (or physisorption). During physisorption, the adsorbed molecule may be stretched and van de Waal forces may bind it to the surface, however, it retains its own identity. Physisorbed molecules are mobile on the surface except at cryogenic temperatures and can be shown hopping along the surface atomic sites. If the molecules gain sufficient energy while doing so, they may desorb into the surrounding after a while. Alternatively, the physisorbed molecules may undergo further interactions and form chemical bonds with the surface atoms. This is known as *chemical adsorption*, or chemisorption.
- *Chemical adhesion* (chemisorption) occurs if the molecule changes its identity through ionic or covalent bonding with the atoms or molecules of the other material. It involves the sharing of electrons in new molecular orbitals and is much stronger than physisorption, which involves only dipole interactions. The strongest joints are obtained for ionic and covalent bonding between the materials. The adhesive strength in these cases is directly related to the nature of the chemical bond formed and varies depending on the bond energies involved. Typical examples here would be the bonding of thiols to gold, silanes to silicon-oxide surfaces and peptide (amide) bond formation in nature. Such adsorption

mechanisms play an important role in protein adsorption, cell adhesion and cell proliferation.

Much weaker bonding is observed between two materials if their oxygen, nitrogen or fluorine atoms share a hydrogen nucleus, leading to hydrogen bonding. In Nature, hydrogen bonding is a widespread mechanism to achieve relatively strong bonds between two biological entities and ultimately leads to the quaternary structure, stability and function of proteins and DNA.

Provided there is enough intimate surface adhesive contact, any 'weak' forces can produce significant amounts of adhesion. The attractive interactions must be much larger than the repulsive forces for adhesion to be effective. In order to achieve this, Nature has shown how a hierarchial surface structure over a range of size scales can influence adhesive properties of surfaces. The most well-known example of this is probably shown by the gecko, which has developed an extremely high level of control allowing it to master any kind of surface [1–3]. This control is based on highly refined 200 nm hair-like structures that maximize van der Waals interactions with the surface [4]. Microfabrication techniques today make use of such structures to influence adhesion between materials. Examples of this are the artificial structures from silicon dioxide covered with polymeric nanorods that behave like a synthetic dry adhesive modeled after the fine-hair adhesives found in Nature [5, 6]. While it is in principle possible to mimic these designs from Nature, modern synthetic techniques still show difficulties in coping with a comparatively low stability and unpredictable interactions that often lead to a reduction or even loss of the desired characteristics.

Following the discussion above it can be said that the strength of adhesion between two materials is dependent on:

1. which type of mechanism is taking place and
2. the surface area over which the contact occurs.

The inherent importance of adhesion in virtually all aspects of technology has led to a tremendous variety of surface-modification techniques. As such, much of the research investigating the technology of adhesion, adhesion improvement and surface pretreatment is concerned with optimizing one or more of the adhesion processes discussed above.

The contact area is in turn dependent on the surface roughness and the relative wettability of the two materials. The term wetting generally describes the contact between a liquid and a solid surface that results from intermolecular interactions once the two are brought together. Similarly, when considering the wettability of two solid materials we refer to the contact that results from intermolecular interactions taking place between the two materials. The extent of wetting depends on the surface energy of the interfaces involved. Materials that wet against each other tend to have a larger contact area than materials that do not. This can be seen by a drop wetting a hydrophilic (high wettability) surface versus a drop wetting a hydrophobic (low wettability) surface, see Figure 1. The wettability of a material is dependent on the surface energy, which can be conveniently measured in terms

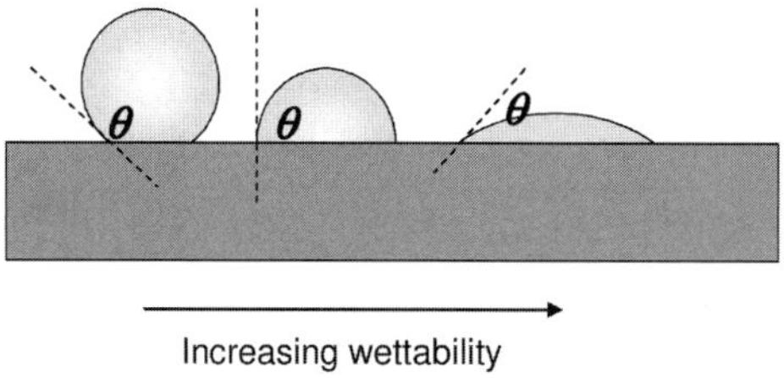

Figure 1 Different degrees of wetting of a surface.

of the contact angle of, for example, a water droplet on the surface [7–10]. High wettability will lead to a low contact angle such that the fluid will spread to cover a larger area of the surface. If the wettability is low, the fluid will form a compact droplet on the surface.

In many engineering applications it is equally important to recognize that the bulk properties of the polymeric materials also play a significant role and control the types of interfacial forces that govern the adhesion between the two materials. Polymers typically have different deformation processes from other materials, obtaining their toughness from their long-chain nature. They generally have a lower modulus and deformation resistance than, for example, metals and inorganics. Thus, both the nature of the polymer and the interface chemistry determine the strength of adhesion under a given environment. Under mechanical stress, for example, the deformation at a polymer/inorganic interface will normally occur within the polymer. Failure of the joint will generally occur within the polymer or at the interface. Situations where adhesion failure occurs within the inorganic material include, for example, adhesion via weak oxides or hydroxides on a metal surface. Excellent adhesion can only be obtained if the interface can sustain a certain amount of stress induced by deformation, for example as a result of changes in the surrounding environment.

1.3.2
Chemical Methods of Adhesion Promotion

Adhesives that are used in everyday life include epoxies and cyanoacrylates whose adhesive mechanisms can be explained by adsorption and chemical reactions [11–13]. Other commonly used adhesives include polyurethanes, silicones, epoxides and polyacrylates. In pharmaceuticals certain polymers such as poly(vinyl pyrrolidone) and cellulose are used as adhesives. Nature has designed its own adhesives, examples would be the sticky adhesives secreted by the venus fly trap, or the protein-based adhesion molecules involved in cell adhesion, bacterial adhesion and biofilm formation or mussels adhesion to rocks.

Under most circumstances, the strength of adhesion is dependent on the density of the covalent bonds at the interface together with an interlocking network and the entanglement of chains. In particular, for polymer–polymer adhesion, the polymer chains may form an entangled network strengthened by a high density of covalent bonds, such that the materials cannot be pulled apart. Under stress, the whole network deforms and fracture only occurs if scission of the covalent bonds takes place.

Thus, to form strong adhesion between two polymeric materials it is necessary for this interlocking and bonded network to be continuous across the interface. Such continuity can be optimized by utilizing

1. sufficiently miscible materials to allow for chain interdiffusion, see Figure 2(a),
2. coupling chains at the interface that chemically bind to both materials, see Figure 2(b), or
3. chemical reactions of reactive groups within the polymer that react at the interface, see Figure 2(c).

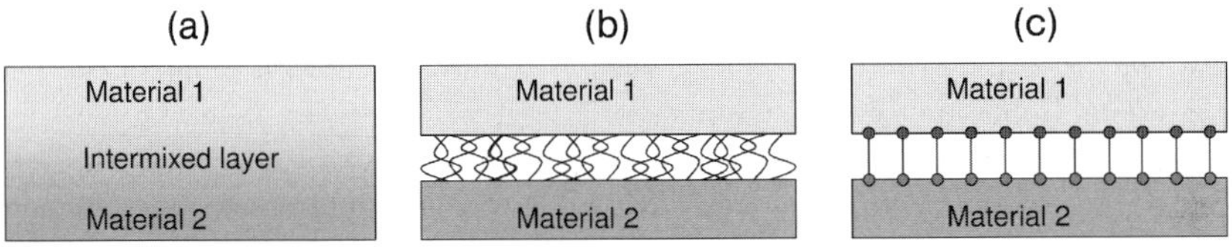

Figure 2 Schematic showing adhesion promotion as a result of (a) chain interdiffusion at the interface, (b) adhesion molecules between the layers and (c) chemical reactions at the interface between the two materials.

Coupling by chain interdiffusion can occur if the two polymeric materials are miscible in each other, or at least sufficiently miscible to form a broad interface. During interface coupling a chain end initially crosses the interface and then slowly more of the chain follows it across into the other material. Failure of such a joint is thus expected to be by chain pullout. As the diffusion distance increases it is chain scission that will lead to joint failure.

Another common technique to couple two bulk polymers is to introduce a small percentage of chemically modified chains into one or both of the materials. These modified chains can react with the other polymer to form coupling chains at the interface. A classic example of this technique is the introduction of maleic anhydride into polypropylene to induce coupling with a polyamide such as Nylon 6. The maleic anhydride functionality can be used to react with the polyamide chain leading to a graft, or block copolymer at the interface. Failure of adhesion for such systems can either be by pullout or by scission of the coupling chains. Pullout can occur if the chains of the interfacial copolymer are short, or if they do not entangle as effectively. This would, for example, be the case if there were too many coupling chains at the interface that become so densely packed that they prevent other chains from diffusing effectively. Unfortunately, it is generally difficult to distinguish between coupling of two materials by reaction and by entanglement.

Simple Van der Waals forces are generally not sufficient to cause significant adhesion between a flat, smooth inorganic substrate and a glassy or semicrystalline polymer. Chain interdiffusion and/or a chemical reaction between the materials is a clear prerequisite for a strong interface. Since interdiffusion is impossible at a polymer/inorganic interface, strong adhesion will require interfacial bonds of higher energy. The typical types of bonds that become relevant are hydrogen bonds, acid–base interactions or covalent bonds between the materials.

Two adhesive systems for inorganic–organic adhesion that have been studied extensively are based on silanes and thiols for the surface modification of inorganic surfaces and membranes. Today, both of these systems are of significant technological importance as they have shown significant potential for providing strong bonds between different materials [14]. Silanes are well known to self-assemble into monomolecular and multimolecular layers on glass or silicon dioxide surfaces [15–17]. One end of the silane molecule typically has functional groups such as di- or tri- methoxy, -ethoxy or -chloro- functionality, whilst the other end normally has other reactive groups or specific molecular chains, as suggested in Figure 3 [18–20]. The Si-bound functionality is allowed to condense with the hydroxyl functionality on the surface of the glass, as shown in Figure 4, whilst the amine functionality at the other end can react with a second material.

This simplified and idealized picture suggests a single molecular layer of silane adhesion promoter between, for example an epoxide and glass. The real situation is, however, is much more complex, in particular since silanes with multiple methoxy or ethoxy functionalities can self-condense and polymerize on the surface.

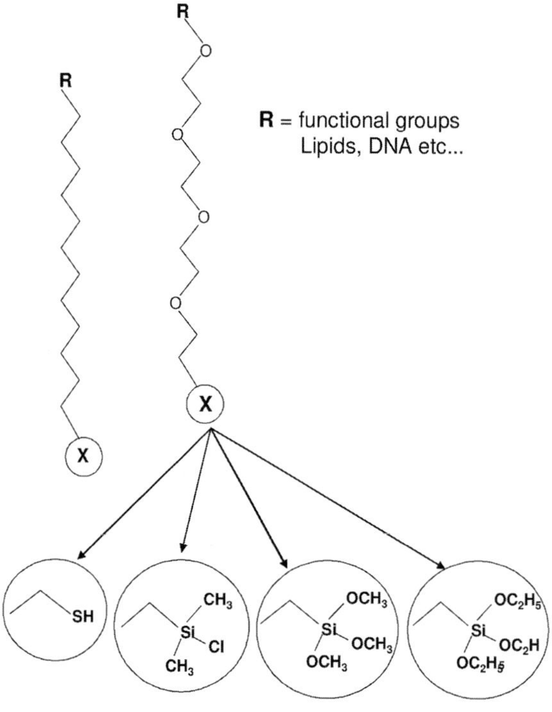

Figure 3 Typical structure of thiol- and silane-based adhesion molecules.

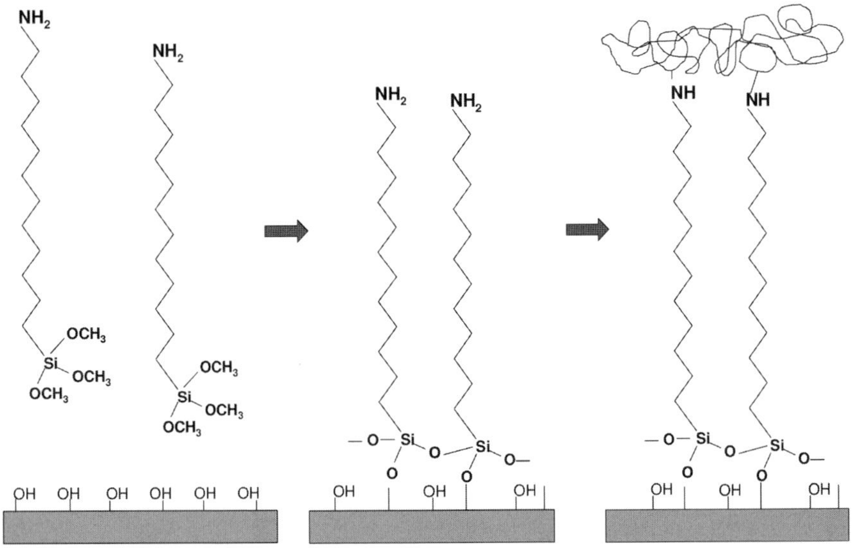

Figure 4 Simplified schematic of the adhesion mechanisms taking place between a polymeric material and a glass surface using a tri-methoxy functionalized silane for coupling.

Even though silanes clearly cause covalent bonding between the glass and the epoxy, there is no way known today to estimate the actual density of coupling bonds at the interface. Furthermore, the adhesion between the materials is markedly reduced in the presence of water. Silanes are used as adhesion promoters for many life-science applications [21–23], membrane technology [24, 25], colloid surface modification, filler and composite materials [18, 26, 27]. See also Chapter 3.3.

Thiols and disilfides are molecules that show a strong affinity for gold and are used as adhesion promoters that bond via the thiol at one end and via another reactive group at the other end [28–31]. Because of their ease of use they have found a wide range of applications in life-science applications and devices based on gold. Molecules to be bonded to a gold surface can be modified to contain a thiol (–S–H) end group, which is used as a relatively easy method to immobilize molecules such as DNA strands [32–35], proteins and sugars [36, 37], bilayer assemblies [38, 39] and polyethylene glycols to gold-coated surfaces [40, 41]. Thiols have been deposited using either dipping methods or by microcontact printing. Today, they are well established in sensor technology and many nanotechnological devices [42–45].

1.3.3
Physicochemical Methods of Surface Modification

Physicochemical methods of surface modification that are relevant to polymer interfaces and that will be discussed here include:

1. photochemical treatment, and
2. plasma-assisted processes using nonpolymerizable gases (e.g. nitrogen, oxygen, argon, helium, ammonia, carbon dioxide, etc.), polymerizable gases or plasma-assisted grafting of reactive molecules.

Photochemical treatment such as UV/ozone and UV laser treatment has been a standard method to modify polymer surfaces by crosslinking polymer chains, to pattern surfaces as well as to clean metal surfaces from organic contaminants (see also Chapter 3.3) [46–49]. UV lamps, such as Hg lamps ($\lambda = 254$ nm and 187 nm) [50], Xe_2^* excimer lamps [51] ($\lambda = 172$ nm) and different excimer lasers (200–400 nm depending on the laser gas) [52, 53] are described in the literature.

UV light of wavelength 185 nm decomposes oxygen molecules to generate ozone, which in turn is decomposed by irradiation at 254 nm to produce high-energy, activated oxygen (O^*).

$$O_2 \rightarrow O + O \text{ (185 nm)}$$
$$O_2 + O \rightarrow O_3$$
$$O_3 \rightarrow O^* + O_2 \text{ (254 nm)}$$

Organic molecules are decomposed and oxidized by the UV irradiation with the formation of CO_2 and H_2O, which desorb from the surface. At the same time reactive groups such as –OH, –COO$^\bullet$, C = O and –COOH are formed, which increase the hydrophilic nature of the surface treated. The radicals formed at the surface react with their environment as well as with each other leading to the incorporation of oxygen containing functional groups and a crosslinking of the surface. UV irradiation is commonly used not only to modify, but also to remove contaminants from a surface in preparation for adhesion and bonding. The types of contaminants that can be removed are organic compounds, in particular fatty substances. For the removal of thick layers of contamination the UV treatment must be preceded by a wet-cleaning process using water or other solvents. Photomodification has the advantage of allowing processing of surfaces under normal atmosphere. Even geometrically complex objects may be sufficiently modified by means of arranging lamp shapes or carrier units, so that it is suitable even for molded products. It is a remarkably effective process for engineered plastics, allowing a low-temperature processing. The effectiveness, however, depends also on the selected materials. Under clean conditions the modifying effect can last for weeks, shorter if contamination is allowed to resume.

The wavelengths of < 200 nm lead to photon energies of > 6 eV, which cause the dissociation of virtually all chemical bonds at the surface of the material. Depending on the wavelengths used, it can be used as a method of ablation and surface modification. Different variations of UV treatment include the process in air as well as different reactive atmospheres such as ammonia [51] or post-treatment reaction with reagents such as acrylamide (AAm), acrylic acid (AAc), azides and different amides [54, 55]. Under reactive gas atmosphere the irradiation results in the introduction of new hydrophilic nitrogen- or oxygen-containing groups and is generally accompanied by a roughening of the surface. Such photochemical processes are, for example, used to improve the wettability and adhesion properties of polymers such as PEEK, PC, PP, PET and PBT in printing, dyeing and metallization processes [56–61].

1.3.3.1 Plasma-Assisted Surface Modification

Plasma-assisted surface modification for the improvement of adhesion between two materials has been the subject of investigation since the 1970s. The plasma state is also known as the fourth state of matter and can be exploited to modify the chemical structure of materials in a number of different ways.

The ionization of the gas is accomplished by applying an electromagnetic field using specific frequencies. The frequencies are regulated by the government and controlled by international agreements to prevent interference with worldwide communication bands. They can be divided into three main categories: low frequency – less than 100 kHz, radio frequency (13.56 MHz) and microwave frequency (2.45 GHz). Today, the literature describes many examples of low-pressure nonequilibrium plasmas for the modification of many types of surfaces using both radio-frequency and microwave excitation.

Depending on the experimental conditions chosen, such as the nature of the gas, pressure, treatment time and input power, a variety of effects can be obtained, such as:

1. Ablation of surface material leading to a cleaning effect.
2. Deposition of material if polymerizable gases are used. Here, in particular, a wide variety of materials and surfaces can be obtained that are useful for adhesion improvement on different levels and applications.
3. A modification, or restructuring of the surface as a result of bombardment of the surface with chemically active species such as O, N, H and OH in their ground state, in higher excited states and as ions. These species interact with surfaces and effectively etch and chemically modify the upper monolayers.

The cleaning effect of oxygenated and many fluorinated gas plasmas is well known; it is a rapid and very effective procedure to remove mostly organic material from any surface. The oxygen gas is broken down into monatomic oxygen (O), as well as different oxygen ions (O^-, O_2^-) by the electromagnetic field [62, 63]. At reduced pressures the oxygen species will readily react with any organic material leading to the formation of water vapor, CO and CO_2, which is carried away by the pumping system [64]. While plasma is very effective in removing organic material it is less efficient in removing inorganics and salts. Today, plasma cleaning is widespread in different industrial processes in particular for optical components and microelectronics. In many optoelectronic devices, for example, plasma-enhanced cleaning procedures are used to prepare surfaces prior to eutectic die attachment and wire bonding. In life-science applications it has been shown to significantly improve sterilization methods by the destruction of spores, the removal of residual proteins and depyrogenation [65–67].

Nonequilibrium plasma processes for the deposition of thin films to be used for adhesion promotion have become of increasing interest [68]. The materials formed are referred to as *plasma polymers*, which differ from their conventional counterparts in the chemical structure and in their chemical and physical properties. While some 20 years ago, plasma deposition was mainly used for hard coatings, optical coatings, diffusion barriers and passivation layers, today it is routinely possible to deposit polymeric layers that show a high density of reactive surface groups. It is now routinely possible to achieve different levels of hydrophilicity that can be used for subsequent bonding to other materials via direct covalent bonding as well as by interdiffusion and entanglement of polymer chains. The advantages of plasma-assisted processes today are still the low quantities of reagents necessary, the absence of solvents, fast and effective treatments resulting in an economically and ecologically favorable process technology for many branches. In modern nanotechnological processing it offers unique ways to modify the surface of temperature-sensitive and highly fragile components. Disadvantages of plasma-assisted processes are often associated with the availability of equipment suitable for specialized applications, lack of specialized personnel required to run the equipment as well as the inherent costs.

Examples of well-researched functional plasma polymer films are those containing amines ($-NH_2$) [69–71], carboxylic acids ($-COOH$) [72–74], alcohols ($-OH$) [75–78], anhydrides [79, 80] epoxides [81] and active ester groups [82, 83]. Depending on the starting material used they can be highly reactive materials. The reactivity, of course, depends on the nature of the chemical groups they contain. Because of their reactive nature, the materials are prone to rapid aging, which leads to a loss of reactivity within days or weeks. If stored under ambient conditions the reactive groups will readily react with their environment. However, if used while still reactive, many of these films are excellent adhesives and adhesion-promoting layers. The literature over recent years shows, for example, that thin films deposited by nonequilibrium plasma can be used as adhesion promoters making the link between materials science and life sciences [84–86].

Plasma-assisted processing of surfaces offers some unique advantage. For example, the plasma process can, in principle, be guided to perform different functions sequentially; firstly cleaning/sterilization [87, 88], followed by the deposition of one or more functional layers. This approach is particularly interesting, since it can be designed for and applied to almost all substrate materials as well as offering economical, ecological and effective solutions in comparison to other more conventional methods of surface modification. This multistep processing is particularly interesting for biomaterial applications and has shown excellent results for *in-vitro* cell-culture tests [89]. In our laboratory a 1-minute modification/sterilization step using an oxygen plasma run at 100 W was sufficient to (i) activate and (ii) sterilize commercially available polystyrene cell-culture dishes. The monomers can be exchanged without exposure of the cell-culture dishes to ambient air. Subsequent plasma-assisted polymerization allows for further functionalization of the culture dish for cell adhesion tests. After this treatment the cell-culture dishes required *no further cleaning or sterilization* even after brief exposures to air, such as might be required during transfer to a cell laboratory. Standard laboratory procedures for cell cultures (flow box, sterile containers for transport, gloves while handling, etc.) are generally sufficient to allow for tests lasting several weeks.

Plasma-deposited layers have already shown promising results for the immobilization of proteins, antibodies and living cells. Since biological applications almost always require an aqueous medium, major efforts have concentrated on the stabilization and adhesion of the biofunctional plasma polymer films on different substrates. Solutions that were found to be reliable include:

1. wet-chemical procedures involving self-assembled monolayers based on thiols and silanes (see above), and
2. physiochemical methods involving plasma-deposited layers from hexamethly disiloxane (HMDSO) in excess oxygen.

The thiols are particularly suited to mediate between a gold surface and a plasma polymer, see Figure 5. While the HS-end forms strong bonds with the gold substrate the hydrocarbon end protruding from the surface can be activated by the plasma and is believed to participate in the plasma-initiated polymeriza-

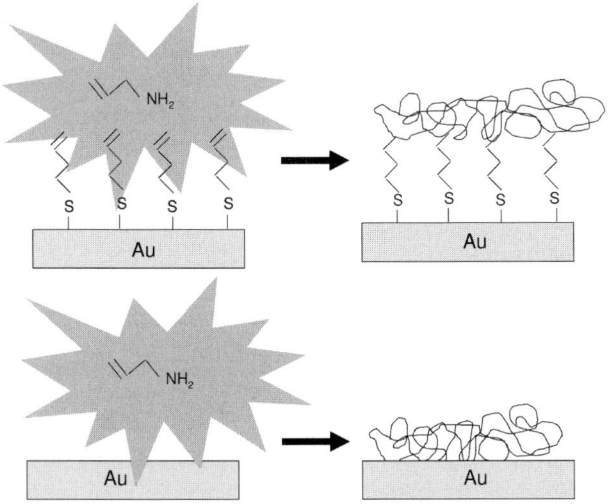

Figure 5 Schematic of adhesion promotion with and without a thiol SAM. The SAM is believed to participate in the plasma polymerization reaction acting as a linker molecule between a plasma polymer and a gold surface.

tion. *In-situ* infrared absorption reflection spectroscopic (IRRAS) studies previously showed that the thiol SAM is not destroyed by the plasma polymerization. The improved adhesion observed in all cases studied suggests strong bonding between the alkyl-thiol and the plasma polymer [79, 97].

The immobilization of proteins on a surface previously coated with an adhesion layer and a functional plasma polymer containing amino ($-NH_2$) groups or carboxylic acid (–COOH) groups has demonstrated that the functional group density at the surface is a major factor determining the immobilization. Kurosawa et al showed that cationic surfaces show a higher affinity for protein adsorption than anionic surfaces [90, 91]. Figure 6(a) shows a schematic of such a multilayer system for the adsorption of IgG on plasma polymer films deposited from allylamine under continuous wave plasma conditions using input powers (P_{peak}) of 90 W, 50 W and 10 W.

Previous work by this group and others has shown that a decrease in the input power is associated with an increase in the functional group density [70, 92–94]. In the case of plasma-polymerized allylamine this means a higher density of amino groups, which will protonate in solution and at pH 7.4 (standard PBS buffer) and show characteristics of a cationic polymer surface [95, 96]. Thus, the polymers deposited at a P_{peak} of 90 W typically have less amino groups available for bonding than those films deposited at a P_{peak} of 10 W. The graph in Figure 6(b) shows typical results obtained for the equivalent IgG thickness observed on plasma polymer films of thickness 30 ± 5 nm using surface plasmon resonance spectroscopy (SPR – see Glossary). For these measurements, refractive indices were previously determined using waveguide mode spectroscopy (see Glossary). The results show that with decreasing plasma input power a surface is obtained

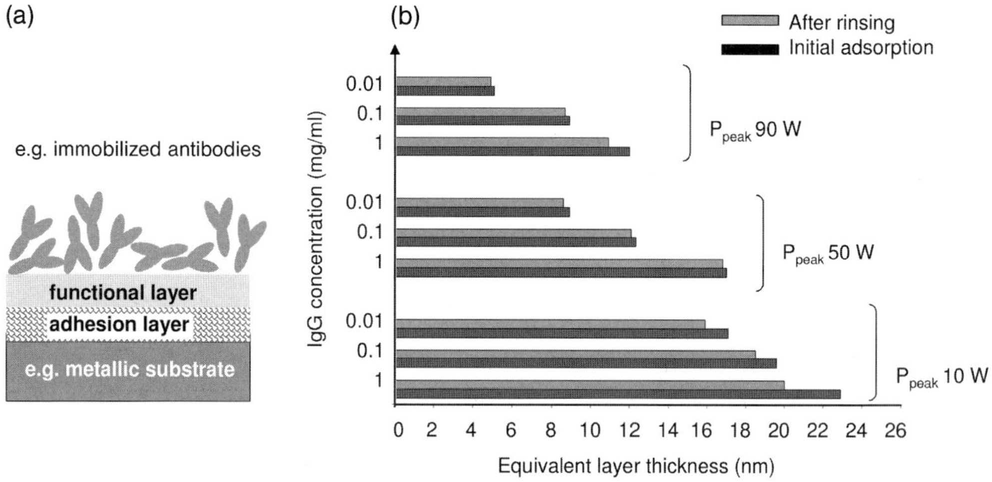

Figure 6 (a) Schematic of a multilayer system for the adsorption of proteins and (b) SPR results of IgG adsorption on plasma-polymerized allylamine using a refractive index *n* of 1.39 for the protein and *n* of 1.47 for the plasma polymer.

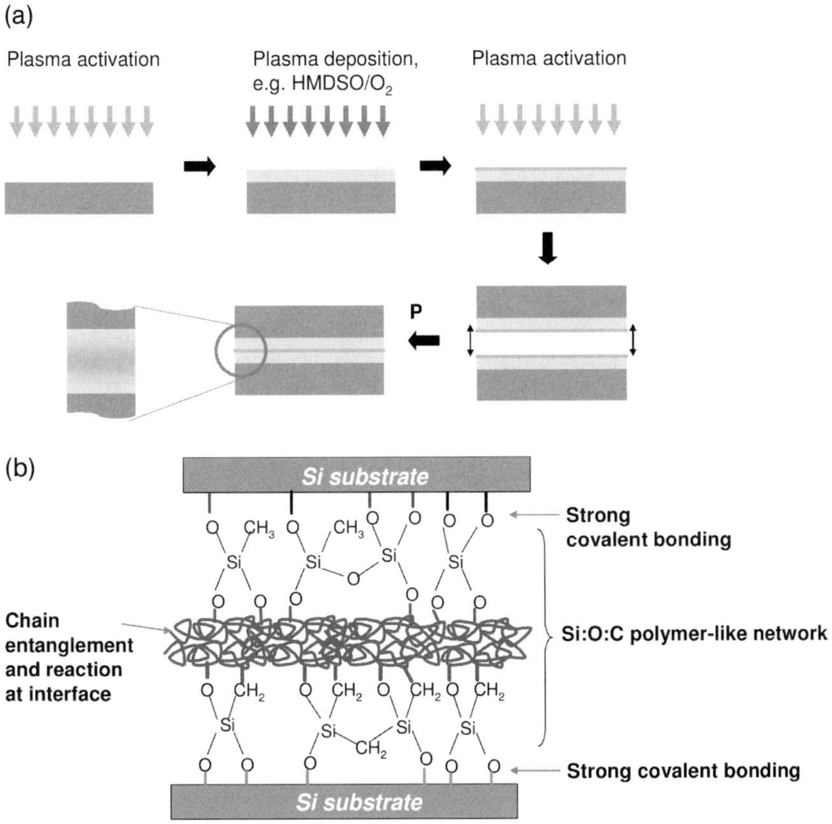

Figure 7 Schematic of silicon bonding using plasma-assisted processes.

that allows for the increased adsorption of IgG. The arrangement of the IgG molecules on the surface is only idealized in Figure 6(a).

Another interesting example in which a process may benefit from a multistep procedure is the bonding of silicon wafers. The procedure for such a process is schematically shown in Figure 7. Silicon bonding is of particular interest in the design and manufacture of different sensor devices. Unmodified silicon wafers will not adhere to one another; however, the adhesion between two Si wafers can be improved by the deposition of an organo-silane layer by plasma deposition followed by pressing of the two surfaces under a load of a few kilograms for at least 2 h.

In order to accomplish this, the two Si wafers were subjected to a series of steps as schematically shown in Figure 8. The wafers to be bonded were firstly wet cleaned and then activated in an oxygen plasma, then coated with a polymeric film derived from hexamethyl disiloxane (HMDSO) in oxygen at a ratio of (1:4) and an rf power of 150 W. These process conditions typically yield a polymer-like adhesive material consisting of silicon, oxygen and carbon in approximately the following proportion: Si_1:O_1:C_1. This material has previously been reported to be hydrophobic and was thus activated during a brief oxygen-plasma exposure [97].

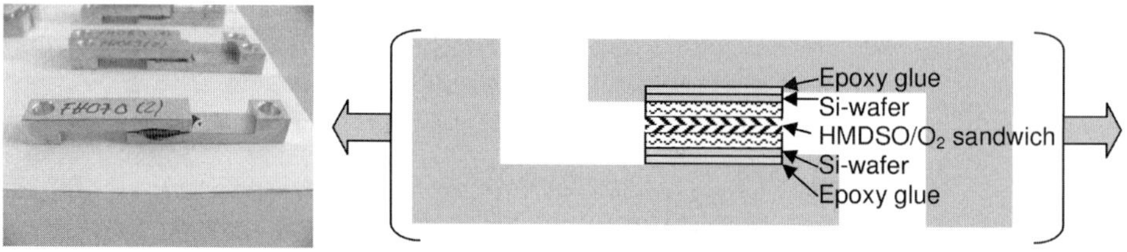

Figure 8 Schematic of the samples holder for testing shear strength using the Instron 6022 tensile testing device.

When pressing the two surfaces together to form an Si-HMDSO-Si sandwich, the reactive chains of the polymer-like ceramic are allowed to diffuse, entangle and react with one another at the interface. Mechanical shear tests were carried out by gluing the Si-HMDSO-Si sandwich onto a specialized holder designed for the Instron strain test device as shown in Figure 8. The above procedure led to a significant improvement in the bonding strength between the two wafers. For a series of 100 samples, results showed maximum strains ranging between 4–8 MPa. The main results are briefly summarized in Table 1 [98].

Table 1 Summary of mechanical test performed on differently treated pieces of Si wafer, The adhesion strengths given are the average of about 10 separate measurements. The term *activ.* stands for surface activation using a 5-s exposure to a 50-W oxygen plasma.

Plasma surface treatment	Plasma conditions	Adhesion
Si–activ. / activ.–Si	50 W, 5 s	no adhesion
Si–HMDSO/O_2–activ. / activ.–HMDSO/O_2–Si	30 W, 1 min	adhesion 2 ± 1 MPa
Si–HMDSO/O_2–activ. / activ.–HMDSO/O_2–Si	150 W, 1 min	good adhesion 4 ± 2 MPa
Si–activ.–HMDSO/O_2–activ./activ.–HMDSO/O_2–activ–Si	150 W, 1 min	v. good adhesion 6 ± 2 MPa

The modification of surfaces using nonequilibrium plasma makes use of the reactive nature of the active species of molecules such as oxygen, nitrogen, ammonia or carbon dioxide. In the plasma state these molecules are dissociated into ions, atoms in their ground state and higher excited states, metastables, charges species and electrons. In addition, the relaxation processes in the plasma lead to the emission of UV radiation. All of these interact, react with and bombard the substrate material subjected to the plasma, leading to a multitude of chemical and physical effects at the surface.

In this case, the modification of the material does not involve the deposition of a thin film, but it follows mechanisms in which the reactive species of the plasma colliding with the surface break bonds within the uppermost layer of the material leading to reactive sites at the surface. These reactive sites can then react either with each other, leading to a crosslinking of the surface, or with oxygen to form oxygenated groups such as alcohols (–OH), aldehydes, ketones (–C = O) and acids (–COOH) [99]. In the case of nitrogen and ammonia plasma treatment nitrogen is also incorporated into the surface and groups such as amines, amides and imines have previously been identified [100–102]. The depth of modification by the reactive species of the plasma is generally only a few nanometers. However, the UV that is associated with the plasma is known to modify the material to a depth of a few μm. The surface modification typically manifests itself as a change in the hydrophilicity of the substrate material. This is particularly the case for hydrocarbon polymers such as polystyrene, polyethylene and polypropylene. Each of these can readily be surface modified using plasma surface-modification processes, allowing these materials to be used in environments requiring the reactivity of the surface, yet maintaining the properties of the bulk material. The incorporation of oxygen and nitrogen groups leads to an overall increase in the hydrophilicity and typical water contact angles after modification are in the range of 30° to 60°, depending on the extent of oxygen or nitrogen incorporation [103–105]. The plasma surface modification of polymers such as nylon or polycarbonate involves similar mechanisms, yet the effect on the hydrophilicity are not as dramatic. This is because: (i) these materials have a lower contact angle to start with and (ii) the overall change in oxygen or nitrogen does not change as much as with the pure hydrocarbons.

As such, the process of surface modification using reactive gases is an extremely powerful tool and has found applications in the activation and modification of polymer webs, textiles and fibers. More recently, it has also found applications in biomaterials and biomedical and sensor applications (e.g. application in lab on a chip [86, 106]). Generally, only a fraction of a second is required to modify the surface of typical polymeric materials, whereby only a 5–7% increase in the surface oxygen or nitrogen content is sufficient to provide a significant improvement in reactivity and adhesion.

When the plasma is generated using inert noble gases such as helium or argon, the predominant effects on polymers are the breaking of C–C or C–H bonds by ion or UV bombardment. The free radicals that are formed in turn react with each other to lead to a crosslinking of the surfaces. Many of the radicals formed by the plasma are fairly long lived, such that subsequent exposure to the ambient leads to post-plasma reactions with water and oxygen in the air. This effect can be utilized to activate highly inert materials such as Teflon [107, 108]. Once activated by a helium or argon plasma Teflon may, for example, be further functionalized, e.g. using an ammonia plasma for the incorporation of reactive amine groups at the surface without changing the bulk properties of the Teflon device [109–111].

The activation of a surface in the plasma of an inert gas can be further exploited to subsequently graft specific molecules to the surface, which in turn exhibit a desired property. This is done by exposing the activated surface directly to the vapors of the molecules to be grafted before exposure to the ambient. This approach has been used since the 1980s to modify the surface of polypropylene and polyethylene [112–115]. Today, it is still a widespread approach to modify the surface of many polymers decorating the surfaces treated with amino groups, acid groups, alcohols and anhydride groups [116].

1.3.4
Analytical Tools to Study Adhesion

Despite the fact that thin films, coatings and different methods of adhesion promotion are an indispensable part of modern technology there are no simple global methods available to quantitatively determine adhesive strength. Some of the more common tests available and generally used are described briefly below. In recent years more sophisticated methods have been described in the literature, however, whatever test is chosen, it is generally restricted to only a few types of samples and the data obtained can often only be used to compare between like samples or specimens tested on the same equipment, thus greatly limiting their applicability. In all cases, however, it is always important to note whether the bonding failure was adhesive (i.e. at the coating/substrate interface) or cohesive (i.e. within the coating or the substrate material).

The bulk tensile strength of a cylindrical specimen can be determined by subjecting it to a tensile load. The adhesion between two materials of which at least one is a thin film, however, cannot be tested in a similar way. The 'peel' test and the 'scratch' test as well as the different variations of these methods remain surprisingly simple and are at best semiquantitative. Even so, they are still the most widespread methods of adhesion analysis. Both of these methods have been comprehensively reviewed in a number of articles and books by Mittal [117, 118] and Lacombe [119]. Recognized methods that can be used to estimate how well a coating is bonded to a substrate can be performed with a knife or with a pull-off adhesion tester. A knife and a cutting guide are used to make two cuts with a 30–45° angle between the legs and down to the substrate forming an X (see Figure 9). The knife is used to attempt to lift the coating from the substrate at the vertex. Alternatively, pressure-sensitive tape can be applied and removed over the cuts made. After the tape has been applied and pulled off, the cut area is inspected and

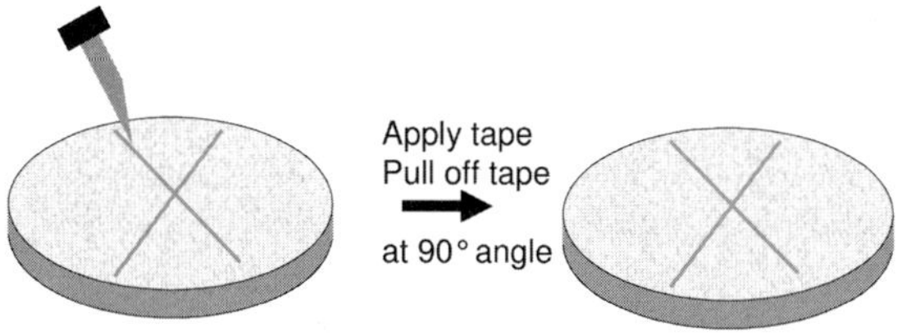

Figure 9 Basic procedure for the pull-off adhesion test.

rated. These tests are highly subjective and greatly dependent on the operator's experience, the types of cuts made, the type of tape used and the materials involved. The standard procedures for the application and performance of such tests are available under ASTM D3359 and ASTM D6677 [120].

Another well-known method that has been used since the 1960s for testing the adhesion between two materials is the blister test [121]. This is based on the formation of a blister by injecting a liquid under pressure between the coating and the substrate. The work of detachment, i.e. the energy needed to detach 1 cm^2 of coating, is determined by recording the liquid pressure as a function of the injected volume and computing the area under this curve. While many of these tests are still sufficient for a relatively rough evaluation of adhesion for macroscopic samples, the advances in nanotechnology and different thin-film coating procedures require much more sensitive and accurate methods to determine adhesion and adhesive forces.

Some of the more sophisticated methods that are currently being investigated include the nanoscratcher, acoustic microcavitation [122] and neutron reflectivity [123, 124]. Acoustic microcavitation, as described by Madanshetty and coworkers [125] is based on surface cavitation occurring near the surface immersed in a liquid subjected to kilohertz acoustic frequencies and can be regarded as a mechanical method for determining adhesion. Inertial cavitation involves the formation and implosion of bubbles between a thin film and the substrate and ultimately leads to the erosion of the surface film, thus potentially providing qualitative information on adhesive strengths between a thin film and any substrate. Since it generally leads to nonuniform erosion it is not a well-regulated process.

Depending on the types of adhesion mechanisms that are to be studied, optical techniques such as ellipsometry, surface plasmon resonance spectroscopy (SPR) and optical waveguide mode spectroscopy (OWS) have also shown promising results and are currently being employed by several groups to study adhesion and binding events of biological systems on different surfaces. These techniques are particularly suited for the investigation of adhesion phenomena on the monolayer and submonolayer scale. A particular advantage, which each of these techniques offer, is the possibility to study binding and adhesion phenomena in real time and in solvents. This makes them very powerful tools for studying the adhesion phenomena between previously modified surfaces with biological molecules and even living cells [126–128]. The optical techniques generally only provide information on changes in the optical constants of the materials involved. However, by careful experimental design and experimental series it is possible to extract and calculate extremely useful information on adhesion phenomena and binding events. As such, it is possible to calculate association and dissociation constants for many surface binding reactions (see Chapter 3.6). Some examples of how these optical techniques find applications in adhesion studies are given in Chapters 1.2, 1.4, and 5.3 of this book.

Another effective way to examine the mechanism of failure at a polymer/polymer interface is to use coupling chains that contain a label. After interface fracture, each of the two fractured surfaces is examined for the labeled section. This is particularly interesting for polymer–polymer adhesion where one polymer has been deuterated. The deuterium can later be identified using ion scattering or secondary ion mass spectrometry (SIMS) providing valuable information of adhesion failure mechanisms. Other alternative ways to study the mechanisms of adhesion failure include SEM, AFM and optical microscopy.

1.3.5
Adhesion Failure – Longevity of Modification

There is a wide range of situations where the adhesion between two materials fails because of environmental- or applications-related influences. These situations include temperature, pH or ionic-concentration changes; they include mechanical stress, aggressive chemical environments or irradiation. All of these external parameters may lead to changes in the structure and the chemical bonding within the materials and subsequently lead to material degradation and device failure. The modification techniques described above involve an activation of the surface that includes the opening of bonds and the formation of radicals on the surface. These reactive species in turn interact with their environment. At the same time, in particular polymeric materials will rearrange at the surface in an effort to reduce surface energy. Thus, much of the modification 'seems' to be lost over time as functional groups turn into the material and away from the surface. The phenomenon of surface aging of plasma-modified surfaces has been reviewed in a number of recent articles [129–132].

Most commercial polymers today contain additives to improve the product stability, flexibility, shelf life and packaging. These additives are often surface active and migrate to the surface where they can easily be detected using techniques such as X-ray photoelectron spectroscopy (XPS) or secondary ion mass spectrometry (SIMS and related techniques). If the materials are modified using the above-mentioned techniques, many of these additives can initially be removed from the surface, or at least they may be embedded or covered by a coating. However, it has been found in particular for methods involving the activation and chemical restructuring of the surface, that over time many of these additives migrate from the bulk material to the surface, where they may or may not interfere with surface properties and adhesion. This has previously been shown for silicon-based stabilizers in polymers such as PE, PS and PP.

Another aspect that influences the adhesion properties of modified surfaces is the presence of low molecular weight material that may be present after UV irradiation, plasma surface modification and even plasma modification [133]. These are fragments of material originating either from etching effects or from reaction termination at an early stage of the polymerization without bonding to the bulk material. These materials are generally soluble and can be washed away from the modified surface. For plasma surface modification processes it has been found that these low weight fragments can be reduced by carrying out the plasma modification at low power or, for moving foils and webs, by working at higher speeds.

Thus, while it is possible to modify polymer surfaces such as PE or PS to contain >> 10% oxygen, optimum adhesion properties are generally already found at around 5–7% oxygen as determined by XPS. Above this surface oxygen concentration it should be assumed that the oxygen originates from etching products that may interfere with adhesion processes.

1.3.6 Summary

Surface modification and adhesion are two surface-related phenomena that are closely linked and interrelated. Many adhesion processes depend strongly on preceding surface-modification techniques and would not be possible without the pretreatment or a modification of at least one of the materials. There are a number of possible strategies that can be applied; many of these are described in this book and some of them are briefly reviewed in this chapter. In all cases, the choice of methodologies used for adhesion improvement depends on:

1. the chemistry of the materials involved,
2. the surface morphologies,
3. the application environment to which the adhering surfaces are subjected.

The present chapter aims to review a few of the methods available today for the surface modification of some inorganic and polymeric materials taking examples from the authors' own work as well as referring to work described in other chapters within this book. There are a number of different types of phenomena that lead to adhesion forces between two materials ranging from mechanical, electrostatic to dispersive and diffusive mechanisms. Optimum adhesion is generally achieved when more than one of these processes takes place at the interface. Testing and quantifying adhesive forces is today still a major challenge and there is to date no universal method available. Some conventional methods to determine adhesion forces macroscopically and some that can be used to estimate (mostly qualitatively) the adhesive forces on the micro- and nanometer scale are introduced and reviewed.

References

1 Autumn, K., Liang, Y.A., Hsieh, S.T., Zesch, W., Wai Pang, C., Kenney, T.W., Fearing, R. and Full, R.J. (2000) Adhesive forces of a single gecko foot-hair. *Nature*, **405** (6787), 681–685.

2 Persson, B.N.J. (2003) On the mechanisms of adhesion in biological systems. *Journal of Chemical Physics*, **118** (16), 7614–7621.

3 Spolenak, R., Gorb, S. and Arzt, E. (2005) Adhesion design maps for bio-inspired attachment systems. *Acta Biomaterialia*, **1**, 5–13.

4 Hildebrand, M. (1988) *Analysis of Vertebrate Structures*, 3rd edn, John Wiley & Sons, Inc., N.Y., p. 701.

5 Northen, M.T. and Turner, K.L. (2006) Meso-scale adhesion testing of inte-

grated micro- and nano-scale structures. *Sensors and Actuators A: Physical,* **130–131**, 583–587.

6 Northen, M.T. and Turner, K.L. (2006) A batch fabricated biomimetic dry adhesive. *Nanotechnology,* **16**, 1159–1166.

7 Berg, J.C. (1993) *Wettability,* Marcel Dekker, New York.

8 Schrader, M.E. and Loeb, G. (1992) *Modern Approach to Wettability,* Plenum Press, New York.

9 Wu, S. (1982) *Polymer Interface & Adhesion,* Marcel Dekker, New York.

10 Andrade, J.D. (1985) in *Surface & Interfacial Aspects of Biomedical Polymers,* Vol. 1, Plenum Press, New York.

11 Veselovsky, R.A. and Kestelman, V.N. (2002) *Adhesion of Polymers,* McGraw-Hill.

12 Packham, D.E. (2005) *Handbook of Adhesion,* 2nd edn, Wiley & Sons, 10: 0471808741.

13 Pocius, A.V. (2002) *Adhesion and Adhesives Technology,* 2nd edn, Hanser Verlag, 10: 1569903190.

14 Ulman, A. (1998) *Thin Films: Self-Assembled Monolayers of Thiols,* 1st edn, Academic Press, 978-0125330244.

15 Singh, Rajinder Pal, Jha, Praveen, Kalpakci, Kerem and Way, J. Douglas (2007) Dual-Surface-Modified Reverse-Selective Membranes. *Industrial & Engineering Chemistry Research,* **46**, 7246–7252.

16 Pham, Khoa N., Fullston, Damian and Sagoe-Crentsil, Kwesi (2007) Surface modification for stability of nano-sized silica colloids. *Journal of Colloid and Interface Science,* **315**, 123–127.

17 Senkevich, J.J., Mitchell, C.J., Yang, G.-R. and Lu, T.M. (2002) Surface chemistry of mercaptan and growth of pyridine short-chain alkoxy silane molecular layers. *Langmuir,* **18** (5), 1587–1594.

18 Bula, K., Jesionowski, T., Krysztafkiewicz, A. and Janik, J. (2007) The effect of filler surface modification and processing conditions on distribution behaviour of silica nanofillers in polyesters. *Colloid and Polymer Science,* **285**, 1267–1273.

19 Atanasov, V., Atanasov, P.P., Vockenroth, I.K., Knorr, N. and Köper, I. (2006) A molecular toolkit for highly insulating tethered bilayer lipid membranes on various substrates. *Bioconjugate Chemistry,* **17**, 631–637.

20 Bothun, Geoffrey D., Peay, Katif and Ilias, Shamsuddin (2007) Role of tail chemistry on liquid and gas transport through organosilane-modified mesoporous ceramic membranes. *Journal of Membrane Science,* **301**, 162–170.

21 Tang, Haiying, Cao, Ting, Wang, Anfeng, Liang, Xuemei, Salley, Steven O, McAllister, James P. II, and Ng, K.Y. Simon (2006) Effect of surface modification of silicone on Staphylococcus epidermidis adhesion and colonization. *Journal of Biomedical Materials Research Part A,* **80** (4), 885–894.

22 http://www.farfield-scientific.com/products_anachip_nano.asp.

23 http://arrayit.blogspot.com/2007/10/microarray-surface-chemistry.html.

24 Szczepanski, V., Vlassiouk, I. and Smirnov, S. (2006) Stability of silane modifiers on alumina nanoporous membranes. *Journal of Membrane Science,* **281**, 587–591.

25 Atanasov, V., Knorr, N., Duran, R.S., Ingebrandt, S., Offenhäuser, A., Knoll, W. and Köper, I. (2005) Membrane on a chip: a functional tethered lipid bilayer membrane on a silicon oxide surface. *Biophysical Journal,* **89**, 1780–1788.

26 Subramani, Sankaraiah, Choi, Sung-Wook, Lee, Jun-Young and Kim, Jung Hyun (2007) Aqueous dispersion of novel silylated (polyurethane-acrylic hybrid/clay) nanocomposite. *Polymer,* **48**, 4691–4703.

27 Girones, J., Mendez, J.A., Boufi, S., Vilaseca, F. and Mutje, P. (2006) Effect of Silane Coupling Agents on the Properties of Pine Fibers/Polypropylene Composites. *Journal of Applied Polymer Science,* **103** (6), 3706–3717.

28 McNally, H., Janes, D.B., Kasibhatla, B. and Kubial, C.P. (2002) Electrostatic investigation into the bonding of poly-(phenylene) thiols to gold. *Superlattices and Microstructures,* **31** (5), 239–245.

29 Rocha, T.A.P., Teresa, M., Gomes, S.R., Duarte, A.C. and Oliveira, J.A.B.P. (1998) Quartz crystal microbalance with gold electrodes for monitoring gas-phase adsorption/desorption of short

chain alkylthiols and alkyl sulphides. *Analytical Communications*, **35**, 415–416.
30 Tlili, A., Abdelghani, A., Hleli, S. and Maaref, M.A. (2004) Electrical characterisation of a thiol SAM ob gold as a first step for the fabrication of immunosensors based on quartz crystal microbalance. *Sensors*, **4**, 105–114.
31 Khire, V.S., Lee, Tai Yeon and Bowman, C.N. (2007) Surface Modification Using Thiol-Acrylate Conjugate Addition Reactions. *Macromolecules*, **40**, 5669–5677.
32 Zhang, Z., Chen, Q., Knoll, W., Foerch, R., Holcomb, R. and Roitman, D. (2003) Plasma polymer film structure and DNA probe immobilization. *Macromolecules*, **36**, 7689–7694; Zhang, Z., Knoll, W., Förch, R., Holcomb, R. and Roitman, D. (2005) DNA hybridization on plasma polymerized allylamine. *Macromolecules*, **38** (4), 1271–1276.
33 Hasan, W., Lee, J., Henzie, J. and Odom, T.W. (2007) Selective Functionalization and Spectral Identification of Gold Nanopyramids. *The Journal of Physical Chemistry C*, **111** (46), 17176–17179.
34 Aoki, H. and Umezawa, Y. (2002) Ion channel sensors for electrochemical detection of DNA based on self assembled PNA monolayers. *Nucleic Acid Research Supplement*, **2**, 131–132.
35 Bietsch, A., Zhang, Jiayun, Hegner, M., Lang, H.P. and Gerber, C. (2004) Rapid functionalization of cantilever array sensors by inkjet printing. *Nanotechnology*, **15**, 873–880.
36 Seo, Jeong Hyun, Adachi, Kyouichi, Lee, Bong Kuk, Kang, Dong Gyun, Kim, Yeon Kyu, Kim, Kyoung Ro, Lee, Hea Yeon, Kawai, Tomoji and Cha, Hyung Joon (2007) Facile and Rapid Direct Gold Surface Immobilization with Controlled Orientation for Carbohydrates. *Bioconjugate Chemistry*, **18**, 2197–2201.
37 Gwenina, C.D., Jones, J.P., Kalaji, M., Lewis, T.J., Llewellyn, J.P. and Williams, P.A. (2007) Viscoelastic change following adsorption and subsequent molecular reorganisation of a nitroreductase enzyme on a gold surface: A QCM study. *Sensors and Actuators B: Chemical*, **126**, 499–507.
38 Schiller, S.M., Naumann, R., Lovejoy, K., Kunz, H. and Knoll, W. (2003) Archaea Analogue Thiolipids for Tethered Bilayer Lipid Membranes on Ultrasmooth Gold Surfaces. *Angewandte Chemie – International Edition*, **42** (2), 208–211.
39 Naumann, R., Schiller, S.M., Gieß, F., Grohe, B., Hartman, K.B., Kärcher, I., Köper, I., Lübben, J., Vasilev, K. and Knoll, W. (2003) Tethered Lipid Bilayers on Ultra Flat Gold Surfaces. *Langmuir*, **19**, 5435–5443.
40 Shenoy, Dinesh, Fu, Wei, Li, Jane, Crasto, C., Jones, G., Dimarzio, C., Sridhar, Srinivas and Amiji, Mansoor (2006) Surface functionalization of gold nanoparticles using hetero-bifunctional poly(ethylene glycol) spacer for intracellular tracking and delivery. *International Journal of Nanomedicine*, **1** (1), 51–58.
41 Piehler, J., Brecht, A., Valiokas, R., Liedberg, B. and Gauglitz, G. (2000) A high-density poly(ethylene glycol) polymer brush for immobilization on glass-type surfaces. *Biosensors & Bioelectronics*, **15** (9–19), 473–481.
42 Vockenroth, I.K., Atanasova, P.P., Knoll, W., Jenkins, A.T.A. and Köper, I. (2005) Functional tethered bilayer membranes as a biosensor platform. in IEEE Sensors 2005 – The 4th IEEE Conference on Sensors. Irvine, CA: IEEE Sensors, pp. 608–610.
43 Busse, S., Käshammer, J., Krämer, S. and Mittler, S. (1999) Gold and thiol surface functionalized integrated optical Mach–Zehnder interferometer for sensing purposes. *Sensors and Actuators B: Chemical*, **60** (2–3), 148–154.
44 Thery-Merland, F., Méthivier, C., Pasquinet, E., Hairault, L. and Pradier, C.M. (2006) Adsorption of functionalised thiols on gold surfaces: How to build a sensitive and selective sensor for a nitroaromatic compound? *Sensors and Actuators B: Chemical*, **114** (1), 223–228.
45 Aoki, Hiroshi, Umezawa, Yoshio, Vertova, A. and Rondinini, S. (2006) Ion-channel Sensors Based on ETH 1001 Ionophore Embedded in Charged-alkanethiol Self-assembled Monolayers on Gold Electrode Surfaces. *Analytical Sciences*, **22** (12), 1581.

46 Blach-Watson, J.A., Watson, G.S., Brown, C.L. and Myhra, S. (2004) UV patterning of polyimide: differentiation and characterization of surface chemistry and structure. *Applied Surface Science*, **235** (1–2), 164–169.

47 Chen, M.-S., Dulcey, C.S., Chrisey, L.A. and Dressick, W.J. (2006) Deep-UV Photochemistry and Patterning of (Aminoethylaminomethyl)phenethylsiloxane Self-Assembled Monolayers. *Advanced Functional Materials*, **16** (6), 774–783.

48 Nicolau, D.V., Taguchi, T., Tanguchi, H. and Yoshikawa, S. (1996) UV-and E-Beam patterning of bioactive molecules. *Journal of Photopolymer Science and Technology*, **9** (4), 645–652.

49 Vig, J. and LeBus, J. (1976) UV/Ozone Cleaning of Surfaces. *IEEE Transactions*, **12** (4), 365–370.

50 Foerch, R., Izawa, J. and Spears, G. (1991) A comparative study of the effects of remote nitrogen plasma, remote oxygen plasma and corona discharge treatments on the properties of polyethylene. *Journal of Adhesion Science and Technology*, **5** (7), 549–564.

51 Olbrich, M., Punshon, G., Frischauf, I., Salacinski, H.J., Rebollar, E., Romanin, C., Seifalianm, A.M. and Heitz, J. (2007) UV surface modification of a new nanocomposite polymer to improve cytocompatibility. *Journal of Biomaterials Science. Polymer Edition*, **18** (4), 453–468.

52 Mortaigne, B., Feltz, B. and Laurens, P. (1997) Study of unsaturated polyester and vinylester morphologies using excimer laser surface treatment. *Journal of Applied Polymer Science*, **66**, 1703–1714.

53 Jahani, Hamid R., Moffat, B., Mueller, R.E., Fumo, D., Duley, W., Northm, T. and Gu, Bo. (1998) Excimer laser surface modification of coated steel for enhancement of adhesive bonding. *Applied Surface Science*, **127–129**, 767–772.

54 Wang, Tie, Kang, E.T., Neoh,K.G., Tan, K.L., and Liaw, D.J. (1998) Surface Modification of Low-Density Polyethylene Films by UV-Induced Graft Copolymerization and Its Relevance to Photolamination. *Langmuir*, **14** (4), 921–927.

55 Sonntag, F., Pietzsch, M., Poll, R., Rabenau, M. and Jäger, M. (2007) Surface Modification of Polymers by using Excimer Laser for Biomedical Applications. *Plasma Processes and Polymers*, **4** (S1), 416–418.

56 Breuer, J., Metev, S., Sepold, G., Hennemann, O.D., Kolleg, H. and Krueger, G. (1990) Laser induced photochemical adherence enhancement. *Applied Surface Science*, **46**, 336–341.

57 Bahners, T. (1995) Eximer laser irradiation of synthetic fibres as a new process for the surface modification of textiles – a review. *Optical and Quantum Electronics*, **27**, 1337–1348.

58 Beil, S., Horn, H., Windisch, A., Hilgers, C. and Pochner, K. (1999) Photochemical functionalization of polymer surfaces for subsequent metallization. *Surface & Coatings Technology*, **116–119**, 1195–1203.

59 Petit, S., Laurens, P., Amouroux, J. and Arefi-Khonsari, F. (2000) Excimer laser treatment of PET before plasma metallization. *Applied Surface Science*, **168**, 300–303.

60 Laurens, P., Sadrasa, B., Decobert, F., Arefi-Khonsari, F. and Amouroux, J. (1998) Enhancement of the adhesive bonding properties of PEEK by excimer laser treatment. *International Journal of Adhesion and Adhesives*, **18**, 19–27.

61 Thomas, D.W., Foulkes-Williams, C., Rumsby, P.T. and Gower, M.G. (1992) Surface modification of polymers and ceramics induced by excimer laser radiation, in *Laser Ablation of Electronic Materials* (eds E. Fogarassy and S. Lazare), Elsevier Science Publishing B.V.

62 Diamy, A.-M., Legrand, J.-C., Rybkin, V.V. and Smirnov, S.A. (2005) Experimental Study and Modelling of Formation and Decay of Active Species in an Oxygen Discharge. *Contributions to Plasma Physics*, **45** (1), 5–21.

63 Saloum, S. and Naddaf, M. (2008) Diagnostic study of low pressure Ar-O_2 remote plasma generated in HCD-L 300 system: relative density of O atom. *Vacuum*, **82**, 66–71.

64 Krstulovic, N., Labazan, I., Milosevic, S., Cvelbar, U., Vesel, A. and Mozetic, M. (2006) Optical emission spectroscopy characterisation of oxygen plasma dur-

ing treatment of a PET foil. *Journal of Physics D-Applied Physics*, **39**, 3799–3804.

65 Lerouge, S., Fozza, A.C., Wertheimer, M.R., Marchand, R. and Yahia, L.H. (2000) *Plasma Polymers*, **5**, 31–46.

66 Moisan, M., Barbeau, J., Moreau, S., Pelletier, J., Tabrizian, M. and Yahia, L'H. (2001) *International Journal of Pharmaceutics*, **226**, 1–21.

67 Rossi, F., Kylian, O. and Hasiwa, M. (2007) in *Advanced Plasma Technology* (eds R. d'Agostino, P. Favia, Y. Kawai, H. Ikegami, N. Sato and F. Arefi-Khonsari), Wiley VCH, Weinheim, pp. 319–340.

68 Chu, P.K., Chen, J.Y., Wang, L.P. and Huang, N. (2002) Plasma-surface modification of biomaterials. *Materials Science and Engineering Reports*, **36** (5–6), 143–206.

69 van Os, M.T., Menges, B., Förch, R., Knoll, W., Timmons, R.B. and Vancso, G.J. (1999) Pulsed plasma deposition of allylamine: aging and effects of solvent treatment. *Materials Research Society Symposium Proceedings*, **544**, 45–50.

70 van Os, M.T., Menges, B., Förch, R., Knoll, W. and Vancso, G.J. (1999) Characterisation of plasma polymerised allylamine using optical waveguide mode spectroscopy. *Chemistry of Materials*, **11** (11), 3252–3257.

71 Choukourov, A., Biedermann, H., Slavinska, D., Trchova, M. and Hollander, A. (2003) The influence of pulse parameters on film composition during pulsed plasma polymerization of diaminocyclohexane. *Surface & Coatings Technology*, **174–175**, 863–866.

72 O'Toole, L., Beck, A.J., Ameen, A.P., Jones, F.R. and Short, R.D. (1995) Radio frequency induced plasma polymerization of propenoic acid and propanoic acid. *Journal of the Chemical Society-Faraday Transactions*, **91** (21), 3907–3912.

73 Alexander, M.R. and Duc, T.M. (1998) The chemistry of deposits formed from acrylic acid. *Journal of Materials Chemistry*, **8** (4), 937–943.

74 Swindells, I., Voronin, S.A., Fotea, C., Alexander, M.R. and Bradley, J.W. (2007) Detection of negative molecular ions in acrylic acid plasma: Some implications for polymerization mechanisms. *The Journal of Physical Chemistry. B*, **111** (30), 8720–8722.

75 Gombotz, W.R. and Hoffman, A.S. (1988) Functionalization of polymeric films by plasma polymerization of allyl alcohol and allylamine. *Journal of Applied Polymer Science: Applied Polymer Symposium*, **42**, 285–303.

76 Rinsch, C.R., Chen, X., Panchalingam, V., Eberhart, R.C., Wang, J.-H. and Timmons, R.B. (1996) Pulsed Radio frequency plasma polymerization of allyl alcohol: controlled deposition of surface hydroxyl groups. *Langmuir*, **12**, 2995–3002.

77 Friedrich, J., Kühn, G., Mix, R. and Unger, W. (2004) Formation of Plasma Polymer Layers with Functional Groups of Different Type and Density at Polymer Surfaces and their Interaction with Al Atoms. *Plasma Processes and Polymers*, **1** (1), 28–50.

78 Oran, U., Swaraj, S., Friedrich, J.F. and Unger, W.E.S. (2005) Surface analysis of plasma deposited polymer films, 5: Tof-SSIMS characterisation of plasma deposited allyl alcohol films. *Plasma Processes and Polymers*, **2**, 563–571.

79 Jenkins, A.T.A., Hu, J., Wang, Y.Z., Schiller, S., Förch, R. and Knoll, W. (2000) Pulsed plasma deposited maleic anhydride thin films as supports for lipid bilayers. *Langmuir*, **16** (16), 6381–6384.

80 Schiller, S., Hu, J., Jenkins, A.T.A., Timmons, R.B., Sanchez-Estrada, F.S., Knoll, W. and Förch, R. (2002) Plasma polymerisation of maleic anhydride: film chemical structure and properties. *Chemistry of Materials*, **14**, 235–242.

81 Tarducci, C., Kinmond, E.J., Badyal, J.P.S., Brewer, S.A. and Willis, C. (2000) Epoxide-functionalized solid surfaces. *Chemistry of Materials*, **12** (7), 1884–1889.

82 Francesch, L., Borros, S., Knoll, W. and Förch, R. (2007) Surface Reactivity of pulsed-plasma polymerised pentafluorophenyl methacrylate (PFM) towards amines and proteins in solution. *Langmuir*, **23**, 3927–3931.

83 Francesch, L., Garreta, E., Balcells, M., Edelman, E.R. and Borros, S. (2005)

Fabrication of bioactive surfaces by plasma polymerization techniques using a novel acrylate derived monomer. *Plasma Processes & Polymers*, **2**, 605–611.

84 Harsch, A., Calderon, J., Timmons, R.B. and Gross, G.W. (2000) Pulsed plasma deposition of allylamine on polysiloxane: a stable surface for neuronal cell adhesion. *Journal of Neuroscience Methods*, **98**, 135–144.

85 Sardella, E., Gristina, R., Ceccone, G., Gilliland, D., Papadopoulou-Bouraoui, A., Rossi, F., Senesi, G.S., Detomaso, L., Favia, P. and d'Agostino, R. (2005) Control of cell adhesion and spreading by spatial microarranged PEO-like and pdAA domains. *Surface & Coatings Technology*, **200**, 51–57.

86 Martin, I.T., Dressen, B., Boggs, M., Liu, Y., Henry, C.S. and Fisher, E.R. (2007) Plasma modification of PDMS microfluidic devices for control of electroosmotic flow. *Plasma Processes & Polymers*, **4**, 414–424.

87 Sun, Y.Z., Qiu, Y.C., Nie, A.L. and Wang, X.D. (2007) Experimental research on inactivation of bacteria by using dielectric barrier discharge. *IEEE Transaction on Plasma Science*, **35** (5), 1496–1500.

88 Moisan, M., Barbeau, J., Moreau, S., Pelletier, J., Tabrizian, M. and Yahia, L.H. (2001) Low-temperature sterilization using gas plasmas: a review of the experiments and an analysis of the inactivation mechanisms. *International Journal of Pharmaceutics*, **226** (1–2), 1–21.

89 Förch, R., Chifen, A.N., Bousquet, A., Khor, H.L., Chu, L., Zhang, Z., Osey-Mensah, I., Sinner, E.-K. and Knoll, W. (2007) Recent and expected roles of plasma polymerized films for biomedical applications. *Chemical Vapor Deposition*, **13**, 80–294.

90 Kurosawa, S., Kamo, N., Aizawa, H. and Muratsugu, M. (2007) Adsorption of I-labelled immunoglobulin G, its F(ab')2 and Fc fragments onto plasma-polymerized films. *Biosensors & Bioelectronics*, **22**, 2598–2603.

91 Kurosawa, S., Kamo, N., Minoura, N. and Muratsugu, M. (1997) Dose response in solid-phase radioimmunoassay to human IgG on plasma-polymerized films coated with F(ab')2 anti-human IgG antibody. *Materials Science & Engineering*, **C4**, 291–296.

92 Zhang, Z. (2004). PhD Thesis, Johannes Gutenberg University.

93 Förch, R., Zhang, Z. and Knoll, W. (2005) Soft-Plasma treated surfaces: tailoring of structure and properties. *Plasma Polymers & Processes*, **2** (5), 351–372.

94 Choukourov, A., Biedermann, H., Kholodkov, I., Slavinska, D., Trchova, M. and Hollander, A. (2004) Properties of amine-containing coatings prepared by plasma polymerization. *Journal of Applied Polymer Science*, **92**, 979–990.

95 Schönherr, H., van Os, M.T., Hruska, Z., Kurdi, J., Förch, R., Arefi-Khonsari, F., Knoll, W. and Vancso, G.J. (2000) Towards mapping of functional group distributions in functional polymers by AFM force titration measurements. *Chemical Communications*, 1303–1304.

96 Schönherr, H., Van Os, M.T., Förch, R., Timmons, R.B., Knoll, W. and Vancso, G.J. (2000) Distribution of functional groups in plasma polymerised allylamine films by scanning force microscopy using functionalised probe tips. *Chemistry of Materials*, **12**, 3689–3694.

97 Chifen, A.N., Jenkins, A.T.A., Knoll, W. and Förch, R. (2007) Adhesion improvement of plasma polymerized maleic anhydride films on gold using HMDSO/O_2 adhesion layers. *Plasma Processes and Polymers*, **4**, 815–822.

98 Pihan, S.A., Tzukruk, T. and Förch, R. (2009) Plasma polymerized hexamethyl disiloxane in adhesion applications. *Surface & Coatings Technology*, **203**, 1856–1862.

99 Foerch, R., McIntyre, N.S. and Hunter, D.H. (1990) Oxidation of polyethylene surfaces by remote plasma discharge: a comparison study with alternative oxidation methods. *Journal of Polymer Science, Polymer Chemistry Edition*, **28**, 193–204.

100 Foerch, R., McIntyre, N.S. and Hunter, D.H. (1992) Remote nitrogen plasma treatment of polymers: polyethylene, nylon 6,6, polyethylene, poly(ethylene vinyl alcohol) and poly(ethylene terephthalate). *Journal of Polymer Science, Polymer Chemistry Edition*, **30**, 279–286.

101 Foerch, R. and Johnson, D. (1991) XPS and SSIMS analysis of polymers: the effect of remote nitrogen plasma treatment on polyethylene, poly(ethylene vinyl alcohol) and poly(ethylene terephthalate). *Surface and Interface Analysis*, **17**, 847–854.

102 Foerch, R., McIntyre, N.S., Sodhi, R.N.S. and Hunter, D.H. (1990) Nitrogen plasma treatment of polyethylene in a remote plasma reactor. *Journal of Applied Polymer Science*, **40**, 1903–1915.

103 Chen, Zhijun, Lu, Xiaolin, Chan, C.-M. and Mi, Yongli (2006) Manipulating the surface properties of polyacrylamide with nitrogen plasma. *European Polymer Journal*, **42** (11), 2914–2920.

104 Vandencasteele, N., Merche, D. and Reniers, F. (2006) XPS and contact angle study of N2 and O2 plasma-modified PTFE, PVDF and PVF surfaces. *Surface and Interface Analysis*, **38** (4), 526–530.

105 Wagner, A.J., Fairbrother, D.H. and Renier, F. (2003) A Comparison of PE Surfaces Modified by Plasma Generated Neutral Nitrogen Species and Nitrogen Ions. *Plasmas and Polymers*, **8** (2), 119–134.

106 Hui, A.Y.N., Wang, G., Lin, B. and Chan, W.-T. (2005) Microwave plasma treatment of polymer surface for irreversible sealing of microfluidic devices. *Lab on Chip*, **5**, 1173–1177.

107 Dasilva, W., Entenberg, A., Kahn, B., Debies, T. and Takacs, G.A. (2006) Surface modification of Teflon® PFA with vacuum UV photo-oxidation. *Journal of Adhesion Science and Technology*, **20**(5), 437–455.

108 Liang, R.Q., Su, X.B., Wu, Q.C. and Fang, F. (2000) Study of the surface-modified Teflon/ ceramics complex material treated by microwave plasma with XPS analysis. *Surface and Coatings Technology*, **131**(1–3), 294–299.

109 Chen, Y. and Momose, Y. (1999) Reaction of argon plasma-treated teflon PFA with aminopropyltriethoxysilane in its n-hexane solution. *Surface and Interface Analysis*, **27**(12), 1073–1083.

110 Momose, Y., Tamura, Y., Ogino, M., Okazaki, S. and Hirayama, M. (1992) Chemical reactivity between Teflon surfaces subjected to argon plasma treatment and atmospheric oxygen. *Journal of Vacuum Science & Technology A: Vacuum, Surfaces and, Films*, **10**(1), 229–238.

111 Shi, M.K., Martinu, L., Sacher, E., Selmani, A., Wertheimer, M.R. and Yelon, A. (2004) Angle-resolved XPS study of plasma-treated teflon PFA surfaces. *Surface and Interface Analysis*, **23**(2), 99–104.

112 Mastihuba, M., Blecha, J., Kamenicka, V., Lapcik, L. and Topolsky, I. (1985) *Acta Physica Slovaca*, **35**(6), 355–362.

113 Suziki, M., Kishida, A., Iwata, H. and Ikada, Y. (1986) Graft copolymerization of acrylamide onto polyethylene surface pretreated with a glow discharge. *Macromolecules*, **19**, 1804–1808.

114 Dogue, I.L.J., Mermilliod, N. and Foerch, R. (1995) Grafting of acrylic acid onto polypropylene – comparison of two pretreatments: gamma irradiation and argon plasma. *Nuclear Instruments and Methods in Physics Research*, **B105**, 164–167.

115 Novak, I., Steviar, M., Chodak, I., Krupa, I., Nedelcev, T., Sprkova, M., Chehimi, Mohamed M., Mosnacek, J. and Kleinova, A. (2007) Study of adhesion and surface properties of low-density poly(-ethylene) pre-treated by cold discharge plasma. *Polymers for Advanced Technologies*, **18**, 97–105.

116 Zou, X.P., Kang, E.T., Neoh, K.G., Cui, C.Q. and Lim, T.B. (2000) Surface Modification of Poly(tetrafluoroethylene) Films by Plasma Pre-Activation and Plasma Polymerization of Glycidyl Methacrylate. *Plasmas and Polymers*, **5** (3–4), 219–234.

117 Mittal, K.L. (2001) *(EDT) – Adhesion Measurement of Films & Coatings*, Brill Academic Publishers, 13: 9789067643375.Mittal, K.L. (2004) *Polymer Surface Modification : Relevance to Adhesion*, V.S.P. Intl Science, 13: 9789067644037.Mittal, K.L. (2007) Adhesion Aspects of Thin Films, Brill Academic Pub, 13: 9789067644556.

118 Mittal, K.L. Pizzi, A. Mittal Mittal (1999) *Adhesion Promotion Techniques: Technological Applications (Materials Engineering)*, Marcel Dekker Ltd, 13: 978-0824702397.

119 Lacombe, R. (2005) *Adhesion Measurement Methods: Theory and Practice*, CRC Books – Taylor & Francis Ltd, 13: 978-0824753610.

120 http://www.astm.org/cgi-bin/SoftCart.exe/index.shtml?E+mystore.

121 Dannenberg, H. (2003) Measurement of adhesion by a blister method. *Journal of Applied Polymer Science*, **5** (14), 125–134.

122 Madanshetty, S.I., Wanklyn, K.M. and Ji, H. (2005) *Testing thin film adhesion strength acoustically*. Acoustic research letters online, **6** (1) (DOI: 10.1121/1.1828106).

123 Stamm, M. and Schubert, D.W. (1995) *Annual Review of Materials Science*, **25**, 325–356.

124 Schnell, R., Stamm, M. and Creton, C. (1999) *Macromolecules*, **32**, 3420–3425.

125 Madanshetty, S.I., Wanklyn, K.M. and Ji, H. (2005) *Testing thin film adhesion strength acoustically*. Acoustic research letters online **6** (1) (DOI: 10.1121/1.1828106).

126 Zhang, Z. and Foerch, R. (2005) Amino-functionalized plasma polymer films for DNA immobilization and hybridisation, in *Surface & Coatings Technology, Special issue: Plasma Surface Engineering (PSE 2004)*, Vol. 200/1–4 (eds M. Albu, U. Helmersson, H. Holleck, C. Oehr, K. Reichel, V. Rigato, H. StÃri and R. Suchentrunk), pp. 993–995.

127 Bender, K., Fraser, S., Förch, R., Jenkins, A.T.A., Köper, I., Naumann, R., Schiller, S.M.S. and Knoll, W. (2005) Plasma Polymer supported Lipid bilayers, in *Plasma Polymers & Related Materials* (eds M. Mutlu, G. Dinescu, R. Förch, J.M. Martin-Martinez and J. Vyskocil), Hacettepe University Press, pp. 32–42.

128 Jenkins, A.T.A., Buckling, A., McGhee, M. and Ffrench-Constant, R.H. (2005) *Journal of the Royal Society Interface*, **20**, 1233–1236.

129 Ling, Ao, Jin-zhen, Niu, Yu-ming, Liu, Jian-fang, Hu and Hui-ying, Chen (2001) The Aging Study on Polyethylene Terephthalate with Surface Modification by Water Vapor Plasma. *Plasma Science and Technology*, **3**, 761–764, 10.1088/1009-0630/3/3/002.

130 Gengenbach, T.R., Vasic, Z.R., Li, Sheng, Chatelier, R.C. and Griesser, H.J. (1997) Contributions of restructuring and oxidation to the aging of the surface of plasma polymers containing heteroatoms. *Plasmas and Polymers*, **2** (2), 91–114.

131 Shyong Siow, K., Britcher, L., Kumar, S. and Griesser, H.J. (2006) Plasma methods for the generation of chemically reactive surfaces for biomolecule immobilization and cell colonization: A review. *Plasma Processes and Polymers: Plasma Processes for, Biomedical Applications*, **3** (6–7), 392–418.

132 Yun, Yo Il, Kim, Kwang Soo, Uhm, Sung-Jin, Khatua, Bhanu Bhusan, Cho, Kilwon, Kim, Jin Kon and Park, Chan Eon (2004) Aging behavior of oxygen plasma-treated polypropylene with different crystallinities. *Journal of Adhesion Science and Technology*, **18** (11), 1279–1291.

133 Foerch, R., Kill, G. and Walzak, M.J. (1993) Plasma surface modification of polyethylene: short-term vs. long-term plasma treatment. *Journal of Adhesion Science and Technology*, **7** (10), 1077–1089.

1.4
Tutorial Review: Modern Biological Sensors

A. Tobias A. Jenkins

1.4.1
Analytical Concepts in Biosensor Design

It is important to define what is meant by a biosensor system. Generally, biosensors are regarded as distinct from chemical sensors if they use a biological element for analyte recognition or signal transduction. The definition therefore means that sensors used in a biological system, e.g. that measure calcium, pH or potassium concentrations are not biosensors unless they employ a biological element for sensing.

Using the definition given above, a microbiologist identifying and reporting on bacteria seen under a microscope is a biosensor system: she/he (a biological element) recognizes an analyte and converts that information via further neural transduction to a verbal or written output. In fact, this approach to sensing is still of vital importance in clinics and hospitals worldwide. However, for the purpose of this discussion, we define a sensor to be a *discrete* system that gives a response on recognition of an analyte. Göpel and Schierbaum described five essential features that describe a biosensor: (i) the detected / measured parameter (the analyte); (ii) the mode of operation of the signal transducer; (iii) the physical chemical or biochemical model used in design of the sensor; (iv) the application of the sensor, and (v) the materials and technology used in the sensor fabrication [1].

Before looking in more detail at specific sensor systems, it is important to define some terms that are commonly used in the literature to define important sensor parameters. These will be based on IUPAC definitions, where possible.

1.4.1.1
Units of Concentration

Standard chemical definitions of concentration are mol dm^{-3}, mol L^{-1} or mg ml^{-1}. However, many sensors that either use enzymes in their signal recognition/transduction or directly detect enzymes as an analyte use the term International Unit, or IU. This is a measure of enzyme activity:

Surface Design: Applications in Bioscience and Nanotechnology
Edited by Renate Förch, Holger Schönherr, and A. Tobias A. Jenkins

ISBN: 978-3-527-40789-7

1 IU of enzyme will catalyze 1 μmol dm^{-3} of its substrate per minute under the specified assay conditions.

To convert this to concentration, the term IU per unit volume is often given, i.e. IU ml^{-1}. Moreover, the assayed activity of an enzyme per unit mass is generally given, allowing conversion, if desired into SI units of concentration. For example, commercial glucose oxidase purchased from Sigma-Aldrich might be assayed to have an activity of 16 000 units g^{-1}. The important thing to note is that activity can vary between different suppliers, between batches or, if expressing the enzyme itself, between different experiments.

1.4.1.2
Sensitivity

Sensor sensitivity is a frequently used term to describe sensor performance, but is often used imprecisely. The sensitivity of a sensor is the gradient of the response–-concentration calibration curve. The important thing to note here is that most sensors have a sensitivity that varies with analyte concentration. This can be clearly seen in Figure 1, where a sensor response to human chorionic gonadtrophin (hCG), measured as a fluorescence response, is plotted against the log of analyte concentration. Three distinct regions can be seen: at high analyte concentration, sensitivity is low, in mid concentration range, between 10^2 and 10^4 mIU ml^{-1} (hCG), sensitivity is high, and at low concentrations, sensitivity is again low. It is therefore important, when quoting an assay or sensor sensitivity to also quote over what analyte concentration range this refers to.

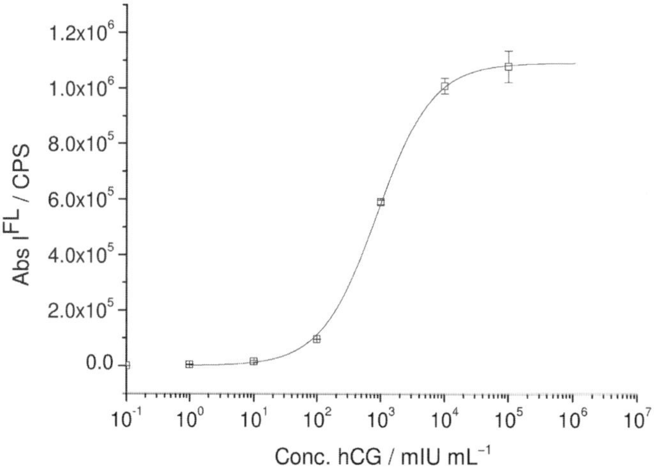

Figure 1 Plot of an assay response against log of analyte concentration (in mIU ml^{-1} of protein (human chorionic gonadotropin).

1.4.1.3
Analyte Selectivity

The selectivity of a sensor for a given analyte can be of crucial importance in sensor design, since false-positive signals can be obtained if other analytes present in the sample being studied interact with the sensor recognition element. A good example of this problem is where fructose can interfere with measurement of glucose in sensors that use phenyl boronic acid as an analyte-recognition element [2]. Designing sensors with high analyte specific recognition elements drives much current biosensor research and will be discussed in more detail later in this chapter.

1.4.1.4
Limit of Detection (LoD)

The limit of detection is a frequently quoted measure of sensor performance. It has a precise meaning: it is the lowest detectable concentration of analyte that gives statistical confidence [3]. The LoD of a sample is the response that relates to the concentration on the calibration curve equivalent to the zero analyte concentration response plus three standard deviations. Therefore, provided that there is a normal distribution of the sensor response at a given concentration, this theoretically equates to being 99.73% confident that the lowest concentration response is due to analyte recognition and not noise. However, due to uncertainty in the origin of the zero analyte response, IUPAC recommend to treat the LoD as having 90% confidence [4]. Figure 2 illustrates a graphical method for determination of LoD.

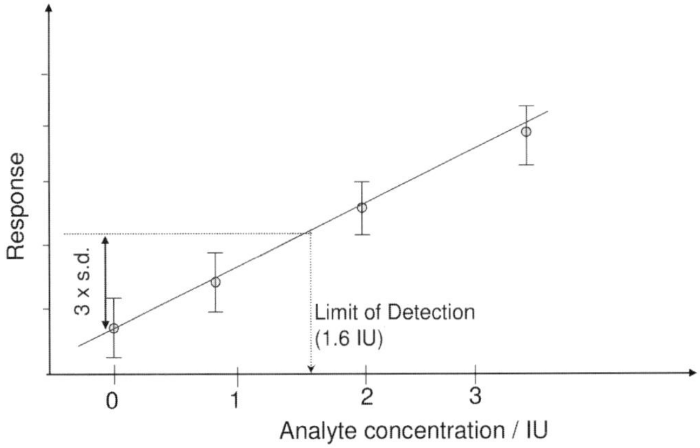

Figure 2 Schematic of methodology for estimating limit of detection from sensor response–concentration curve at lower limiting concentrations.

1.4.2
Signal Amplification

In order to provide high levels of analyte sensitivity and limits of detection at very low concentrations, some degree of signal amplification is built into many biosensors. Specific examples are given in the overview of modern biosensing strategies, later in this chapter. In general, there are three principal methods for amplifying a biosensor signal: to increase analyte concentration in the sensor; to amplify the signal in the detection assay; to improve the detection mode, e.g. by improvement in instrumentation such as a more effective photomultiplier tube in a fluorescent detector.

1.4.2.1
Signal Amplification by Increasing the Effective Analyte Concentration

This methodology is the method used since the nineteenth century by bacteriologists culturing bacteria on agar plates. A biological sample containing the analyte is grown in growth media to a point where its concentration is high enough to be detected by eye or a microscope. A more modern approach is used in many nucleic-acid based assays, where a technique known as the polymerase chain reaction (PCR) is used to exponentially amplify DNA to high concentrations. The method requires alternating heating / cooling cycles of a polymerase enzyme and single nucleotides to build up replicate copies of a target sequence. PCR was originally invented by Kary Mullis, who won the Nobel prize for the work in 1993. It can be used to amplify both small oligonucleotides and larger sequences [5]. The PCR process can be miniaturized and included on a chip-based platform within a sensor device [6]. Methods for detecting the amplified DNA are discussed later.

1.4.2.2
Signal Amplification by Assay Design

A further methodology for amplifying a signal is not the direct amplification of the analyte, but rather amplification of molecules that report the existence of the analyte via a catalytic response. This methodology utilizes the fact that a single enzyme can react with many times its own number of a specific substrate per second. For example, glucose oxidase when purchased commercially can convert thousands of molecules of glucose to gluconic acid per second. The reduced enzyme then needs to be regenerated back to its active form by further reaction with an oxidant mediator such as oxygen or Fe^{2+}. Hence, detection of the rate of mediator production or loss can give direct information on the rate of substrate reaction by the enzyme. Modern glucose sensors use organo-metallic mediators rather than oxygen, which are detected electrochemically [7]. Since the mediator is regenerated (reduced) following oxidation on an electrode, the resultant electrochemical signal is catalytic, thus providing signal amplification (see Section 1.9). Other applications of catalysis are discussed later, but include, for example where an enzyme-labeled antibody in an HIV test, converts a colorless substrate to a colored product, with the product being many times in excess of the enzyme concentration.

1.4.2.3
Signal Amplification using High-Sensitivity Instrumentation

Advances in instrumentation offer attractive methods for signal amplification in biosensors. Many sensors utilize a fluorescent output detected by a charge coupled device (CCD) element for their primary signal transduction following the recognition of an analyte. Advances in CCD technology mean that the construction of simple, sensitive and relatively cheap fluorescent sensors is now possible. Fluorescent signals themselves can be amplified by using enhanced electric fields created by surface plasmon resonance to excite fluorophores, rather than direct light excitation, as demonstrated in a number of prototype sensing applications by Knoll *et al.* This methodology uses the intense electric field created on a thin metal (usually gold or silver) during surface plasmon resonance to excite fluorophores situated close to the metal (within ca. 200 nm) in a technique known as surface plasmon enhanced fluorescence (SPEFS) [8].

1.4.3
Strategies for Attaching Functional Biomolecules to Surfaces

A common feature of most biosensor systems, both commercially available or at a research level is the need to create a structured sensor surface. This surface may need to perform several functions, for example: it may need to be conducting for electrochemical measurement, it may need to allow light in or out in an optical-sensor format and, in most cases it will need to couple the analyte recognition unit: protein / enzyme / antibody, oligonucleotide such that the functional activity of the unit is retained. The retention of activity is critical, especially in the case of antibodies and enzymes, which may need highly specific environments in which to work and may easily denature. This discussion is divided into considering the requirements for biomolecule attachment in an R& D environment and in a 'real' sensor system – although the distinction is perhaps arbitrary, and in any cases there can be considerable overlap.

In a research environment, gold is often used material as the underlying base material. Gold can be evaporated or sputtered to form reasonably homogenous films that can be deposited on a range of backing material such as silicon, glass, mica or even plastic. Gold films can be made very flat using methods such as template stripping [9]. Gold has the further advantage that it is electrically conducting and is optically translucent as a thin film (< 50 nm). This property is used in optical sensors based on surface plasmon resonance, whereby incident light is coupled into a thin gold film, via a prism or grating, causing excitation of surface plasmons in the gold, which can be measured and used to either excite fluorophores as in the SPEFS method or the angle of incident light required to create the resonant conditions can be followed in real time, giving information on protein binding on the surface. The final advantage of gold is that it is relatively inert in many conditions, but will react readily with thiolated compounds such as alkyl mercaptans, allowing self-assembled monolayers (SAMs) to be created, which can be used as support architecture for the attachment of biomolecules [10]. Other substrates such as silicon and glass can also be modified with silanized molecules to form SAMs [11]. Further discussion of this subject is provided in Chapter 1.4

Commercial sensors tend to use a disposable or semidisposable analyte interface, which avoids problems with contamination and cleaning. Polymers or modified glass, often with a nitrocelluose or polylysine layer onto which biomolecules such as antibodies can be immobilized [12]. Surfaces can be modified such that they specifically adsorb the biomolecule of interest, whilst having relatively low nonspecific protein binding. Many of the commercially successful biosensors, such as the hCG pregnancy test, use this technology. Electrochemical biosensors in commercial operation often utilize screen-printing technology for the fast, cheap and reproducible fabrication of electrodes, for example in the blood-glucose self-test system [13].

Once a support layer has been developed, the biorecognition element, i.e. antibody or oligonucleotide, may need to be conjugated to the support layer. This can be achieved by simply loading an absorbent film such as nitrocellulose with the molecule, or spotting oligonucleotide onto clean glass or silicon, as is often practised in DNA arrays [14]. However, it is often desirable to specifically conjugate the biomolecule of interest to the support layer. There are a number of chemical

Table 1 Summary of common immobilization strategies for biomolecules. Further conjugation methods for carboxylic acid groups are shown in Figure 4.

Functional group on (bio)molecule	Target on surface	Mechanism	References
Thiol, R-SH	Au, Ag and some other metals	Au-SR bond formation	[15]
Amine $R\text{-}NH_2$	R′-COOH	EDC, DCC or SSMCC mediated peptide coupling	[16–18]
His-tagged $R\text{-}(His)_6$	NTA – Ni^{2+} ligands	Ni^{2+} complexation	[19]
Biotin tagged R-biotin	Streptavidin	Biotin-streptavidin affinity	[20, 21]
Carboxylic acid R-COOH	R′-NH2	Mediated peptide coupling	[22]
Silanyl chloride $R\text{-}Si(OH)_nCl_y$	SiO_2 (glass, silicon)	Si-O-Si (uncertain)	[11, 23]
Antibody Fc binding protein, Protein A	SiO_2	Protein / protein (noncovalent)	[24]

and biochemical noncovalent coupling strategies that can be readily employed for this purpose. Some common strategies are summarized in Table 1.

1.4.3.1
Biotin-Streptavidin

The biotin-streptavidin method utilizes the very high noncovalent binding of biotin to the protein streptavidin. Streptavidin is a tetramer with four equivalent domains, all of which can bind biotin, and hence it can be used as a protein linker. A biotinylated SAM surface can be formed and streptavidin bound to this surface, with biotin-tagged biomolecules then being attached (Figure 3). This strategy has been used for attaching oligonucleotides and proteins to surfaces. Oligonucleotides can be purchased from commercial sources with the biotin tag already attached at the 5′ or 3′ end [25, 26], and biotin-labeling kits for proteins are also widely available.

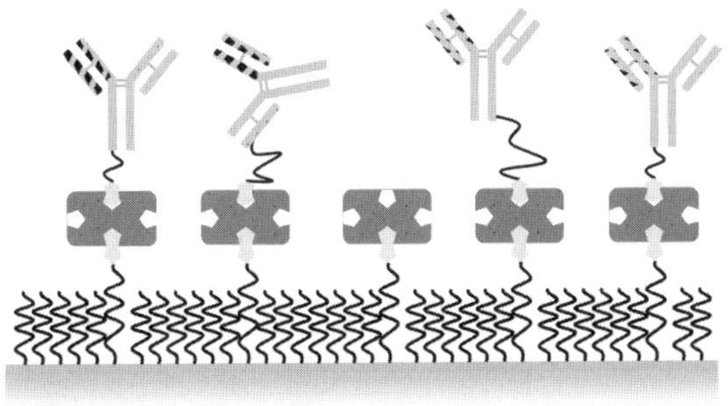

Figure 3 Schematic of a streptavidin matrix, in this case with biotin-tagged antibodies attached.

1.4.3.2
Direct Covalent Coupling of Biomolecules Using EDC–NHS

Methods for direct attachment of proteins to surfaces (or linker molecules / fluorescent tags) generally utilize coupling to free primary amines found in amino acids such as lysine, although coupling to thiol-containing amino acids such as cysteine using maleimide and other functional groups are also possible [27] Oligonucleotides can be immobilized using this strategy, normally by purchasing with a terminal amine at the desired end.

This strategy uses an active N-hydroxysuccinimide (NHS) ester group, which spontaneously reacts with free amines to form an amide linkage. Many common protein tags such as biotin or fluorophores are supplied commercially in the active NHS ester form. In many cases, a carboxylic acid group exists that has first to be converted to an NHS ester. The carboxylic acid group is activated by reaction with a component such as EDC (1-Ethyl-3-(3-dimethylaminopropyl)-carbodiimide) and N-hydroxysuccinimide to form the NHS ester [16]. For example, in work by Chifen *et al.*, lipid vesicles containing phosphtidylethanolamine head groups were immobilized on carboxylic-acid-containing polymer using this coupling methodology:

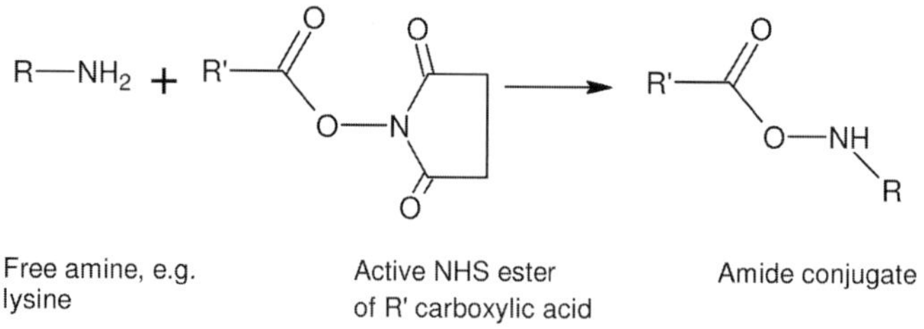

Figure 4 Schematic of coupling active NHS ester to primary amine for conjugation of biomolecule to surface, linker molecule or fluorescent / colorimetric tag.

the polymer was first activated with EDC / NHS to produce a reactive surface, conjugation of the amine on the lipid head group then proceeded [28].

Once the activated ester has been made, a number of coupling strategies become available. In Figure 4 the formation of the amide by reaction of a primary amine (such as a lysine amino acid in a protein) is shown, but a number of other possibilities also present themselves, which allow for direct binding to thiol residues (found in cysteines) or aldehydes. Figure 5 summarizes these strategies [31].

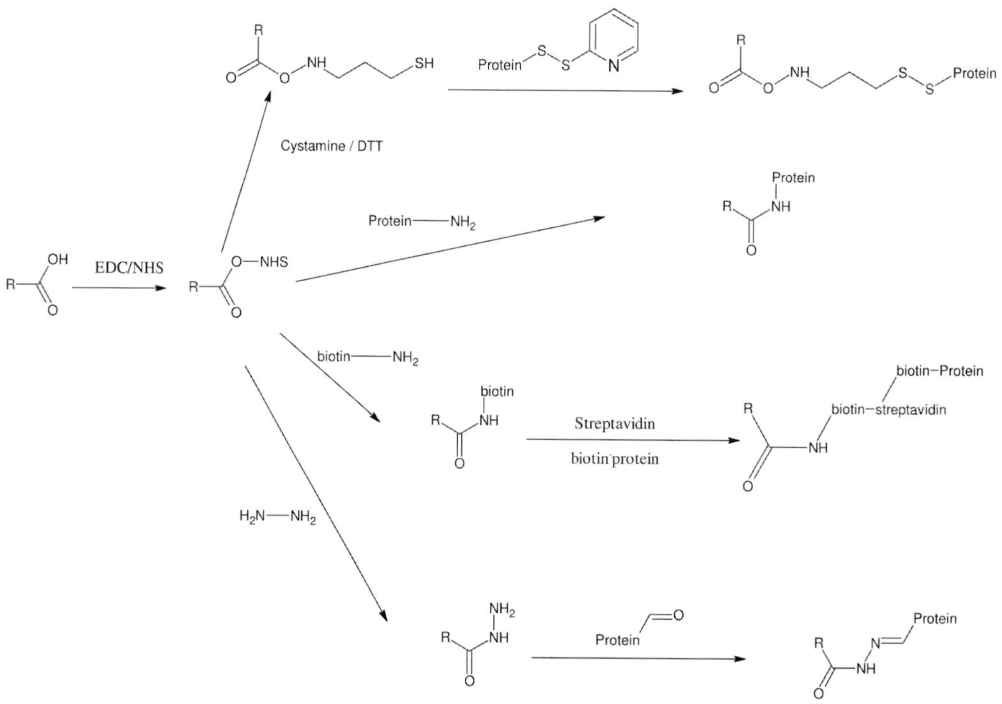

Figure 5 An overview of methods for coupling proteins to acid groups present on carboxymethyldextran brushes. These methods, in principal, work equally well for coupling proteins to any free acid group.

1.4.3.3
His Tag

The introduction of a multihistidine (*His*) tag (normally 6 *His* or greater) into proteins inserting a DNA vector that codes for 6-*His* into a plasmid that contains the gene for the protein of interest. This method confers a number of advantages for molecular biologists and biochemists, as the *His* tag has a very high affinity for Ni^{2+}. This means that when proteins are expressed in a bacterium or yeast, they can be easily separated from other cell constituents in a column containing Ni^{2+} immobilized on the stationary phase [29]. The advantage of *His*-tagged proteins from the perspective of persons working on sensor research, is that the purified protein, can be surface immobilized using surface bound Ni^{2+} chelating the hexahistidine tag on the protein. This is generally achieved by modifying surfaces with nitrilotriacetic acid (NTA) and adding Ni^{2+}. Surfaces including silicon, gold and biopolymers such as agarose can all be modified with NTA [30].

1.4.3.4
Three-Dimensional Sensor Surface Structuring

Two further issues that relate to improving sensor sensitivity must now be considered. The first is how more biorecognition sites, such as antibodies, can be attached to a sensor surface and the second is to consider the role of orientation of receptor molecules in improving sensitivity and LoD [50].

1.4.3.5
Brush Surfaces

Much effort has been put into creating 3-dimensional sensor chips over the last two decades. This methodology utilizes a hydrogel matrix, which swells in water to form a 'brush' morphology, but, by weight is mainly composed of water. Most 3D chips of this nature utilize the polysaccharide carboxymethyldextran, which is itself tethered to the surface. Carboxylic acid groups on the dextran can then be modified using EDC–NHS coupling to make specific chemical conjugations either to aminoacids directly, or via biotin-streptavidin, pyridyl disulphide linkages or hydrazine-activated acid, as shown in Figure 6 [31].

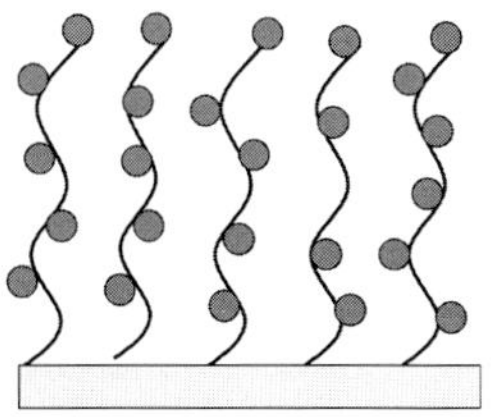

Figure 6 Schematic of a dextran 'brush' chip, with carboxylic acid groups represented as circles. In aqueous electrolyte, the brush structure swells essentially creating a noncrosslinked hydrogel matrix. Carboxylic groups can be readily conjugated to biomolecule receptors.

It can be observed that use of such a 3-dimensional brush structure can achieve high loadings of biorecognition molecules on sensor surfaces. However, mass-transport considerations can become an important consideration, as significant concentration gradients of analyte between the bulk solution and the surface hydrogel can build up. In some cases, especially where binding kinetics are being measured, a more open gel matrix should be used.

1.4.3.6
Specific Orientation of Bound Proteins

Most protein receptors have specific sites where they bind analytes. A good example are antibody immunoglobulins (IGG), which have two important recognition regions. The generic IGG structure is a Y shape, with the two Fab fragments uppermost and the Fc region at the bottom end. The Fab is the primary binding site in monoclonal antibodies, and in many sensors that utilize antibodies, it can be advantageous to orientate these sites in apposition where easy, nonsterically hindered analyte access can be achieved. One method for doing this, discussed by Vareiro *et al.*, is to use immobilized Protein A that specifically binds the Fc region, to bind the antibody to the sensor surface [53].

1.4.4
Methods to Prevent Nonspecific Adsorption

The nonspecific adsorption of an analyte or other material present in a sample being measured by a biosensor can lead to false-positive or false-negative results. Additionally, it can lead to an impairment of performance and reduction in signal accuracy. A number of approaches are used in sensor systems to reduce this problem, including the addition of a filter membrane over the sensor surface to adding surfactant molecules to the sample to be tested. The severity of the problem is very dependent on the nature of the sample, for example, blood is a particularly difficult system in this regard as it contains large amounts of proteins (such as albumin), cells and plasma serum. A detailed study of the physical chemistry that underlies the nonspecific binding of the protein β-lactoglobulin, which may well be of relevance to other protein systems, is given by Rabe *et al.* [32].

The problem of blood products fouling blood-glucose sensors has largely been solved by coating the electrode with a thin polymer membrane that allows transfer of glucose to the electrode but blocks the access of larger proteins. One situation where the problem of protein fouling is particularly acute is in implantable electrodes used for *in-vivo* blood monitoring. Churchhouse *et al.* utilized a two-component polymer layer on their *in-vivo* needle electrodes for glucose monitoring [33]. A polyethersulphone membrane was applied directly to the electrode, which itself was coated with a low-permeability polyurethane film. This effectively excluded large proteins from interfering with the measured signal, while allowing access of glucose to the electrode.

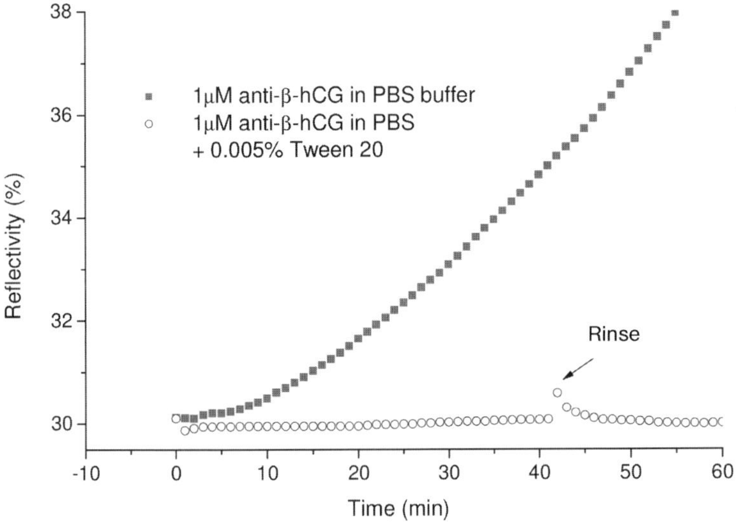

Figure 7 SPR measurements of the nonspecific binding of the anti-hCG antibody in the presence (■) or absence (○) of Tween 20.

Another approach to the problem of nonspecific adsorption is to incorporate material onto the sensor surface, either as a support for or coadsorbed with the biorecognition element, that is very 'slippery' to proteins and other potential contaminants. The 3D dextran brush surfaces discussed in Section 1.4.3.5 are also reported to demonstrate low nonspecific protein adsorption [34]. Ethylene glycols, either as short oligomers or longer polymers (PEGs) have been shown to be effective in resisting nonspecific protein adsorption. Two strategies present themselves: surface derivatization of surfaces with PEGs and/or addition of detergents such as Tween 20 (polyoxyethylene (20) sorbitan monolaurate), which contain nonionic surfactants based on ethylene glycol [35].

The addition of surfactants to the analyte is a commonly used method for preventing nonspecific binding. Figure 7 shows results from the authors' own work and Vareiro *et al.* studying the nonspecific adsorption of anti-hCG antibodies on a surface [53]. The reflectivity units on the *y*-axis are proportional to the mass density adsorbed on the surface. The reduction in nonspecific adsorption of antibodies in the presence of Tween is clearly demonstrated.

1.4.5
Analyte Recognition

The need for analyte recognition has been discussed above. Strategies for building recognition sites into a biosensor platform depend on the nature of the analyte, and also to some extent, on the medium in which the analyte is exposed to the sensor-recognition element. The discussion in this chapter will be confined to sensors where the recognition element is surface immobilized. Different sensing strategies will be grouped by nature of analyte, arbitrarily starting with large complex systems such as bacteria and viruses, before considering proteins, nucleic acids and finally sugars.

1.4.6
Overview of some Biosensing Strategies

1.4.6.1
Bacterial and Viral Sensing

The diagnosis of infection by culturing bacteria, colorimetric staining and observation under a light microscope was pioneered by the German scientist, Robert Koch in the nineteenth century [36]. This basic methodology is still used as a fundamental approach to diagnosis by many microbiologists today. However, in the twentieth century, new methods for rapid detection of bacteria have been developed. This is especially important for diagnosing viral infection, since viruses are many times smaller than bacteria. There are three fundamental biosensing approaches to the detection of bacteria and viruses: (i) Immobilization on antibody coated surfaces and reporting via modified reporter antibodies. (ii) Detection of antibodies produced by the host animal in response to a primary infection. (iii) Detection of nucleic acids unique to the pathogen (DNA or RNA). The discussion in this section will confine itself to point (i) above – direct detection of bacteria using antibodies.

1.4.6.2
Antibody 'Sandwich' Assay Sensing Platform for Detection of Group B Streptococcus Bacterium

Group B *streptococcus (Streptococcus agalactiae)* is the principal cause of neonatal meningitis and infection [37]. The bacteria are passed from an infected mother to the baby during the birth process, normally as a consequence of the baby swallowing fluid in the birth canal. A general review of various tests for Group B strep versus laboratory culture, is provided by Honest *et al.* [38]. Group B strep testing provides a good example of the development of clinical microbiological sensors over a relatively short time. One of the first sensor systems for Group B strep was developed by Morrow *et al.* in 1984 [39]. However, more recently and following the sequencing of the genome of *Streptococcus agalactiae* in 2002, nucleic-acid-based assays have largely superseded antibody tests for this pathogen [40].

Antibody-based sensors use the fact that antibodies can be generated with a very high degree of antigen specificity: so-called monoclonal antibodies, using standard immunological methods. Antibodies are proteins produced in response to infection by mammals. They bind to an infective agent, in doing so labeling it for destruction by a number of mechanisms. This immunological response is

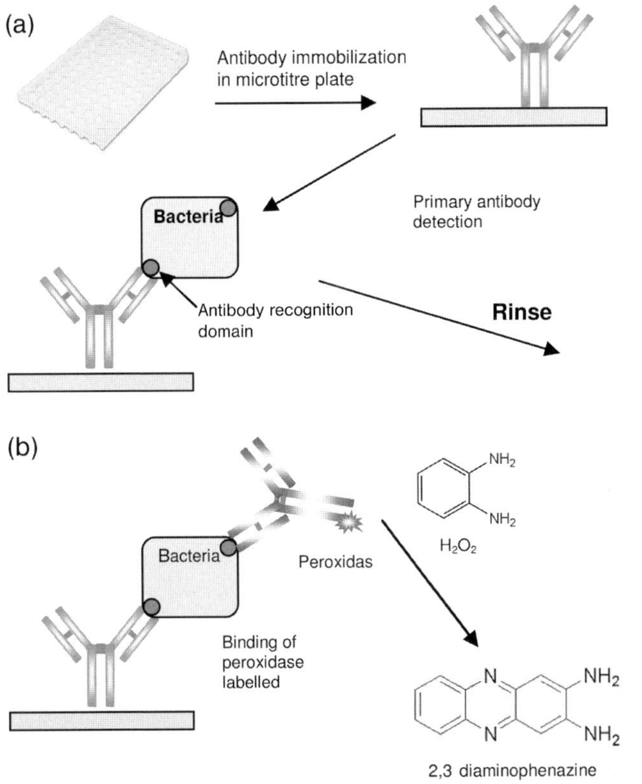

Figure 8 (a) Target recognition: Immobilization of antibodies in microtitre plate, and binding of bacterial target. (b) Reporting binding event and signal transduction via color change on binding of secondary reporter antibody.

exploited in antibody sensors, since antibodies can be raised against a specific bacteria or viral target. Moreover, antibodies can be raised against antibodies, this being exploited by sensors for the human immunodeficiency virus (HIV).

The antibody 'sandwich' assay described by Morrow *et al.* uses monoclonal antibodies against Group B *strep* [39]. The approach described is reasonably generic, with the methodology being used (with variations) in a variety of bacterial sensors. Monoclonal antibodies against Group B *strep* were made by immunizing rats with Group B *strep* bacteria, sacrifice of the rats and fusing antibody-forming cells from their spleens with tumor cells to produce an immortal cell line – a hybridoma that expresses the anti-Group B *strep* antibody [41]. The purified antibodies were then immobilized on a microtitre plate, and solution containing potential analyte added. Figure 8 provides a schematic view of the assay. Following incubation, the microtitre plate was rinsed and a secondary antibody added containing a peroxidase label. The peroxidase is used to convert *o*-phenylenediamine to a colored product, diaminophenazinine.

Protein attachment to microtitre plates can be performed in a number of ways. Microtitre plates containing streptavidin can be purchased from companies such as Sigma-Aldrich and used with biotin-tagged proteins. Other methods approached used polylysine as a noncovalent coating on a polystyrene plate. This can then be activated by gluataraldehyde to bind proteins to the surface [42]. Other approaches include the attachment of reactive 1-fluoro-2-nitro-4-azidobenzene by photochemical activation of polystyrene microtitre plates [43]. The polymer-bound fluoro-compound undergoes nucleophilic attack by amine residues in the protein (i.e. Lysine), resulting in covalent protein coupling and loss of fluorine. Other chemical grafting methods are detailed by various authors [44]. Plasma activation has also been explored by Hayat *et al.* [45].

1.4.6.3
Biosensors for Detecting Viruses: Antibody Recognition and the HIV Test

Most animals produce specific antibodies against a pathogen species following infection. Since these antibodies specifically recognize the infective species, a reasonable commonly used biosensing strategy is to make a biosensor that recognizes not the primary infective agent, but instead antibodies produced following infection. The primary disadvantage of this strategy is that it can take time for sufficient antibody to be made in the host animal, so-called seroconversion, limiting the accuracy in the period immediately following initial infection. For example, the antibody HIV test that will be discussed here is only considered accurate after 25 days following initial infection [46]. In most countries, both an antibody ELISA test (as discussed here) and Western Blotting is used to confirm diagnosis.

Infection by HIV in humans induces the production of a range of antibodies including anti-HIV against core HIV proteins and antibodies against core (p24) and envelope (gp160, gp120, gp41) HIV proteins. There are many actual and putative antibody tests discussed in the literature and on the market. This discussion focuses on a generic approach to testing for the presence of the primary anti-HIV antibody, although this technology can also be used for the detection of other associated antibodies, e.g. anti-p24. An overview of the clinical reliability of number of rapid HIV test strategies is given by Makuwa *et al.* [47]. The description of the assay below is based on reports by Kwon *et al.* on three of the most widely used commercial HIV biosensors on the market in 2006 [48].

The basis of most assays is similar to the test system shown in Figure 8. Antigens to the specific HIV proteins are immobilized either on a solid surface, or on polymer beads. This is generally achieved via EDC coupling of carboxylic-acid-functionalized polymers (1.4.2) [49]. Analyte containing HIV antibodies is added (e.g. blood / saliva samples). Following incubation, the bound antigen–target antibody conjugate is rinsed and the presence of the bound antigen revealed via a labeled secondary antibody. The label can be a fluorescent tag such as acrinidinium ester, an enzyme, e.g. peroxidise, or another reporting and signal amplification system [50].

1.4.6.4
Biosensors for Detecting Enzymes: Sensors for Human Pregnancy Hormone

A further variation of the antibody sandwich assay sensors described above is the system for detecting enzymes. In this format, two antibodies – a capture antibody that binds to one domain of the target enzyme and a secondary antibody with a fluorescent or colorimetric tag are used. The most common example of the wide spread application of this technology is the human home pregnancy test kit, commercially available in most countries. This sensor has a line of capture antibodies raised against the β domain of human chorionic gonadotrophin (hCG), an enzyme produced by women shortly following conception. hCG production increases geometrically over the first ten weeks of pregnancy. Typical hCG levels 28 days into pregnancy are around 100 mIU ml^{-1} in serum, plasma or urine [51].

The consumer pregnancy test kit has a line of immobilized anti-β-hCG immobilized on a hydrophilic substrate. Urine is applied to the substrate, where it is carried up over the absorbent substrate by capillary action (wicking), carrying hCG (if present) with it. As the hCG travels along the absorbent substrate they bind to nonsurface attached labeled anti-α-hCG antibodies. These antibodies are either labeled with a polymer bead, or gold nanoparticles. The hCG, thus labeled then continues to travel over the surface until the enzyme come across a line of capture antibodies, where they become immobilized. After sufficient labeled enzyme is immobilized, a colored line appears, due to the concentrating of the polymer or nanoparticle labels in one area. Current commercial sensors can detect as little as 25 mIU ml^{-1} in often highly suboptimal analytical conditions, sufficient to detect pregnancy up to 4 days before the due menstrual period [52].

This well-characterized assay has also been used as a benchmark test by researchers looking into methods for improving sensitivity and limit of detection of immunoassays. Vareiro *et al.* used ultra-sensitive fluorescence measurements (surface plasmon field-enhanced fluorescence) to quantify the effect of steric crowding of antibodies on the surface on LoD, and to study the effect of using antibody fragments rather than whole antibodies for the surface immobilization of hCG. A limit of detection under laboratory conditions of 0.3 mIU ml^{-1} (6×10^{-13} mol dm^{-3}) hCG was achieved on an optimized surface architecture [53]. A related study by Yu *et al.*, using a 3-dimensional dextran matrix (Biacore CM5 chip) coupled to antiprostate specific antigen (anti-PSA) in a sandwich format gave a LoD of 8×10^{-14} mol dm^{-3} for the detection of PSA in human serum [54]. This enhanced LoD reflects the higher antibody loading on a dextran matrix. A discussion of 3-dimensional capture surfaces for sensing is provided by Xu *et al.* [55].

1.4.6.5
Nucleic-Acid-Based Sensors

The attraction of nucleic acids for biosensing is that with fast genome sequencing and bioinformatic software such as the basic local alignment search tool (BLAST) now widely available, it is possible to obtain information on the genome of many pathogens and determine which parts of the genome are unique to that particular species. Thus, nucleic-acid sensors can in principal have a very high degree of analyte recognition capability – even recognizing differences between closely related subtypes of, for example, a bacteria (provided the gene fragment being detected is different in the two subtypes). Bacteria and some viruses contain deoxyribonucleic acid (DNA) while some viruses contain only ribonucleic acid (RNA). DNA is more robust than RNA, and hence most viral RNA is detected by first converting it to DNA using an enzyme, reverse transcriptase. DNA and RNA can be easily extracted from cells by lysing the cellular membrane. Once extracted, the sensor needs to recognize the specific gene fragment and report its presence. Classically, DNA has been detected by running an electrophoresis gel (a Southern blot) [56]. This method falls outside the scope of this book and the discussion here will focus on the direct surface hybridization of single-strand DNA fragments (oligonucleotides) with surface-immobilized complementary sequences [57].

One of the challenges in detecting nucleic acids is the reporting / signal transduction step. Bioinformatics software is good at identifying *target* oligonucleotide sequences, and a number of strategies exist for attaching *probe* oligonucleotides onto sensor surfaces such as biotin-streptavidin, and other architectures discussed in more detail in other chapters. There are two principle approaches commonly employed: electrochemical detection and fluorescence detection. This is a vast area of research, with many reviews available in the literature [58]. The purpose of this short overview is to show that the fundamentals of DNA biosensors are the same as any other sensor: analyte recognition, signal transduction and signal output. However, because the concentration of the target in nucleic-acid sensors is often very low, signal amplification is also very important (see Section 1.4).

An electrochemical detection system has been proposed by Williams *et al.* [59]. A graphite–epoxy composite electrode was modified with streptavidin: a tetrameric protein that has a very high affinity for the small molecule, biotin. Biotin tagged probe oligonucleotides were immobilized on the electrode. The probe oligonucleotide was chosen to be complementary to half of the target probe (Figure 9). Following probe–target hybridization, a digoxigenin labeled probe was added that was complementary to the target sequence that had not hybridized to the electrode attached probe sequence. Digoxigenin is a small molecule steroid, for which a monoclonal antibody is commercially available. Horseradish peroxidase (HRP) labeled antidigoxigenin then binds to digoxigenin on the hybridized oligonucleotide. The HRP catalyses the reduction of hydrogen peroxide (H_2O_2) to water and is itself reduced by reaction with hydroquinone (HQ), a redox couple. The resultant oxidized hydroquinone is then reduced at the electrode surface, giving a catalytic current [60].

Knoll *et al.* used SPEFS as a method for detecting on-chip DNA hybridization in real time. A methodology similar to that shown in Figure 4 is used, but with some key differences. A biotin-thiol self-assembled monolayer (SAM) is formed on a gold surface with control of biotin thiol density to ensure optimal attachment of streptavidin [61]. A probe surface of both DNA and peptide nucleic acid (PNA) was investigated and long products of the polymerase chain reaction (PCR) ampli-

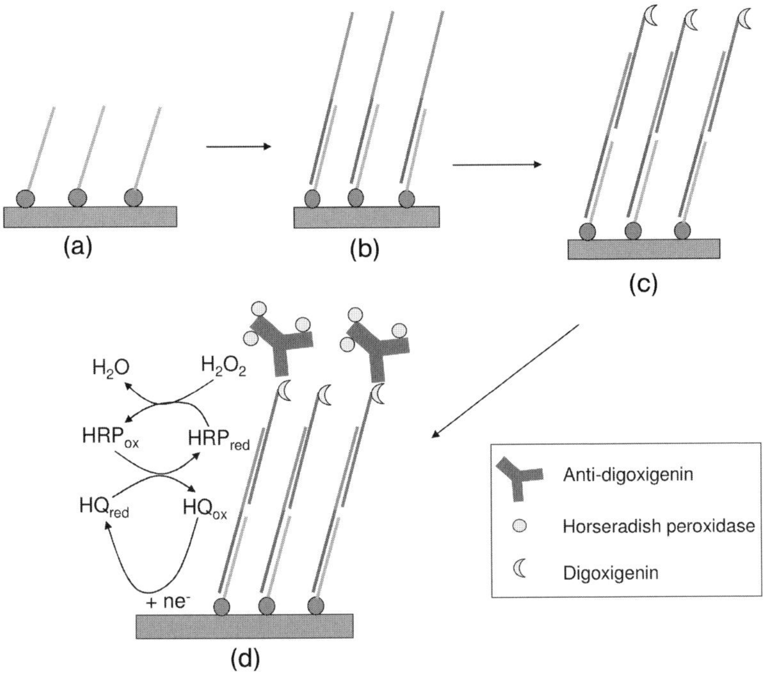

Figure 9 Schematic strategy for electrochemical detection of surface-hybridized oligonucleotide: (a) attachment of biotin-tagged probe oligonucleotide to streptavidin on electrode; (b) hybridization with target oligonucleotide; (c) hybridization with digoxigenin labeled probe; (d) binding of HRP labeled antibody and detection of catalytic current using a hydroquinone redox couple.

fication of target sequences were detected. The PCR product targets were all labeled prior to hybridization with a Cy5 fluorescent dye. The fluorescent increase, measured by SPEFS allowed real-time quantification of binding and showed the assay sensitive to single base-pair mismatches in the probe – target sequence. LoDs down to 100 fmol dm^{-3} (1×10^{-13} mol dm^{-3}) were achieved using PNA probes.

1.4.6.6
Electrochemical Measurement of Blood Glucose

The measurement of blood glucose by persons suffering from diabetes is essential for prevention of side effects caused by high blood glucose concentration or too low blood glucose concentration, such as blindness, gangrene / tissue necrosis and death. The most successful electrochemical point-of-care biosensor to reach the market is the blood-glucose meter. This system, based on the pioneering work of Clarke and Lyons in 1962, who showed that enzymes could be immobilized on an electrode, used glucose oxidase (GOD) to oxidize glucose to gluconolactone

[62]. The early GOD-based sensors operated by the electrochemical oxidation of hydrogen peroxide produced by the reaction of the reduced enzyme with oxygen [63]. The difficulty with this system is that variations in oxygen concentration could affect the measurement accuracy. In 1984 Cass *et al.* reported the use of ferrocene carboxylic acid (Fc) as a mediator in the amperometric detection of glucose. The Fc replaced oxygen in removing electrons from reduced GOD, and had the added advantage that following oxidation at the electrode it was recycled back to the form (the ferrococinium ion) required to react with GOD, thus giving a catalytic response, illustrated in Figures 9 and 10 and the scheme below [60, 64]:

$$\text{glucose} + \text{GOD} \rightarrow \text{gluconolactone} + \text{GOD}_{\text{(reduced)}} \qquad \text{C}'$$
$$\text{GOD}_{\text{(reduced)}} + \text{Fc}^{+} \rightarrow \text{GOD} + \text{Fc}$$
$$\text{Fc} \rightarrow \text{Fc}^{+} + \text{e}^{-} \text{ (at electrode)} \qquad \text{E}$$

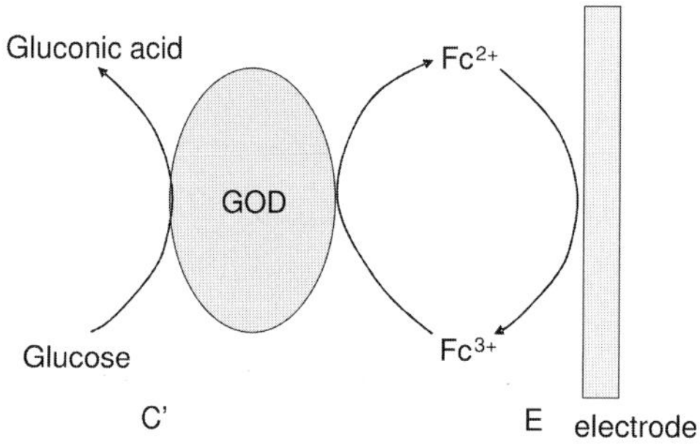

Figure 10 Schematic of GOD-catalyzed oxidation of glucose and ferrocene-mediated electron transfer to an electrode. E refers to electron transfer, C′ to the catalytic reaction.

There has been also been much interest in 'real-time' continuous glucose measurement by use of electrodes planted *in vivo*. The advantage of such a system is that it could be connected to an insulin- and glucose-delivery system, thus improving the health of diabetics [65]. A recent review of *in-vitro* and *in-vivo* glucose biosensors is provided by Koschwanez, and Reichert, who also consider much of the current regulatory framework that affects testing and development of such sensors [66].

1.4.7 Summary and Conclusions

There are two emerging trends in sensors: miniaturization and point of care applications. The rapid development of microfabrication technology that has allowed computer microprocessors to get ever faster has crossed over into sensor design and research, with particular interest in so-called 'lab on the chip' technology. The concept here is simple: many current diagnostic procedures require multiple steps prior to acquiring a readout. Lab on the chip technology seeks to miniaturize on patterned silicon all, or most of the routine sample-preparation steps current undertaken, such as PCR in DNA diagnostics and integrate with a sensor element. Hence, a lab on a chip DNA sensor for a bacterial infection would be supplied with whole blood, the cells in the sample would be lysed, DNA extracted, amplified and detected [67]. The advantages of such an approach include a great simplification of diagnostic approach, a need for a less-skilled operator, higher sample throughput and reduced assay cost.

The other important trend in diagnostics is a gradual move away from centralized diagnostic labs to a point of care environment, such as a doctor's surgery, or a patient's bedside or home. Advances in miniaturization mentioned above, coupled with a general need to drive down the cost of diagnostics means that increasingly, diagnostics are being performed at point of care. The advantages in reduced personnel costs and time are obvious [68].

References

1 Göpel, W. and Schierbaum, K.D. (1991) *Sensors – a comprehensive survey, in Chemical and Biochemical Sensors*, Vol. 2 (eds W. Göpel, T.A. Jones, M. Kleitz, J., Lundsstrom and T. Seiyama), VCH, New York.

2 Dean, K.E.S., Horgan, A.M., Marshall, A.J., Kabilan, S. and Pritchard, J. (2006) *Chemical Communications*, 3508–3509.

3 Thomsen, V., Schatzlein, D. and Mercuro, D. (2003) *Spectroscopy*, **18**, 112–114.

4 Gilfrich, J.V. and Birks, L.S. (1984) *Analytical Chemistry*, **56**, 77–79.

5 Cheng, S., Fockler, C., Barnes, W.M. and Higuchi, R. (1994) *Proceedings of the National Academy of Sciences of the United States of America*, **9**, 5695–5699.

6 Poulsen, C.R., El-Ali, J., Perch-Nielsen, I.R., Bang, D.D., Telleman, P. and Wolff, A. (2005) *Journal of Rapid Methods and Automation in Microbiology*, **13**, 111–126.

7 Cass, A.E.G., Davis, G., Francis, D.G., Hill, H.A.O., Aston, W.J., Higgins, L.J., Plotkin, E.V., Scott, L.D.L. and Turner, A.P.F. (1984) *Analytical Chemistry*, **56**, 667–671.

8 Neumann, T., Johansson, M.-L., Kambhampati, D. and Knoll, W. (2002) *Advanced Functional Materials*, **12**, 575–586.

9 Mosley, D.W., Chow, B.Y. and Jacobson, J.M. (2006) *Langmuir*, **22**, 2437–2440.

10 Ulman, A. (1991) *Ultrathin Organic Films*, Academic Press.

11 Onclin, S., Ravoo, B.J. and Reinhoudt, D.N. (2005) *Angewandte Chemie – International Edition*, **44**, 6282–6304.

12 Joos, T.O., Schrenk, M., Hopfl, P., Kroger, K. *et al.* (2000) *Electrophoresis*, **21**, 2641–2650.

13 Renedo, O.D., Alonso-Lomillo, M.A. and Martinez, M.J.A. (2007) *Talanta*, **73**, 202–219.

14 Trevino, V., Falciani, F. and Barrera-Saldaña, H.A. (2007) *Molecular Medicine*, **13**, 527–541.

15 Love, J.C., Estroff, L.A., Kriebel, J.K., Nuzzo, R.G. and Whitesides, G.M. (2005) *Chemical Reviews*, **105**, 1103–1169.

16 Staros, J.V., Wright, R.W. and Swingle, D.M. (1986) *Analytical Biochemistry*, **156**, 220–222.

17 Grabarek, Z. and Gergely, J. (1990) *Analytical Biochemistry*, **185**, 131–135.

18 Peelen, D. and Smith, L.M. (2005) *Langmuir*, **21**, 266–71.

19 Nieba, L., Nieba-Axmann, S.E., Persson, A. *et al.* (1997) *Analytical Biochemistry*, **252**, 217–228.

20 Qi, K., Zhou, C., Walker, A.V., Wooley, K.L., Jhaveri, S.B., Sogah, D.Y., Malkoch, M., Beinhoff, M., Carter, K.R.

and Hawker, C.J. (2005) *Polymer Preprints*, **46**, 363.
21 Rogers, K.R. and Mulchandani, A. (1998) *Methods in Biotechnology. 7: Affinity Biosensors: Techniques and Protocols*, Humana Press.
22 Lin, Z., Strother, T., Cai, W., Cao, X., Smith, L.M. and Hamers, R.J. (2002) *Langmuir*, **18**, 788–796.
23 Finklea, H.O., Robinson, L.R., Blackburn, A., Richter, B., Allara, D. and Bright, T. (1986) *Langmuir*, **2**, 239–244.
24 Dutra, R.F., Castro, C.M.H.B. and Azevedo, C.R. (2000) *Biosensors and Bioelectronics*, **15**, 511–514.
25 Ladd, J., Boozer, C., Yu, Q., Chen, S., Homola, J. and Jiang, S. (2004) *Langmuir*, **20**, 8090–8095.
26 Liebermann, T., Knoll, W., Sluka, P. and Herrmann, R. (2000) *Colloids and Surfaces A*, **169**, 337–350.
27 Wong, S.S. (1991) *Chemistry of Protein Conjugation and Cross-Linking*, CRC Press.
28 Chifen, A.N., Förch, R., Knoll, W., Cameron, P.J., Khor, H.L., Williams, T.L. and Jenkins, A.T.A. (2007) *Langmuir*, **23**, 6294–6298.
29 Hengen, P.N. (1995) , **20**, 285–286.
30 Paik, H.-J., Kim, Y.-R., Orth, R.N., Ober, C.K., Coates, G.W. and Batt, C.A. (2005) *Chemical Communications*, 1956–1958.
31 Johnsson, B., Löfås, S., Lindquist, G., Edström, Å.A., Müller Hillgren, R.-M. and Hansson, A. (1995) *Journal of Molecular Recognition*, **8**, 125–131.
32 Rabe, M., Verdes, D., Rankl, M., Artus, G.R.J. and Seeger, S. (2007) *ChemPhysChem*, **8**, 862–872.
33 Churchouse, S.J., Battersby, C.M., Mullen, W.H. and Vadgama, P.M. (1986) *Biosensors*, **2**, 325–342.
34 Akkoyun, A. and Bilitewski, U. (2002) *Biosensors and Bioelectronics*, **17**, 655–664.
35 Bi, H., Meng, S., Li, Y., Guo, K., Chen, Y., Kong, J., Yang, P., Zhong, W. and Liu, B. (2006) *Lab Chip*, **6**, 769–775.
36 Kaufmann, S.H.E. and Schaible, U.E., (2005) *Trends in Microbiology*, **13**, 469–475.
37 Center for Disease Control. (2000) Early-onset group B streptococcal disease – United States, 1998–1999. *MMWR*, **49**, 793–796.
38 Honest, H., Sharma, S. and Khan, K.S. (2007) *Pediatrics*, **117**, 1055–1066.
39 Morrow, D.L., Kline, J.B., Douglas, S.D. and Polin, R.A. (1984) *Journal of Clinical Microbiology*, **19**, 457–459.
40 Tettelin, H., Masignani, V., Cieslewicz, M.J. *et al.* (2002) *PNAS*, **99**, 12391–12396.
41 Kohler, G., Howe, S.C. and Milstein, C. (1976) *European Journal of Immunology*, **6**, 292–295.
42 Fujiwara, K. and Kitagawa, T. (1993) *Journal of Biochemistry*, **114**, 708–713.
43 Bora, U., Chugh, L. and Nahar, P. (2002) *Journal of Immunological Methods*, **268**, 171–177.
44 Larsson, P.H., Johansson, S.G.O., Hult, A. and Gothe, S. (1987) *Journal of Immunological Methods*, **98**, 129.
45 Hayat, U., Tinsley, A.M., Calder, M.R. and Clarke, D.J. (1992) *Biomaterials*, **13**, 801–806.
46 Food and Drugs Administration (2001) Talk paper number T01–42.
47 Makuwa, M., Souquiere, S., Niangui, M., Rouquet, P., Apetrei, C., Roques, P. and Simon, F. (2002) *Journal of Virological Methods*, **103**, 183–190.
48 Kwon, J.A., Yoon, S.-Y., Lee, C.-K., Seung Lim, C., Le, K.N., Sung, H.J., Brennan, C.A. and Devare, S.G. (2006) *Journal of Virological Methods*, **133**, 20–26.
49 Chersi, A., Rosano, L. and Tanigaki, N. (2000) *Human Immunology*, **61**, 1298–1306.
50 Hart, R.C. and Taaffe, L.R. (1987) *Journal of Immunological Methods*, **101**, 91–96.
51 Brody, S. and Carlstrom, G. (1962) *The Journal of Clinical Endocrinology*, **22**, 564–568.
52 Lenton, E.A., Neal, L.M. and Sulaiman, R. (1982) *Fertility and Sterility*, **37**, 773–8.
53 Vareiro, M.M.L.M., Liu, J., Knoll, W., Zak, K., Williams, D., Toby, A. and Jenkins, A. (2005) *Analytical Chemistry*, **77**, 2426–2431.
54 Yu, F., Persson, B., Löfås, S. and Knoll, W. (2004) *Analytical Chemistry*, **76**, 6765–6770.

55 Xu, F., Persson, B., Löfås, S. and Knoll, W. (2006) *Langmuir*, **22**, 3352–3357.

56 Southern, E.M. (1975) *Journal of Molecular Biology*, **98**, 503–517.

57 Szunerits, S., Bouffier, L., Calemczuk, R., Corso, B., Demeunynck, M., Descamps, E., Defontaine, Y., Fiche, J.-B., Fortin, E., Livache, T., Mailley, P., Roget, A. and Vieil, E. (2005) *Electroanalysis*, **17**, 2001–2017.

58 Khanna, V.K. (2007) *Biotechnology Advances*, **25**, 85–98, Elsevier, Amsterdam.

59 Williams, E., Pividori, M.I., Merkoc, A., Forster, R.J. and Alegret, S. (2003) *Biosensors and Bioelectronics*, **19**, 165–175.

60 Bard, A.J. and Faulkner, L.R. (2001) *Electrochemical Methods: Fundamentals and Applications*, 2nd edn, Wiley.

61 Yao, D., Yu, F., Kim, J., Scholz, J., Nielsen, P.E., Sinner, E.-K. and Knoll, W. (2004) *Nucleic Acids Research*, **3**, e177.

62 Clark, L.C. and Lyons, C. (1962) *Annals of the New York Academy of Sciences*, **102**, 29–45.

63 Iannlello, R.M. and Yacynych, A.M. (1981) *Analytical Chemistry*, **53**, 2090–2095.

64 Cass, A.E.G., Davis, G., Francis, G.D. and Hill, H.A.O. (1984) *Analytical Chemistry*, **56**, 667–671.

65 Klueh, U. and Kreeutzer, D.L. (2005) *Diabetes Technology and Therapeutics*, **7**, 727–737.

66 Koschwanez, H.E. and Reichert, W.M. (2007) *Biomaterials*, **28**, 3687–3703.

67 Liao, J.C., Mastali, M., Li, Y., Gau, V., Suchard, M.A., Babbitt, J., Gornbein, J., Landaw, E.M., McCabe, E.R.B., Churchill, B.M. and Haake, D.A. (2007) *The Journal of Molecular Diagnostics*, **9**, 158–168.

68 McLaughlin, J.A.D. (2005) Point of Care biomedical sensors in Studies in Health Technology and Informatics, Vol. 117, Personalised Health Management Systems – The Integration of Innovative Sensing, Textile, Information and Communication Technologies, eds Chris D. Nugent, Paul J. McCullagh, Eric T. McAdams, Andreas Lymberis.

2
Functional Thin Film Architecture and Platforms Based on Polymers

Surface Design: Applications in Bioscience and Nanotechnology
Edited by Renate Förch, Holger Schönherr, and A. Tobias A. Jenkins

ISBN: 978-3-527-40789-7

2.1
Controlled Block-Copolymer Thin-Film Architectures

Monique Roerdink, Mark A. Hempenius, Ulrich Gunst, Heinrich F. Arlinghaus, and G. Julius Vancso

2.1.1
Introduction

The self-organization of block-copolymers [1–3] in thin films has attracted much interest since the resulting nanoperiodic structures have potential for applications as high-density data storage [4, 5], or photonic or plasmonic waveguides [6], in sensing and other technologically important areas [7] due to the intrinsic sizes of tens of nanometers and the convenient processing of block-copolymers. The two microphase-separated blocks are typically immiscible and thus phase separate to form various morphologies, such as spheres of the minority phase in the matrix formed by the majority component, of which the length scales can be tuned by varying the molar mass of the individual blocks. In bulk, the spherical domains are typically found to adopt a body-centered cubic (bcc) packing, while in thin films of one monolayer of domains an inplane hexagonal array is formed [8–10]. The inplane order is typically short range, extending over a small number of inter-domain spacings, after which the directionality of the hexagonal packing is changed. This causes an overall inplane appearance reminiscent of the grain structure found in metals. The positional control of *block*-copolymer domains is a key issue for many technological applications and much work has been directed towards obtaining order and understanding of the mechanisms involved. A variety of methods has been explored to enlarge the grain sizes in block-copolymer thin films, such as using substrate interactions [11, 12], solvent evaporation [13–15] and crystallization [16, 17], or topographical patterns [18–22].

Kramer and coworkers[23] demonstrated surface-induced ordering in films of poly(styrene-*block*-2-vinylpyridine) up to one micrometer in thickness perpendicular to the surface, although the inplane ordering was short range. Inplane order was more recently obtained by the same group using topographically structured substrates on which thin films consisting of one monolayer of domains were cast, resulting in single grains of domains in the wells and on top of the mesas that extended over micrometers [18, 24, 25]. This work on the templating effect of edges on domain ordering in block-copolymer thin films was the first example of

Surface Design: Applications in Bioscience and Nanotechnology
Edited by Renate Förch, Holger Schönherr, and A. Tobias A. Jenkins

ISBN: 978-3-527-40789-7

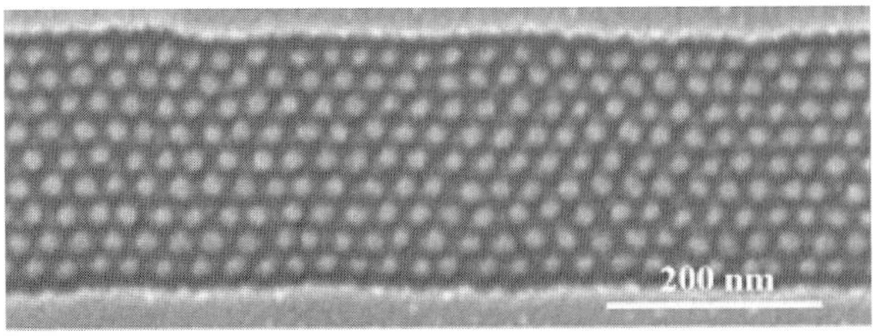

Figure 1 Polystyrene-*block*-poly(ferrocenyldimethylsilane) (PS-*b*-PFS) block-copolymer confined within grooves in a silica substrate, leading to well-ordered rows of PFS spheres. (Reprinted with permission from [26]).

graphoepitaxial ordering with amorphous block-copolymers. Another study of graphoepitaxy was reported by Cheng *et al.* [20, 26], aligning domains of poly(ferrocenylsilane) in a polystyrene matrix (Figure 1). Poly(ferrocenylsilane) (PFS) block-copolymers, pioneered by Manners and coworkers [27–29], are of special interest, since, due to the presence of iron and silicon in the main chain, PFSs display unique properties such as catalytic activity [30–32], redox responsiveness [33, 34], and etch resistance [35, 36]. Thin films of PFS block-copolymers can, for example, be used for direct pattern transfer of the microphase-separated morphology to underlying materials such as silicon, silicon nitride, or cobalt to form dense arrays of nanoparticles on flat substrates [37, 38] or in grooves [20].

Graphoepitaxy studies on PS-*b*-PFS showed adjustment of the ideal, 'commensurable' number of rows with the groove width by changes in the interrow spacing (stretching or compressing) or changes in domain size directly near the edges. The formation of ordered patterns in grooves was described in terms of elastic strain in the array of self-assembled microdomains [39]. Based on this argument, precisely modulated block-copolymer domains could be obtained through the design of grooves with uniform modulated groove widths. Although the system described by Cheng has a high tolerance to imperfections in templating geometry or edge roughness [39, 40], only a maximum of 11 to 12 rows could be successfully guided by templating grooves.

Amorphous PI-*b*-PFS (Scheme 1) was reported to have intrinsic long-range order on flat substrates extending up to 2 μm and has, in addition, the advantage of glass-transition temperatures (T_g) of both blocks that lie below room temperature [41], making time-consuming annealing steps to improve ordering redundant. Amorphous poly(ferrocenylsilane)s are in principle obtained by polymerizing unsymmetrically substituted, silicon-bridged ferrocenophanes [42]. Living anionic polymerization, a method of choice for preparing well-defined PFS homo- and block-copolymers [28, 43], however, may in case of the unsymmetrical [1]ethylmethylsilaferrocenophane or [1]methylphenylsilaferrocenophane lead to corresponding amorphous PFSs with higher polydispersities (M_w/M_n = 1.2–1.3) [44] than typically found for the [1]dimethylsilaferrocenophane monomer (M_w/M_n = 1.05–1.1) although examples of well-defined amorphous PFSs have been reported [45]. Anionic copolymerization of symmetrically and unsymmetrically substituted silicon-bridged ferrocenophanes, such as [1]dimethylsilaferrocenophane **1** and [1]ethylmethylsilaferrocenophane **2**, produces well-defined amorphous PFS blocks. Suppression of crystallization of the PFS blocks prevents the formation of hedrite-like structures in thin films that destroy the microdomain morphology [46].

The low-T_g amorphous block-copolymers are of interest for studying the possibilities of expanding the grain size by topographic templating and to investigate to what extent similar constraints, such as the maximum number of rows susceptible to templating, apply. Further, the effect of groove geometry on the templating ability of topological structures is of interest, to fabricate addressable microphase-separated structures. The orientational effects of spherical domains in thin films of PS-*b*-PFS on silica substrates were attributed to the presence of a PFS brush layer at the substrate, which was supported by dynamic secondary ion mass spectrometry (DSIMS) data and the decrease of groove width by twice a brush-layer thickness after spin casting a film [26, 47]. A study of the wetting layer of PI-*b*-PFS has not been presented so far but since we expect the polyisoprene block in our block-copolymer to be more apolar than PS, we are curious to know if we have a similar case consisting of a PFS wetting layer on the substrate or if a different situation, such as symmetric wetting by both blocks, exists. A detailed study of the wetting layer in both PS-*b*-PFS as PI-*b*-PFS is presented here, demonstrating a qualitatively different case of wetting of the substrate surface for these two block-copolymers. Furthermore, we describe experimental evidence of a structural transition from hexagonally packed domains in thin films to bcc morphology in bulk, and the influence of the surface and substrate on this transition. Finally, we demonstrate successful alignment of PI-*b*-PFS in linear and hexagonal grooves, and discuss the special case of circular pits, reflecting on the results obtained by dynamic SIMS.

2.1.2
Results and Discussion

A viable route to amorphous, low-T_g organic-organometallic block-copolymers involves living anionic copolymerization of symmetrically and unsymmetrically substituted silicon-bridged ferrocenophanes, such as [1]dimethylsilaferrocenophane **1** and [1]ethylmethylsilaferrocenophane **2**, to form amorphous PFS blocks. A series of statistical copolymers of **1** and **2** with varying compositions (Table 1) were synthesized and characterized by ^{1}H- and ^{13}C NMR spectroscopy and gel permeation chromatography (GPC). The amounts of **2** incorporated in the PFS chains, as gauged by ^{1}H NMR, agreed well with the monomer ratios used. Clearly, polydispersity values did not increase by copolymerizing **1** with **2**. The copolymers constitute orange, gummy substances, with T_gs in the range of 11–20 °C, depending on molar mass and composition. Their thermal behavior was studied by allowing isothermal crystallization at 95 °C, followed by DSC heating scans (Figure 2, top). For poly(ferrocenyldimethylsilane) (M_n = 15200 g/mol, M_w/M_n = 1.05), trace A, two endothermic peaks associated with the melting–recrystallization behavior of PFDMS were observed [48]. Upon introducing ferrocenylethylmethylsilane

(EM) units in the chains, the melting transitions shift to lower temperatures as anticipated for statistical copolymers. At 12 mol% EM the high-temperature melting transition disappeared. Melting transitions were no longer observed upon increasing the EM percentage to 23%. A similar trend of decreasing, and subsequently disappearing melting transitions was found for the corresponding PI-*b*-PFS block-copolymers (Figure 2, bottom) [49].

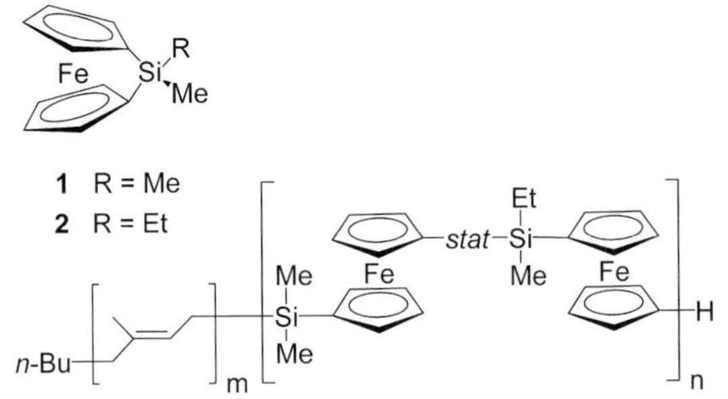

Scheme 1 Organic-organometallic block-copolymer featuring an amorphous organometallic block.

Table 1 PFS statistical copolymer and PI-*b*-PFS block-copolymer characteristics.[a)]

Polymer	$\overline{M}_n$ [kg/mol]	$\overline{M}_w$ [kg/mol]	$\overline{M}_w/\overline{M}_n$	EM [mol%]	PFS [vol%]
F 13($DM_{95}EM_5$)	12.8	13.5	1.05	5	
F 16($DM_{88}EM_{12}$)	15.9	17.7	1.11	12	
F 15($DM_{77}EM_{23}$)	15.3	16.7	1.09	23	
IF 58/22	79.8	83.3	1.04	0	22
IF 54/18($DM_{94}EM_6$)	71.8	74.2	1.03	6	19
IF 35/11($DM_{89}EM_{11}$)	45.1	46.8	1.04	11	18
IF 52/20($DM_{89}EM_{11}$)	71.5	74.8	1.04	11	22
IF 44/15($DM_{86}EM_{14}$)	58.2	61.2	1.06	14	20
IF 56/21($DM_{77}EM_{23}$)	76.7	80.8	1.05	23	22
IF 47/13($DM_{55}EM_{45}$)	60.3	61.5	1.02	45	16
SF65/20	84.0	89.0	1.07		20

a) Compositions measured by ^{1}H NMR spectroscopy. Molar masses obtained by GPC in THF, relative to polystyrene standards. The notation used to identify the diblock-copolymers includes the corresponding molar masses (M_n, in 10^3 g/mol) of the two blocks, i.e. IF 58/22 is a 58000 g/mol polyisoprene-*b*-22000 g/mol poly(ferrocenylsilane) diblock. DM denotes incorporated **1**, EM denotes incorporated **2**. (Reprinted with permission from [41], copyright 2005 by American Chemical Society).

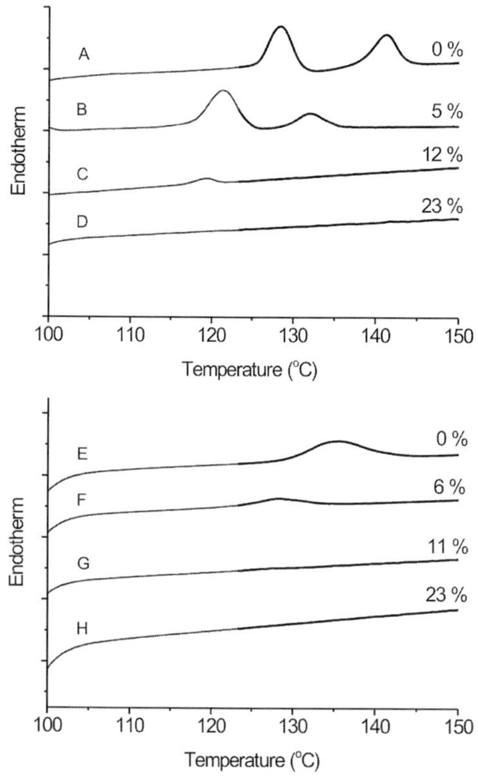

Figure 2 DSC heating scans of PFS copolymers (left) with increasing amounts of EM: (A) PFDMS, (B) F 13($DM_{95}EM_5$), (C) F 16 ($DM_{88}EM_{12}$) and (D) F 15($DM_{77}EM_{23}$). DSC heating scans of PI-*b*-PFS block-copolymers (right) with increasing amounts of EM: (E) IF 58/22 [EM = 0%], (F) IF 54/18($DM_{94}EM_6$), (G) IF 52/20($DM_{89}EM_{11}$) and (H) IF 56/21 ($DM_{77}EM_{23}$). All samples were kept at 95 °C for 2 h, prior to heating. (Reprinted with permission from [41], Copyright 2005 by American Chemical Society).

2.1.2.1
Wetting Layer

We performed dynamic TOF-SIMS depth-profiling measurements on thin films of a 35 000 $g mol^{-1}$ PI-*b*- 11 000 $g mol^{-1}$ PFS with 14 mol% incorporated **2** (IF44/15, Scheme 1) and 67 000 $g mol^{-1}$ PS-*b*-21 000 $g mol^{-1}$ PFS (SF65/20, Table 1) on both silica and silicon (with native oxide) substrates. The thickness of the silica layer can influence the stability of a nonpolar liquid on a solid substrate, as demonstrated for PS on silicon with oxide layers of various thickness.[50] The interplay of both short- and long-range interfacial forces can influence the film stability in the case of our nonpolar block-copolymer, and may also affect the substrate wetting. The PFS is the minority phase and forms spherical domains arranged in a bcc phase in the bulk [51]. In DSIMS, bombardment-induced erosion from a sample surface down-

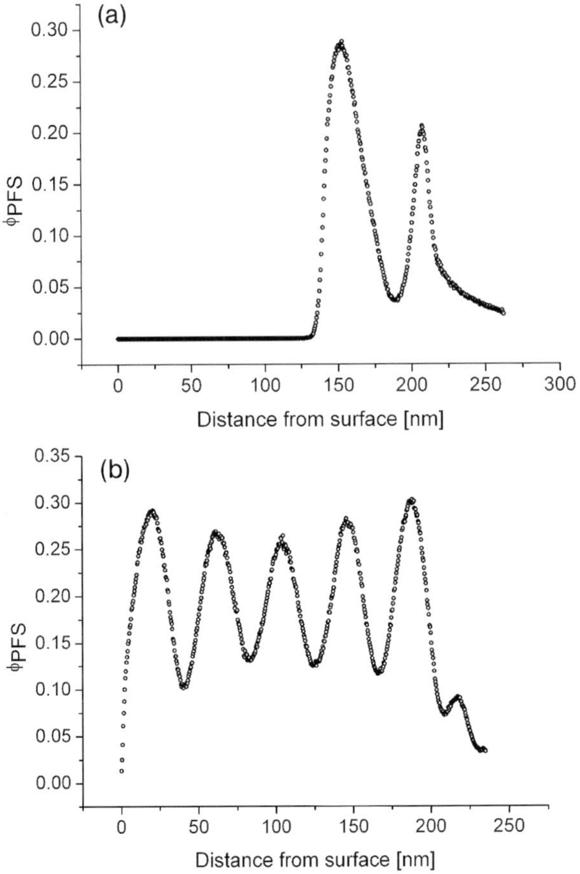

Figure 3 Depth profiles established with TOF-SIMS, represented as the PFS volume fraction, for SF65/20. (a) 69-nm film on silicon; (b) 215-nm film on silicon oxide. (Reprinted with permission from [54]).

wards occurs and one or more mass-over-charge intensities are recorded as a function of time, which can be converted to a concentration–depth profile [52, 53]. By detecting the Fe ions that are present in PFS, the compositional depth profile in thin block-copolymer films was obtained in terms of PFS volume fraction (ϕ_{PFS}).

Figure 3 shows the PFS depth profile of the Fe ion intensity converted to ϕ_{PFS} in SF65/20 on Si and SiO_2, with film thicknesses of 69 and 215 nm, respectively. The location of the block-copolymer surface in the sample of Figure 3(a) was determined from the Au signal (not shown) of a thin gold layer evaporated on the polymer surface, which featured a sacrificial 125-nm thick PS top layer. Na and K moieties could also be used to determine the block-copolymer interface (Figure 3b and Figure 4) without the use of Au and a sacrificial polymer layer. The first maximum of ϕ_{PFS} is located at a distance from the surface, denoting the first layer of spherical domains located within the film. The peak in ϕ_{PFS} corresponds to a layer

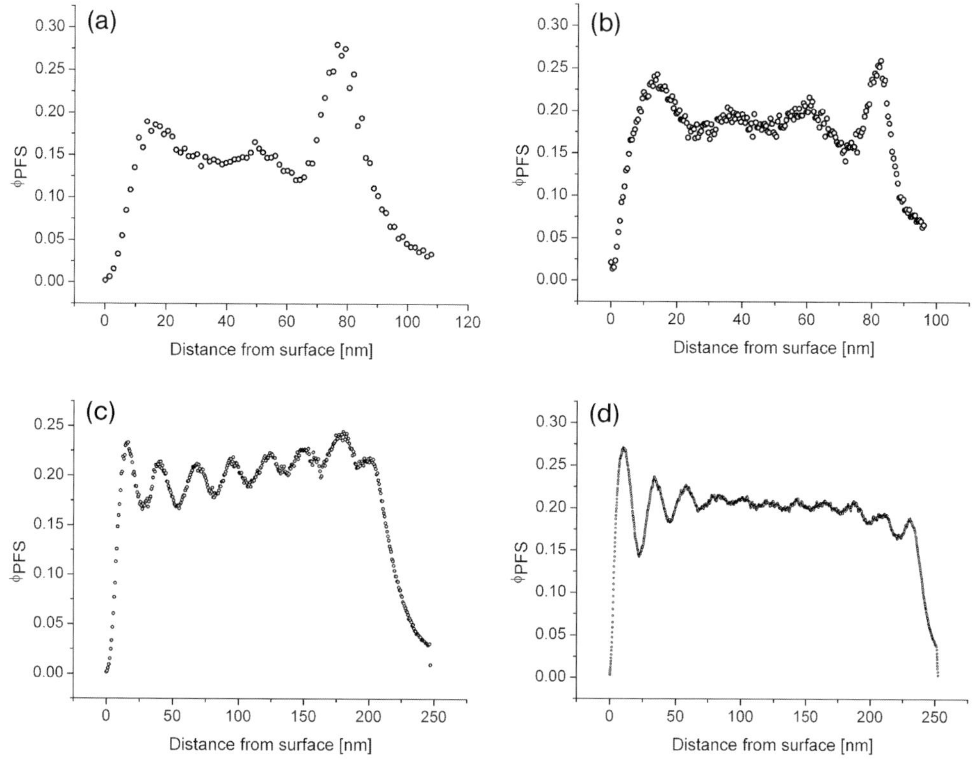

Figure 4 TOF-SIMS depth profiles of IF44/15, convoluted to the PFS volume fraction, for films of (a) 73 nm on silica; (b) 84 nm on silicon; (c) 204 nm on silicon; (d) 233 nm on silica. (Reprinted with permission from [54]).

of PFS spherical domains with a half-peak width of 30 nm, which we take as the domain diameter. The next maximum in ϕ_{PFS} is located directly at the silicon substrate. The width of this PFS layer is 13 nm, which is approximately half the domain diameter. The PFS volume fraction at the substrate is considerably lower compared to the layer of spheres. These results indicate that in PS-*b*-PFS block-copolymer films the PFS preferentially resides at the silicon substrate, whereas the PS exhibits an affinity for the free surface. Figure 3(b) shows a similar trend of a somewhat thicker PS-*b*-PFS film but on a silicon-oxide substrate. At the substrate, a PFS layer is present, as expected based on previous work [20, 26], with half the thickness of the domain size within the specimen. The oscillating signal clearly indicates layers of spherical domains ordered laterally with respect to the substrate over the entire analysis area of 30 μm^2. There are five layers of spherical PFS domains present in this film with an average domain diameter of 22 nm. The layer periodicity (l), that is, the spacing between two adjacent maxima in ϕ_{PFS}, is 42 nm. This corresponds well with the spacing of 39.2 nm for the first allowed reflection in the bcc structure (l_{bcc}), which is the (110) plane, determined by small-angle X-ray scattering (SAXS). Similar results where spherical domains were found to order on the substrate with the (110) plane parallel to the substrate were found by others for PS-*b*-PVP [55]. In Table 2, the average domain diameter (2*r*) and layer periodicity for the layers are summarized.

Table 2 Number of layers (*n*) observed with dynamic TOF-SIMS depth profiling and the deviation from the natural film thickness (h_0) for asymmetric wetting of the substrate and surface. (Reprinted with permission from [54]).

	h [nm]	*n*	*l* [nm]	2*r* [nm]	h_0 [nm]	$(h-h_0)/h_0$ [%]
SF65/20	69	1	53	30	59	17
	215	6	42	22	216	< 1
IF44/15EM14	73	3	31	15	63	16
	84	4	23	11	88	5
	204	8	26	14	188	9
	233	11	24	12	263	8

Figure 4 shows the volume fraction of PFS in thin films of IF44/15 on silicon and silica. First, the depth profile of the PI-*b*-PFS has a weaker variation in the PFS volume fraction (ϕ_{PFS}) than observed for PS-*b*-PFS. The weaker oscillation is likely due to the weaker phase segregation in PI-*b*-PFS [56]. From contact-angle measurements it is known that PI wets the air interface [46]. In all films, maxima in the PFS volume fraction are found at the substrate interface and there is no noticeable difference in wetting behavior on silicon or silicon oxide. Contrary to the copolymer film with a polystyrene majority phase, there is no notable decrease in PFS volume fraction directly near the substrate, nor are the widths of these peaks smaller compared to other depth values in the plots. Comparison of the TOF-SIMS depth profile data of SF65/20 and IF44/15 leads us to conclude that while in PS-*b*-PFS films a layer of PFS is present at the substrate, due to the significant decrease in PFS volume and layer thickness at the surface, both blocks are present at silicon or silica substrates in the case of PI-*b*-PFS. The presence of both blocks directly at the substrate in PI-*b*-PFS is supported by scanning electron microscopy (SEM) images of O_2 reactive ion etching exposed films on substrates with 100-nm deep grooves, where domains are observed directly on the groove walls.

Although the manner in which PFS is present at the substrate differs for PS-*b*-PFS and PI-*b*-PFS, the natural film thickness h_0 that is commensurate with the periodicity is for both polymers consistent with asymmetric wetting for surface and substrate, and is given by:

$$h_0 = [(n-1) + 0.5]l_{(bcc)} \quad (1)$$

where n is an integer. The number of layers found by TOF-SIMS depth profiling in the PI-*b*-PFS thin films and the corresponding average layer thickness and domain sizes are given in Table 2. First, the deviation from the natural film thickness is less than 1% for the thicker film of SF65/20, which denotes that either the film thickness was coincidentally very well chosen or that the copolymer responds easily to small deviations from commensurability with the natural film thickness by chain stretching or compression, distributed throughout the entire film. For PI-*b*-PFS films of four layers or more (Figures 4(b)–(d)) the average layer thickness corresponds well with the equilibrium bulk value l_{bcc} for the (110) plane in a bcc morphology, which was determined by SAXS to be 25.1 nm. The planar (110) oriented bcc morphology found in DSIMS does not correspond to the inplane morphology of spheres in a monolayer, which are hexagonally packed (HP), as can be easily imaged, for instance, by atomic force microscopy (AFM). The (110) plane of a bcc structure is a distorted hexagonal packing of domains. The (111) plane in face-centered cubic (fcc) and the (111) plane in bcc structures exhibits hexagonal packing of domains. However, if one assumes either one of these planes to be parallel to the substrate, the corresponding layer periodicities are not compatible with the TOF-SIMS depth-profile results. Based on our TOF-SIMS results, we explain the discrepancy with the HP monolayer and the bcc bulk morphology by a rearrangement of the microdomains in thin films, a similar process to surface relaxations and relocations in atomically thin films of metals [57]. The first layer of domains forms an HP array of domains. When a second or even third layer of domains is present, a repositioning towards a more bcc-like morphology occurs, presumably via a base-centered orthorhombic arrangement. The position of the domains in both layers is shifted slightly with respect to the hexagonal packing. In the following layers of domains, the HP morphology is abandoned and a bcc structure is adopted. Our results of a larger layer spacing in thinner films (see Table 2) agree well with such a transition. A more close-packed arrangement of microdomains in a monolayer, as is the case for HP arrays compared to a bcc arrangement, will necessarily induce a mechanical stress normal to the plane of organization on the chains, which will result in chain stretching and larger domain sizes and interdomain spacings. This is supported by the film thickness of 41 nm for one monolayer of domains ($n = 1$) in PI-*b*-PFS, which is almost three times the calculated h_0, and close to h_0 for 1.5 layers ($n = 2$). However, just one layer of domains exists directly at the substrate ($n = 1$) [58]. Similarly, the thickness of the monolayer in SF65/20 deviates by 17% from the natural film thickness. For films of more than one monolayer, a positional reorganization of the domains from HP to the bcc phase most probably induces chain stretching both inplane and out-of-plane. In the film of three layers of domains of IF44/15 (Figure 4(a)) there is a considerable difference in the distance between adjacent layers and the bulk value. Moreover, the first layer directly next to the substrate is 27 nm, slightly larger than l_{110}, while the second layer is 34 nm. The increase in domain sizes, as well as the increased layer thickness in the thinner films of both SF65/20 and IF44/15 compared to the bulk, support our idea of a transition in the morphology going from films of a few layers of microdomains to thicker films in the copolymer studied. In thicker films, this difference in layer periodicity throughout the film is absent. Clearly, the rearrangement from HP to bcc strongly affects the orientation in films up to three layers in this polymer, but for thicker films the bcc (110) morphology planar to the free surface dominates the order down to the substrate.

Small variations in periodicity and domain size are observed in the three thicker polyisoprene copolymer films. In the case of incommensurability of the film thickness with the periodicity of the block-copolymer the formation of islands or holes is typically observed. However, for some copolymer systems, deviations from the ideal domain size and periodicity occur in films of more than a few layers of domains, which is apparently more favorable than the rupture of the film [23]. This also seems to be the case in thin films of PI-*b*-PFS block-copolymer, which seems to exhibit a large degree of elasticity, as seen from the deviation from the natural thickness given in Table 2, ranging from 5 to 9%. For the film thickness of 233 nm (Figure 4(d)), layers of domains are present near both the surface and the substrate but the ordered layering does not propagate to the middle of the film. Since the SAXS scattering profile of a bulk specimen of this copolymer results in sharp, well-defined higher-order peaks, it appears unlikely that this is due to the absence of long-range order. Further, since the two blocks of the copolymer are above T_g, rendering the chains mobile, the absence of order in the film interior does not seem likely to be caused by kinetic trapping. Therefore, the absence of order is most likely the result of frustration due to incommensurability of the block-copolymer repeat length with the film thickness. The surface and substrate strongly direct the position of the layers of domains directly next to the interfaces, and the frustration is spread over the interior layers of domains.

2.1.2.2
Ordering on Planar Substrates

Thin films of the amorphous block-copolymers on silicon wafers were imaged by tapping-mode AFM (Figure 5a), showing a relatively well-ordered monolayer of PFS spheres. To establish the degree of ordering more quantitatively, a Voronoi construction was employed, which aids in the visualization of packing defects and grain boundaries occurring in the block-copolymer thin film [59]. Spheres surrounded by 6 neighbors appear as hexagons, while 5- and 7-fold-coordinated sites are displayed as pentagons and heptagons, respectively. A dislocation core is seen as a bound 5–7 pair, a grain boundary appears as a row of dislocations [24, 25]. A simple sphere-finding algorithm was used to identify the sphere centers, enabling the construction of a Voronoi diagram that was superimposed on the original image (Figure 5b). A perfect match between the domains and the polygons was found. Defects are shaded in the Voronoi diagram in Figure 5(c). The amorphous block-copolymer self-organized at room temperature to a well-ordered monolayer of organometallic spheres, with a single hexagonal grain extending over almost the whole scanned area of 1 μm × 1 μm. A pair-distribution function [59] associated with the AFM image (Figure 5d) confirmed the ordering quality and the

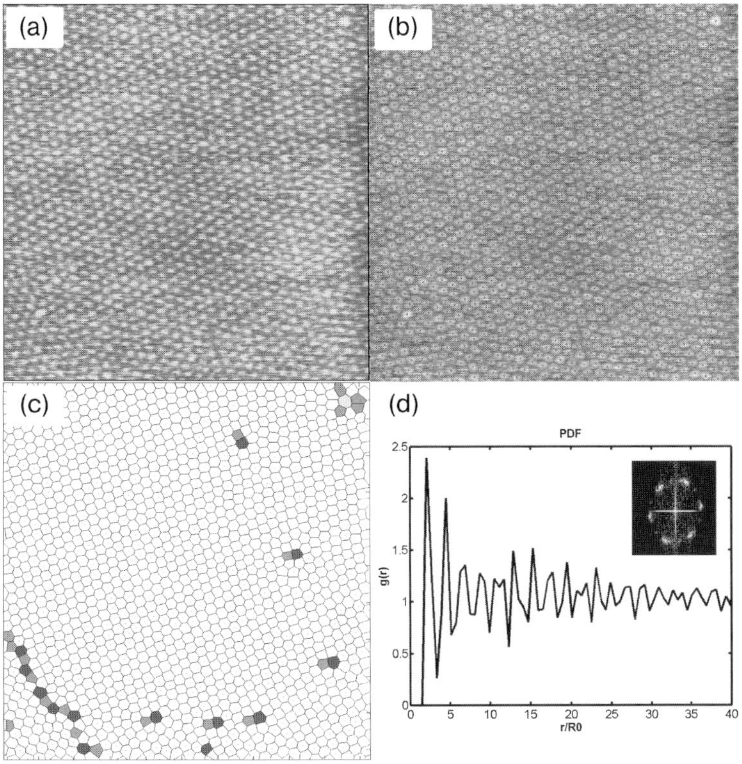

Figure 5 (a) AFM TM height image of a IF 35/10($DM_{89}EM_{11}$) block-copolymer thin film, scan size 1 μm × 1 μm. (b) Voronoi diagram of image 4A, projected on top of the image. (c) Voronoi diagram, where sites surrounded by 6 neighbors are unshaded, 5-fold-coordinated sites are magenta and 7-fold-coordinated sites are blue. (d) Pair-distribution function and Fourier transform of image 5A. (Reprinted with permission from [41], Copyright 2005 by American Chemical Society).

relatively large single-grain size. In comparison, the average grain size found in PS-*b*-PFDMS diblock-copolymer thin films is typically around 280 nm.[26] Similar correlation length values were observed for PI-*b*-PFDMS diblock-copolymers. PS-*b*-P2VP diblock-copolymer thin films on flat silicon substrates showed single-grain sizes of about 400 nm [18]. The increased correlation length for the amorphous PI-*b*-PFS block-copolymer monolayers is likely a result of the high molecular mobility, allowing for the removal of point defects and grain boundaries by sphere rearrangement. The high molecular mobility is due to both the low-T_g polyisoprene block, a lowered T_g of the PFS block [46, 60] combined with the suppression of crystallization of the PFS domains.

Figure 6(a) shows an AFM TM image, scan size 2 μm × 2 μm, of an amorphous PI-*b*-PFS block-copolymer thin film exposed to an O_2 plasma. In this treatment, the organic matrix is removed and arrays of Fe/Si/O domains remain. The relatively high degree of order of the etched film is demonstrated by the Voronoi

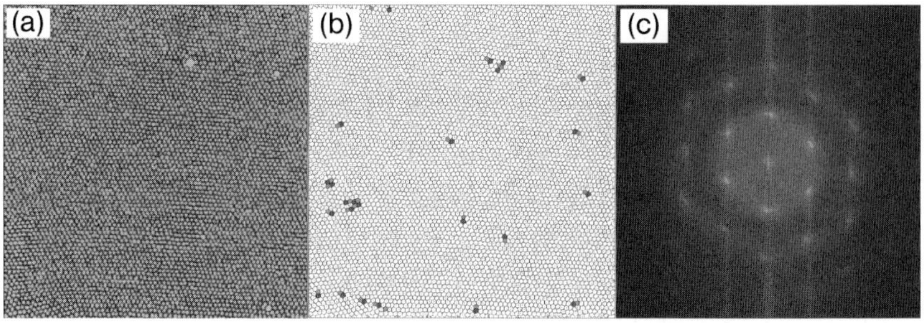

Figure 6 (a) AFM TM height image of a IF 35/10($DM_{89}EM_{11}$) block-copolymer thin film after an O_2 plasma treatment, scan size 2 μm × 2 μm. (b) Voronoi diagram, where sites surrounded by 6 neighbors are unshaded, 5-fold-coordinated sites are magenta and 7-fold-coordinated sites are blue. (c) Fourier transform of image A, showing the high degree of order of the etched film. (Reprinted with permission from [41], Copyright 2005 by American Chemical Society).

diagram, which shows the scanned area to consist of a single grain containing approximately 5 defects/μm^2 (Figure 6b). The six sharp first-order Bragg peaks and twelve second-order peaks observed in the Fourier transform (Figure 6c) provide further evidence for the high degree of order. This image also demonstrates that large, well-ordered areas can be transferred to underlying substrates by reactive ion etching, using these block-copolymer patterns as nanolithographic masks.

By incorporating unsymmetrically substituted ferrocenylsilane units in PFDMS blocks, crystallization in PI-*b*-PFDMS block-copolymer thin films can be effectively suppressed. In thin films, the amorphous low-T_g diblock-copolymers self-assemble to form well-ordered monolayers of PFS spheres, with single-grain sizes larger than 1 μm × 1 μm, without thermal annealing. The ordering of these amorphous block-copolymers on topographically patterned substrates is discussed in the following section.

2.1.2.3
Ordering on Topographically Patterned Substrates

The graphoepitaxy effect of side-wall constraint on the domains with PI-*b*-PFS block-copolymer was evaluated next since the difference in substrate wetting between PS-*b*-PFS and PI-*b*-PFS presents a qualitatively different situation for the graphoepitaxial templating of domains along edges in these polymers. The polymer at the mesa directly next to the side walls flows into the grooves immediately after spin coating. The area around the grooves is therefore depleted from polymer except for a brush layer of adhered chains, as can be seen from the remaining contrast in both SEM and AFM phase images. It is probable that the rapid transport of material into the grooves proceeds over this adhered brush [61]. Figure 7 shows PFS domains in a linear groove after O_2-RIE exposure [62]. There are 38 rows accommodated in the 1.1-μm wide groove, aligned parallel to the groove

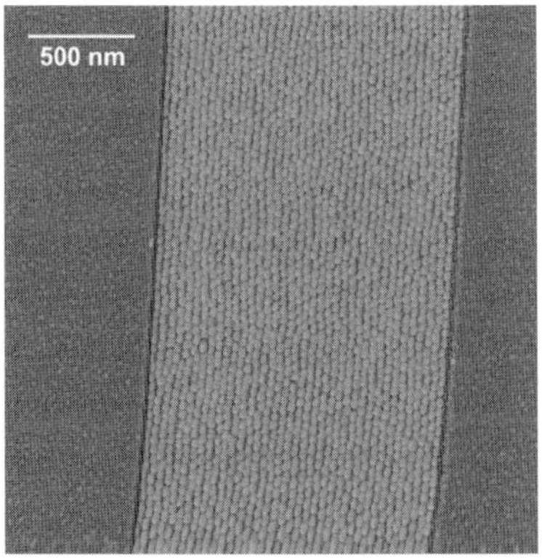

Figure 7 AFM-TM phase image of a IF47/13EM45 thin film in a linear groove of 1.1 μm. All 38 rows of domains are aligned parallel to the groove edges. (Reprinted with permission from [54]).

edge. In grooves up to 1.3 μm, alignment is observed, but in grooves of 2 μm the alignment is lost. The number of rows aligned in the grooves by far exceeds the 12 rows that can be aligned in PS-*b*-PFS [39]. The width of the grooves that can successfully guide block-copolymer domains for PI-*b*-PFS is comparable to the grain size of the polymer on flat substrates, as expected. Therefore, there is apparently no qualitative difference in topographically induced ordering of block-copolymer domains based on symmetric or asymmetric wetting of the substrate. Recent work on cylindrical microdomains in a PS-*b*-PMMA block-copolymer on topographically patterned substrates that were covered by a neutral brush agrees with our findings [22].

Figure 8 shows two rectangular groove endings. Next to all three side walls, rows align parallel to the edge but the two parallel side walls dominate the overall ordering. The 90° angle is incommensurate with the hexagonal packing of domains. The result is a semicircular row of dislocations perpendicular to the groove length, located close to the end of the 300-nm wide groove. The row of dislocations was found to advance more to the interior of the groove for wider grooves (not shown). The grain boundary formed by the dislocations absorbs

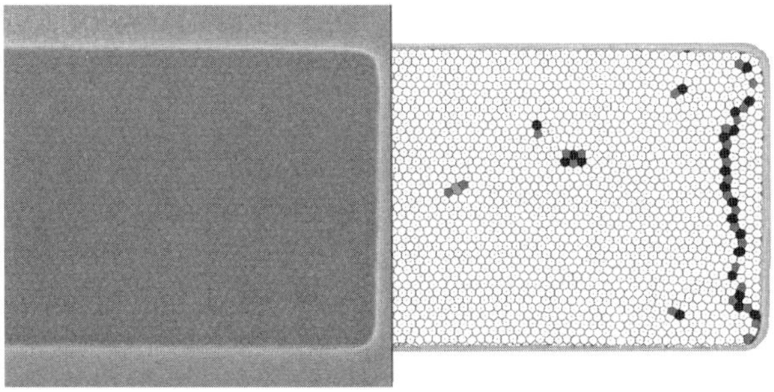

Figure 8 Alignment of PFS domains in square corners of grooves with widths of 1.1 μm. Across the groove, 38 rows of domains are accommodated.

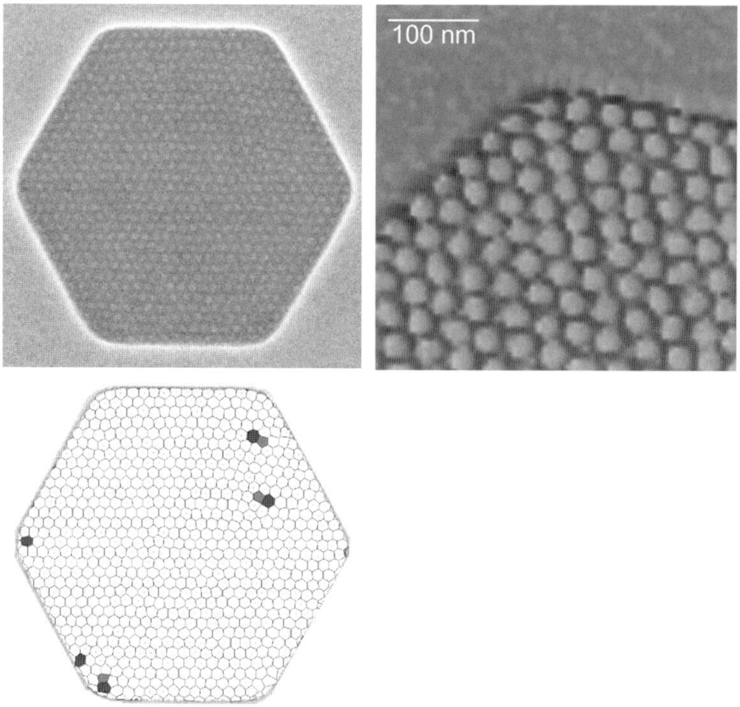

Figure 9 Left: SEM image of PI-*b*-PFS block-copolymer in an 800-nm wide hexagonal groove, and below, a Voronoi representation of the domains. Right: AFM-TM phase image of PFS domains in a 120° corner. One domain is located exactly in the middle of the corner, allowing for precise positional control of the domains over the entire hexagon. (Reprinted with permission from [54]).

internal stresses induced by the incommensurability between the 90° side wall and the block-copolymer lattice and tends to minimize its length, resulting in the semicircular shape observed.

The combination of wall-induced alignment and large-area correlation is expected to be optimal in hexagonal grooves that match the HP lattice of the copolymer. In series of hexagonal pits with different widths we observed single grains where the block-copolymer matched its grain lattice to the angle of registration of 120°. In the smaller hexagonal pits, variations in row spacing and domain size were observed close to the edges. Figure 9 shows an SEM image of a hexagonal groove approximately 800 nm in diameter. The domains are aligned with respect to the six side walls over the entire groove, resulting in 2D ordering over a μm^2 area. No apparent differences in domain sizes or row spacing values near the edges were observed, indicating that deviations from incommensurability are distributed more easily over a large number of domains. The hexagonal grooves open up the possibility of accurate positioning of the domains, since one single domain is located in each corner, as shown in Figure 9, right-hand side.

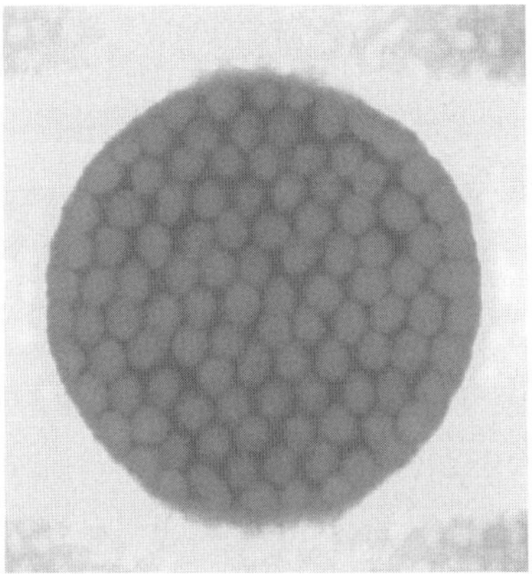

Figure 10 AFM-TM height image of a PI-*b*-PFS block-copolymer thin film in a circular pit. (Reprinted with permission from [54]).

The results of ordering in linear and hexagonal grooves suggest that the ordering mechanism is a combination of the aligning effect of the wall and the interaction between microdomains over a correlation length that extends over larger areas. An interesting case is therefore presented by the study of graphoepitaxial alignment in a series of circular pits with varying radii and, therefore, varying curvature. Figure 10 shows a typical example of PFS domains in a circular pit with a diameter of 300 nm. Although the row directly next to the edge follows the curvature, the other rows are not aligned with respect to the edge but organize in one or a few well-ordered grains, independent of the side wall. The block-copolymer system adjusts to the side-wall curvature by a strong compression or expansion of the row spacing and strong variation in the domain size directly near the side wall. Side walls in circular pits do not align more than one row of domains, which is in contrast to the directing effect of the side walls in linear and hexagonal grooves. This displays the considerable ability to absorb stress by local adjustments, and demonstrates the high degree of elasticity in the phase-separated PI-*b*-PFS that was also found in response to incommensurate film thickness.

2.1.3 Conclusions

The domain diameters and layer thicknesses in thin films of PFS copolymers with a polystyrene or polyisoprene block were found to be larger in films consisting of one to three layers of spherical domains, indicating chain stretching normal to the substrate. This stretching is thought to originate from a transition in microdomain packing from HP arrays of domains in a monolayer, probably via a transition from a base-centered orthorhombic to a bcc arrangement. In thicker films, a bcc phase is adopted with the (110) plane parallel to the surface, which persists through the entire film, including the layers directly near the substrate. Deviations from the natural film thickness, dictated by the asymmetric wetting conditions of substrate and surface, are distributed over the layers of domains. Particularly in the case of the polyisoprene copolymers, a highly elastic response to incommensurability was observed, rather than the typically observed island or hole formation. Despite the presence of both blocks of PI-*b*-PFS at the substrate, alignment of domains in grooves by the side walls was observed, even over larger areas than found for PS-*b*-PFS. This demonstrates that the nature of the substrate wetting is not a key issue in graphoepitaxial alignment of microdomains. Successful alignment was achieved as long as the groove width did not extend beyond the natural grain size of the copolymer, which in the case of PI-*b*-PFS is between 1 and 2 μm. 2D alignment was achieved in hexagonal grooves, where the HP coordination axes are registered with the side walls. A complete mismatch between the HP-ordered microdomain and the side walls, as in the case of circular pits, resulted in a complete absence of edge-induced order beyond more than one row of domains. Instead, the copolymer formed a single grain and absorbed the stress induced by the side walls by variation in the domain sizes and row spacings near the edge.

2.1.4
Experimental Section

[1]Dimethylsilaferrocenophane **1** and [1]ethylmethylsilaferrocenophane **2** were prepared and purified as described elsewhere [28, 44, 63]. Anionic polymerizations were carried out in tetrahydrofuran (THF) in a MBraun Labmaster 130 glove box under an atmosphere of prepurified nitrogen (< 0.1 ppm H_2O). Amorphous poly(isoprene-*block*-ferrocenylsilane) (PI-*b*-PFS) was synthesized by sequential anionic polymerization. Isoprene polymerizations in ethylbenzene were initiated by *n*-butyllithium. After completion of the isoprene block, **1** and **2** were added, followed by THF, for the formation of the PFS block. Degassed methanol was added after 1 h. The polymers were precipitated in methanol and dried under vacuum. ^1H NMR spectra were recorded in $CDCl_3$ on a Varian Unity Inova (300 MHz) instrument at 300.3 MHz and on a Varian Unity 400 spectrometer at 399.9 MHz. The solvent peak with a chemical shift of $\delta = 7.26$ ppm was used as a reference. GPC measurements were carried out in THF at 25 °C, using microstyragel columns with pore sizes of 10^5, 10^4, 10^3 and 10^6 Å (Waters) and a dual detection system consisting of a differential refractometer (Waters model 410) and a differential viscometer (Viscotek model H502). Molar masses were determined relative to polystyrene standards. Differential scanning calorimetry (DSC) measurements were performed on a Perkin-Elmer DSC-7 instrument. Patterned substrates were prepared by electron-beam lithography. Series of lines, circles, and hexagons with widths of 300 nm, 500 nm, 800 nm, 1 μm, and 1.5 μm were patterned on the wafers. After resist removal, the wafers were cleaned in fuming nitric acid (100%) for 10 min and in boiling nitric acid (70%) for 10 min to destroy any residual organic material. Thin films were spin coated from 1 wt% PI-*b*-PFS solution in toluene on flat or topographically patterned silicon substrates. Film thicknesses

(measured on flat areas of the substrates) ranged from 20 to 30 nm, established by ellipsometry, so as to yield a film with only one monolayer of microphase-separated spherical domains in the grooves. The films were annealed at room temperature for approximately 20 h and subsequently exposed to an oxygen plasma in a reactive-ion etching (O_2-RIE) setup, carried out in an Elektrotech PF 340 apparatus. The pressure inside the etching chamber was 10 mTorr, the substrate temperature was set at 10 °C, and an oxygen flow rate of 20 cm^3 min^{-1} was maintained. Power was set at 75 W and the thin films were exposed to the O_2 plasma for 10 s. The morphology of the thin films and the etched patterns was studied using a NanoScope III multimode AFM instrument (Digital Instruments/Veeco), operated in the tapping mode, and by SEM (LEO 1550 FEG microscope). To establish the degree of order more quantitatively, a Voronoi construction was employed [24, 59]. Spheres surrounded by six neighbors appear as hexagons while, five- and seven-fold-coordinated sites are displayed as pentagons and heptagons, respectively. A dislocation core is seen as a bound 5–7 pair; a grain boundary appears as a row of dislocations. A simple sphere-finding algorithm was used to identify the sphere centers, enabling the construction of a Voronoi diagram. Dynamic SIMS measurements (DSIMS) were performed using a TOF-SIMS IV with 2 keV 20 nA Ar^+, analysis (using noninterlace mode in the case of SiO_2 substrates and interlace mode on Si substrates) was done with 10-keV Ar^+ with an additional $^{18}O_2$ flooding. Positive ions of H, CH, O, Na, Si, SiC, and Fe were monitored as a function of time. The sputter beam was rastered over a 500-μm^2 area; the analysis area was 30 μm^2 centered within the sputtered area. Cooling to –100 °C was possible with a cooling device in the TOF-SIMS instrument. Films of PS-*b*-PFS for DSIMS measurements were annealed at 150 °C for two days. To determine the wetting layer at the air interface, a 5-nm gold layer was evaporated on the annealed film, on top of which a sacrificial polystyrene film was placed by floating. The gold layer indicates the start of the film boundary during sputtering and the sacrificial layer allows one to determine the sputter rate and to stabilize the sputter beam. The PI-*b*-PFS films were cooled to –100 °C during sputtering to ensure the polymer being below the T_g to prevent rearrangements during sputtering. The intensity of the Fe ions as a function of the depth ($I(z)$), directly obtained from dynamic TOF-SIMS depth profiling, was normalized to give the depth profile of the PFS domains in the film, using:

$$\phi_{PFS}(z) = hf_{PFS}I(z) \Big/ \int_0^h I(z)dz \qquad (2)$$

where h is the film thickness and f_{PFS} is the bulk fraction of PFS in the polymer.

Acknowledgments

The University of Twente, MESA[+] Institute for Nanotechnology, the Netherlands Foundation for Chemical Research NWO/CW and NanoImpuls, a Nanotechnology Program of the Dutch Ministry of Economic Affairs are acknowledged for financial support. DIMES Institute for Nanotechnology is acknowledged for the fabrication of the e-beam substrates. C.J. Padberg (University of Twente) is acknowledged for GPC measurements. Prof. R.A. Segalman (University of California, Berkeley) is thanked for the Voronoi script. The SAXS work referred to in this work is based upon research conducted at the Cornell High-Energy Synchrotron Source (CHESS), which is supported by the National Science Foundation and the National Institutes of Health/National Institute of General Medical Sciences under award DMR-0225180.

References

1 (a) Hamley, I.W. (2003) *Nanotechnology*, **14**, R39; (b) Park, C., Yoon, J. and Thomas, E.L. (2003) *Polymer*, **44**, 6725.
2 Bockstaller, M.R., Mickiewics, R.A. and Thomas, E.L. (2005) *Advanced Materials*, **17**, 1331.
3 For general references on block-copolymers see: (a) Hamley, I.W. (ed.) (2004) Developments in Block Copolymer Science and Technology, Wiley, Chichester; (b) Hadjichristidis, N., Pispas, S. and Floudas, G. (2003) Block Copolymers: Synthetic Strategies, Physical Properties, and Applications, Wiley, Hoboken.
4 Ross, C.A. (2001) *Annual Review of Materials Science*, **31**, 203.
5 Naito, K., Hieda, H., Sakurai, M., Kamata, Y. and Asakawa, K. (2002) *IEEE Transactions on Magnetics*, **38**, 1949.
6 Maier, S.A., Brongersma, M.L., Kik, P.G., Meltzer, S., Requicha, A.A.G. and Atwater, H.A. (2001) *Advanced Materials*, **13**, 1501.
7 Ruffolo, R., Evans, C.E.B., Liu, X.H., Ni, Y.Z., Pang, Z., Park, P., McWilliams, A.R., Gu, X.J., Lu, X., Yekta, A., Winnik, M.A. and Manners, I. (2000) *Analytical Chemistry*, **72**, 1894.
8 Thomas, E.L., Kinning, D.J., Alward, D.B. and Henkee, C.S. (1987) *Macromolecules*, **20**, 2934.
9 Bates, F.S. and Fredrickson, G.H. (1990) *Annual Review of Physical Chemistry*, **41**, 525.
10 Fasolka, M.J. and Mayes, A.M. (2001) *Annual Review of Materials Science*, **31**, 323.
11 Kim, S.O., Solak, H.H., Stoykovich, M.P., Ferrier, N.J., de Pablo, J.J. and Nealey, P.F. (2003) *Nature*, **424**, 411.
12 Edwards, E.W., Stoykovich, M.P., Müller, M., Solak, H.H., de Pablo, J.J. and Nealey, P.F. (2005) *Journal of Polymer Science Part B*, **43**, 3444.
13 Mansky, P., Liu, Y., Russell, T.P. and Hawker, C. (1997) *Science*, **275**, 1458.
14 Kimura, M., Misner, M.J., Xu, T., Kim, S.H. and Russell, T.P. (2003) *Langmuir*, **19**, 9910.
15 Kim, S.H., Misner, M.J., Xu, T., Kimura, M. and Russell, T.P. (2004) *Advanced Materials*, **16**, 226.
16 Reiter, G., Castelein, G., Hoerner, P., Riess, G., Sommer, J.U. and Floudas, G. (2000) *The European Physical Journal E*, **2**, 319.
17 Park, C., De Rosa, C. and Thomas, E.L. (2001) *Macromolecules*, **34**, 2602.
18 Segalman, R.A., Yokoyama, H. and Kramer, E.J. (2001) *Advanced Materials*, **13**, 1152.
19 Segalman, R.A., Cochran, E., Fredrickson, G.H. and Kramer, E.J. (2005) *Macromolecules*, **38**, 6575.

20 Cheng, J.Y., Ross, C.A., Smith, H.I. and Vancso, G.J. (2002) *Applied Physics Letters*, **81**, 3657.

21 Sundrani, D., Darling, S. and Sibener, S. (2004) *Nano Letters*, **4**, 273.

22 Xiao, S., Yang, X., Edwards, E.W., La, Y.H. and Nealey, P.F. (2005) *Nanotechnology*, **16**, S324.

23 Yokoyama, H., Mates, T.E. and Kramer, E.J. (2000) *Macromolecules*, **33**, 1888.

24 Segalman, R.A., Hexemer, A., Hayward, R.C. and Kramer, E.J. (2003) *Macromolecules*, **36**, 3272.

25 Segalman, R.A., Hexemer, A. and Kramer, E.J. (2003) *Macromolecules*, **36**, 6831.

26 Cheng, J.Y., Ross, C.A., Thomas, E.L., Smith, H.I. and Vancso, G.J. (2003) *Advanced Materials*, **15**, 1599.

27 Rulkens, R., Ni, Y. and Manners, I. (1994) *Journal of the American Chemical Society*, **116**, 12121.

28 Ni, Y., Rulkens, R. and Manners, I. (1996) *Journal of the American Chemical Society*, **118**, 4102.

29 For reviews on poly(ferrocenylsilane)s see: (a) Manners, I. (1999) *Chemical Communications*, 857; (b) Kulbaba, K. and Manners, I. (2001) *Macromolecular Rapid Communications*, **22**, 711; (c) Whittell, G.R. and Manners, I. (2007) *Advanced Materials*, **19**, 3439.

30 Hinderling, C., Keles, Y., Stöckli, T., Knapp, H.F., De Los Arcos, T., Oelhafen, P., Korczagin, I., Hempenius, M.A., Vancso, G.J., Pugin, R. and Heinzelmann, H. (2004) *Advanced Materials*, **16**, 876.

31 Lastella, S., Jung, Y.J., Yang, H.C., Vajtaj, R., Ajayan, P.M., Ryu, C.Y., Rider, D.A. and Manners, I. (2004) *Journal of Materials Chemistry*, **14**, 1791.

32 Roerdink, M., Pragt, J., Korczagin, I., Hempenius, M.A., Stöckli, T., Keles, Y., Knapp, H.F., Hinderling, C. and Vancso, G.J. (2007) *Journal of Nanoscience and Nanotechnology*, **7**, 1052.

33 Nguyen, M.T., Diaz, A.F., Dement'ev, V.V. and Pannell, K.H. (1993) *Chemistry of Materials*, **5**, 1389.

34 Rulkens, R., Lough, A., Manners, I., Lovelace, S., Grant, C. and Geigger, W. (1996) *Journal of the American Chemical Society*, **118**, 12683.

35 Lammertink, R.G.H., Hempenius, M.A., Chan, V.Z.H., Thomas, E.L. and Vancso, G.J. (2001) *Chemistry of Materials*, **13**, 429.

36 Korczagin, I., Lammertink, R.G.H., Hempenius, M.A., Golze, S. and Vancso, G.J. (2006) *Advances in Polymer Science*, **200**, 91.

37 Lammertink, R.G.H., Hempenius, M.A., van den Enk, J.E., Chan, V.Z.H., Thomas, E.L. and Vancso, G.J. (2000) *Advanced Materials*, **12**, 98.

38 Cheng, J.Y., Ross, C.A., Chan, V.Z.H., Thomas, E.L., Lammertink, R.G.H. and Vancso, G.J. (2001) *Advanced Materials*, **13**, 1174.

39 Cheng, J.Y., Mayes, A.M. and Ross, C.A. (2004) *Nature Materials*, **3**, 823.

40 Cheng, J.Y., Zhang, F., Smith, H.I., Vancso, G.J. and Ross, C.A. (2006) *Advanced Materials*, **18**, 597.

41 Roerdink, M., Hempenius, M.A. and Vancso, G.J. (2005) *Chemistry of Materials*, **17**, 1275. The bulk T_g of PFDMS is 33.5 °C, see ref 48, the bulk T_g of PI is around 70 °C.

42 Foucher, D., Ziembinski, R., Petersen, R., Pudelski, J., Edwards, M., Ni, Y., Massey, J., Jaeger, C.R., Vancso, G.J. and Manners, I. (1994) *Macromolecules*, **27**, 3992.

43 Rulkens, R., Lough, A.J. and Manners, I. (1994) *Journal of the American Chemical Society*, **116**, 797.

44 Temple, K., Massey, J.A., Chen, Z., Vaidya, N., Berenbaum, A., Foster, M.D. and Manners, I. (1999) *Journal of Inorganic and Organometallic Polymers*, **9**, 189.

45 Rider, D.A., Power-Billard, K.N., Cavicchi, K.A., Russell, T.P. and Manners, I. (2005) *Macromolecules*, **38**, 6931.

46 Lammertink, R.G.H., Hempenius, M.A. and Vancso, G.J. (2000) *Langmuir*, **16**, 6245.

47 Lammertink, R.G.H., Hempenius, M.A., Vancso, G.J., Shin, K., Rafailovich, M.H. and Sokolov, J. (2001) *Macromolecules*, **34**, 942.

48 Lammertink, R.G.H., Hempenius, M.A., Manners, I. and Vancso, G.J. (1998) *Macromolecules*, **31**, 795.

49 Isothermal crystallization experiments performed for several days at larger undercoolings (at 40 °C) led to the same observations. Thin films of PI-*b*-PFS, containing 11 or 23 mol% of EM, showed no sign of crystallization after being kept at room temperature for several days.

50 Seemann, R., Herminghaus, S. and Jacobs, K. (2001) *Physical Review Letters*, **86**, 5534.

51 The bcc morphology in the bulk was established by means of small-angle X-ray scattering experiments.

52 Benninghoven, A., Rudenauer, F. and Werner, H. (1987) *Secondary Ion Mass Spectrometry*, John Wiley & Sons, New York.

53 Zalm, P.C. (2000) *Mikrochimica Acta*, **132**, 243.

54 Roerdink, M., Hempenius, M.A., Gunst, U., Arlinghaus, H.F. and Vancso, G.J. (2007) *Small*, **3**, 1415.

55 Yokoyama, H., Kramer, E.J., Rafailovich, M.H., Sokolov, J. and Schwarz, S.A. (1998) *Macromolecules*, **31**, 8826.

56 The order–disorder transition (ODT) temperature was measured using SAXS. The ODT temperatures were significantly lower for the block-copolymers of PI and statistical PFS compared to PS-*b*-PFS (unpublished data).

57 Morrison, S.R. (1990) *The Chemical Physics of Surfaces*, 2nd edn, Plenum, New York.

58 The existence of more than one layer of domains would result in superimposed images after O_2-RIE.

59 Allen, S.M. and Thomas, E.L. (1999) *The Structure of Materials*, John Wiley & Sons, New York.

60 Glass–rubber transitions of the amorphous PI-*b*-PFS block-copolymers, with relatively low molar mass PFS blocks, were in the range of 10–20 °C. Poly(ferrocenylethylmethylsilane) has a reported bulk T_g value of 15 °C; see ref [44].

61 Segalman, R.A., Schaefer, K.E., Fredrickson, G.H. and Kramer, E.J. (2003) *Macromolecules*, **36**, 4498.

62 The imaging by AFM or SEM is facilitated by exposing the copolymer to an oxygen plasma, which removes the organic majority phase.

63 Lammertink, R.G.H., Hempenius, M.A., Thomas, E.L. and Vancso, G.J. (1999) *Journal of Polymer Science Part B*, **37**, 1009.

2.2 Stimuli-Responsive Polymer Brushes

Edmondo Benetti, Melba Navarro, Szczepan Zapotoczny, and G. Julius Vancso

2.2.1 Introduction

Polymer brushes [1, 2] (Figure 1c) refer to an assembly of macromolecules attached with one end to a surface and stretched away from it. The density of the anchoring sites should be high enough to ensure extended conformation of the crowded chains (in the swollen state) with the end-to-end distance larger than for the free chains in the same solvent. At the smaller densities of the tethered chains they adopt rather 'pancake' or 'mushroom' conformations (Figure 1a and b). Polyelectrolytes may often form brushes at lower grafting density than neutral polymers due to intra- and intermolecular electrostatic repulsions, in addition to the excluded-volume effect that drives expansion of the grafted layer. While polymer chains may form brushes on different substrates the majority of the studies focus on solid surface supports.

The stretched conformation of the tethered chains leads to unique properties of the brushes and specific applications. Among others, they are able to efficiently stabilize colloids [3] and significantly reduce friction between the surfaces [4]. Using brushes as biointerfaces may have advantages over conventional polymeric films and self-assembled monolayers (SAMs). Polymeric brushes provide control

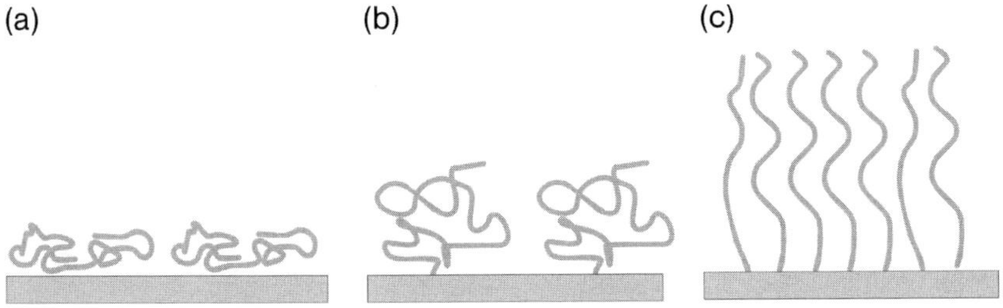

Figure 1 Scheme of the conformations of the surface tethered polymeric chains: (a) 'pancake' (the chains are adsorbed on a surface), (b) 'mushroom' and (c) 'brush'.

Surface Design: Applications in Bioscience and Nanotechnology
Edited by Renate Förch, Holger Schönherr, and A. Tobias A. Jenkins

ISBN: 978-3-527-40789-7

over both thickness and density of the interface layer, and can therefore influence its mechanical properties, wetting and adhesion characteristics. They can be synthesized on a variety of metallic and nonmetallic substrates. Chemical functionality can be incorporated in specific positions embedding desired groups at a defined depth from the top surface. In this way, the brushes enable selective attachment of biologically active molecules and allow the loading capacity of the surface layer to be easily tuned.

Thus, polymer brushes are versatile platforms for the molecular engineering of surfaces with complex functionality. Their design determines many important parameters of biointerfaces that are responsible for enhanced cell adhesion, antifouling behavior [5] or enable biosensing [6]. Polymeric brushes find numerous applications in biotechnology and medicine such as immobilization of cells or biocatalysts, drug-delivery and bioseparation systems, as biomimetic actuators and chemical valves [7].

2.2.2 Synthesis

A number of methods may be used to fabricate surface-tethered macromolecules but only a few of them are suitable for the synthesis of high-density polymer brushes. In the 'grafting to' approach (Figure 2a) preformed end-functional macromolecules are attached to a surface via physisorption or chemical reactions with appropriate groups on a surface [8]. Due to steric hindrance (slow diffusion of the macromolecules through the already grafted chains to the surface) only low grafting densities can be achieved. At most, 'semidilute' brushes may be formed for low molecular weight polymers. 'Grafting from' (Figure 2b) utilizes surface-tethered initiating sites from which polymeric chains may be grown [9]. The active sites can be generated *in situ* via, e.g., plasma or UV/ozone treatment of surfaces [10]. As an alternative, polymerizations initiators may be immobilized on surfaces, by embedding them in SAMs, and subsequently activated [11]. This method allows to control the surface concentration of the active sites and leads to synthesis of high-density polymer brushes.

Many synthetic routes have been developed so far for the fabrication of polymer brushes such as: conventional [12, 13] and controlled radical polymerization (CRP) [14], cationic [15] and anionic [16] living polymerizations as well as ring-opening polymerizations [17]. CRP has proven to be the most popular method for surface-initiated polymerization (SIP) on model surfaces (gold, silicon) but also on, e.g., colloidal particles. The controlled polymerizations used include nitroxide-mediated and iniferter-based (initiator-transfer-terminator agent) polymerizations, atom transfer radical polymerization (ATRP), and reversible addition–fragmentation chain transfer (RAFT) polymerization that resulted in synthesis of homo- and copolymer brushes. The SIP methods and general properties of the polymeric brushes have been summarized in several recent reviews [1, 9, 14, 18, 19].

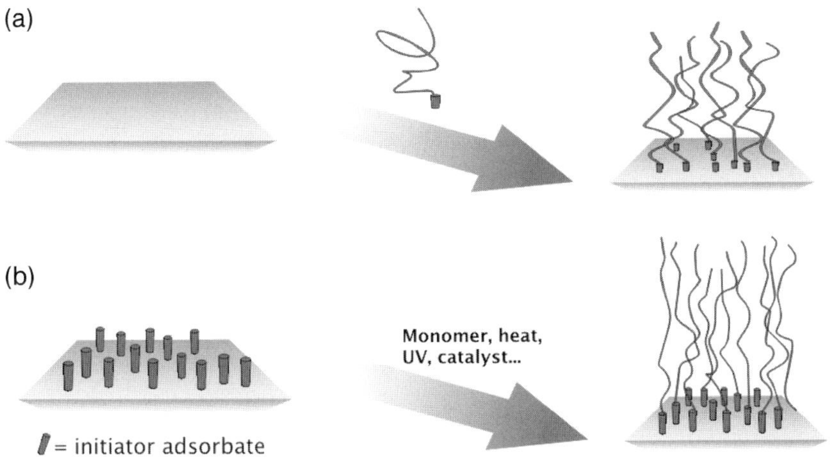

Figure 2 Scheme of 'grafting to' (a) and 'grafting from' (b) techniques for fabricating surface-tethered macromolecules.

2.2.3 Stimuli-Responsive Brushes

The group of brushes responsive to different stimuli is of special interest due to promising bio-related applications [20]. So-called 'smart' or 'intelligent' brushes respond highly nonlinearly to a stimulus or multiple stimuli (pH, temperature, ionic strength, solvent quality, etc.) by cooperative conformational change of the macromolecules [21]. While responsiveness of the tethered chains in the 'mushroom' regime is similar to that of the free chains, the response of the brushes is different (usually enhanced) due to constrains of the densely packed chains [22].

Stimuli-responsive brushes sensitive to temperature changes have been widely synthesized and their behavior studied both experimentally and theoretically. 'Smart' poly(N-isopropylacrylamide) (PNIPAM) brush is the most common example, while other (co)polymer brushes were also found to undergo large conformational changes responding to small variations of temperature around the critical one. This type of brush is described below in more details.

Special group of stimuli-responsive brushes are polyelectrolyte brushes that may be sensitive to changes of ionic strength and pH [23, 20]. Brushes consisting of weak polyelectrolytes, undergo reversible expansion and shrinking at pH around their pK values. Strong polyelectrolytes are insensitive to changes of pH, but in a given regime the brushes respond to changes of ionic strength. Polymer brushes containing photoisomerizable groups, e.g. azobenzene, were synthesized and their behavior studied [24]. Solvent quality is also a stimulus that may drive reorganization of the tethered chains especially of block amphiphilic copolymers [25]. Macromolecules containing redox active groups assembled in brushes were found to react cooperatively in response to reduction and oxidation processes [26].

2.2.4
Engineering of Surfaces with Stimuli-Responsive Polymer Brushes

The chemical structuring of stimuli-responsive polymer brushes via copolymerization, incorporation of nano-objects or simple chemical modification represent a versatile tool for synthesizing multifunctional tunable surfaces that can act as biological platforms.

Great interest has been vested in using PNIPAM as prototype 'smart' polymer, as it exhibits a lower critical solution temperature (LCST) in aqueous environments. Upon passing the LCST by heating, PNIPAM precipitates at physiologically relevant temperatures (ca. 32 °C), which forms the basis for responsive applications.

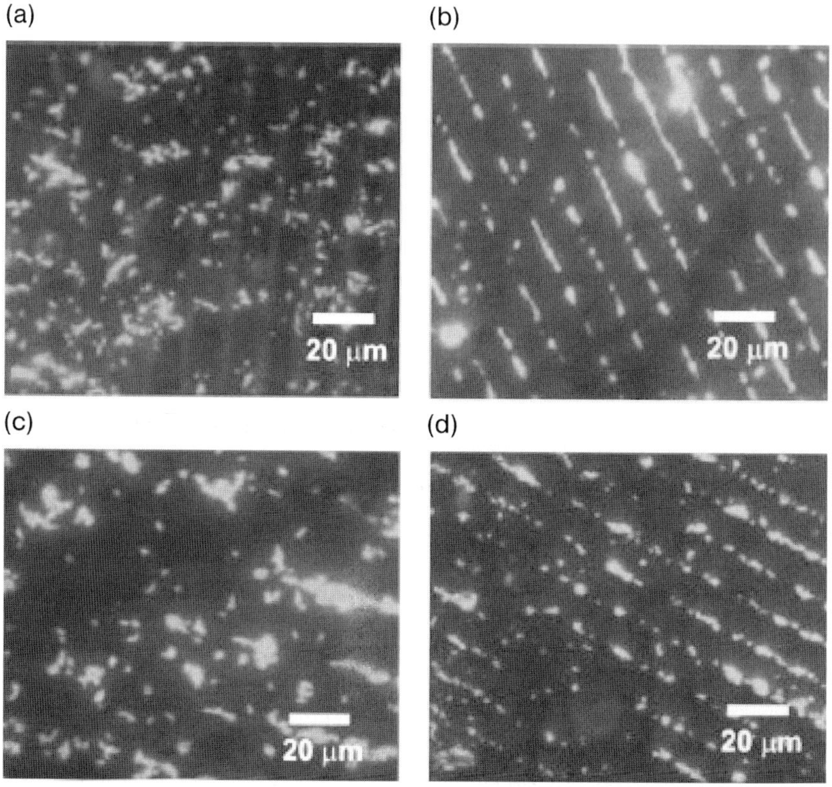

Figure 3 Fluorescence microscopy images of micropatterned PNIPAM brushes showing two cycles of incubation in suspensions of *Streptococcus mutans* at 37 °C (a and c) followed by incubation at 4 °C (b and d). The hydrophobic–hydrophilic transition from above to below the LCST of PNIPAM caused the selective adhesion of the bacteria in correspondence of the interpattern regions. (With permission from [35]).

The use of PNIPAM in the form of a brush layer has been studied by several research groups, and it has been employed to modify nanoparticles and flat substrates of different nature [27–33]. A fundamental study, in this sense, was done by Huck *et al.* [34] who successfully grafted PNIPAM brushes from self-assembled monolayers of initiators on gold using ATRP in aqueous medium. This controlled polymerization technique, developed by Matyjaszewski, was shown to be effective, although not fully controlled in such a polar medium, for the grafting of up to 100-nm thick brush layers. These thermoresponsive surfaces were subsequently employed for the synthesis of bioplatforms that can regulate the adhesion of bacteria in function of the temperature of the medium in which they are immerged [35] (Figure 3).

In addition to ATRP other polymerization techniques have been proved to be suitable synthetic routes for the grafting of PNIPAM brushes. Matsuda *et al.* explored an effective photopolymerization method based on photoiniferters immobilized on glass surfaces [36]. These initiators bear dithiocarbamate moieties that upon UV irradiation form reactive carbon radicals paired with less-reactive sulfur-centered radical species that can act as capping agents [37–39]. The existence of these reversible chain terminators was shown to control the growing of the grafted chains and, furthermore, allowed the chemical modification of the chain ends by tailoring of the functionalities bounded to the dithiocarbamates moieties. As a demonstration of this notion, in the latter work, an albumin-modified iniferter was immobilized on the glass surface in the form of a trichlorosilane adsorbate. After photoirradiation in the presence of NIPAM, polymer brushes bearing albumin at the chain ends were synthesized. The polymeric brushes thus synthesized were shown to have potential applications as temperature-induced switchable fouling–antifouling surfaces towards the adsorption of proteins which selectively bind albumin.

The use of iniferter-based polymerization for the synthesis of thermosensitive polymer brushes was recently proved as an effective technique to also modify metallic surfaces like gold [40]. The choice of such substrates has several advantages as compared to the already reported photoiniferter-based grafting from silicon/glass surfaces [13]. Initiators in the form of thiols or disulfides forming SAMs on gold surfaces are well ordered, their compositions may be easily varied, the deposition procedure is compatible with soft-lithographic techniques and AFM-tip-assisted (atomic force microscopy) nanolithographic methods. Moreover, this chemistry is not sensitive to humidity variations [41, 42].

In the work by Vancso *et al.* [40] a disulfide-containing photoiniferter was shown to be easily deposited on gold using a PDMS stamp, thus forming regular patterns of SAMs of initiators. After just 5 min of irradiation with UV light in the presence of aqueous solution of the NIPAM monomer, brush layers of 10 ± 2 nm in height were obtained (Figure 4a).

In order to confirm the presence of diethyldithiocarbamil groups at the chain ends that can reinitiate the polymerization upon further irradiation, a chain expansion test was performed. Following the reported procedure, after interruption of the polymerization, the samples were placed back in the monomer solu-

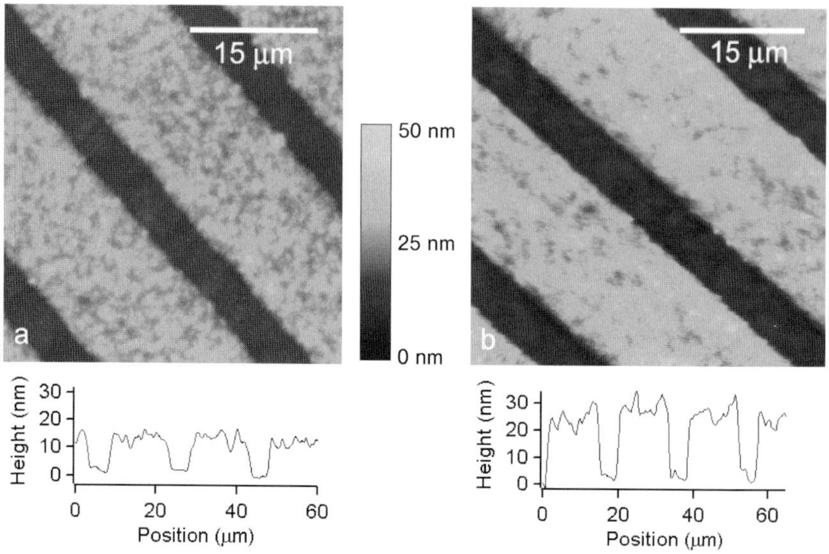

Figure 4 Height images from AFM measurements in tapping mode of the patterned surface with the grafted PNIPAAM chains (in the dry state) after (a) 5 min and (b) an additional 5 min of irradiation. The respective cross-sections are displayed below each image. (With permission from [40]).

tion and a further UV-irradiation cycle was performed. As can be seen in Figure 4(b), the height of the brushes doubled after the second irradiation step, reaching values of 20 ± 3 nm and confirming their activity as tethered macroinitiators.

As it has been already mentioned, the iniferter-based technique allows the modification of the chain ends via tuning the chemistry of the carbamate species forming the starting initiator layer. Alternatively, the chemical modification of the brush ending groups can be achieved by radical exchange with, e.g., stable 2,2,6,6-tetramethylpiperidyl-1-oxyl (TEMPO) radicals [43]. In a simple experiment, the tethered polymer chains were irradiated in the presence of 4-amino-TEMPO radicals that irreversibly couple at room temperature with the *in-situ* formed macroradicals. The effectiveness of the radical exchange method can be confirmed by labeling the amine moieties present in the newly attached end-groups with a fluorophore (Figure 5). The contrast shown in the fluorescent microscope image in Figure 5(b) comes from the fluorescamine adducts formed at the chain ends of the polymers grafted from the μCP patterns.

The synthetic routes presented here enabled efficient fabrication of tunable, temperature-responsive polymeric layers based on PNIPAM brushes. The development of SIP techniques featuring dormant initiators that turn into active radicals responding to a physical (UV) or a chemical trigger, allowed full control over the thickness of the so-formed brush layers. Furthermore, the chemistry involved in these processes allowed one to functionalize the tethered macromolecules via simple synthetic steps.

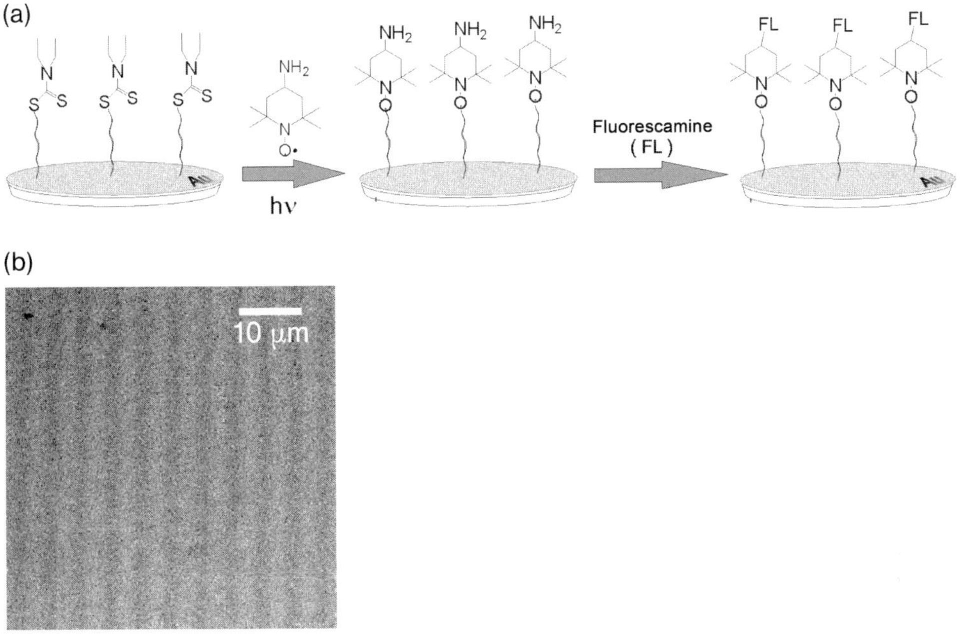

Figure 5 (a) Scheme of the exchange procedure of the PNIPAM polymer brush end groups with the stable 4-amino-TEMPO radicals and the subsequent fluorescent labeling of the amine groups with fluorescamine. (b) A fluorescent microscope image of the patterned PNIPAM brushes modified according to the mentioned procedure. (With permission from ref. [40]).

Polyelectrolytes are another important class of macromolecules that form brushes with very interesting smart properties. Weak polyelectrolytes can vary the charge density along their backbone by varying the pH of the solution they are dissolved in. The latter property makes weak polyelectrolytes pH-responsive materials since variations in the electrostatic interactions can cause a variation of the hydrodynamic volume of the polymer. The physical responses of pH-responsive brushes have been exhaustively described in several fundamental papers [44–48]. Here, we will rather concentrate on the manufacturing of functional materials exploiting the intrinsic properties of the platforms based on polyelectrolyte brushes.

An example is given by polymer films constituted by multistimuli-responsive triblock copolymer brushes bearing both cationic and anionic groups along the polymer chain, which were fabricated using sequential aqueous ATRP in the group of Huck [49]. These model platforms incorporate blocks that respond to changes in pH and ionic strength (polymethacrylic acid) together with other segments that are exclusively sensitive to ionic strength (poly-[2-(methacryloyloxy)ethyl]trimethylammonium chloride) and a neutral block of poly(methyl methacrylate). The obtained surfaces possessed well-defined compositional characteristics, nanostructured in the vertical direction from the underlying substrate to the interface.

The last model brings up an appealing feature typical of brush layers made of charged polymers: the possibility to incorporate counter ions of different nature via adsorption from solution. Through easy procedures, salts containing metallic ions can be incorporated inside the polymeric matrixes and later act as precursors for the synthesis of metal nanoparticles embedded in the polymer brushes. In the first reported attempt, silver nanoparticles were successfully synthesized inside poly(acrylic acid) brushes by first adsorbing Ag^+ ions from aqueous solution followed by reduction of them to metallic silver using hydrogen at elevated temperatures [50]. In order to avoid the use of hydrogen at elevated pressures and temperatures, a recent report demonstrated the efficient synthesis of gold nanoparticles from $AuCl_4^-$ ion inside a polycation brush using $NaBH_4$ solutions as a reducing agent [51].

Another original example of using polyelectrolyte brushes as templates for the synthesis of functional surfaces was recently reported by Klok *et al.* [52]. They utilized poly(methacrylic acid) (PMAA) brushes in order to mediate the crystallization of thin calcite films according to the step-by-step fabrication methodology reported in Figure 6. First, micropatterned PMAA brushes were synthesized on glass substrates, then they were exposed to supersaturated solutions of $CaCO_3$ and finally, by temperature-induced crystallization, regular structures made of polycrystalline calcite films were formed following closely the shape of the precursor brush layers (Figure 5). This synthetic procedure showed how functional platforms based on tethered macromolecules can mimic natural environments directing the mineralization of biomaterials.

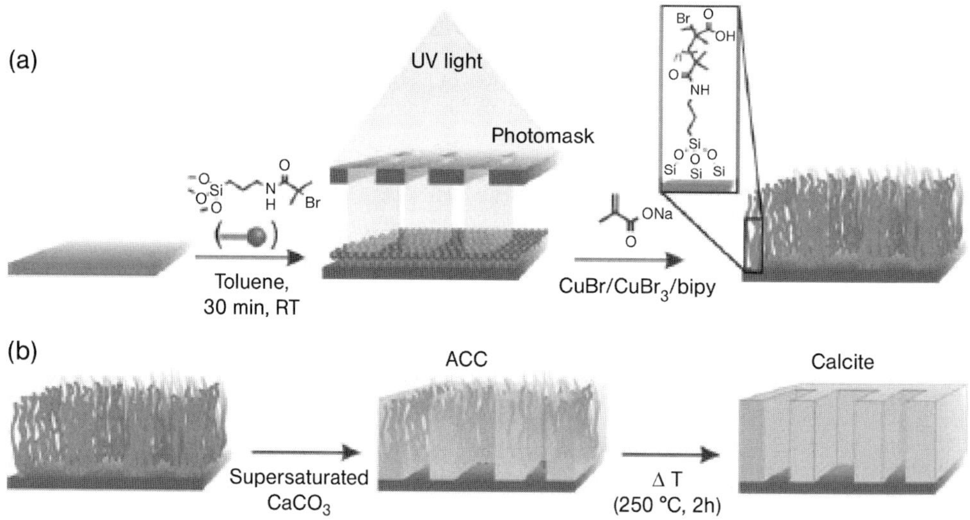

Figure 6 Stepwise fabrication of microstructured calcite films as reported in [52]. (a) Grafting of PMAA brushes via surface-initiated ATRP from photopatterned SAMs. (b) Temperature-induced formation of polycrystalline calcite film using the PMAA brushes as templates. (With permission from [52]).

As demonstrated above, the versatility of chemistry and possible morphology of polymer brush systems justify their usage as starting materials for nanotechnology. The broad range of different features and properties typical of these platforms makes them an extraordinary tool for the engineering of surfaces suitable also for biomedical applications.

2.2.5
Polymer Brushes Across the Length Scales: A Tool for Functional Nanomaterials

'Grafting-from' approaches that combine microcontact printing of initiators and subsequent controlled polymerization to obtain patterned stimulus responsive polymer brushes have been reported by several research groups [34, 40] However, these approaches still do not enable full control of surface-tethered macromolecules on the level of single chains to be obtained. Control of the overall dimensions and the exact position of single polymer graft chains on multiple length scales is an issue of central importance in macromolecular nanotechnology.

Some proteins, as examples of biopolymers, were shown to be easily immobilized on substrates by AFM-assisted lithographic methods forming isolated macromolecular nanostructures. These biomolecules can thus play a role as starting blocks for the synthesis of complex macromolecular architectures that can function as selective biorecognition sites precisely positioned on a flat surface in a single molecular fashion [53–58]

The nanomodifications of the surfaces of synthetic polymers were shown to represent an even more fascinating field of research. Also in this case, AFM-based lithographic techniques were shown to play a fundamental role for the deposition of macromolecules in a 'grafting-to' manner [59, 60], or for the promotion of the direct growth of polymers in locally defined areas following a 'grafting from' approach [61–64].

The combination of tip-assisted nanolithography to position initiator molecules on surfaces and SIP methods would allow one to form polymeric nanostructures keeping the control over the properties and the morphology of the macromolecular components. Moreover, this approach would represent a first step towards the controlled growth of single macromolecules from prefunctionalized sites.

A classical dip-pen nanolithography (DPN) experiment demonstrated the capability to deliver initiator molecules on a surface in predetermined patterns from which localized polymer brushes could be grafted [63]. In other reports, the AFM tip was used as a 'destructive' tool to locally damage already formed SAMs on silicon or gold surfaces forming exposed areas of the bare substrate that can be subsequently filled up via deposition from solution of functional adsorbates [65]. This approach, named as 'nanoshaving', was successfully used to deliver thiol-functionalized initiators for ATRP on the gold surface, forming regular patterns from scratches drawn on a pre-existing organic monolayer [64]. Using subsequent SIP, temperature-responsive polymer brushes were grown forming line patterns with typical widths ranging from 300 to 500 nm.

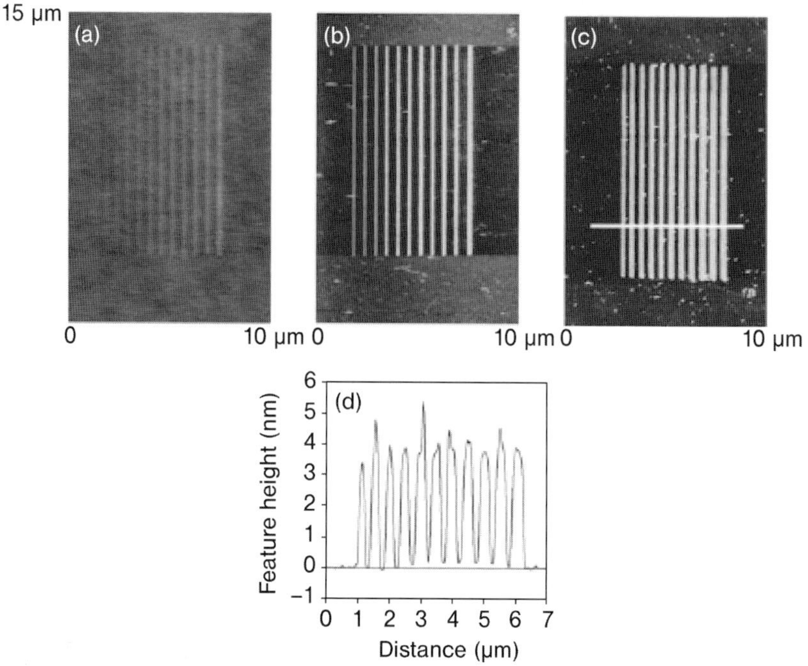

Figure 7 AFM micrographs showing: (a) contact mode height image of silicon oxide patterns generated by anodization lithography; (b) lateral force micrographs emphasizing silicon oxide features; (c) tapping mode height image depicting poly(cyclooctatetraene) brushes grafted from functionalized SiO_2 patterns after 12 h of polymerization; (d) cross-section of the image (c) along the drawn line. (With permission from [62]).

Local modification of functionalized surfaces can also be achieved by applying a potential to the AFM tip positioned in close proximity to the surface [66, 67]. Zauscher *et al.* [62] used this anodization potential lithography technique to form silicon oxide nanofeatures via local oxidation of an organic SAM on silicon substrate (Figure 7). Siliconoxide patterns could act as platforms towards the selective chemisorption of Ru-based metathesis catalysts through the reaction of organosilanes linkers. After the subsequent ring-opening polymerization in the presence of cyclooctatetraene in solution or 5-ethylidene-norborene in vapor phase, polyconjugated brushes were successfully grafted forming regular patterns reaching sub-100-nm widths.

The synthesis of a nanostructured platform based on gold nanowires on silicon substrates was recently reported as an alternative highly selective system towards the adsorption of initiators for the grafting of functional polymer brushes [68]. In the first step, gold nanowires were deposited on hydride-terminated silicon followed by the selective immobilization of disulfide-based adsorbates bearing iniferter groups. Finally, the UV-initiated grafting of poly(methacrylic acid) was performed using the functionalized nanowires as platforms (Figure 8).

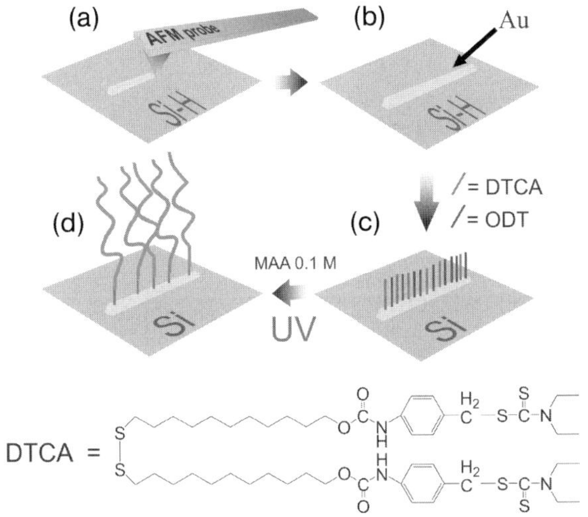

Figure 8 Preparation of polymer brushes grafted from immobilized precursors on gold nanowires. (a) and (b) tip-assisted deposition of gold nanowires on hydride-terminated silicon; (c) selective immobilization of functional adsorbates on the gold structures; (d) UV-initiated grafting of PMAA brushes using the functionalized nanowires as platforms. (With permission from [68]).

In this work, the height and width of the polymer brush nanostructures obtained were reported to be easily controlled by tuning the size of the underlying nanowires. These were deposited on silicon by DPN, using the AFM tip to deliver a precursor ion ($AuCl_4^-$) that is locally reduced to metallic gold by the previously hydrogenated silicon surface [69]. Varying the environmental humidity during the deposition experiments, Au lines with average thicknesses varying from 240 nm to 20 nm were formed (Figure 9a and c).

The subsequently polymer brush grafted from the wider gold lines reached a height value of 5 ± 1 nm and the corresponding width showed an average increase by 8% reaching 260 ± 10 nm. Surprisingly, in the case of the thinnest, 20 nm wide, functionalized gold wires the height observed after the same polymerization time did not show a remarkable change. At the same time the width of the grafted brush was doubled with respect to the precursor Au lines, reaching an average value of 42 ± 4 nm (widening by 110% with respect to the starting gold wires).

The reason why from very thin patterns the subsequently grown polymer chains do not extend markedly in the vertical direction can be explained if one takes into account the clear widening of the features after polymerization and caused by extension of the polymer chains beyond the edges of the grafting region. This effect was recently theoretically described by Patra *et al.* [70] who studied the conformation of polymer brushes grafted from patterns with variable width. The relevant widening of the patterns after the polymer grafting is a consequence of the release of osmotic pressure inside the brush system. Such relative widening effect, due to chain flattening, is obviously much more pronounced with decreasing pattern width.

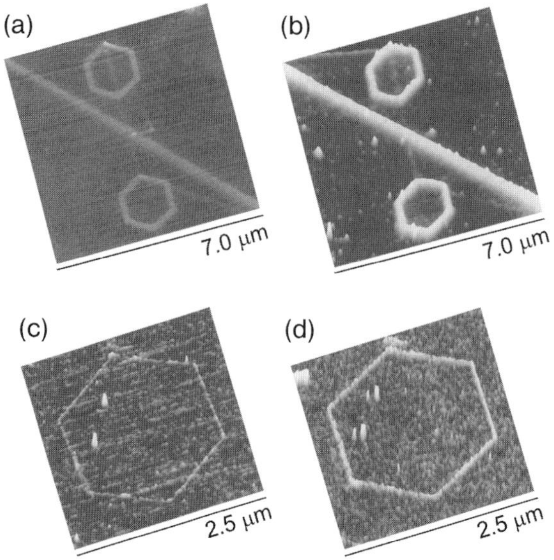

Figure 9 Height images (vertical scale from black to white 2 nm) from AFM tapping mode measurements of 240 nm (a) and 20 nm (c) wide gold nanowires deposited on hydride-terminated silicon and the subsequently grafted PMAA brushes (b) and (d), respectively, from 240 and 20 nm wide patterns. (With permission from [68]).

The examples presented here, can give an idea of how the combination of DPN and SIP allows the nanoengineering of various surfaces using tethered polymers. These newly developed stepwise 'bottom-up' approaches enabled one to synthesize brush-based platforms ranging from hundreds of nanometers down to sizes comparable with only a few arrays of grafted macromolecules. Thus, such platforms represent promising models for the controlled growth of isolated functional macromolecules, opening new horizons in surface chemistry and biology.

2.2.6
Polymer Brushes and Cell Adhesion

It is well known that cell response is affected by the physicochemical parameters of the biomaterial surfaces [71]. Thus, tailoring surface characteristics such as surface energy, surface charges, topography and chemical composition among others may modulate cell behavior and provide adequate signaling for the activation of specific cellular events. The chemical approach and specifically, functionalization of surfaces with biomolecules is one of the most promising alternatives to provide chemical signals that trigger and stimulate biological cascades at the molecular level.

Most eukaryotic cells rely on a natural scaffold that consists of a wide variety of biomolecules and is known as extracellular matrix (ECM). The ECM is a combination of specific proteins that are secreted by cells to form a complex network. This network is a supporting framework where cells grow and differentiate giving rise and shape to new tissues. The ECM provides the environment within which migratory cells can move and interact between them and immobile cells are anchored. Interactions between cells and the extracellular matrix are crucial to initiate and mediate biological events that control cell adhesion, growth, migration, differentiation, survival, tissue organization, and matrix remodeling among others. Therefore, the development of interfaces that mimic the natural ECM and its interaction with cells are of great interest for tissue engineering and regenerative medicine [72].

The cell-adhesion process is one of the most important cellular events since it plays a main role in cells development and behavior. It consists of a cascade of four steps: (i) cell attachment, (ii) cell spreading, (iii) cell organization of cytoskeleton through actin filaments and (iv) formation of focal contacts [73]. The process is mediated by protein–cell interactions based on the biological recognition of peptide sequences by cell-surface receptors known as integrins. These are transmembrane proteins that possess two domains (α and β) that contain units for the specific recognition of cell-adhesive peptide motifs such as RGD (arginine-glycine-aspartic acid). This sequence is found in many ECM proteins and enhances cell attachment.

Some of the most commonly employed strategies to mimic cells/ECM adhesion interactions rely on the functionalization of biomaterial surfaces using proteins and peptides present in the ECM, in particular, the RGD sequence that has been shown to be the most effective for stimulation of cell adhesion on synthetic materials [73]. Within this context, several techniques including proteins and peptides immobilization by physisorption methods or via SAMs have been exploited, among them the use of polymer brushes has risen as an interesting approach to engineer and modify the surface of synthetic materials [7].

Polymer brushes offer valuable advantages in the development of engineered biointerfaces: they provide multiple binding sites for biomolecules [74], their flexible nature allows reorganization of the chemical functionalities upon adhesion of cells [75], their thickness and grafting density can be readily tuned and, they can be easily synthesized on a variety of substrates [9].

Functional polymer brushes have been used as platforms for the immobilization of cell-adhesion peptides [75–77]. In general, nonadherent polymer chains are tethered to the surface and subsequently functionalized with cell-adhesive ligands. These modified surfaces are intended to be used in cell/surface interactions studies that are usually carried out under *in-vitro* conditions. Once functionalized surfaces are in contact with cell-culture medium mimicking the physiological environment, the mixture of proteins present in the medium tends to absorb on the surface, hindering the effect of the grafted ligands. Polymers such as poly(ethylene glycol) (PEG), poly(methacrylic acid)(PMAA), poly(2-hydroxyethyl methacrylate)(PHEMA) and poly(poly(ethylene glycol)methacrylate)(PPEGMA) have been used as antifouling surfaces to reduce nonspecific protein adhesion. Surfaces functionalized with these nonadherent polymer brushes act as an inert platform where specific peptide sequences can be introduced to enhance cell adhesion and study the specific integrin–surface interactions of different cell types [78, 79].

There are several parameters such as grafting density/thickness of the polymer chains, density of the RGD motifs, type of the polymer chains regime, polymer used and RGD vertical position within the brushes layer, that affect the cell-adhesion process on RGD-functionalized polymer brushes.

Polymer grafting density seems to be one of the most relevant aspects to avoid protein adsorption [80]. It has also been shown that there is a grafting density threshold that depends on the polymer used, above this critical density the resistance to protein adsorption decreases [81, 13]. Furthermore, grafting density affects the conformation of the tethered chains. It has been shown that the ability of proteins to absorb on the modified surface depends on the configuration acquired by the grafted chains. Mei *et al.* [82] reported that PHEMA layers in the mushroom regime were able to absorb proteins and therefore cells could adhere to them, whereas the brush regime (higher grafting density) avoided protein and cell adhesion. Another example of the effect of chains configuration has been reported by Singh *et al.* [76]. They studied the effect of grafting density (independently from layer thickness) on MC3T3 mouse fibroblast cells for PPEGMA nanolayers and reported that RGD peptide absorption and cell attachment took place mostly in the mushroom regime rather than in the brushes one.

Moreover, the concentration of RGD units may also be modulated by means of a controlled grafting density. It is expected that a high density of polymer chains leads to a higher number of available binding sites and therefore higher concentrations of peptide units on the surface may be achieved. Harris and coworkers [83] have prepared PMMA brushes gradients varying grafting density, chains length and peptide concentration and evaluated MC3T3 fibroblasts adhesion across these gradients. According to their study, cell density increases with RGD density. However, there is a critical concentration above which cell adhesion is not improved. Other studies involving GGG*RGD*S (glycine-glycine-glycine-arginine-aspartic acid-glycine-serine) functionalized brushes made of nonadherent polymers such as PPEGMA and PHEMA has been evaluated using HUVEC endothelial cells, and corroborate that there is indeed a critical concentration of the adhesive sequence for the improvement of cell adhesion. They also found that the critical threshold depends on the polymer used [75]. Besides RGD density, the lateral spacing of these cell adhesive units also influences cell spreading and the formation of focal adhesion points [84].

The last step of the cell-adhesion process is the formation of focal adhesions that are the connecting points between cell membrane and the ECM. Focal adhesions comprise of the formation of integrin clusters and more than 50 other transmembrane, membrane-associated and other cytosolic molecules [73]. The flexibility and mobility of peptide sequences coupled to polymer brushes promote the formation of such clusters. Depending on the polymer structure, the flexibility and rigidity of the tethered chains vary, leading to differences in the size and morphology of cell focal points. A study carried out by Tugulu *et al.* [75] showed that depending on the polymer used (PPEGMA or PHEMA), large mature focal adhesions at the cell periphery or fibrillar adhesions toward the nucleus of the cell are formed. This is a clear example of the effect of polymer structure on the flexibility of the brushes and their influence on cell behavior.

Another key aspect that influences the RGD/integrins interaction is the vertical position of the adhesive ligand within the polymer layer. In order to mimic ECM conditions, the RGD sequence should be positioned in such a way that resembles the natural structure of extracellular matrix proteins. Thus, the RGD unit must be separated from the synthetic surface to be reached by the recognition site of integrins. In fact, it has been shown that certain spacing between the RGD motif and the surface is crucial for cell attachment [85]. In this sense, methods allowing a controlled length tuning of the tethered chains play an important role. In addition to the spacing between the anchoring moieties and the RGD peptide, there seems to be a certain distance on the surface of the polymer layer, which is available for interactions with the cell-membrane receptors [86].

Recently, Navarro *et al.* [77] have studied the effect of RGD vertical position within PMAA brushes. They synthesized bioadhesive platforms using the iniferter-mediated polymerization technique, which enables maintaining active chemical functionalities at the chain ends. Thus, it is possible to reinitiate the polymerization mechanism in subsequent stages. Given the versatility of this technique, it was used for an indepth chemical structuring of the polymer chains. PMAA polymer brushes were grown, functionalized with the RGD sequence and extended with a second layer of brushes, the procedure used to develop the functionalized surface is illustrated in Figure 10.

MG63 osteoblastic cells were used to assess the effect of the position of the adhesive motifs (on the surface of the brushes or buried within the polymer layer) using immunofluorescence methods to evaluate cell spreading and the formation of focal adhesions. Immunofluorescence results showed remarkable differences with respect to cell morphology. Cells spread well with marked focal adhesion points at the periphery of the cytoplasm on samples with RGD motifs coupled on

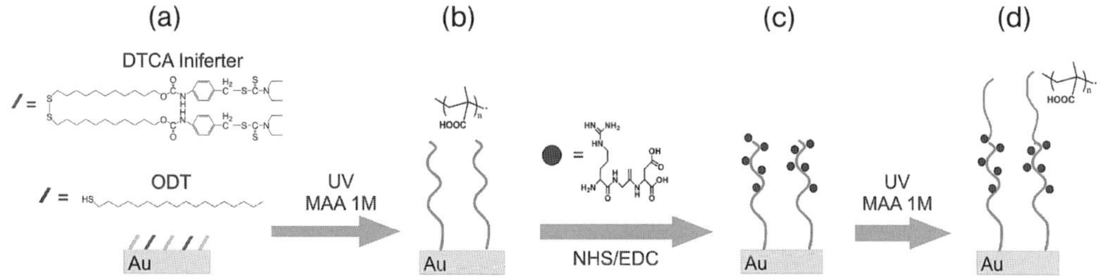

Figure 10 Preparation of RGD-modified polymer brushes grafted from immobilized precursors on gold by photopolymerization: (a) and (b) photografting of PMAA brushes from immobilized photoiniferter DTCA (dithiodiundecane-11,1diylbis[4({[diethylamino)carbonothioyl]thioethyl)phenyl]carbamate}; (c) immobilization of RGD peptides; (d) chain extension via photografting of a top PMAA brush layer. (With permission from [77]).

the surface, whereas in the case of the samples where RGD was buried, cells were found to adopt a rounded morphology and focal adhesions concentrated toward the internal part of the cell (Figure 11).

Figure 12 describes the cell–surface interaction in both cases, RGD exposed or buried within the PMAA brushes. In the former case, RGD units are highly concentrated at the exposed surface of the layer and the cells can access them easily. In the latter, when RGD is not directly accessible but buried beneath a thin but uniform layer of nonfunctionalized PMAA brushes, the cells cytoskeleton acquired a different configuration showing a more rounded morphology and different focal adhesions.

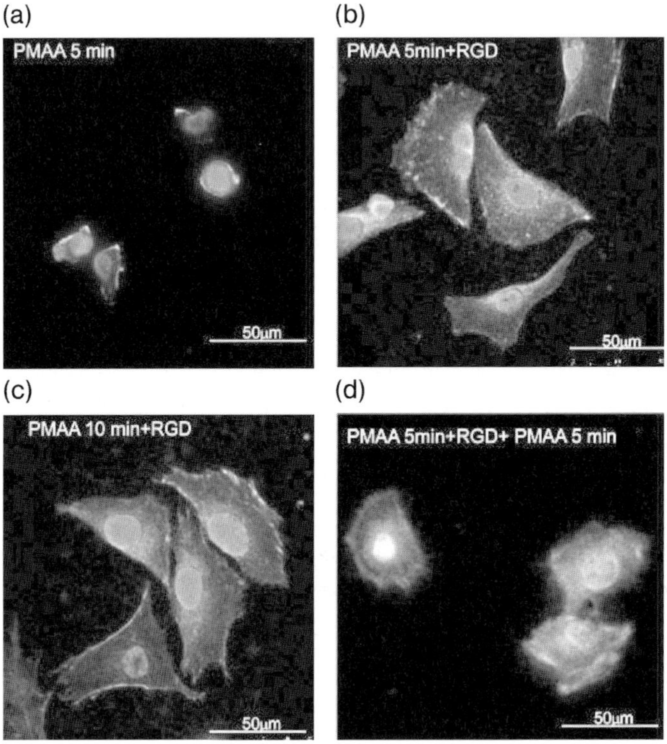

Figure 11 Immunofluorescence images of MG63 cells on the studied surfaces: (a) PMAA, (b and c) PMAA-RGD, (d) PMAA-RGD-PMAA after 6 h of contact. (With permission from [77]).

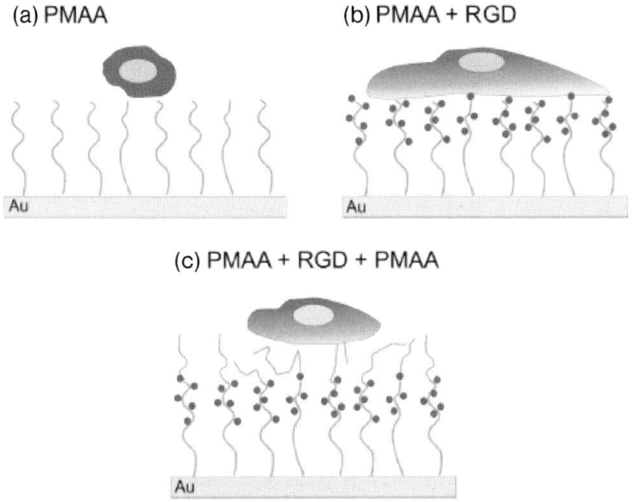

Figure 12 A schematic representing the morphological changes of MG-63 cells on a PMAA brush without RGD (a) and with RGD at different depths in the brush layer (b and c). (With permission from [77]).

Thus, polymer brushes versatility makes them an attractive approach to engineer surfaces of synthetic materials. Tethered polymer chains represent a valuable tool for the development of well-defined biochemically structured platforms for the study of specific cellular events such as cell adhesion.

Acknowledgments

The Commission of the European Union (Marie Curie RTN Contract Number MRTN-CT-2004-005516 BioPolySurf) and the Catalan Government are acknowledged for financial support.

References

1 Advincula, R.C., Brittain, W.J., Caster, K.C. and Rühe, J. (eds), (2004) *Polymer Brushes: Synthesis, Characterization, Applications*, Wiley-VCH, Weinheim.
2 Milner, S.T. (1991) *Science*, **251**, 905.
3 Jusufi, A., Likos, C.N. and Ballauff, M. (2004) *Colloid and Polymer Science*, **282**, 910.
4 Raviv, U., Giasson, S., Kampf, N., Gohy, J.-F., Jérôme, R. and Klein, J. (2003) *Nature*, **425**, 163.
5 Alarcon, C.D.H., Farhan, T., Osborne, V.L. and Huck, W.T.S. (2005) *Journal of Materials Chemistry*, **15**, 2089.
6 Tugulu, S., Arnold, A., Sielaff, I., Johnsson, K. and Klok, H.-A. (2005) *Biomacromolecules*, **6**, 1602.

7 Senaratne, W., Andruzzi, L. and Ober, C.K. (2005) *Biomacromolecules*, **6**, 2427.
8 Mansky, P., Liu, Y., Huang, E., Russell, T.P. and Hawker, C. (1997) *Science*, **275**, 1458.
9 Zhao, B. and Brittain, W.J. (2000) *Progress in Polymer Science*, **25**, 677.
10 Ito, Y., Ochiai, Y., Park, Y.S. and Imanishi, Y. (1997) *Journal of the American Chemical Society*, **119**, 1619.
11 Prucker, O. and Rühe, J. (1998) *Macromolecules*, **31**, 602.
12 Prucker, O. and Rühe, J. (1998) *Macromolecules*, **31**, 592.
13 De Boer, B., Simon, H.K., Werts, M.P.L., van Der Vegte, E.W. and Hadziioannou, G. (2000) *Macromolecules*, **33**, 349.
14 Tsujii, Z., Ohno, K., Yamamoto, S., Goto, A. and Fukuda, T. (2006) *Advances in Polymer Science*, **197**, 1.
15 Zhao, B. and Brittain, W.J. (1999) *Journal of the American Chemical Society*, **121**, 3557.
16 Jordan, R., Ulman, A., Kang, J.F., Rafailovich, M.H. and Sokolov, J. (1999) *Journal of the American Chemical Society*, **121**, 1016.
17 Jordan, R. and Ulman, A. (1998) *Journal of the American Chemical Society*, **120**, 243.
18 Luzinov, I., Minko, S. and Tsukruk, V.V. (2004) *Progress in Polymer Science*, **29**, 635.
19 Edmondson, S., Osborne, V.L. and Huck, W.T.S. (2004) *Chemical Society Reviews*, **33**, 14.
20 Zhou, F. and Huck, W.T.S. (2006) *Physical Chemistry Chemical Physics*, **8**, 3815.
21 Minko, S. (2006) *Journal of Macromolecular Science: Part C: Polymer Reviews*, **46**, 397.
22 de Gennes, P.G. (1980) *Macromolecules*, **13**, 1069; Alexander, S. (1997) *Journal of Physiology, Paris*, **38**, 983.
23 Rühe, J., Ballauff, M., Biesalski, M., Dziezok, P., Gröhn, F., Johannsmann, D., Houbenov, N., Hugenberg, N., Konradi, R., Minko, S., Motornov, M., Netz, R.R., Schmidt, M., Seidel, C., Stamm, M., Stephan, T., Usov, D. and Zhang, H. (2004) *Advances in Polymer Science*, **165**, 79.
24 Uekusa, T., Nagano, S. and Seki, T. (2007) *Langmuir*, **23**, 4642.
25 Granville, A.M., Boyes, S.G., Akgun, B., Foster, M.D. and Brittain, W.J. (2004) *Macromolecules*, **37**, 2790.
26 Ito, Y., Nishi, S., Park, Y.S. and Imanishi, Y. (1997) *Macromolecules*, **30**, 5856.
27 Lista, L.K., Mendez, S., Perez-Luna, V.H. and Lopez, G.P. (2001) *Langmuir*, **17**, 2552.
28 Huber, D.L., Manginell, R.P., Samara, M.A., Kim, B.-I. and Bunker, B.C. (2003) *Science*, **301**, 352.
29 Shan, J., Chen, J., Noupponen, M. and Tenhu, H. (2004) *Langmuir*, **20**, 4671.
30 Tu, H., Heitzman, C.E. and Braun, P.V. (2004) *Langmuir*, **20**, 8313.
31 Wang, X., Tu, H., Braun, P.V. and Bohn, P.W. (2006) *Langmuir*, **22**, 817.
32 He, Q., Kuller, A., Grunze, M. and Li, J.B. (2007) *Langmuir*, **23**, 3981.
33 Li, D., Cui, Y., Wang, K., He, Q., Yan, X. and Li, J.B. (2007) *Advanced Functional Materials*, **17**, 3134.
34 Jones, D.M., Smith, J.R., Huck, W.T.S. and Alexander, C. (2002) *Advanced Materials*, **14**, 1130.
35 Alarcon, C.D.H., Farhan, T., Osborne, V.L. and Huck, W.T.S. (2005) *Journal of Materials Chemistry*, **15**, 2089.
36 Matsuda, T. and Ohya, S. (2005) *Langmuir*, **21**, 9660.
37 Otsu, T. and Yoshida, M. (1982) *Makromolekulare Chemie. Rapid Communications*, **3**, 127.
38 Otsu, T., Yoshida, M. and Tazaki, T. (1982) *Makromolekulare Chemie. Rapid Communications*, **3**, 133.
39 Otsu, T. and Matsumoto, A. (1998) *Advances in Polymer Science*, **136**, 75.
40 Benetti, E.M., Zapotoczny, S. and Vancso, G.J. (2007) *Advanced Materials*, **19**, 268.
41 Bain, C.D., Evall, J. and Whitesides, G.M. (1989) *Journal of the American Chemical Society*, **111**, 7155.
42 Love, J.C., Estroff, L.A., Kriebel, J.K., Nuzzo, R.G. and Whitesides, G.M. (2005) *Chemical Reviews*, **105**, 1103.
43 Beckwith, A.L.J., Bowry, V.W. and Ingold, K.U. (1992) *Journal of the American Chemical Society*, **114**, 4983.

44 Pincus, P. (1991) *Macromolecules*, **24**, 2912.

45 Israels, R., Leermakers, F.A.M. and Fleer, G.J. (1994) *Macromolecules*, **27**, 3087.

46 Zhulina, E.B., Birshtein, T.M. and Borisov, O.V. (1995) *Macromolecules*, **28**, 1491.

47 Bielanski, M., Johannsmann, D. and Rühe, J. (2002) *Journal of Chemical Physics*, **117**, 4988.

48 Zhang, H. and Rühe, J. (2005) *Macromolecules*, **38**, 4855.

49 Osborne, V.L., Jones, D.M. and Huck, W.T.S. (2002) *Chemical Communications*, **17**, 1838.

50 Boyes, S.G., Akgun, B., Brittain, W.J. and Foster, M.D. (2003) *Macromolecules*, **36**, 9539.

51 Azzaroni, O., Brown, A.A., Cheng, N., Wei, A., Jonas, A.M. and Huck, W.T.S. (2007) *Journal of Materials Chemistry*, **17**, 3433.

52 Tugulu, S., Harms, M., Fricke, M., Volkmer, D. and Klok, H.-A. (2007) *Angewandte Chemie – International Edition*, **45**, 7458.

53 Wadu-Mesthrige, K., Xu, S., Amro, N.A. and Liu, G.-Y. (1999) *Langmuir*, **15**, 8580.

54 Kenseth, J.R., Harnisch, J.A., Jones, V.W. and Porter, M.D. (2001) *Langmuir*, **17**, 4105.

55 Liu, G.Y. and Amro, N.A. (2002) *Proceedings of the National Academy of Sciences of the United States of America*, **99**, 5165.

56 Hyun, J., Ahn, S.J., Lee, W.K., Chilkoti, A. and Zauscher, S. (2002) *Nano Letters*, **2**, 1203.

57 Lee, K.-B., Park, S.-J., Mirkin, C.A., Smith, J.C. and Mrksich, M. (2002) *Science*, **295**, 1702.

58 Case, M.A., McLendon, G.L., Hu, Y., Vanderlick, T.K. and Scoles, G. (2003) *Nano Letters*, **3**, 425.

59 Yu, M., Nyamjav, D. and Ivanisevic, A. (2005) *Journal of Materials Chemistry*, **15**, 649.

60 Lee, S.W., Sanedrin, R.G., Oh, B.-K. and Mirkin, C.A. (2005) *Advanced Materials*, **17**, 2749.

61 Woodson, M. and Liu, J. (2006) *Journal of the American Chemical Society*, **128**, 3760.

62 Lee, W.-K., Caster, K.C., Kim, J. and Zauscher, S. (2006) *Small*, **2**, 848.

63 Liu, X., Guo, S. and Mirkin, C.A. (2003) *Angewandte Chemie – International Edition*, **115**, 4933.

64 Kaholek, M., Lee, W.-K., LaMattina, B., Caster, K.C. and Zauscher, S. (2004) *Nano Letters*, **4**, 373.

65 Liu, G.Y., Xu, S. and Qian, Y.L. (2000) *Accounts of Chemical Research*, **33**, 457.

66 Dgata, J.A., Schneir, J., Harary, H.H., Evans, C.J., Postek, M.T. and Bennett, J. (2002) *Applied Physics Letters*, **80**, 2592.

67 Mamin, H.J., Guethner, P.H. and Rugar, D. (1990) *Physical Review Letters*, **65**, 2418.

68 Zapotoczny, S., Benetti, E.M. and Vancso, G.J. (2007) *Journal of Materials Chemistry*, **17**, 3293.

69 Maynor, B.W., Li, Y. and Liu, J. (2001) *Langmuir*, **17**, 2575.

70 Patra, M. and Linse, P. (2006) *Nano Letters*, **6**, 133.

71 Stevens, M.M. and George, J.H. (2005) *Science*, **310** (5751), 1135.

72 Lutolf, M.P. and Hubbell, H.A. (2005) *Nature Biotechnology*, **23** (1), 47.

73 Hersel, U., Dahmen, C. and Kessler, H. (2003) *Biomaterials*, **24**, 4385.

74 Li, X., Wei, X.L. and Huson, S.M. (2004) *Biomacromolecules*, **25**, 677.

75 Tugulu, S., Silacci, P., Stergiopulos, N. and Klok, H.A. (2007) *Biomaterials*, **28** (5), 763.

76 Singh, N., Cui, X.F., Boland, T. and Husson, S.M. (2007) *Biomaterials*, **28**, 763.

77 Navarro, M., Benetti, E.M., Zapotoczny, S., Planell, J. and Vancso, G.J. (2008) *Langmuir*, **24** (19), 10996.

78 Maheshwari, G., Brown, G., Lauffenburger, D.A., Wells, A. and Griffith, L.G. (2000) *Journal of Cell Science*, **113**, 1677.

79 Ruoshlahti, E. and Piersbacher, M.D. (1987) *Science*, **268**, 491.

80 McPherson, T., Kidane, A., Szleifer, I. and Park, K. (1998) *Langmuir*, **14**, 176.

81 Unsworth, L.D., Sheardown, H. and Brash, J.L. (2005) *Langmuir*, **21**, 1036.

82 Mei, Y., Wu, T., Xu, C., Langebach, K.J., Elliot, J.T. and Vogt, B.D. (2005) *Langmuir*, **21**, 12309.

83 Harris, B.P., Kutty, J.K., Fritz, E.W., Webb, C.K., Burg, K.J. and Metters, A.T. (2006) *Langmuir*, **22**, 4467.

84 Cavalcanti-Adam, E., Micoulet, A., Blümmel, J., Auernheimer, J., Kessler, H. and Spatz, J.P. (2006) *European Journal of Cell Biology*, **85**, 219.

85 Beer, J.H., Springer, K.T. and Coller, B.S. (1992) *Blood*, **79**, 117.

86 Elbert, D.L. and Hubbell, J.A. (2001) *Biomacromolecules*, **2**, 430.

2.3
Cyanate Ester Resins as Thermally Stable Adhesives for PEEK

Basit Yameen, Matthias Tamm, Nicolas Vogel, Arthur Echler, Renate Förch, Ulrich Jonas, and Wolfgang Knoll

2.3.1
Introduction

Polyether ether ketone (PEEK) is a high-temperature-resistant, semicrystalline thermoplastic polymer [1]. Excellent thermomechanical properties, chemical stability, and fire resistance have made PEEK an attractive material for applications in various fields such as automotive, electrical engineering, home appliances and aircraft parts [2, 3]. Due to the increased need for PEEK with complex shapes and geometries, the joining of separate components in an assembly by adhesive bonding is often inevitable. Epoxies have been the most widely used adhesives in this context [4–8], which are generally processable at ambient temperatures. They form a thermoset resin at a relatively low curing temperature (< 180 °C) and are therefore suitable for industrial processing. The major drawback, however, is their poor performance at elevated temperatures. With a general recommended application temperature below < 180 °C, epoxies lose their adhesive properties as a result of their poor thermal stability above this temperature [9]. Conversely, PEEK exhibits a continuous working temperature of 250 °C with retention of its excellent mechanical properties, solvent resistance, and chemical resistance [10]. Thus, in order to benefit from the superior thermal properties of PEEK, more sophisticated adhesives with high thermal stability are needed.

Cyanate ester resins (CERs), first developed by Grigat *et al.* in the 1960s at Bayer AG, is another class of thermosetting polymers that have been under intense investigation and development in recent years [11]. Cyanate ester monomers (CEM) have multiple reactive cyanate groups (–OCN) that can crosslink thermally by cyclotrimerization to give polycyanurate (PC) thermosets (Scheme 1). PCs are constituted by a network of 1,3,5-triazine rings and are known for their high T_g and high thermal stability.

Beside their thermal stability, CERs possess other interesting properties such as low dielectric constants, radar transparency and low water absorption, which, like PEEK, make them attractive materials for microelectronics, aircraft, and spacecraft technologies [12]. In the light of these enhanced properties we hereby pro-

Surface Design: Applications in Bioscience and Nanotechnology
Edited by Renate Förch, Holger Schönherr, and A. Tobias A. Jenkins

ISBN: 978-3-527-40789-7

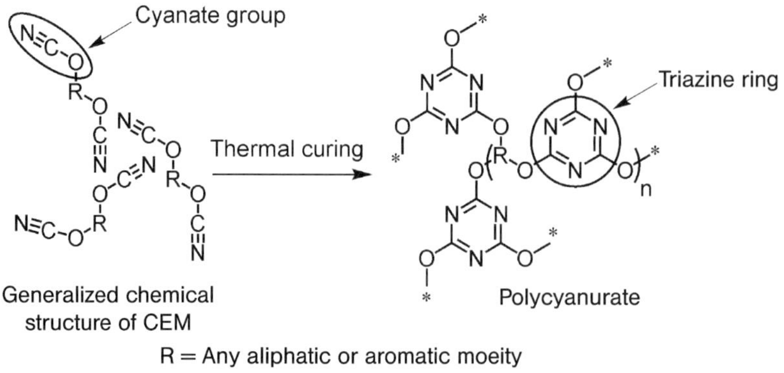

Scheme 1 A generalized scheme for thermal curing by cyclotrimerization of the cyanate groups in cyanate ester monomers (CEMs) to give a thermosetting polycyanurate. For simplicity only three triazine rings are shown, while the real crosslinked polycyanurate network (marked with *) consists of many triazine rings.

pose CEMs as thermally stable adhesives for PEEK. For the proof-of-concept, a range of experiments using a commercial novolac-based PT-30 cyanate ester monomer (CEM) are presented here as an example (Figure 1).

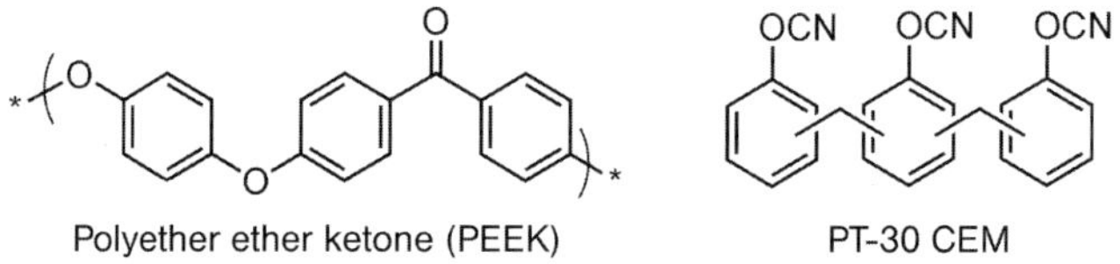

Figure 1 Chemical structure of PEEK and PT-30 CEM.

To improve the PEEK-PEEK joint strength while using PT-30 CEM as adhesive wet-chemical and plasma-assisted surface activations were applied. The wet-chemical surface activation involved the functionalization of the PEEK surface with –OCN groups, which are reactive towards the CEM adhesive. This will result in higher joint strengths due to interfacial covalent bonding between the PEEK surface and the PT-30 adhesive. Surface treatment of polymers by plasma techniques has proved to be one of the possible methods for improving the adhesion properties without affecting the bulk characteristics (see also Chapter 1.4) [13]. Plasma treatments using different mixtures of N_2 and O_2 were investigated here to activate the PEEK surface. The PEEK-CER-PEEK joint strength after wet chemical and plasma activations were tested at room temperature and at 200 °C.

2.3.2 Experimental

2.3.2.1 Materials and Methods

PEEK samples were purchased from Victrex Europa GmbH, Hofheim/Taunus, Germany. The novolac-based cyanate ester monomer (CEM) Primaset PT-30, used as model adhesive, was kindly provided by Lonza, Switzerland. A two-component epoxy resin adhesive was obtained from UHU GmbH, Bühl, Germany (Art. No. 45705). The two components of epoxy were mixed together in 1:1 ratio before use. Dry tetrahydrofuran (THF), sodium hydride, sodium borohydride, iron acetylacetonate, and bromocyan were used as received from Sigma-Aldrich, Schnelldorf, Germany. Dimethylsulfoxide (DMSO) was distilled prior to use. All the other reagents were used as received unless otherwise described. Sand paper of 600 grit, ERSTA, was obtained from Dieter Schmid Fine Tools, Berlin, Germany.

The average roughness (R_a) of PEEK surfaces was determined using a TENCOR P-10 surface profiler (Tencor Instruments, San Jose, CA, USA). X-ray photoelectron spectroscopy (XPS) measurements were carried out using a Physical Electronics 5600 A instrument. The MgK_α (1253.6 eV) X-ray source was operated at 300 W. A pass energy of 117.40 eV was used for the survey spectra. The spectra were recorded using a 45° take-off angle relative to the surface normal. The XPS scans were analyzed using the MultiPak 5.0 software. Attenuated total reflection infrared (ATR-IR) spectra were recorded on a Nicolet FT-IR 730 spectrometer. Differential scanning calorimetric (DSC) analyses were performed on a DSC 822 (Mettler-Toledo, Greifensee, Switzerland) under a nitrogen purge of 30 $cm^3\ min^{-1}$. Thermogravimetric analyses (TGA) were performed on TGA 851 (Mettler-Toledo, Greifensee, Switzerland) at a heating rate of 10 K min^{-1} under a N_2 or N_2:O_2 (80:20) purge of 30 $cm^3\ min^{-1}$. Rheological measurements were performed on an advanced rheometric expansion system (ARES, Rheometric Scientific Inc., New Jersey, NJ, USA). The PEEK-CER-PEEK joint strength of half-cut tensile bone joined together by PT-30 CEM was tested on a universal material-testing machine (Instron 6022, Instron Co., Buckinghamshire, UK) equipped with a 10-kN load cell and an oven (Brabender Realtest, GmbH, Moers, Germany) to control the sample temperature.

2.3.2.2 Sample Preparation

The as-received PEEK was cut into bone-shaped structures (Standard: Din 52455 No. 4) with the dimensions as illustrated in Figure 2. For tensile testing after surface modification and gluing, these bones were cut again into two equal halves. The surfaces to be glued were roughened with sand paper, thoroughly rinsed with THF, and dried in a stream of N_2 gas before gluing. For ATR-IR and XPS measurements, small slices of $1 \times 1 \times 0.5\ cm^3$ and $0.5 \times 0.5 \times 0.1\ cm^3$ were cut out of

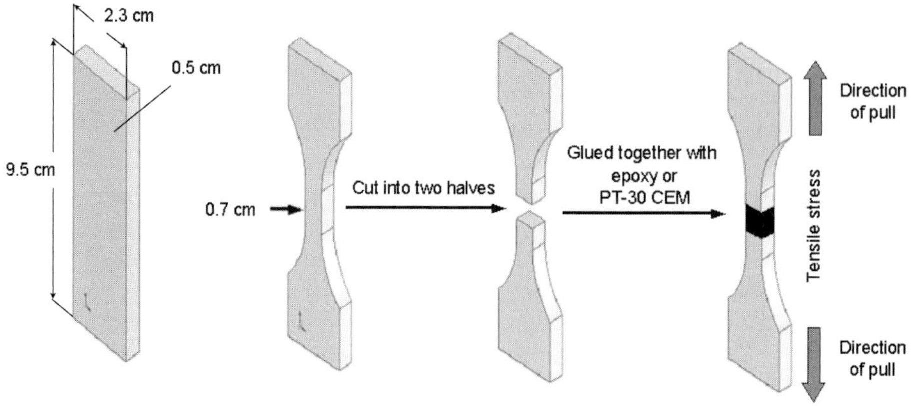

Figure 2 Bone-shaped samples for mechanical testing.

the commercial PEEK blocks and treated in the same manner as the bone-shaped samples.

2.3.2.3
Wet-Chemical Surface Activation – Reduction of Surface Carbonyl Groups to Hydroxy Groups (PEEK-OH)

A 1-L three-neck round-bottom flask was equipped with a reflux condenser and argon inlet and outlet. 500 mL of freshly distilled DMSO and 1.2 g (0.032 mole) of sodium borohydride were added to the flask. The reaction mixture was heated at 120 °C under stirring until dissolution occurred. Five half-cut PEEK tensile bones along with samples for ATR-IR and XPS measurements were immersed in the stirred reaction mixture and heated at 120 °C for 3 h under argon. After removal from the reaction mixture the PEEK samples were successively immersed in stirred methanol for 15 min, in distilled water for 10 min, in 0.5 N HCl for 10 min, in water for 10 min and in ethanol for 10 min. The samples were dried at 60 °C under vacuum for 2 h and stored under N_2. These samples will hereafter be referred to as PEEK-OH.

2.3.2.4
Wet-Chemical Surface Activation – Transformation of Surface Hydroxy Groups into Cyanate Groups (PEEK-OCN)

To a 500-mL three-neck round-bottom flask, equipped with condenser and argon inlet and outlet, 360 mL of THF, and 3.2 g (0.13 mole) of sodium hydride were added. The mixture was stirred for 30 min at room temperature. Five half-cut PEEK-OH tensile bones along with the samples for ATR-IR and XPS measurements were immersed in the reaction mixture and refluxed for 5 h. After cooling to room temperature all the samples were thoroughly rinsed with dry THF and transferred to another 500-mL round-bottom Schlenk flask under argon. The Schlenk flask was cooled in a salt-ice bath and the temperature was kept below 0 °C. 13.2 g (0.128 mole) of bromocyan dissolved in 350 mL of dry THF was added into the Schlenk flask. The reaction mixture was stirred below 0 °C for 1 h and at room temperature for 12 h. The thus obtained PEEK-OCN samples were rinsed twice with dry THF. The residual solvent was removed under vacuum at room temperature for 3 h and stored under N_2.

2.3.2.5
Plasma-Assisted Surface Activation

The plasma reactor consisted of a 30-cm long and 10-cm diameter cylindrical flow reactor equipped with two concentric metal braid rings (separated by about 10 cm), which delivered the 13.56 MHz radio-frequency to the chamber. The reactor is described in detail elsewhere [19]. The typical base pressure achieved in the system using a Leybold Trivac 16BCS rotary vane pump was 10^{-4} mbar. Half-cut tensile bones were stacked (in packs of 3–4) on the glass substrate holder placed half way between the electrodes with the surfaces to be treated and pointing in the direction of the incoming gases. Treatment was carried out using different mixtures of O_2 and N_2, both with a purity of 99.9%. The different gas ratios and plasma conditions used during the PEEK treatment are given in Table 2.

2.3.2.6
Gluing and Adhesion Tests

PT-30 was applied to the surface of one half of the tensile bone of PEEK and two halves were then glued at 200 °C for 12 h. A similar procedure was used for the epoxy glue with a curing temperature of 70 °C. The joint strengths were determined by applying a pull force acting perpendicular to the plane of the joint (Figure 2). The nature of stress that develops in such a joint geometry is tensile stress. Structural adhesives perform worse under tension, therefore, any improvement of joint strength under tension can be translated into higher strengths under shear, cleavage, or peel stress. During gluing, the two halves of the tensile bone were held together with the help of a steel clip that induces a physical restraint to ensure constant axial loading and avoids any undesirable cleavage or peel stresses. For the sake of simplicity, the values of tensile stress determined in the experiment will be termed 'joint strength' in this work. The strength of the joint was measured by pulling apart the glued halves of the PEEK tensile bone perpendicular to the joint cleft at room temperature and at 200 °C. The joint strengths reported are the average of 5 to 10 individual tensile tests.

2.3.3
Results and Discussion

2.3.3.1
Choice of PT-30 CEM as Adhesive for PEEK

In comparison to the widely used epoxies, CEMs generally suffer from poor pre-curing processability due to their usually solid character at room temperature. Besides recent efforts to develop CEMs [14, 17] that are processable at ambient temperature, some highly processable CEMs have already been commercialized. The PT-30 CEM used in the present study is a commercially available CEM with a room temperature viscosity of 2047 Pa s, which decreases substantially to 9 Pa s at 50 °C. This reflects a high epoxy-like ambient temperature processability of the PT-30 CEM.

Thermal curing of CEMs generally yields PC by cyclotrimerization of the –OCN groups to triazine rings. This cyclotrimerization is an exothermic process, which can be monitored by DSC. The DSC thermogram of PT-30 CEM showed the exothermic transition corresponding to a curing reaction at 287 °C (Figure 3a). This exothermic curing transition of PT-30 CEM occurs at a temperature that is 37 K higher than the recommended continuous working temperature of PEEK, quoted at 250 °C. In order to use PT-30 as an adhesive for PEEK it is, however, necessary to carry out curing below 250 °C.

It is known that the curing temperature of CEM can be lowered by catalysis of the cyanate groups cyclotrimerization by a transition-metal acetylacetonate [15]. The inset in Figure 3(a) shows the concentration dependence (in wt.%) for iron acetylacetonate as catalyst on the curing of PT-30 CEM. With increasing concentration of the catalyst the exothermic curing transition shifted to lower temperatures. Thus, for 0.1 wt.% catalyst the exothermic transition was substantially lowered from 287 °C to 135 °C, which is already 152 K below the continuous working temperature of PEEK. The exothermic transition shifted further down to 112 °C if 0.4 wt.% of the catalyst was used. Increasing the catalyst concentration from 0.4 wt.% to 0.8 wt.% had no apparent effect in the DSC, but concentrations beyond 0.8 wt.% resulted in spontaneous curing at room temperature.

While using a catalyst has the advantage of lowering the curing temperatures for CEMs, several disadvantages have to be considered: (i) when mixed with the CEM the shelf live of the adhesive may be substantially lowered. (ii) Any residual solvent (acetone in present case) used to homogenize the CEM–catalyst mixture can adversely affect the mechanical strength of the resulting thermoset. (iii) Another major disadvantage of the catalyst, which will remain in the thermoset PC, is its activation of hydrolysis reactions of the PC network and hence accelerated ageing [11]. The included metal ions may also negatively influence the absorption behavior of the PC with respect to radiation (radar, light, or X-rays). These combined effects will ultimately reduce the joint strength when employed as adhesive. Therefore, the use of a curing catalyst for CEM is not encouraged for practical applications.

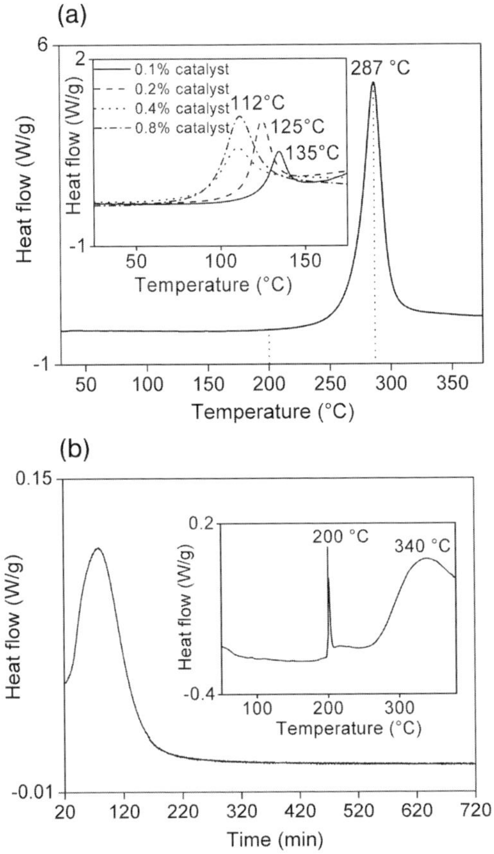

Figure 3 (a) DSC thermogram of neat PT-30 CEM. The exothermic transition peaking at 287 °C corresponds to the thermal curing by cyclotrimerization of cyanate groups (–OCN). The inset shows the catalytic effect of iron acetylacetonate (in wt.%) on thermal curing. (b) Inset: DSC thermogram of PT-30 following the heating program: heated dynamically at 10 K/min to 200 °C where it was annealed isothermally for 12 h (represented by the very sharp exothermic transition at 200 °C) followed by heating to 400 °C at 10 K/min. Beside the exothermic transition at 200 °C, a 100-K wide shoulder related to a diffusion controlled cure occurs around 340 °C. Figure (b) is the isothermal part of the same heating scan depicting heat flowing (W/g) as a function of time (min) at 200 °C.

These observations and problems motivated the study of neat PT-30 CEM curing at lower temperatures. As the DSC thermogram of neat PT-30 CEM (Figure 3a) indicates the first onset of the exothermic transition for curing via cyclotrimerization already around 200 °C, curing may be induced at this lower temperature even without employing a catalyst. For this purpose isothermal DSC at 200 °C was employed for PT-30 CEM curing. The temperature program followed in DSC was as follows: PT-30 CEM was first heated to 200 °C dynamically at 10 K/min, where it was annealed isothermally for 12 h followed by heating to 400 °C at 10 K/min.

The resulting DCS thermogram is depicted in Figure 3(b) as an inset. The heat flow (W/g) during the 200 °C isothermal annealing step of the whole DCS scan is depicted as a function of time (min) in Figure 3(b). It took about 20 min to reach to 200 °C, therefore, the annealing curve starts from 20 min. An exothermic transition related to curing can be observed that rises sharply at 20 min, peaks around 1 h and ceases at about 180 min. The isothermal heating was continued for another 9 h but no significant further change was observed. After isothermal heating at 200 °C for 12 h the temperature was raised dynamically to 400 °C at 10 K/min during the same DSC scan. Apart from an exothermic curing transition at 200 °C, a 100-K wide shoulder related to diffusion-controlled curing centred at 340 °C [16]. No exothermic transition at 287 °C, as found in dynamic DSC of untreated PT-30 CEM (Figure 3a), was observed, which reflects a reasonably high extent of curing by chemical conversion of –OCN groups to the triazine ring at 200 °C even without adding a catalyst.

Beside DSC, FT-IR spectroscopy is another powerful tool for monitoring the thermal curing of CEMs by cyclotrimerization of –OCN groups to a network of triazine rings. –OCN groups have a characteristic stretching band in the IR spectrum, which in the case of PT-30 CEM appeared as a bifurcated band at 2260 cm^{-1} and 2240 cm^{-1} (Figure 4). The intensity of the band was observed to decrease substantially when PT-30 CEM was subjected to curing at 200 °C for 12 h with the appearance of new bands at 1556 cm^{-1} and 1360 cm^{-1}, which are characteristic of the triazine rings of PC (Figure 4).

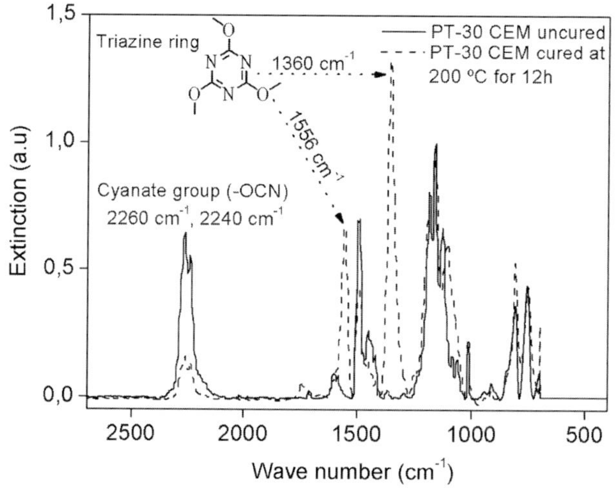

Figure 4 Normalized FT-IR spectra of uncured and cured PT-30 CEM. Two new bands for the cured PT-30 CEM appeared at 1556 cm^{-1} and 1360 cm^{-1} for the triazine ring at the expense of the cyanate group –OCN stretching band at 2260 cm^{-1}. The consumption of cyanate groups while curing at 200 °C for 12 h was calculated from the normalized areas of cyanate absorbance before $(A_{OCN})_{t=0\,h}$ and after $(A_{OCN})_{t=12\,h}$ the thermal treatment and was found to be 70%.

In order to determine the extent of curing the IR spectra presented in Figure 4 were normalized to the out-of-plane deformation of CH groups in the substituted aromatic ring at 754 cm^{-1}. The extent of curing was quantified by calculating the percent conversion factor $\alpha(t)$ [17] for the residual –OCN groups from the normalized area of –OCN group absorbance before $(A_{OCN})_{t=0\,h}$ and after $(A_{OCN})_{t=12\,h}$ the thermal treatment according to the following equation:

$$\alpha(t) = \frac{(A_{OCN})_{t=12\,h}}{(A_{OCN})_{t=0}} \times 100 \quad (1)$$

It was found that curing PT-30 CEM at 200 °C for 12 h resulted in 70% conversion of –OCN groups. These experiments also highlight that the DSC and FT-IR are very convenient tools for qualitative and quantitative analysis of the thermal transformations during CEM curing.

Thermal properties such as the T_g and thermo-oxidative stability of a material are critical for any potential high-temperature application, therefore, PT-30 CEM cured for 12 h at 200 °C was subjected to DSC, rheological analysis and TGA. It was found that the thermal curing for 12 h at 200 °C was enough for the resulting PC to show a rheological T_g of 187 °C (Figure 5a, inset), which is 44 K higher than the T_g (143 °C) of PEEK and 7 K higher than the maximum recommended use temperature of epoxies, i.e. 180 °C. The T_g of 200 °C/12 h cured PT-30 CEM was deduced from its tanδ curve because no transition was observed in its DSC thermogram (not shown here for simplicity). The DSC thermogram for PEEK is also depicted in Figure 5(a) for comparison. Two endothermic transitions for T_g and melting T_m can be observed at 143 °C and 342 °C. The thermal stability of 200 °C/12 h cured PT-30 CEM was evaluated from the TGA shown in Figure 5b. The TGA thermograms of PEEK are also plotted in the same figure for comparison. The TGA revealed a high thermal stability of 200 °C/12 h cured PT-30 CEM as no weight loss was observed below 424 °C under N_2 atmosphere and 392 °C under a mixture of N_2:O_2 in the ratio of 80:20. PEEK on the other hand showed a higher thermal stability as no weight loss was observed before 550 °C, irrespective of gaseous atmosphere used in TGA. The comparison of TGA reflects a slightly higher thermal stability of PEEK but the stability of 200 °C/12 h cured PT-30 is still ~150 K higher than the continuous working temperature of PEEK. An interesting feature observed in the TGA under N_2 was a higher char yield (68%) for 200 °C/12 h cured PT-30 than PEEK (52%) at 900 °C, which suggests that 200 °C/12 h cured PT-30 possesses a better flame retardancy and fire resistance than PEEK. It was also observed during oxidizing TGA (N_2:O_2 80:20) that both 200 °C/12 h cured PT-30 and PEEK exhibited a two-step degradation process with the main degradation occurring during the second step where 75% weight loss was observed. The onset temperatures for the second step of degradation of 200 °C/12 h cured PT-30 and PEEK were quite close, i.e. 575 °C, and 589 °C respectively, and differ only by 14 K. It can thus be inferred from the TGA that 200 °C/12 h cured PT-30 possesses thermal stability comparable to PEEK. A comparison of thermal properties of PEEK and 200 °C/12 h cured PT-30 CEM is presented in Table 1.

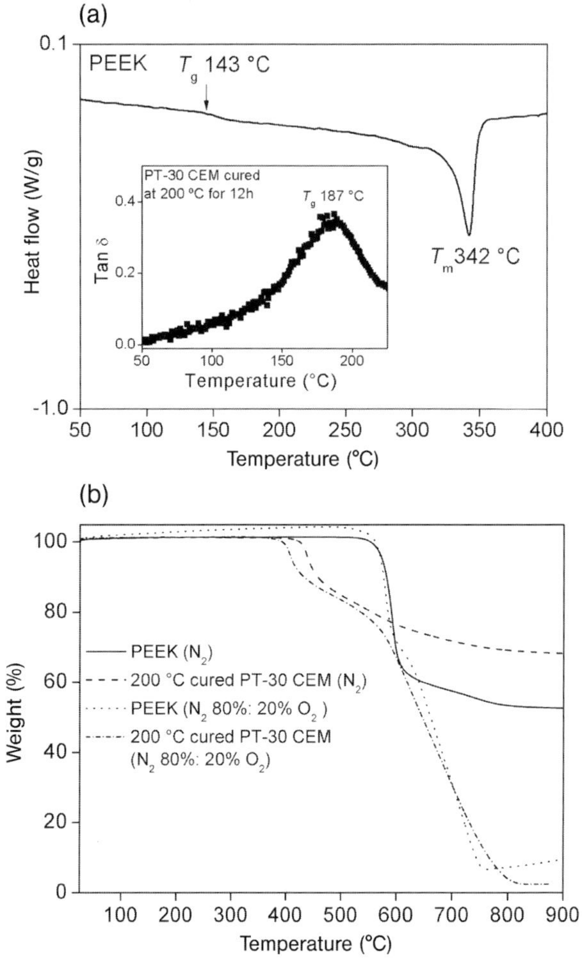

Figure 5 (a) DSC thermogram of PEEK. The endothermic transitions at 143 °C and 342 °C correspond to T_g and T_m, respectively. The inset is a tanδ plot of the thermoset of 200 °C/12 h cured PT-30 wherein the maxima is the T_g (187 °C). (b) TGA thermogram of PEEK and of the 200 °C/12 h cured PT-30 thermoset. No weight loss was observed below 400 °C.

The investigation of thermal properties of PT-30 CEM and their comparison with the thermal properties of PEEK demonstrates that PT-30 CEM is a highly promising material for application in combination with PEEK under high temperatures. In order to investigate this experimentally, two halves of a half-cut bone (see materials and methods) were glued together using PT-30 CEM as adhesive and the assembly was kept for 12 h at 200 °C without a special surface treatment of the PEEK material. The resulting PEEK-CER-PEEK joint was subjected to a pull-apart test. At room temperature the average PEEK-CER-PEEK joint strength

Table 1 Comparison of thermal properties of PEEK and 200 °C/12 h cured PT-30 and joint strengths of unmodifierd PEEK using epoxy and PT-30 CEM adhesives.

Material	T_g (°C)	T_m (°C)	T_0–N_2 (°C)	T_0–N_2/O_2 (80:20) (°C)	Joint strength (MPa)
PEEK	143[a)]	342[a)]	550	550	–
200 °C/12 h cured PT-30	187[b)]	No melting	424	392	–
PEEK-epoxy-PEEK	–	–	–	–	7[c)]
PEEK-CER-PEEK	–	–	–	–	13[c)] and 6[d)]

a) DSC, b) Rheological, c) at room temperature,
d) at 200 °C, T_0 = onset of decomposition in TGA

obtained was 13 MPa. For comparison a commercial two-component epoxy was also applied as an adhesive for the same joint geometry. PEEK-epoxy-PEEK joints showed an average room-temperature strength of 7 MPa. This indicates that under the same joint geometry, PEEK-CER-PEEK joints using the PT-30 CEM adhesive exhibit a higher average room-temperature strength than the joints using the epoxy adhesive.

The PEEK-adhesive-PEEK joint strengths were also measured at 200 °C for the PT-30 CEM- and epoxy adhesives. Interestingly, at 200 °C the joints with PT-30 CEM could retain an average strength of 6 MPa, whereas at the same temperature epoxy simply failed to hold together the two halves of the glued tensile bone and consequently no joint strength could be measured. At this temperature PT-30 CEM shows a room-temperature epoxy-like adhesive performance, while the epoxy completely fails under these conditions. The reduced joint strength for PT-30 CEM at 200 °C is partially due to a softening of the adhesive above its T_g of 187 °C. These experiments clearly demonstrate the suitability of CEMs as thermally stable adhesives for PEEK. In order to further improve the PEEK-adhesive-PEEK joint strength, the PEEK surface in contact with the adhesive was subjected to different activation treatments and the result obtained are discussed below.

2.3.3.2
Surface Activation of PEEK for Adhesion Improvements

The chemical nature of polymer surfaces govern a wide range of interfacial properties [18]. The chemical inertness of PEEK is highly attractive for many applications, however, this inertness also leads to immense difficulties with the adhesion of PEEK or to other materials. In order to glue PEEK, the adhesion must be optimized without losing the bulk properties that make PEEK such an attractive material. This can be achieved by modifying the chemical nature of the surface either in wet-chemical processes or plasma-assisted processes. Several methods of surface modification have been reported for adhesion improvements of PEEK-PEEK joints with epoxies as adhesives [4–8]. The use of CEM as adhesive for PEEK has previously not been reported. Here, we discuss two means of surface activation, namely wet chemical and plasma treatment, applied to the PEEK surface in order to improve the PEEK-adhesive-PEEK joint strength with CEMs as adhesives.

Wet-Chemical Surface Activation

During wet-chemical surface activation, the surface of PEEK was functionalized with –OCN groups by exposure to reagents in organic solvents in a two-step process, as outlined in Scheme 2. The surface carbonyl groups of PEEK were first reduced to hydroxy groups by treating with $NaBH_4$ in DMSO at 120 °C for 3 h. The modified PEEK with surface hydroxy groups was designated as PEEK-OH.

The surface hydroxy groups of PEEK-OH were subsequently transformed into –OCN groups under the conditions slightly modified as compared to those generally applied for CEM synthesis from phenols [11f]. The procedure involved the activation of surface hydroxy groups of PEEK-OH by treatment with NaH. This activation step was necessary because the oxygen in alcohols is not as nucleophilic as that of phenols to give a reasonable yield during the nucleophilic attack on BrCN. The activated surface alkoxide groups –ONa were transformed into –OCN

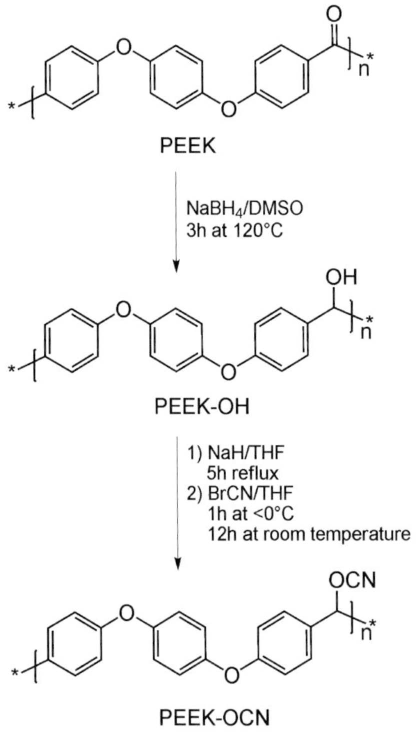

Scheme 2 Wet-chemical functionalization of the PEEK surface with cyanate groups (–OCN).

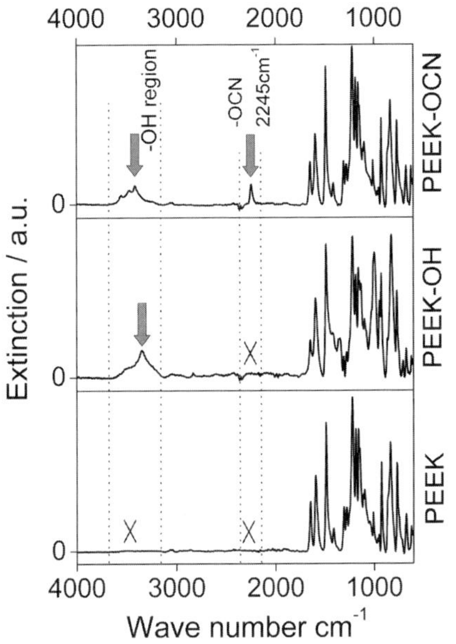

Figure 6 ATR-IR monitoring of wet-chemical surface modifications of PEEK. –OH stretching bands at around 3400 cm^{-1} can be observed after reduction of carbonyl groups at the PEEK surface to hydroxy groups. The –OCN stretching band appeared at 2245 cm^{-1} after transformation of surface –OH groups to –OCN groups.

groups by reacting with BrCN in THF first at 0 °C for 1 h and then at room temperature for 12 h. The progress of these wet-chemical transformations at the PEEK surface were monitored by XPS and ATR-IR spectroscopy.

The reduction of surface carbonyl groups of PEEK to hydroxy groups was evident from the ATR-IR spectra (Figure 6). The characteristic stretching band for hydroxy groups was observed around 3400 cm^{-1} in the ATR-IR spectrum of PEEK-OH, while the band was absent in the ATR-IR spectrum of untreated PEEK. The transformation of surface hydroxy groups to –OCN groups was confirmed by the appearance of a band at 2245 cm^{-1} in the ATR-IR spectrum of PEEK-OCN, which is characteristic for –OCN groups. The presence of chemically bound nitrogen at the PEEK surface was confirmed in subsequent XPS measurements. It must be noted, that the conversion of surface –OH groups to –OCN groups was never 100% and an –OH stretching band was always present along with the –OCN stretching band in the IR spectrum of PEEK-OCN.

The C1s and N1s XPS spectra for PEEK and PEEK-OCN are depicted in Figure 7(a) and (b), respectively. Unmodified PEEK (inset in Figure 7a) shows a main C1s peak centred around 285 eV with a high energy tail that leads directly into a broad, low intensity peak around 294 eV. The main peak can be associated with the aromatic rings of PEEK, while the high-energy tail reflects carbon bonded to oxygen, presumably as the ether and ketone groups present in unmodified PEEK. The peak at 294 eV represents the π–π^* shake up transition of the aromatic rings in PEEK. After wet-chemical modification the high-energy tail became more prominent, indicating the formation of the –O–C≡N group. The successful functionalization of the PEEK surface with –OCN groups was more evident from the

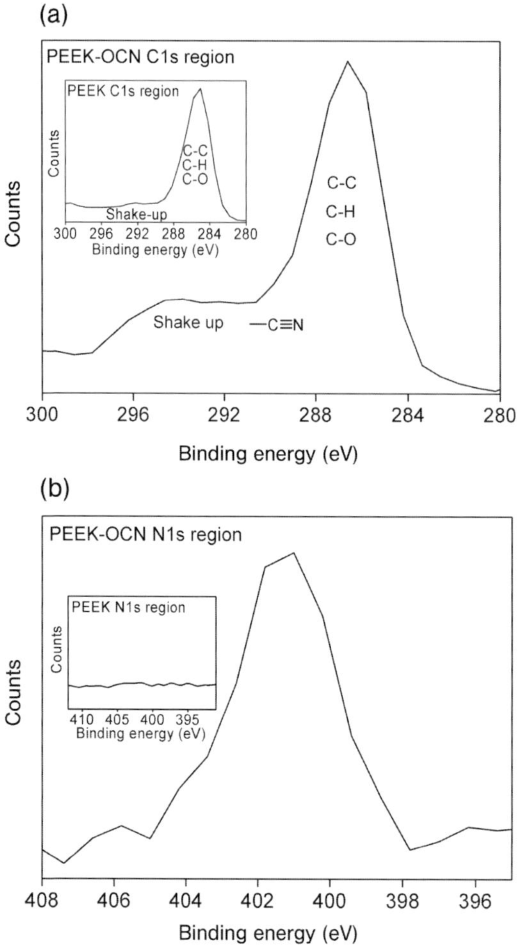

Figure 7 (a) C1s XPS spectra of PEEK-OCN showing the change in the spectrum after wet-chemically functionalization of the PEEK surface. For comparison a C1s spectrum of unmodified PEEK is shown in the inset. (b) N1s spectrum of PEEK-OCN with a N1s spectrum of unmodified PEEK in the inset.

comparison of the N1s spectral region of PEEK and PEEK-OCN (Figure 7b). While the unmodified material revealed no evidence for chemically bound nitrogen, the PEEK-OCN showed a signal at 401 eV, very characteristic for nitrogen. The XPS analysis fully supports the FTIR data showing the chemical incorporation of nitrogen into the PEEK surface and the formation of –OCN surface groups.

While XPS spectroscopy is sensitive down to a surface layer depth of ~5 nm, ATR-IR spectroscopy can sample a depth of about 2.0 μm. Since the present samples possessed a high surface roughness (R_a) of around 9 μm on average, no attempts were made to quantify the surface transformations by XPS or ATR-IR

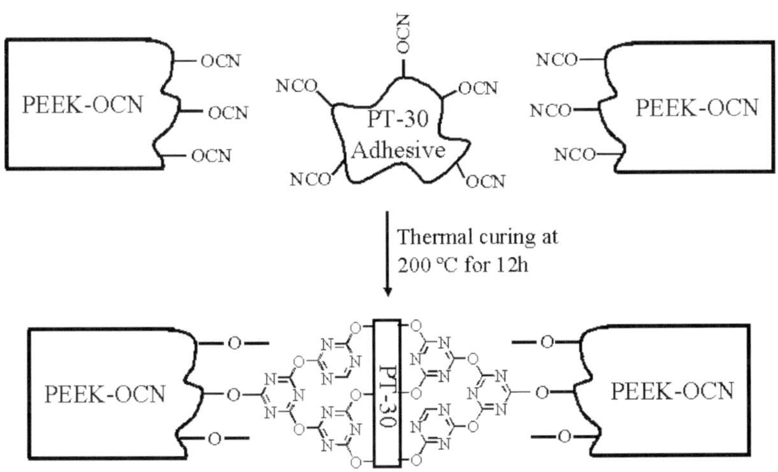

Figure 8 Schematic illustration of covalent bonding at the PEEK/adhesive interface after introduction of cyanate functionalities on PEEK surface.

spectroscopy. Probably because of the high surface roughness it was possible to monitor the chemical changes by ATR-IR, which has previous been reported as rather difficult [19].

After wet-chemical surface activation, the half-cut PEEK tensile bones with surface –OCN groups were glued together with PT-30 CEM, yielding an average room-temperature PEEK-CER-PEEK joint strength of 28 MPa. This joint strength is almost double compared to untreated PEEK (13 MPa). Apparently, the increased joint strengths with PEEK-OCN is a result of the participation of surface –OCN groups of PEEK in the cyclotrimerization of PT-30 CEM –OCN groups during thermal curing. This results in strong covalent bonding between the PEEK surface and the PT-30 CEM adhesive through triazine ring formation at the PEEK/PT-30 CEM interface (Figure 8). The results show that the combined effect of surface functionalization with –OCN groups and using CEMs as adhesives allows for a significant increase in PEEK-to-PEEK bonding strength.

PEEK-CER-PEEK joints involving PEEK-OH were also fabricated. Interestingly, these joints exhibited an average joint strength of 26 MPa, which is not much different from PEEK-OCN. The high joint strength shown by PEEK-OH is probably due to the chemical bonding between surface –OH groups and –OCN groups of PT-30 CEM. This chemical bonding involves the nucleophilic attack of the –OH group oxygen on the electrophilic carbon of the –OCN group resulting in the formation of a triazine ring passing through an imidate intermediate. A probable mechanism [20] is depicted in Scheme 3. These results show that surfaces with –OH groups can also be used with CEMs as adhesives to substantially improve the joint strength.

The high temperature joints strength of PEEK-OH and PEEK-OCN were measured at 200 °C, where epoxies fail, and were found to range between 8–10 MPa.

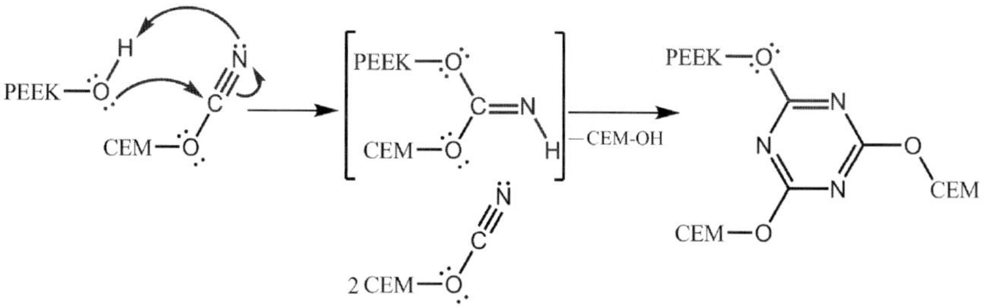

Scheme 3 Proposed mechanism for the reaction of surface hydroxy groups of PEEK-OH with PT-30 CEM resulting in adhesion improvement.

The average values were always slightly higher than PEEK-CEM-PEEK joints of untreated PEEK. In summary, it was found that wet-chemical surface activation significantly improved the PEEK-CEM-PEEK joint strength at room temperature and at 200 °C.

Plasma-Assisted Surface Activation

The modification of surfaces using plasma-assisted processes has been intensively studied since the 1960s and a large number of today's industrial processes are based and rely on plasma-assisted modification and deposition processes [21]. Cleaning, ablation, and surface activation are the major effects of plasma treatment. One or more of these effects can dominate and can be controlled by tuning the nature of gas and plasma operating parameters [22]. To date the available literature demonstrates the effect of plasma treatment on PEEK-PEEK adhesion improvement using epoxy adhesives [5–9], but only few references report on the functionalities introduced at the PEEK surface after plasma treatment using different gases. Therefore, half-bones of PEEK were subjected to different N_2/O_2 plasma conditions in order to optimize the PEEK-CER-PEEK joint strength using PT-30 CEM as adhesive. Even though PEEK is known to be chemically rather inert towards many reagents, its surface chemistry can be modified using an oxygen plasma [23]. The extent to which the surface is oxidized with oxygen as the reactive gas depends on the input power, P_{peak} (or the plasma density) and leads to the dissociation of bonds at the material surface with the formation of new oxygen-based functional groups. This process involves for PEEK the dissociation of the aromatic rings and the C–O–C bonds at the polymer surface (Figure 1). The breaking of aromatic rings in oxygen or nitrogen plasmas has previously been demonstrated and results in a number of different free radical sites that spontaneously react with oxygen to form alcohols, ketones, aldehydes and carbonyl groups [24]. The quantification of the new functional groups on the surface is difficult and has not yet been achieved satisfactorily. If the gas used during plasma treatment is nitrogen the radical sites will also with the nitrogen species in the plasma. It is thus also possible to form various amines, amides, and imines on the surface, that are available for subsequent reaction [25].

The different plasma conditions applied in this work for PEEK treatment are summarized in Table 2. From the data in Table 2 it can be concluded that increasing input power during oxygen plasma treatment leads to an increase in the PEEK-CER-PEEK joint strength. Similar values as those obtained by wet-chemical surface modification are achieved (i.e. about 26 MPa). When using pure nitrogen at a high input power of 180 W a similarly high PEEK-CER-PEEK joint strength was also observed. For N_2:O_2 mixtures the adhesion strength improved with increasing input power, treatment time, and nitrogen flow rate.

The highest room-temperature adhesion strength of 48 MPa was obtained using a N_2:O_2 mixture of 150 sccm:50 sccm flow rate at an input power of 210 W for 10 min and a pressure of 0.898 bar. It can be assumed that under these conditions the PEEK structure at the uppermost surface is largely destroyed by the plasma

Table 2 PEEK-CEM-PEEK joint strengths for different surface activations

Nature surface activation	Joint strength at room temperature (MPa)	Joint strength at 200 °C (MPa)
Without any treament:	13	6
Wet chemical surface activation:		
PEEK-OH	26	10
PEEK-OCN	28	8
Plasma-assisted surface activation:		
O_2 10 sccm* Power: 10 W 2 min	16	
O_2 10 sccm* Power: 100 W 7 min	26	
O_2 10 sccm* Power: 180 W 7 min	25	
N_2 10 sccm* Power: 180 W 2 min	28	
N_2:O_2 mixtures:		
Flow 10:10 sccm, P_{peak}: 10 W, 2 min, P: 0.898 mbar	21	
Flow 10:10 sccm, P_{peak}: 100 W, 7 min, P: 0.898 mbar	27	
Flow 10:10 sccm, P_{peak}: 160 W, 7 min, P: 0.898 mbar	33	
Flow 20:10 sccm, P_{peak}: 10 W, 2 min, P: 0.898 mbar	23	
Flow 20:10 sccm, P_{peak}: 100 W, 7 min, P: 0.898 mbar	21	
Flow 20:10 sccm, P_{peak}: 160 W, 7 min, P: 0.898 mbar	38	
Flow 100:10 sccm, P_{peak}: 10 W, 2 min, P: 0.898 mbar	25	13
Flow 100:10 sccm, P_{peak}: 100 W, 7 min, P: 0.898 mbar	36	
Flow 100:50 sccm, P_{peak}: 210 W, 10 min, P: 0.898 mbar	48	

* sccm = standard cubic centimetre, given in each case as the flow of N_2:O_2
P is the proces pressure

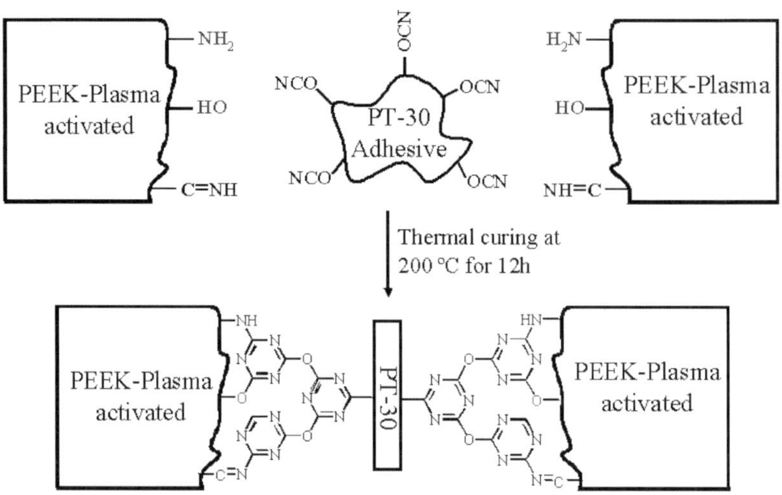

Figure 9 Proposed scheme showing possible interfacial bonding occurring between plasma-treated PEEK and the CEM adhesive.

and that the surface is marked by a high density of different oxygen and nitrogen functional groups, which undergo chemical bonding with the CEM adhesive as shown schematically in Figure 9. At the same time the polymer molecules at the surface of the modified PEEK are shortened due to extensive bond scission, such that the uppermost layer of the polymer no longer resembles the original material, but consists of a multitude of short- and medium-length molecules that are (i) highly reactive and (ii) are mobile and can diffuse partially into the CEM adhesive. This has been discussed in Chapter 1.4 in more detail. It is possible that the combined effect of a high reactive-group density and the diffusion of molecular chain ends from the PEEK surface into the adhesive leads to the high observed joint strengths.

The strength of the PEEK-CER-PEEK joints obtained under these optimized plasma conditions was also determined at 200 °C and was found to be 13 MPa. This high-temperature joint strength is slightly higher than that for the wet-chemically modified PEEK-CER-PEEK joint (8–10 MPa, see above) and represents a significant improvement compared to epoxies.

2.3.4 Conclusion

We have demonstrated the high potential of CEMs as thermally stable adhesives for PEEK. Commercially available PT-30 CEM was used as a model CEM adhesive. Even though the maximum of the exothermic transition for the curing reaction was observed at 287 °C in the DSC of PT-30 CEM, curing could be induced at a substantially lower temperature (200 °C), as inferred from the onset of curing exothermic transition. The thermal properties, like a high T_g (187 °C) and thermal stability, of the PC obtained after curing PT-30 CEM at 200 °C for 12 h are comparable to PEEK.

The potential of PT-30 as an adhesive was evaluated by fabricating PEEK-CER-PEEK joints and subjecting them to a tensile test by applying a pull force perpendicular to the plane of the joint (tensile stress geometry). Compared to a two-component epoxy PT-30 CEM exhibited better adhesive properties at room temperature and at 200 °C, where epoxies generally fail. The PEEK-CER-PEEK joint strength could be improved substantially by wet-chemical and plasma-assisted activation of the PEEK surface. During wet-chemical activation the PEEK surface was successfully functionalized with –OH and –OCN groups. These surface transformations could be monitored with XPS and ATR-IR spectroscopy. The room temperature joint strengths involving PEEK with surface –OH and –OCN groups (26 MPa and 28 MPa) were double compared to joints involving untreated PEEK surfaces (13 MPa).

Furthermore, plasma treatment was applied to activate the PEEK surface, varying the gas composition, input power, and treatment time. The highest room-temperature joint strength of 48 MPa was achieved for optimized plasma treatment conditions of N_2:O_2 150 sccm:50 sccm 210 W, 10 min, and a pressure of 0.898 bar. The joint strength of 8 to 13 MPa at 200 °C involving PEEK surface activation showed a slight improvement over the untreated surfaces (6 MPa). In summary, it can be concluded that CEMs possess a favorable combination of thermal and mechanical properties that make them a very attractive material class as a structural adhesives at high temperatures.

Acknowledgments

B. Yameen gratefully acknowledges financial support from Higher Education Commission (HEC) of Pakistan and Deutscher Akademischer Austauschdienst DAAD (Code # A/04/30795). The authors also acknowledge Dr. Lenz and Dr. Lauer from Siemens for technical discussions. Thanks to Andreas Hanewald, MPIP, for assistance with tensile testing, and to Daniela Mössner, IMTEK, University of Freiburg, Germany, for the XPS analysis of our samples.

References

1 May, R. (1988) in *Encyclopedia of Polymer Science and Engineering*, 2nd edn, Vol. 12 (eds H.F. Mark, N.M. Bikales, C.G. Overberger, G. Menges and J.I. Kroschwitz), John Wiley & Sons, New York, p. 313, 0-471-88786-2.

2 Petillo, O., Peluso, G., Ambrosio, L., Nicolais, L., Kao, W.J. and Anderson, J.M. (1994) *Journal of Biomedical Materials Research*, **28**, 635.

3 Morrison, C., Macnair, R., McDonald, C., Wykman, A., Goldie, I. and Grant, M.H. (1995) *Biomaterials*, **16**, 987.

4 Davies, P., Courty, C., Xanthopoulos, N. and Mathieu, H.J. (1991) *Journal of Materials Science Letters*, **10**, 335.

5 Comyn, J., Mascia, L. and Xiao, G. (1996) *International Journal of Adhesion and Adhesives*, **16**, 97–104.

6 Comyn, J., Mascia, L., Xiao, G. and Parker, B.M. (1996) *International Journal of Adhesion and Adhesives*, **16**, 301–304.

7 Laurens, P., Sadras, B., Decobert, F., Arefi-Khonsari, F. and Amouroux, J. (1998) *International Journal of Adhesion and Adhesives*, **18**, 19.

8 Mathieson, I. and Bradley, R.H. (1998) *International Journal of Adhesion and Adhesives*, **16**, 29.

9 Nair, C.P.R., Mathew, D. and Ninan, K.N. (2001) *Advances in Polymer Science*, **155**, 1.

10 McGrail, P.T. (1996) *Polymer International*, **41**, 103.

11 (a) Fan, J., Hu, X. and Yue, C.Y. (2003) *Journal of Polymer Science Part B – Polymer Physics*, **41**, 1123; (b) Lakshmi, M.S. and Reddy, B.S.R. (2002) *European Polymer Journal*, **38**, 795; (c) Herr, D.E., Nikolic, N.A. and Schultz, R.A. (2001) *High Performance Polymers*, **13**, 79; (d) Hamerton, I. and Hay, J.N. (1998) *Polymer International*, **47**, 465; (e) Hamerton, I. and Hay, J.N. (1998) *High Performance Polymers*, **10**, 163–174; (f) Hamerton, I. (1994) *Chemistry and Technology of Cyanate Ester Resins*, Blackie Academic, Glasgow.

12 Fang, T. and Shimp, D.A. (1995) *Progress in Polymer Science*, **20**, 61.

13 Jama, C., Dessaux, O., Goudmand, P., Gengembre, L. and Grimblot, J. (1992) *Surface and Interface Analysis*, **18**, 751.

14 (a) Laskoski, M., Dominguez, D.D. and Keller, T.M. (2005) *Journal of Materials Chemistry*, **15**, 1611; (b) Laskoski, M., Dominguez, D.D. and Keller, T.M. (2006) *Polymer*, **47**, 3727.

15 Mathew, Dona, Reghunadhan Nair, C.P., Krishnan, K. and Ninan, K.N. (1999) *Journal of Polymer Science Part A – Polymer Chemistry*, **37**, 1103.

16 Mondragonb, Iñaki, Solar, L., Recalde, I.B. and Gómez, C.M. (2004) *Thermochimica Acta*, **417**, 19.

17 Yameen, B., Duran, H., Best, A., Jonas, U., Steinhart, M. and Knoll, W. (2008) *Macromolecular Chemistry and Physics*, in press, 10.1002/macp.200800155.

18 Henneuse, C., Goret, B. and Marchand-Brynaert, J. (1998) *Polymer*, **39** (4), 835.

19 Mirabella, F.M. (1982) *Journal of Polymer Science – Polymer Physics Edition*, **20**, 2309.

20 Grenier-Loustalot, M.F. and Lartigau, C. (1997) *Journal of Polymer Science A – Polymer Chemistry*, **35**, 1245.

21 Förch, R., Zhang, Z. and Knoll, W. (2005) *Plasma Processes and Polymers*, **2**, 351.

22 (a) Friedrich, J., Loeschcke, I., Frommelt, H., Reiner, H.D., Zimmermann, H. and Lutgen, P. (1991) *Polymer Degradation and Stability*, **31**, 97; (b) Liston, E.M. (1989) *Journal of Adhesion*, **30**, 199.

23 Shard, A.G. and Badyal, J.P.S. (1992) *Macromolecules*, **25**, 2053.

24 Shard, A.G. and Badyal, J.P.S. (1991) *The Journal of Physical Chemistry*, **23**, 95.

25 Foerch, R., Mcintyre, N.S., Sodhi, R.N.S. and Hunter, D.H. (1990) *Journal of Applied Polymer Science*, **40**, 1903.

2.4
Structured and Functionalized Polymer Thin-Film Architectures

Holger Schönherr, Chuan Liang Feng, Anika Embrechts, and G. Julius Vancso

2.4.1
Introduction

In the context of advanced materials and tailored interfaces for application in the areas of (bio)sensors and biologically oriented research involving biointerfaces, structured and functionalized thin films received increasing attention in recent years [1, 2]. Prime examples are the so-called DNA arrays that were pioneered in the 1990s [3]. By spotting and binding single-strand probe DNA of known nucleotide sequence in particular locations of a raster-like pattern, the identification of unknown target DNA is rendered possible in a massively parallel manner. This approach has been further developed and refined for the commercialization of the technology. In addition, alternative fabrication schemes have been developed, most notably based on on-chip DNA synthesis using, e.g. advanced photochemistry [4, 5]. More recently, DNA arrays have been exploited as a route to obtain protein arrays using protein–DNA constructs [6].

Even though DNA array technology is considered a mature technology, it is interesting to note that the first tests were FDA approved only in 2004 [7]. Furthermore, reproducibility, variations in surface activity due to ageing, and reliability of the coupling chemistry, etc., have been critically discussed by some authors [8, 9]. The shortcomings identified are mostly related to surface design and warrant a closer look into the physics and chemistry of such sensor chips.

The chemistry and surface engineering involved in the fabrication of (bio)sensor surfaces possess very similar requirements as those used for the generation of biointerfaces to experimentally address and interrogate, e.g., cell–surface interactions. In addition to a sophisticated and optimized surface chemistry to facilitate efficient and controllable bioconjugation (see Chapter 1.1) and to implement the function of suppressing nonspecific adsorption of proteins, these requirements include compatibility with micro- and nanopatterning methodologies. For the specific case of substrates for cell work, topographical structures and the control of the substrate modulus are important as well [10].

These mandatory requirements are met with advanced surface-engineering approaches. Among the possible formats for the immobilization of antibodies or

Surface Design: Applications in Bioscience and Nanotechnology
Edited by Renate Förch, Holger Schönherr, and A. Tobias A. Jenkins

ISBN: 978-3-527-40789-7

DNA for biosensing (see Chapter 1.4), thin organic films (self-assembled monolayers, SAMs) on solid supports, polymeric substrates or polymer thin films, as well as tailored polymer brushes (see Chapter 2.2) have been reported. Compared to SAMs [11], polymer films may offer the advantage of a quasi-3D structure (high loading per unit surface area) and the possibility to exploit the rich structural hierarchy of ordering on different length scales. The advantages further comprise robustness and stability, high molecular loading, high reactivity, and the inherent possibility to introduce simultaneously chemical (compositional), as well as topographical patterns and structures. In order to use polymeric films for chemical patterning, reproducible coverages or activity quantification of molecules immobilized in patterns through mere physical interactions are often difficult to achieve because the molecules can be washed away during the processing [12]. Covalent linkages, on the other hand, are thought to provide more specific and stronger attachment [13, 14]. In general, covalent attachment utilizes bifunctional linker chemistry that involves, e.g, amide or imine bond formation, gold–thiol interactions, or silane chemistry [15].

While biosensors in the already mentioned array-based sensing format possess spot sizes with dimensions on the order of several micrometers to 100 micrometer (Figure 1) [16], nanostructured and nanopatterned sensors may offer several added advantages. As demonstrated by the group of Rossi, Colpo and coworkers [17], the immunoreaction efficiency of uniformly functionalized polyacrylic acid (PAA) surfaces and chemically nanopatterned PAA surfaces differ significantly for antigen–antibody interactions. Using a combination of colloidal lithography and plasma-enhanced chemical vapor deposition (PE-CVD), these authors fabricated nanopatterned surfaces that exposed COOH-functionalized nanoareas in a passivating poly(ethylene oxide) (PEO)-like matrix. The immunoreaction efficiency was found to be enhanced on the nanopatterned film in the absence of protein A (pA) as orienting protein.

Cells possess typical lateral dimensions that exceed several to 10 micrometers, thus controlled patterning on these length scales is important. In early work by Whitesides, Ingber *et al.* patterned SAMs obtained by microcontact printing [18, 19] were employed to unravel the effect of feature dimensions in patterns on cell behavior [10c]. The size of cell-adhesive areas with micrometer-scale dimensions was found to control cell behavior, e.g. proliferation or apoptosis (Figure 3a).

Interestingly, there are other length scales involved that show a clear relation to adhesion protein clustering in the cell membrane. As has also been highlighted by the work of Spatz and coworkers, distances between individual RGD-(arginine-glycine-aspartate)-peptides [20] of ≤ 58 nm and ≥ 73 nm the presence or absence, respectively, of focal adhesions was observed, thus unveiling a characteristic length scale for integrin–receptor clustering (Figure 3b) [21–23]. These reports illustrate the necessity of controlling surface chemical functionality and structures on multiple length scales, i.e. on the length scales of cells down to the molecular level [24].

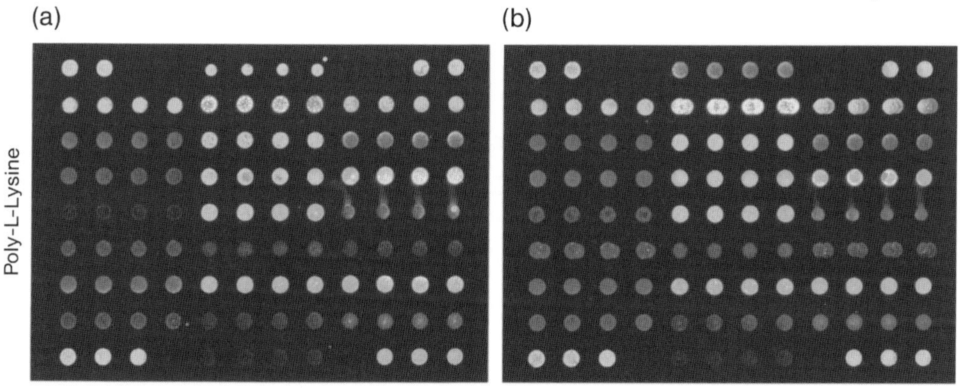

(c)

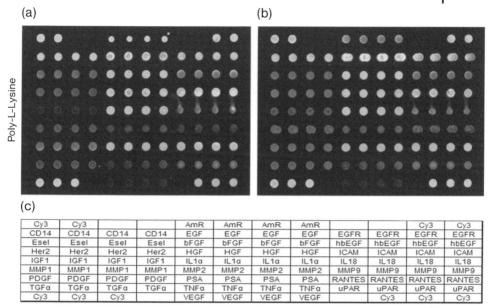

Cy3	Cy3			AmR	AmR	AmR	AmR			Cy3	Cy3
CD14	CD14	CD14	CD14	EGF	EGF	EGF	EGF	EGFR	EGFR	EGFR	EGFR
Esel	Esel	Esel	Esel	bFGF	bFGF	bFGF	bFGF	hbEGF	hbEGF	hbEGF	hbEGF
Her2	Her2	Her2	Her2	HGF	HGF	HGF	HGF	ICAM	ICAM	ICAM	ICAM
IGF1	IGF1	IGF1	IGF1	IL1α	IL1α	IL1α	IL1α	IL18	IL18	IL18	IL18
MMP1	MMP1	MMP1	MMP1	MMP2	MMP2	MMP2	MMP2	MMP9	MMP9	MMP9	MMP9
PDGF	PDGF	PDGF	PDGF	PSA	PSA	PSA	PSA	RANTES	RANTES	RANTES	RANTES
TGFα	TGFα	TGFα	TGFα	TNFα	TNFα	TNFα	TNFα	uPAR	uPAR	uPAR	uPAR
Cy3	Cy3	Cy3		VEGF	VEGF	VEGF	VEGF		Cy3	Cy3	Cy3

Figure 1 Representative fluorescence scanner images of two replicate sensing experiments conducted on different days. Panels (a) and (b) show an example of a slide type with superior performance. Panel (c) shows the array layout used for all experiments. Cy3, cyanine 3-conjugated immunoglobulin G (positive control); AmR, amphiregulin; CD14, CD14; EGF, epidermal growth factor; EGFR, epidermal growth factor receptor; Esel, E-selectin; bFGF, basic fibroblast growth factor; HBEGF, heparin binding epidermal growth factor; Her2, c-erbB-2 extracellular domain; HGF, hepatocyte growth factor; ICAM, intracellular adhesion molecule 1; IGF1, insulin-like growth factor 1; IL1a, interleukin 1 alpha; IL18, interleukin 18; MMP1, matrix metalloprotease 1; MMP2, matrix metalloprotease 2; MMP9, matrix metalloprotease 9; PDGF, platelet-derived growth factor AA; PSA, prostate-specific antigen; RANTES, regulated on activation normal T cell expressed and secreted (CCL5); TGFa, transforming growth factor alpha; TNFa, tumor necrosis factor alpha; uPAR, urokinase-type plasminogen activator receptor; VEGF, vascular endothelial growth factor A. (Reprinted from S. L. Seurynck-Servoss, A. M. White, C. L. Baird, K. D. Rodland, R. C. Zangar, Evaluation of surface chemistries for antibody microarrays, Anal. Biochem., 371, 105–115, Copyright 2007, with permission from Elsevier).

Thin films of polymers on solid supports, as discussed in detail in this chapter, offer the advantage that topographical structures are easily obtained by imprinting [25] or photolithography. As mentioned, polymeric thin films have been shown to possess a number of important advantages compared to SAM-based systems for the mentioned biosensor or biointerface applications. In general, patterned thin films of polymers are already being used in a wide range of applications, for example, as etch resists [26], in biological [27, 28] and chemical sensors [29], and in tissue engineering [30].

Thin films of patterned polymers that incorporate reactive functional groups provide a surface that can be further modified by chemical reactions [31]. Examples include the fabrication of DNA arrays on glassy polymers, such as

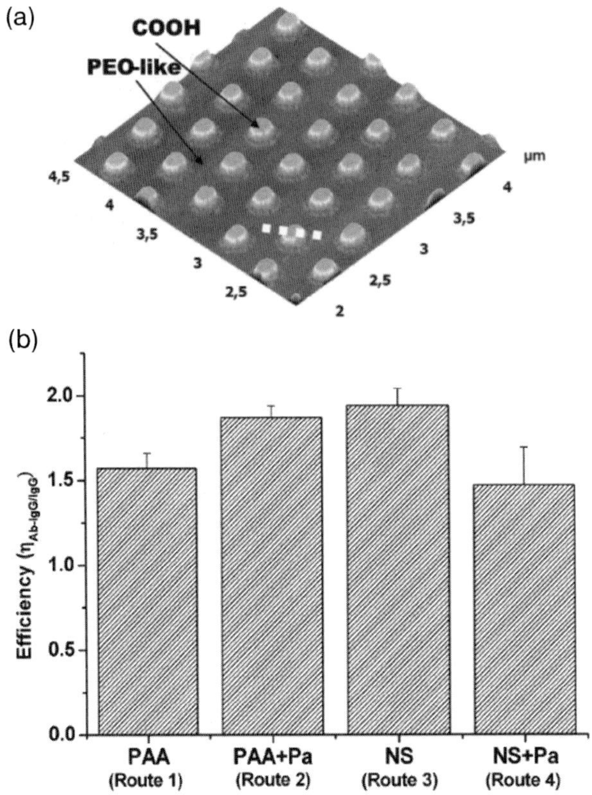

Figure 2 (a) AFM image of the nanopatterned surface. (b) Efficiency for the immunoreaction IgG/Ab-IgG. NS, nanostructures. Routes 1 and 2 refer to unstructured PAA surfaces. (Reproduced with permission from [17]; A. Valsesia, P. Colpo, I. Mannelli, S. Mornet, F. Bretagnol, G. Ceccone, F. Rossi, Anal. Chem. 2008, 80, 1418, Copyright 2008 by American Chemical Society).

poly(methyl methacrylate), PMMA [32], the synthesis of star-shaped poly(ethylene glycols) [33], and the immobilization of proteins and peptide sequences on polyurethanes [34]. In addition, new approaches, including the fabrication of reactive surface by electrografting [35], using click chemistry [36], or surface-initiated polymerizations have been reported [37, 38]. To increase the molecular loading, i.e. the number of molecules immobilized per unit area, hydrogels [39, 40], dendrimers [12, 41, 42, 14], hyperbranched polymers [43], chemical vapor deposition approaches [44], self-assembled polyelectrolyte mutilayers [45], self-assembled polymer layers [46], plasma polymers [47], and comb copolymers [48] have been investigated, among others [49].

This chapter will focus on spin-coated polymer thin films since the film thickness can be tuned and imprint lithography can be applied as a means to topographically structure the films. To establish such polymer-based platforms for the tar-

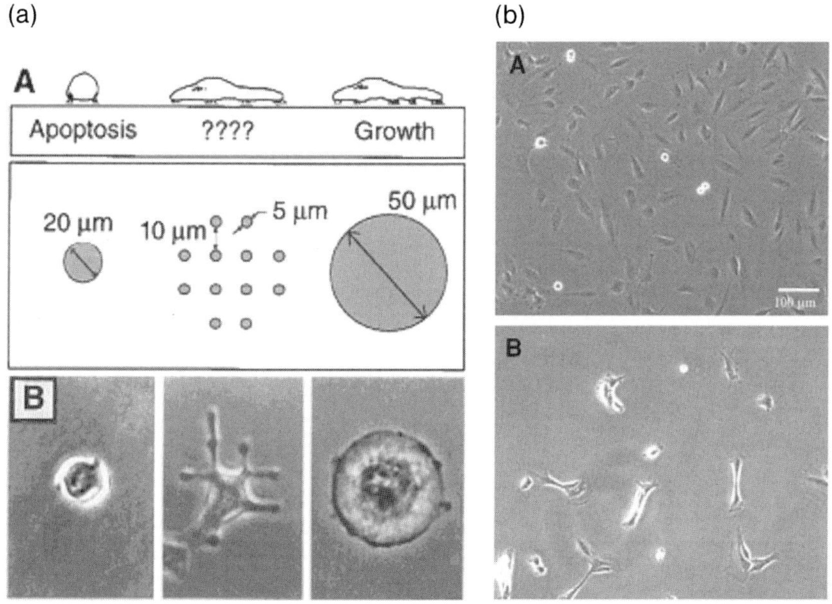

Figure 3 (a) Cell–ECM contact area versus cell spreading as a regulator of cell fate. (a, A) Diagram of substrates used to vary cell shape independently of the cell–ECM contact area. Substrates were patterned with small, closely spaced circular islands (center) so that cell spreading could be promoted as in cells on larger, single round islands, but the ECM contact area would be low as in cells on the small islands. (a, B) Phase-contrast micrographs of cells spread on single 20- or 50-µm diameter circles or multiple 5-µm circles patterned as shown in (A). (From C. S. Chen, M. Mrksich, S. Huang, G. M. Whitesides, D. E. Ingber, Science 1997, 276, 1425. Reprinted with permission from AAAS). (b) Phase-contrast images of fibroblasts (REF52) adhering for 24 h to nanopatterned surfaces of RGD peptides placed at 58 nm (b, A) and 110 nm (b, B) distance. (Reprinted from [21]; E. Cavalcanti-Adam, A. Micoulet, J. Blümmel, J. Auernheimer, H. Kessler, J. P. Spatz, Lateral spacing of integrin ligands influences cell spreading and focal adhesion assembly, Eur. J. Cell Biol., 85, 219–224, Copyright 2006 with permission from Elsevier).

geted applications, it is, however, necessary to also implement efficient surface functionalization and patterning on the one hand, and to maintain film integrity and stability under the processing and application conditions on the other hand. If successful, these methodologies not only promise to enable one to address advanced studies of, e.g., cancer-cell–surface interactions (Chapter 3.4), but also to contribute to analyze single cells in designed local microenvironments and to address the relevant question of cell–surroundings interactions in tailored 3D environments.

2.4.2
Results and Discussion

The fabrication of polymer thin-film-based platforms for the application in biosensing and cell–surface interaction studies will be discussed here starting with the preparation and characterization of spin-coated thin films of functional polymers, followed by bioconjugation, patterning and structuring, and finally biosensing.

In the following we focus on the surface engineering of spin-coated films of functional polymer on solid supports, such as glass and oxide-covered silicon. Alternative approaches comprise among others the deposition of polymer layers by surface-initiated polymerization (Chapter 2.2) [50, 51], by plasma polymerization (Chapters 1.3 and 3.6) [52] or by physisorption (PLL-PEG or related systems) [53, 54]. These approaches possess their respective advantages in terms of chemical functionality, versatility and reaction conditions, however, they all require topographically prestructured supports if one aims at combined topographic *and* chemically patterned biointerfaces. On the other hand, spin-coated films of functional polymers, such as polystyrene-block-poly(*tert*-butyl acrylate) (PS-b-PtBA), provide, in conjunction with soft lithographic and imprinting methodologies, direct access to platforms with the mentioned combined attributes [56–59].

Central to our approach is the combination of covalent and thus very robust attachment of (bio)molecules to wet chemically activated polymer surfaces under well-controlled conditions in buffer, the use of newly developed and refined soft lithographic tools and the versatility and flexibility of a toolbox thus developed.

In the particular example of PS_{690}-b-$PtBA_{1210}$ the microphase separated morphology of the block copolymer with PS cylinders in a matrix of the reactive tert-butyl ester acrylate affords the required stability of the films under all processing steps [55]. Owing to the different surface free energies of the two constituent blocks, one observes symmetric wetting with a ~8-nm thick PtBA skin layer [55]. Central to the fabrication of biointerfaces is the facile deprotection of the pending tert-butyl ester groups by organic acids (trifluoro acetic acid, TFA) [57] or heat [60], followed by the established activation using *N*-hydroxysuccinimide (NHS) chemistry (Chapter 1.1), to yield active esters that are reactive towards primary amines in aqueous medium (Figure 4a) [56, 57]. This is shown in Figure 4(b) for the immobilization of amino-end functionalized PEG with different molar mass. Based on ellipsometry and XPS data, grafting densities of ~2.4, 0.8, and 0.5 PEG molecules per nm^2 were determined for PEG with a molar mass of 500, 2000, and 5000 g/mol. On SAMs of a NHS-ester-terminated disulfide on gold, more than 3 times thinner PEG layers were covalently grafted. Since a SAM is intrinsically 2D in nature, the increased coverage on the polymer platform is attributed to an enhanced availability of NHS ester groups at and near the surface of the film. Thus, the polymer film is functionalized in a quasi-3D manner. The improved coverages of PEG lead to a better functionality of the PEG layer to suppress nonspecific adsorption of proteins and, depending on molar mass, to reduce and eliminate the interaction of cells with correspondingly functionalized surfaces (Chapter 3.4).

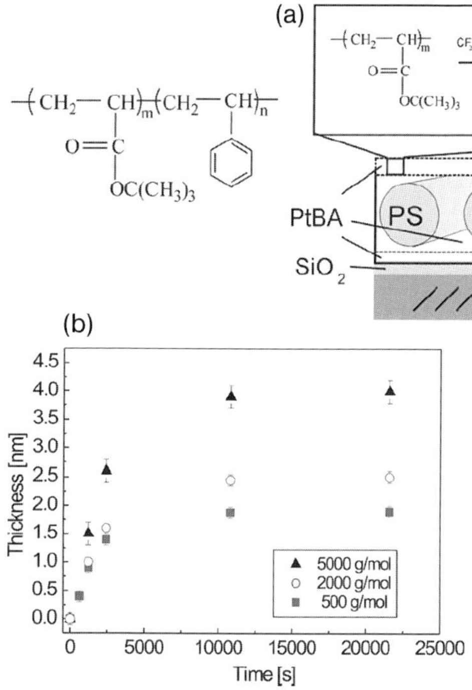

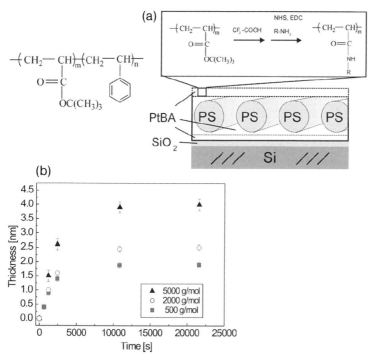

Figure 4 (a) Schematic of the block-copolymer architecture and the surface chemistry employed for immobilization. (b) Thickness changes of activated PS-b-PtBA films after grafting PEG_n–NH_2 (n = 500, 2000, 5000) for various times as determined by *ex-situ* ellipsometry. (Reprinted from [56]; C. L. Feng, A. Embrechts, I. Bredebusch, A. Bouma, J. Schnekenburger, M. García-Parajó, W. Domschke, G. J. Vancso, H. Schönherr, Tailored interfaces for biosensors and cell–surface interaction studies via activation and derivatization of polystyrene-block-poly (tert-butyl acrylate) thin films, Eur. Polym. J., 43, 2177–2190, Copyright 2007 with permission from Elsevier).

Virtually identical grafting densities of PEG were observed on spin-coated films of related homopolymers, namely poly(*N*-hydroxysuccinimidyl methacrylate) (PNHSMA), on oxidized silicon and gold surfaces [61]. In addition, this homopolymer showed increased grafting densities for the immobilization of amino-end functionalized 25-mer probe DNA (Figure 5). In the surface plasmon resonance (SPR) kinetic scan the rapid covalent binding and subsequent equilibration of the DNA-polymer film architecture on gold is shown. Upon rinsing with pure buffer, the layer thickness remains virtually unchanged, which indicates the stable and irreversible attachment of the probe DNA. Compared to SAMs the DNA coverage was increased by a factor of 2.

The immobilization of PEG as a means to efficiently passivate the polymer thin-film surface [62] and the covalent attachment of probe DNA as an active sensing component have thus been demonstrated [63]. Similar to these simple molecules, the immobilization of various proteins, including bovine serum albumin (BSA), fibronectin, laminin, collagen, or protein G, and various antibodies has been

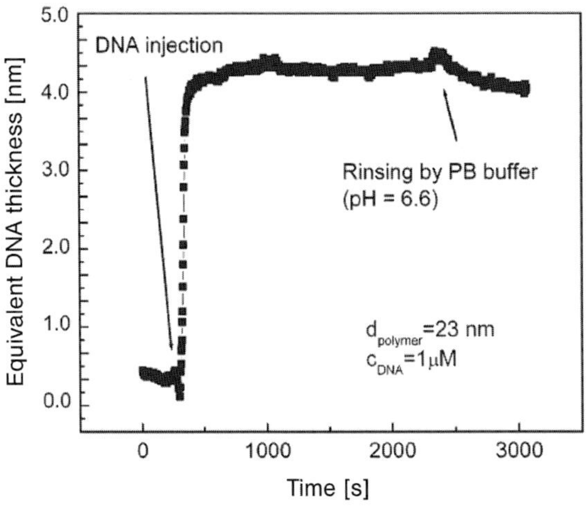

Figure 5 DNA immobilization on PNHSMA film (dry thickness 23 nm) investigated *in situ* by SPR (kinetic scan). (Reproduced with permission from [61]; C. L. Feng, Z. Zhang, R. Förch, W. Knoll, G. J. Vancso, H. Schönherr, Biomacromolecules 2005, 6, 3243, Copyright 2005 by American Chemical Society).

demonstrated. Thus, to within the limits of NHS-ester chemistry (Chapter 1.1), biomolecules can be coupled to the platform introduced. Since the attachment results in the formation of covalent amide bonds, these linkages are stable under the subsequent processing conditions, such as biorecognition and washing.

The films are also amenable to patterning using soft lithographic means. In principle, established photolithography using photoacid is an option for the fabrication of patterns [64], however, soft lithography offers several advantages. Among these advantages are the facile creation of patterns on length scales that approach or bypass the practical limitations of photolithography (e.g. the use of a clean room) and eventually the diffraction limit. Under certain circumstances, soft lithography may also provide a route to enhanced functionality (see below) [58].

In reactive microcontact printing, chemical reactions that are limited to the conformal stamp-polymer film contact area are exploited [58, 59]. These reactions may include transfer and covalent coupling of PEG-NH_2 or biomolecules or the localized deprotection and thus activation by TFA. The patterns shown in Figure 6 were obtained by transfer printing PEG-NH_2 using an UV-ozone oxidized polydimethyl siloxane (PDMS) [65] stamp to NHS-ester-activated PS-b-PtBA films. The reactant is transferred from the stamp to the activated polymer only in the contact areas, leaving the noncontacted circular areas unreacted and thus chemically active. These areas are still equipped with the NHS-ester moieties, to which (bio)-molecules of interest are coupled from buffered aqueous solution. This approach avoids the contact of the biomolecules with the stamp or with air and thus circumvents one shortcoming of conventional microcontact printing [66].

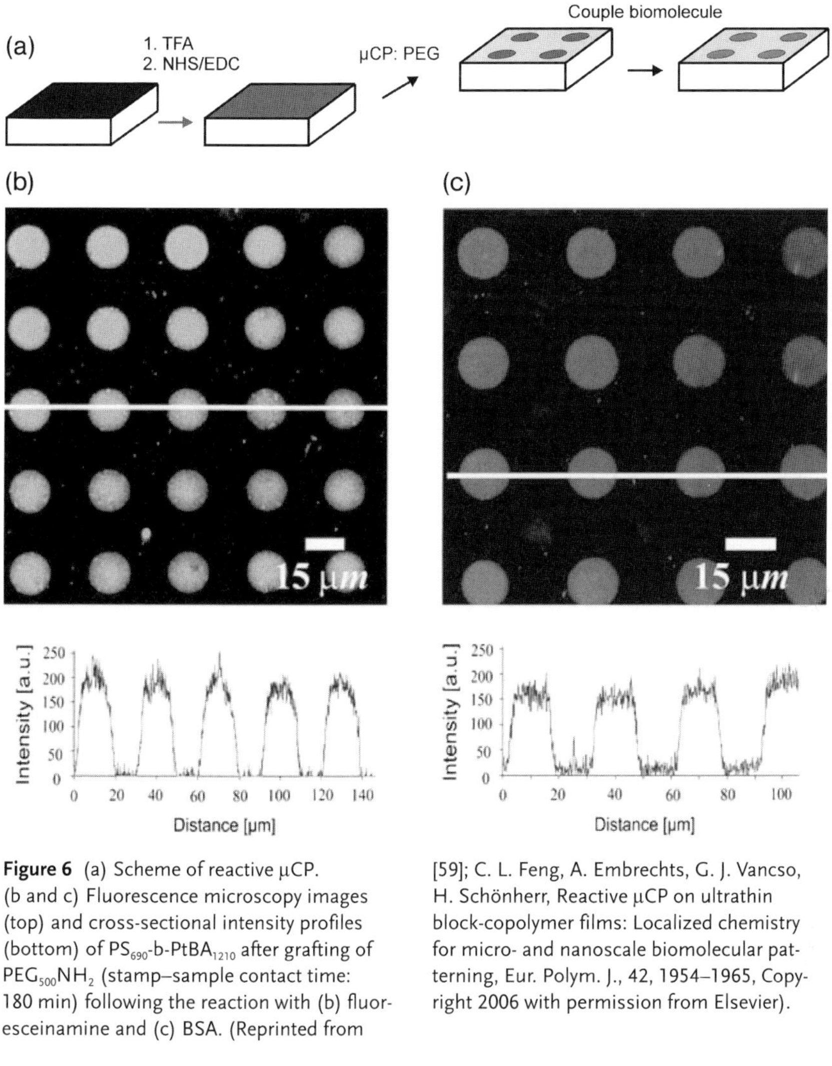

Figure 6 (a) Scheme of reactive μCP. (b and c) Fluorescence microscopy images (top) and cross-sectional intensity profiles (bottom) of PS_{690}-b-$PtBA_{1210}$ after grafting of $PEG_{500}NH_2$ (stamp–sample contact time: 180 min) following the reaction with (b) fluoresceinamine and (c) BSA. (Reprinted from [59]; C. L. Feng, A. Embrechts, G. J. Vancso, H. Schönherr, Reactive μCP on ultrathin block-copolymer films: Localized chemistry for micro- and nanoscale biomolecular patterning, Eur. Polym. J., 42, 1954–1965, Copyright 2006 with permission from Elsevier).

The cross-sectional plots of the fluorescence micrographs illustrate the high definition of the patterns and the absence of irregular spot shapes observed in conventional spotting. Interestingly, it was shown that reactions in the stamp-film contact area may deviate from reactions carried out in solution or at interfaces. The high pressure and the high local concentration (in the absence of solvent) were reported to enhance the efficiency of interfacial peptide synthesis [67], as well as the grafting of PEG-NH_2 [58].

Similar high-definition patterns can be obtained in reactive μCP when TFA is locally administered to deprotect the PtBA ester [58]. In this approach, the ester surface is converted to carboxylic acid only in the contacted areas, while the non-contacted areas remain unreacted, i.e. protected. Subsequently, the carboxylic acid

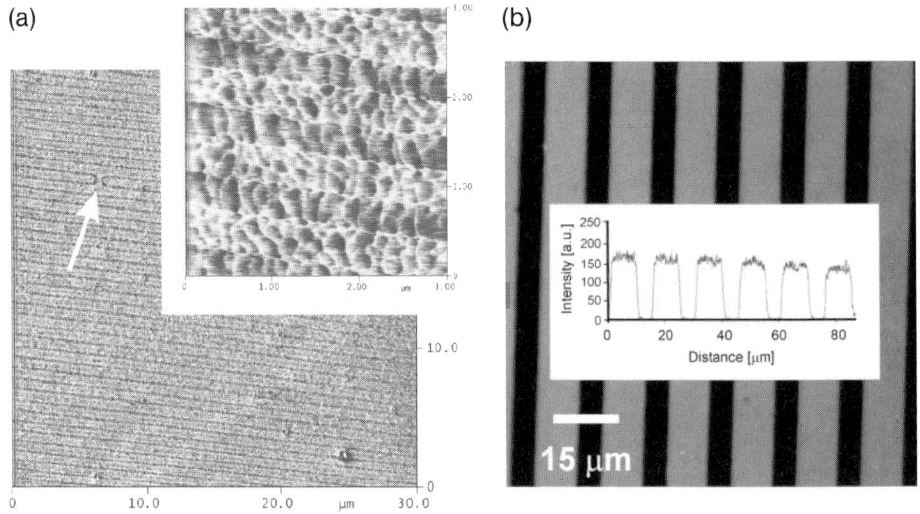

Figure 7 (a) AFM friction image acquired on a PS_{690}-b-$PtBA_{1210}$ film after local hydrolysis by reactive μCP (inset: high-resolution AFM friction image). The stripes with high friction-force contrast (width ca. 500 nm) are ascribed to the hydrophilic part (poly (acrylic acid): PAA), the low friction-force stripes (width ca. 300 nm) are ascribed to the hydrophobic part (PtBA). Note in (a) how a defect in the stamp has been faithfully reproduced (the defect is highlighted by the arrow). (b) Fluorescence microscopy image and corresponding cross-sectional intensity plot (inset) for PS_{690}-b-$PtBA_{1210}$ films patterned with a stamp that possessed line features (10 μm × 5 μm). The block-copolymer film was hydrolyzed by placing a TFA-covered oxidized PDMS stamp in conformal contact for 20 min, followed by activation of the carboxylic acid groups formed in the contact areas with NHS/EDC and grafting of fluoresceinamine from phosphate buffer (PB). (Reproduced with permission from [58]; C. L. Feng, A. Embrechts, I. Bredebusch, J. Schnekenburger, W. Domschke, G. J. Vancso, H. Schönherr, Adv. Mater. 2007, 19, 286, Copyright 2007 by Wiley-VCH).

is chemically activated and functionalized with the (bio)molecules, as desired. This approach provides access to patterns as small as 300 nm × 500 nm (Figure 7).

In this method it is imperative to avoid the presence of liquid TFA on the stamp surface, as this liquid also spreads into the areas not intended for the deprotection [57]. Unlike for SAMs on gold where mere contact of diffusing thiols leads to instantaneous reaction, i.e. SAM formation, and pattern deterioration, the reactive chemistry exploited in μCP with TFA is more tolerant. Although TFA as a strong organic acid deprotects PtBA efficiently, it requires a particular local concentration for the reaction and TFA also partitions into the polymer film, presumably leading to a complex concentration profile in 3D.

To widen the scope of these polymer film-based interfaces, topographical structures were introduced in PS-PtBA films by imprint lithography (Figure 8) [68]. In this technique, a master (stamp) with a topographic pattern is pressed into the polymer film that has been heated above its softening temperature. For the microphase separated PS-PtBA this refers to the glass transition temperature T_g of the

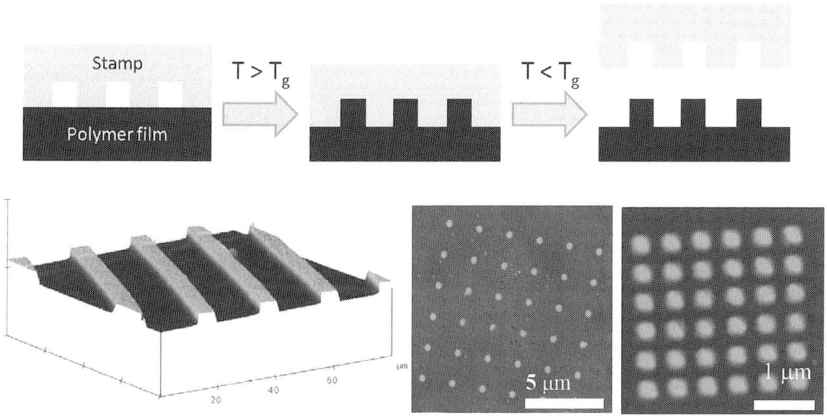

Figure 8 AFM height images of topographic structures imprinted into PS-b-PtBA block-copolymers. The smallest structure shown measures 100 nm in height, 270 nm in width with a period of 1.0 µm. (Reproduced with permission from [68]; A. Embrechts, C. L. Feng, C. A. Mills, M. Lee, I. Bredebusch, J. Schnekenburger, W. Domschke, G. J. Vancso, H. Schönherr, Langmuir 2008, 24, 8841, Copyright 2008 by American Chemical Society).

PS block (T_g = 105 °C). The silicon or silicon nitride masters are coated with an antiadhesive coating. After imprinting the temperature is lowered and the master is separated from the structured polymer film. As shown, this approach provides access to structures on length scales of tens of micrometers to well below 1 micrometer for this system.

As long as the temperature does not exceed the onset of the thermal deprotection of the PtBA block ($T \sim$ 160–180 °C) [69], the structured polymer film exposes a PtBA skin layer (as shown by XPS) and can be covalently functionalized as described above. More sophisticated biointerfaces can be fabricated when imprint lithography and microcontact printing are carried out sequentially (see Chapter 3.4) [70].

Finally, the application of the patterned biointerfaces in sensing and in cell–surface interaction studies will be treated. After the successful immobilization of probe DNA on PNHSMA films, as shown by SPR (see Figure 5), hybridization of complementary DNA was studied [61]. To detect the hybridization of target DNA labeled with a fluorophore, surface plasmon resonance enhanced fluorescence spectroscopy (SPFS) was employed [71]. SPFS is a recently developed very sensitive technique to detect the hybridization of target DNA on probe DNA (Chapter 3.6). In essence, the surface plasmons excite the nearby fluorophore of the bound DNA. In the experiment the emitted photons are detected. Owing to the effects of fluorescence quenching and decaying coupling efficiency between surface plasmons and surface-immobilized fluorophores, there is an optimized film thickness of ~ 30–50 nm for this particular detection method. Since the film thicknesses of spin-coated films can be easily controlled through variations of polymer solution

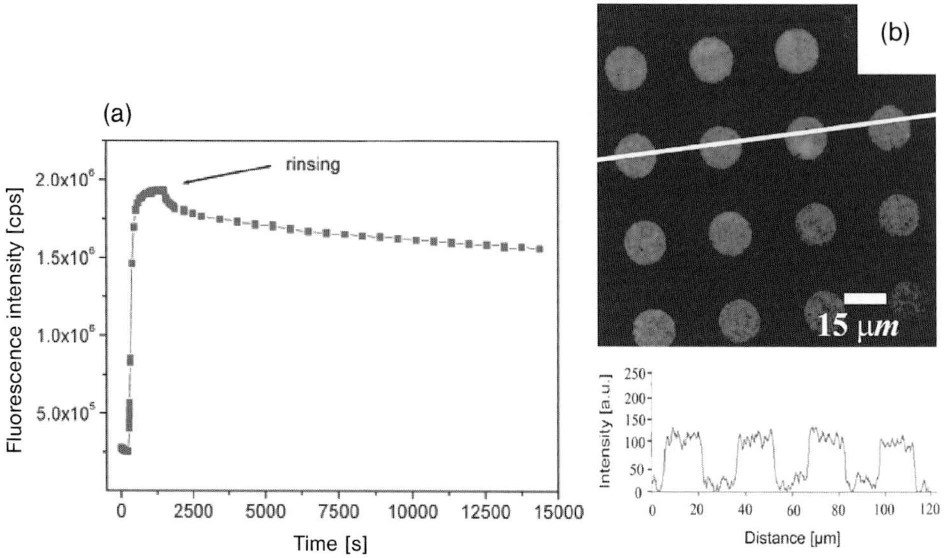

Figure 9 (a) Hybridization of complementary target DNA on probe DNA (4.2 nm) modified PNHSMA film (dry thickness 14 nm) investigated by SPFS (PB buffer; pH 7.4). (Reproduced with permission from [61]; C. L. Feng, Z. Zhang, R. Förch, W. Knoll, G. J. Vancso, H. Schönherr, Biomacromolecules 2005, 6, 3243, Copyright 2005 by American Chemical Society). (b) Fluorescence microscopy image (top) and cross-sectional intensity profile (bottom) of a PEG-probe DNA patterned PS_{690}-b-$PtBA_{1210}$ surface following hybridization with complementary target DNA. (Reprinted with permission from [59]; C. L. Feng, A. Embrechts, G. J. Vancso, H. Schönherr, Reactive μCP on ultrathin block-copolymer films: Localized chemistry for micro- and nanoscale biomolecular patterning, Eur. Polym. J., 42, 1954–1965, Copyright 2006 from Elsevier).

concentration or spinning speed, reactive thin-film systems that show only limited swelling, such as PNHSMA, can be advantageous in this respect.

Upon the addition of the target DNA solution, the fluorescence intensity rises very rapidly and reaches a stable, constant value (Figure 9a). Rinsing with pure buffer affects the intensity only a little. The hybridization was also successfully detected in micropatterns fabricated by reactive μCP on the block copolymer platform [59]. Noteworthy are the homogeneous fluorescence intensity across the spots and the uniformity of the spots.

The controlled modification of the polymer film surface by covalent attachment of passivating PEG or adhesion-promoting poly-L-lysine (PLL) was further exploited to generate surfaces with differential interaction with cells [56]. Figure 10 shows the surface coverage of K562 cells on block copolymer films after applying different surface chemistry. Cell-surface coverages on PLL functionalized films were very high and cells were found almost everywhere (Figure 10b). Cell-surface interaction with the neat PS_{690}-*b*-$PtBA_{1210}$ films was also studied. For both cases, it was found that cell morphology was not affected by the presence of the PS_{690}-*b*-$PtBA_{1210}$ substrate. To prevent K562 cell adhesion via nonspecific interac-

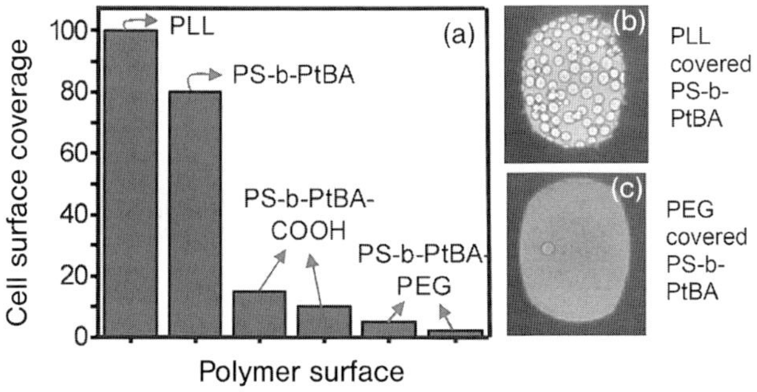

Figure 10 (a) Histogram of cell-surface coverage vs. polymer surface with different composition is obtained from wide field optical microscope images. (b) Wide-field optical micrograph of PLL covered PS_{690}-b-$PtBA_{1210}$ films (scale: 150 μm × 150 μm). (c) Wide-field optical micrograph of PEG_{500}-NH_2 covered PS_{690}-b-$PtBA_{1210}$ film (scale: 150 μm × 150 μm). (Reprinted from [56]; C. L. Feng, A. Embrechts, I. Bredebusch, A. Bouma, J. Schnekenburger, M. García-Parajó, W. Domschke, G. J. Vancso, H. Schönherr, Tailored interfaces for biosensors and cell–surface interaction studies via activation and derivatization of polystyrene-block-poly(tert-butyl acrylate) thin films, Eur. Polym. J., 43, 2177–2190, Copyright 2007 with permission from Elsevier).

tions, hydrolyzed films and PEG_{500}-NH_2 functionalized films were incubated in the cell solution. For both cases, cell-surface coverages were found to be decreased markedly. On the PEG functionalized surface only very few isolated cells were observed, as seen in Figure 10(c). The result demonstrates that PEG_{500}-NH_2 functionalized surfaces can be used as an antifouling layer for preventing the cell interaction with the substrate.

2.4.3 Outlook

The relevance of structured and functionalized polymers and thin-film architectures, as discussed in this chapter, is not at all limited to the application areas of biosensing and controlled cell–surface interactions. With the more recent developments in suppression of unwanted nonspecific adsorption of proteins and efficient bioconjugation / surface decoration, numerous new areas for potential application and usage have been opened. In the specific context of biointerfaces, these developments promise to provide new incentives across the traditional boundaries of the involved disciplines. Tailored microenvironments, as will be introduced in Chapter 3.4, allow one already today to study the behavior of single cells, which is also of relevance for toxicology screening. Similarly important are microenvironments and niches for stem-cell differentiation. The approaches developed now offer control in terms of dimensionality (i.e. 2D vs. 3D) and sophistication, i.e. designed/engineered surfaces and structures versus 'random' trial and error type approaches. Thereby the pathways to systematic studies have been opened and researchers in biology and the biomedical field can expand the pioneering studies mentioned to a wide range of different cell types, on the one hand, and the development of low cost and more efficient biosensors on the other hand, e.g. in the development of personalized point of care applications.

Acknowledgments

This work has been financially supported by the MESA$^+$ Institute for Nanotechnology of the University of Twente and the Council for Chemical Sciences of the Netherlands Organization for Scientific Research (CW-NWO) in the framework of the vernieuwingsimpuls program (grant to HS).

References

1 Kasemo, B. (2002) *Surface Science*, **500**, 656.

2 (a) Lahann, J. (2006) *Chemical Engineering Communications*, **193**, 1457; (b) Yoshida, M., Langer, R., Lendlein, A. and Lahann, J. (2006) *Polymer Reviews*, **46**, 347.

3 (a) Schena, M., Shalon, D., Davis, R.W. and Brown, P.O. (1995) *Science*, **270**, 467; (b) Schena, M. (2003) *Microarray Analysis*, John Wiley and Sons Inc., Hoboken, NJ.

4 Fodor, S.P.A., Read, J.L., Pirrung, M.C., Stryer, L., Lu, A.T. and Solas, D. (1991) *Science*, **251**, 767.

5 (a) Chee, M., Yang, R., Hubbell, E., Berno, A., Huang, X.C., Stern, D., Winkler, J., Lockhart, D.J., Morris, M.S. and Fodor, S.P.A. (1996) *Science*, **274**, 610; (b) Pirrung, M.C. (2002) *Angewandte Chemie – International Edition*, **41**, 1277.

6 He, M., Stoevesandt, O., Palmer, E.A., Khan, F., Ericsson, O. and Taussig, M.J. (2008) *Nature Methods*, **5**, 175.

7 http://www.fda.gov/bbs/topics/news/2004/new01149.html.

8 Gong, P. and Grainger, D.W. (2004) *Surface Science*, **570**, 67; Gong, P., Harbers, G.M. and Grainger, D.W. (2006) *Analytical Chemistry*, **78**, **2342.**

9 Canales, R.D., Luo, Y.L., Willey, J.C. *et al.* (2006) *Nature Biotechnology*, **24**, 1115.

10 (a) Ito, Y. (1999) *Biomaterials*, **20**, 2333; (b) Falconnet, D., Csucs, G., Grandin, H.M. and Textor, M. (2006) *Biomaterials*, **27**, 3044; (c) Chen, C.S., Mrksich, M., Huang, S., Whitesides, G.M. and Ingber, D.E. (1997) *Science*, **276**, 1425; (d) Curtis, A. and Wilkinson, C. (1997) *Biomaterials*, **18**, 1573; (e) Pelham, R.J. and Wang, Y.-L. (1997) *Proceedings of the National Academy of Sciences of the USA*, **94**, 13661; (f) Curtis, A.S., Casey, B., Gallagher, J.O., Pasqui, D., Wood, M.A. and Wilkinson, C.D. (2001) *Biophysical Chemistry*, **94**, 275; (g) Curtis, A.S., Gadegaard, N., Dalby, M.J., Riehle, M.O., Wilkinson, C.D. and Aitchison, G. (2004) *IEEE Transactions on Nanobioscience*, **3**, 61; (h) Dalby, M.J., Silvio, L.D., Harper, E.J. and Bonfield, W. (2002) *Journal of Materials Science-Materials in Medicine*, **13**, 311; (i) Ingber, D.E. (2003) *Journal of Cell Science*, **116**, 1157; (j) Ingber, D.E. (2003) *Journal of Cell Science*, **116**, 1397.

11 Love, J.C., Estroff, L.A., Kriebel, J.K., Nuzzo, R.G. and Whitesides, G.M. (2005) *Chemical Reviews*, **105**, 1103.

12 Pathak, S., Singh, A.K., McElhanon, J.R. and Dentinger, P.M. (2004) *Langmuir*, **20**, 6075.

13 Williams, R.A. and Blanch, H.W. (1994) *Biosensors & Bioelectronics*, **9**, 159.

14 Degenhart, G.H., Dordi, B., Schönherr, H. and Vancso, G.J. (2004) *Langmuir*, **20**, 6216.

15 (a) Wadu-Mesthrige, K., Amro, N.A., Garno, J.C., Xu, S. and Liu, G.Y. (2001) *Biophysical Journal*, **80**, 1891; (b) Kenseth, J.R., Harnisch, J.A., Jones, V.W. and Porter, M.D. (2001) *Langmuir*, **17**, 4105.

16 Seurynck-Servoss, S.L., White, A.M., Baird, C.L., Rodland, K.D. and Zangar, R.C. (2007) *Analytical Biochemistry*, **371**, 105.

17 Valsesia, A., Colpo, P., Mannelli, I., Mornet, S., Bretagnol, F., Ceccone, G. and Rossi, F. (2008) *Analytical Chemistry*, **80**, 1418.

18 Kumar, A. and Whitesides, G.M. (1993) *Applied Physics Letters*, **63**, 2002.

19 (a) Bernard, A., Renault, J.P., Michel, B., Bosshard, H.R. and Delamarche, E. (2000) *Advanced Materials*, **12**, 1067; (b) Falconnet, D., Koenig, A., Assi, T. and Textor, M. (2004) *Advanced Functional Materials*, **14**, 749.

20 Ruoslahti, E. (1996) *Annual Review of Cell and Developmental Biology*, **12**, 697.

21 Cavalcanti-Adam, E., Micoulet, A., Blümmel, J., Auernheimer, J., Kessler, H. and Spatz, J.P. (2006) *European Journal of Cell Biology*, **85**, 219.

22 (a) Maheshwari, G., Brown, G., Lauffenburger, D.A., Wells, A. and Griffith, L.G. (2000) *Journal of Cell Science*, **113**, 1677; (b) Koo, L.Y., Irvine, D.J., Mayes, A.M., Lauffenburger, D.A. and Griffith, L.G. (2002) *Journal of Cell Science*, **115**, 1423.

23 (a) Horbett, T.A. (1994) *Colloids and surfaces B – Biointerfaces*, **2**, 225; (b) Siebers, M.C., ter Brugge, P.J., Walboomers, X.F. and Jansen, J.A. (2005) *Biomaterials*, **26**, 137.

24 Blattler, T., Huwiler, C., Ochsner, M., Stadler, B., Solak, H., Voros, J. and Grandin, H.M. (2006) *Journal of Nanoscience and Nanotechnology*, **6**, 2237.

25 (a) Guo, L.J. (2004) *Journal of Physics D – Applied Physics*, **37**, R123; (b) Charest, J.L., Bryant, L.E., Garcia, A.J. and King, W.P. (2004) *Biomaterials*, **25**, 4767.

26 Xia, Y., Mrksich, M., Kim, E. and Whitesides, G.M. (1995) *Journal of the American Chemical Society*, **117**, 9576.

27 Knoll, W., Matsuzawa, M., Offenhäusser, A. and Rühe, J. (1996) *Israel Journal of Chemistry*, **36**, 357.

28 Murata, H., Chang, B.-J., Prucker, O., Dahm, M. and Rühe, J. (2004) *Surface Science*, **570**, 111.

29 Wells, M. and Crooks, R.M. (1996) *Journal of the American Chemical Society*, **118**, 3988.

30 Langer, R. and Vacanti, J.P. (1993) *Science*, **260**, 920.

31 (a) Bruening, M.L., Zhou, Y., Aguilar, G., Agee, R., Bergbreiter, D.E. and Crooks, R.M. (1997) *Langmuir*, **13**, 770; (b) Chance, J.J. and Purdy, W.C. (1997) *Langmuir*, **13**, 4487.

32 Fixe, F., Dufva, M., Telleman, P. and Chistensen, C.B.V. (2004) *Nucleic Acid Research*, **32**, e9.

33 Heyes, C.D., Groll, J., Möller, M. and Nienhaus, G.U. (2007) *Molecular Biosystems*, **3**, 419.

34 Wang, D.-A., Ji, J., Sun, Y.-H., Shen, J.-C., Feng, L.-X. and Elisseeff, J.H. (2002) *Biomacromolecules*, **3**, 1286.

35 Jerome, C., Gabriel, S., Voccia, S., Detrembleur, C., Ignatova, M., Gouttebaron, R. and Jerome, R. (2003) *Chemical Communications*, 2500.

36 Nandivada, H., Chen, H.-Y., Bondarenko, L. and Lahann, J. (2006) *Angewandte Chemie – International Edition*, **45**, 3360.

37 Dai, J., Bao, Z., Sun, L., Hong, S.U., Baker, G.L. and Bruening, M.L. (2006) *Langmuir*, **22**, 4274.

38 Ma, H., Li, D., Sheng, X., Zhao, B. and Chilkoti, A. (2006) *Langmuir*, **22**, 3751.

39 Malmqvist, M. and Karlsson, R. (1997) *Current Opinion in Chemical Biology*, **1**, 378.

40 (a) Sigal, G.B., Bamdad, C., Barberis, A., Strominger, J. and Whitesides, G.M. (1996) *Analytical Chemistry*, **68**, 490; (b) Gehrke, S.H., Vaid, N.R. and McBride, J.F. (1998) *Biotechnology and Bioengineering*, **58**, 417.

41 (a) Niemeyer, C.M. and Blohm, D. (1999) *Angewandte Chemie – International Edition*, **38**, 2865; (b) Schulze, A. and Downward, J. (2001) *Nature Cell Biology*, **3**, E190.

42 Benters, R., Niemeyer, C.M., Drutschmann, D., Blohm, D. and Wöhrle, D. (2002) *Nucleic Acid Research*, **30** (e10), 1.
43 Rowan, B., Wheeler, M.A. and Crooks, R.M. (2002) *Langmuir*, **18**, 9914.
44 Lahann, J., Balcells, M., Rodon, T., Lee, J., Choi, I.S., Jensen, K.F. and Langer, R. (2002) *Langmuir*, **18**, 3632.
45 Zhou, X., Wu, L. and Zhou, J. (2004) *Langmuir*, **20**, 8877.
46 Park, S.J., Lee, K.B., Choi, I.S., Langer, R. and Jon, S.Y. (2007) *Langmuir*, **23**, 10902.
47 (a) Zhang, Z., Chen, Q., Knoll, W., Foerch, R., Holcomb, R. and Roitman, D. (2003) *Macromolecules*, **36**, 7689; (b) Zhang, Z., Knoll, W., Foerch, R., Holcomb, R. and Roitman, D. (2005) *Macromolecules*, **38**, 1271.
48 Kuhlman, W., Taniguchi, I., Griffith, L.G. and Mayes, A.M. (2007) *Biomacromolecules*, **8**, 8206.
49 Kumar, N., Parajuli, O., Dorfman, A., Kipp, D. and Hahm, J.I. (2007) *Langmuir*, **23**, 7416.
50 Edmondson, S., Osborne, V.L. and Huck, W.T.S. (2004) *Chemical Society Reviews*, **33**, 14.
51 Senaratne, W., Andruzzi, L. and Ober, C.K. (2005) *Biomacromolecules*, **6**, 2427.
52 Förch, R., Zhang, Z.H. and Knoll, W. (2005) *Plasma Processes and Polymers*, **2**, 351.
53 Huang, N.P., Michel, R., Voros, J., Textor, M., Hofer, R., Rossi, A., Elbert, D.L., Hubbell, J.A. and Spencer, N.D. (2001) *Langmuir*, **17**, 489.
54 Feller, L.M., Cerritelli, S., Textor, M., Hubbell, J.A. and Tosatti, S.G.P. (2005) *Macromolecules*, **38**, 10503.
55 Feng, C.L., Vancso, G.J. and Schönherr, H. (2005) *Langmuir*, **21**, 2356.
56 Feng, C.L., Embrechts, A., Bredebusch, I., Bouma, A., Schnekenburger, J., García-Parajó, M., Domschke, W., Vancso, G.J. and Schönherr, H. (2007) *European Polymer Journal*, **43**, 2177.
57 Feng, C.L., Vancso, G.J. and Schönherr, H. (2007) *Langmuir*, **23**, 1131.
58 Feng, C.L., Embrechts, A., Bredebusch, I., Schnekenburger, J., Domschke, W., Vancso, G.J. and Schönherr, H. (2007) *Advanced Materials*, **19**, 286.
59 Feng, C.L., Embrechts, A., Vancso, G.J. and Schönherr, H. (2006) *European Polymer Journal*, **42**, 1954.
60 Duvigneau, J., Schönherr, H. and Vancso, G.J. (2008) *Langmuir*, **24**, 10825–10832.
61 Feng, C.L., Zhang, Z., Förch, R., Knoll, W., Vancso, G.J. and Schönherr, H. (2005) *Biomacromolecules*, **6**, 3243.
62 Michel, R., Pasche, S., Textor, M. and Castner, D.G. (2005) *Langmuir*, **26**, 12327.
63 Ameringer, T., Hinz, M., Mourran, C., Seliger, H., Groll, J. and Möller, M. (2005) *Biomacromolecules*, **6**, 1819.
64 Pan, F., Wang, P., Lee, K., Wu, A., Turro, N.J. and Koberstein, J.T. (2005) *Langmuir*, **21**, 3605.
65 Hillborg, H., Tomczak, N., Olàh, A., Schönherr, H. and Vancso, G.J. (2004) *Langmuir*, **20**, 785.
66 Feng, C.L., Vancso, G.J. and Schönherr, H. (2006) *Advanced Functional Materials*, **16**, 1306.
67 Sullivan, T.P., van Poll, M.L., Dankers, P.Y.W. and Huck, W.T.S. (2004) *Angewandte Chemie – International Edition*, **43**, 4190.
68 Embrechts, A., Feng, C.L., Mills, C.A., Lee, M., Bredebusch, I., Schnekenburger, J., Domschke, W., Vancso, G.J. and Schönherr, H. (2008) *Langmuir*, **24**, 8841.
69 Litmanovich, A.D. and Cherkezyan, V.O. (1984) *European Polymer Journal*, **20**, 1041.
70 Dusseiller, M.R., Schlaepfer, D., Koch, M., Kroschewski, R. and Textor, M. (2005) *Biomaterials*, **26**, 5917.
71 (a) Liebermann, T. and Knoll, W. (2000) *Colloids and Surfaces A*, **171**, 115; (b) Liebermann, T., Knoll, W., Sluka, P. and Herrmann, R. (2000) *Colloids and Surfaces A*, **169**, 337; (c) Yu, F., Yao, D. and Knoll, W. (2003) *Analytical Chemistry*, **75**, 2610.

3
Biointerfaces, Biosensing, and Molecular Interactions

Surface Design: Applications in Bioscience and Nanotechnology
Edited by Renate Förch, Holger Schönherr, and A. Tobias A. Jenkins

ISBN: 978-3-527-40789-7

3.1
Surface Chemistry in Forensic-Toxicological Analysis

Martin Jübner, Andreas Scholten, and Katja Bender

3.1.1
Introduction

Forensic toxicological assignment comprises the investigation of unclear cause of death, driving under the influence of illegal and therapeutically used central active drugs and narcotics, self-poisoning, misuse of therapeutic drugs, abuse of narcotics and alcohol, abuse of narcotics in the context of sexual violations and other delicts. Detecting the abuse of narcotics by road-traffic participants is also part of daily police work. Furthermore, recent sociological developments show that the fast and reliable testing of suspicious materials is becoming more important in the prevention of terrorist attacks. The broad variety of forensic and clinically relevant substances (legal and illegal drugs and narcotics, toxins and other poisons) and samples (body fluids and tissues) that have to be analyzed as well as the different pharmacological and toxicological tasks in forensic science are underlining the importance of novel analytical screening devices. Standard immunochemical screening techniques used in forensic toxicology usually narrow down the eligible substances into substance classes (such as opiates, benzodiazepines, tricyclic antidepressants, cannabinoides, opiates, and amphetamines). For further clarification, time-consuming and expensive analytical methods are used for precise identification and quantification of the individual substance that belong to the substance class tested positive in the screening procedure. The methods are focused mainly on the detection of toxic substances (e.g. parathion, cyanide, botulinum toxin), therapeutic (e.g. benzodiazepines, barbiturates, opioids) and illegally used drugs (e.g. cannabinoids, morphine, amphetamines) and their metabolites. Common detection methods for such substances are based on immunochemical reactions, often coupled with photometric measurements (e.g. Inspec®, Cedia®), or on chromatographic principles (liquid or gas chromatography). This works well when detecting substances that are usually present at substantial concentration in the body, like amphetamines or benzoylecgonin.

Nevertheless, some substances are rapidly metabolized or excreted (e.g. gamma-hydroxybutyric acid (GHB), LSD, suxamethoniumchloride). In these cases, the concentration in blood or tissue can be below the limit of detection before sam-

Surface Design: Applications in Bioscience and Nanotechnology
Edited by Renate Förch, Holger Schönherr, and A. Tobias A. Jenkins

ISBN: 978-3-527-40789-7

pling. In addition, long storage times under unsuitable conditions (room temperature, UV radiation) or postmortem decomposition can result in the loss of a substance. After sampling and even after death, enzymes remain active for a while. During this period a decrease of substances can appear due to enzyme activity. This is well known for cocaine [1–5] and heroin [6] that are rapidly hydrolyzed by blood cholinesterase enzymes. These enzymes can be inhibited by adding NaF to the sample. But the disadvantage of using NaF as an additive is that it causes hemolysis and inhibits also the enzyme reaction, which is a crucial component in enzymatic immunoassays (such as ELISA, EMIT). On the other hand, many substances, e.g. LSD, ricin, botulinum toxin, fentanyl, sufentanil, and digoxin show pharmacological or toxicological activities below nano- or picomolar concentrations in blood [7].

Thus, a central issue of forensic-toxicological analysis is the detection of substances that are present at ultralow concentrations in body fluids and tissues. In the last decade, the limit of detection of such analytes has decreased as a result of the development of new or enhanced analytical methods like coupled chromatographic and spectrometric instrumental devices or immunoassays based on surface-bound antibodies. Typical limits of detection concerning LC/MS analysis of opioids, like fentanyl, or GC/MS analysis of opiates, like morphine or cocaine are in the nanomolar range. However, trace analysis is often affected by the properties of the analytes and the matrix. Samples with complex matrices, such as blood, urine, saliva, cerebrospinal fluid, tissue, and hair, hamper the recovery of the analyte and increase the detection limit of the analytical method. Matrix overloads can cause high background noise. Matrix components and/or analytes may overlap in their analytical behavior, which results in insufficient separation of the substances during chromatography. In both cases the detection at the (ultra-)trace level can be hindered massively. This is a challenge, especially in cases of survived poisoning or poisoning followed by an elongated agonal state when the detection of substances below a nano- or picomolar range has to be possible to prove potential intake. Moreover, the analysis of decomposed sample material from corpses is often affected by such matrix effects.

Postmortem changes due to autolysis, putrefaction, and decomposition can cause certain difficulties in toxicological analysis. Cadaver-associated biogenous substances have a notable influence on the recovery of analytes in body fluids or tissues. In Figure 1 HPLC-chromatograms of two different blood samples are shown, to demonstrate the complexity of samples affected by autolysis and putrefaction in comparison to samples of living individuals. Normally, fresh blood samples show less interference resulting from matrix effects than samples taken from dead bodies. This is represented by the few clear peaks in the left chromatogram of a living person. In contrast, the right chromatogram of a dead body shows markedly more peaks that partly overlap and can hardly be sufficiently separated by chromatography, even if an extraction step was performed before analysis. Peaks with high intensity or broad (multiple) peaks may hide substance peaks with a lower intensity. Consequently, an identification and quantification of low-abundant substances is very difficult in such cases, which means additional ana-

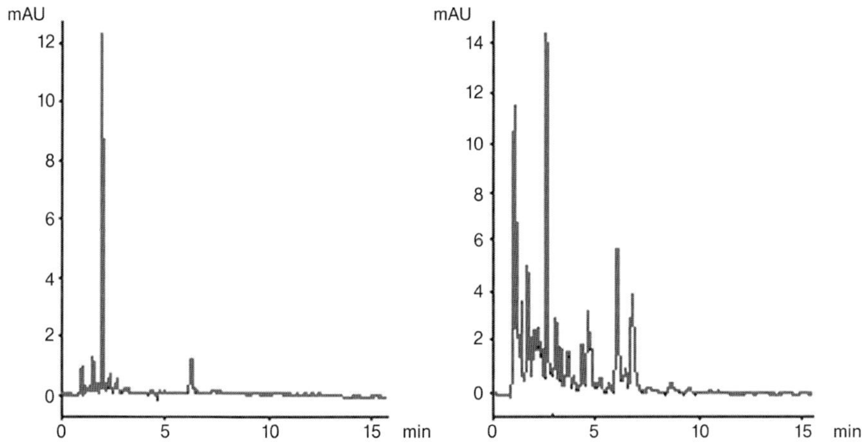

Figure 1 HPLC-chromatograms of two serum-samples after adding Na_2HPO_4 and liquid/liquid extraction with 1-chlorbutane (HPLC-column: VDS optilab nucleosil 100 C8, 8.5 μm, 250 × 4.0 mm, isocratic mobile phase consisting of CH3CN/H2SO4 (1 mM) 1:1.61 V/V, flowrate 1.7 mL min^{-1}). Left: Serum sample of a living person. Right: Serum sample of a dead body.

lytical methods have to be applied. This issue implicates an imperative necessity of a quick and simple extraction of the analytes from matrix components with a high recovery and their separation, followed by very sensitive detection methods.

In the case of total excretion, body fluids or tissues are unsuitable samples, therefore alternative samples have to be analyzed. Due to the fact that many substances get incorporated into hair and nail matrix [8–18], these samples can be used for testing, even if the ingestion was long before the testing procedure [19–21]. Nail and hair testing is also applicable to exhumed bodies [22]. Drug detection is only possible if the length of the available and analyzed hair or nail samples covers the period of substance intake. The amount of such incorporated substances is even lower than the pharmacological or toxicological relevant blood levels would be. The expected concentration in hair after intake of a single dose of a highly potent therapeutically used substance, such as flunitrazepam, is often below ng/mg or pg/mg [23–26]. This is why it is not always possible to detect the substance in hair even within a suitable period after intake [25]. Contamination of hair with sweat and airborne particles are also possible. Therefore, the time of substance intake and the traceability in hair does not always correlate [25]. Furthermore, the incorporated amount in hair of a substance of same dosages is interindividually highly variable. For these reasons, the interpretation of the results achieved by hair testing can be very difficult. With far lower detection limits than those that can be reached with conventional analytical methods like advanced LC/MS- or GC/MS-methods it may be possible to get a better correlation between substance intake and its appearance in hair. This would be very helpful especially when investigating drug-facilitated (sexual) assaults. Biosensor applications using novel surface chemistry or new surface materials provide the potential to complement and to improve the standard analytical methods used in forensic toxicology.

Another challenge in forensic science is the detection of pathological or medically relevant proteins that can be used as biomarkers that indicates a certain illness or pathological event. The detection of such markers is very common in some fields of medical diagnosis, e.g. the detection of carbohydrate deficient transferrine (CDT) as an indicator of alcoholism, C-reactive protein (CRP) as an indicator of inflammatory processes or of cardiac troponins as markers of cardiac disease. Further possibilities for biosensor applications are seen in the detection of specific protein patterns and/or biomarkers, indicating certain causes of death like suffocation, starvation, death of thirst, hypothermia, hyperthermia, and the sudden infant death syndrome (SIDS), or are correlated with the postmortal interval (PMI). In these cases macroscopic findings during autopsy are often very indistinct, therefore analytical methods that allow a far more specified diagnosis in addition to the established procedures would be very helpful.

Moreover, pathologically relevant proteins (e.g. cardiac infarction markers, C-reactive protein) can be rapidly decomposed by autolysis or putrefaction [27–32]. In this case the concentration of marker proteins would decline below the detection limit and these proteins thus would not be detectable. Postmortal cell breakdown can also cause a release of marker proteins. Some intracellular proteins, which are usually markers of pathologically relevant processes like the creatine kinase, would then be detectable at pathologically relevant concentrations in blood samples even if no pathological event has occurred. Rapid test methods would be necessary in order to detect molecules of interest before cell breakdown takes place when performing a postmortem analysis of disease associated marker proteins. Point-of-scene testing devices would be a must to keep the postmortem interval between sampling and analysis as short as possible in order to minimize distortion. Postmortem-associated perturbations in the analysis of toxicological or clinically relevant parameters have to be excluded as much as possible. The key issues therefore are a sufficient extraction, separation and detection of the desired analyte.

3.1.2
Forensic Analysis and Novel Surface-Based Biosensors

Crucial premises for successful sample pretreatment and trace analysis are a sufficient extraction, separation and detection of the analyte. Laboratory-based drug testing is usually done in two steps:

1. screening and
2. confirmation.

The screening test is applied to nearly all samples that go through the laboratory. The confirmation test is only applied to samples that are tested positive during the screening test. Screening tests are usually done by immunoassay, like enzyme multiplied immunoassay technique (EMIT), enzyme-linked immunosorbent assay (ELISA), and radioimmunoassay (RIA), or modifications of these methods, or by chromatographic methods like high-performance liquid chromatography (HPLC), gas chromatography/mass spectroscopy (GC/MS), liquid chromatography/mass spectroscopy (LC/MS) or tandem mass spectroscopy. Commercially available immunoassays for drugs or pathologically relevant marker proteins are predominantly working well, but biogenous substances like amines can cause false-positive results, because of their crossreactivity on the capture/detection antibody used in the immunoassay. Moreover, body fluids and tissues consist of very complex matrices. The high concentrations of fatty acids, lipids, proteins and other components have a strong tendency to adsorb on surfaces [33–35]. The main driving forces of protein adsorption are the formation of hydrogen bonds, hydrophobic, van der Waals, and Coulomb interactions [36]. Polymers, e.g. polyethylene glycol, polyethylene oxide [37] and polysaccharides [38–40] with polar and hydrophilic properties are able to prevent nonspecific protein adsorption [41, 42]. The demands on a biosensor surface can rise exponentially when sensor surfaces are exposed to complex samples, such as whole blood, urine or samples from highly decomposed corpses. Subsequently, the access of the analyte to the recognition site on a conventional sensor surface can be blocked, giving rise to false results and misinterpretations. In modern trace analysis the above-mentioned liquid and gas chromatographic methods coupled with several detection methods are often applied to perform screening tests as well as confirmation tests. In the case of the liquid chromatographic methods intrinsic and extrinsic substances can overlap during chromatographic separation. In the case of detecting the substances by their UV/VIS-absorption spectrum such an overlap may cause another overlap in the absorption spectra. Identification and quantification difficulties then depend on the dimension of the substance overlap. Chromatographic overlaps could also interfere in GC/MS or LC/MS analysis.

One advantage of both methods is that they detect the substances via their molecular mass and mass fragments. Uncharacteristic fragments are always present, originating from sample content (amino acids, proteins, etc.). Frequently, a derivatization of the desired analyte is performed. On the one hand, this aims at a better volatility of the substance in order to reach a more effective transfer to the gas phase of the GC-system, on the other hand, it is also used to get larger and characteristic substance fragments. In the case of LC/MS, substance or matrix overloads can cause ion suppression (quenching of the signal), which means an analyte is not recorded even if it is contained in the sample. Those analytical methods require a high standard of knowledge and experience in order to prove that toxicological analysis has been performed correctly.

3.1.2.1
Principles and Confines of Extraction, Separation and Detection of Analytes and the Role of Surface Chemistry

As mentioned before, in forensic trace analysis the extraction of the analytes from the sample (matrix) is usually performed before the standard chromatographic methods are applied. The chromatographic separation of the analyte mixture as part of the chromatographic method then takes place before detection and quantification. Some special analytical techniques include the extraction step in the analytical method (e.g. SPME, SPDE). To a large extent, the extraction and separation are based on methods that include a surface design and adsorption.

Extraction

These days, a standard extraction method, beside liquid–liquid extraction, is solid-phase extraction (SPE). Some of the above-mentioned difficulties in forensic toxicological analytics caused by matrix effects can be overcome by a precleanup and preconcentration step done with SPE. Widely used are commercial prepacked columns containing stationary phases that may be adsorbents such as silica gel or functionalized silica. Mainly octadecyl (C18)- and octyl (C8)-siloxans used as reversed-phase materials or ion-exchange media (e.g. carboxylpropyl-, (diethyl-) aminopropyl-, propyl- or phenylsulfonic acid-functionalized silica). But also molecular imprinted polymers (MIPs) are usable as column fillings for SPE [43–46]. A good overview of molecular imprinted polymers can be found in the review article of Alexander *et al.* [47].

In some cases, when standard analytical methods are no longer sufficient, highly advanced analytical techniques have to be applied. Crescenzi and coworkers have applied MIPs followed by liquid chromatography/electrospray ion-trap multiple-stage mass spectrometry to determine clenbuterol in bovine liver [48].

Moreover, in recent years an analytical separation method called head-space solid-phase microextraction (HS-SPME) [49–51] has been developed to enable direct separation of analytes from matrix compounds. This works by adsorbing the analyte onto a surface-coated fiber from the vapor phase above the matrix. However, direct immersion into the matrix (DI-SPME) [52] is also possible. The fibers used are generally coated with activated carbon, polydimethylsiloxane (PDMS) and polyvinylalcohol (PVA), polyacrylate (PA), carbowax or composites of these coatings [53]. A further development of SPME was achieved by head-space solid-phase dynamic extraction (HS-SPDE) [54–59]. In contrast to SPME, SPDE (Figure 2) uses a hollow needle. The inside of the needle is coated with one of the above-mentioned materials. In SPME as well as in SPDE (except when the SPME-fiber is directly immersed into the matrix) the sample matrix is filled into a special, gas-tight vial, allowing an atmospheric space above the sample matrix (Figure 2a). An equilibrium between substances in the matrix and the atmospheric space above will develop over time. By gentle heating of the matrix in an oven the equilibrium will be reached faster and a higher concentration of the analyte in the vapor phase can be found than without heating (Figure 2b). A major advance of SPDE is the

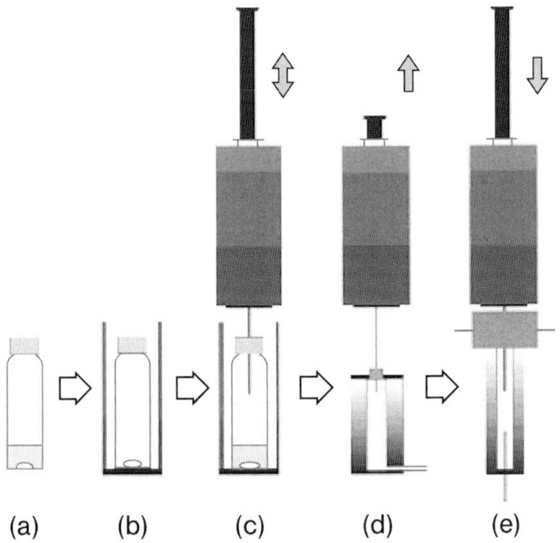

Figure 2 HS-SPDE: After the matrix is placed into a gas tight vial (a), the vial is transferred into an oven (b). The elevated temperature is enforcing fast setting of the equilibrium between vapor phase and matrix. After the extraction step of the analyte by several aspiration steps of the vapor phase and adsorption of the analyte onto the needle coating (c), the analyte is desorbed in the GC-injector by the heat of the injector (d and e). (Provided by D. Lenz, Institute of Legal Medicine, University of Cologne).

multiple active aspiration of vapor containing analyte (Figure 2c). For this reason, the equilibrium between adsorbed analyte and analyte contained in the vapor phase can be obtained more effectively than using SPME. Fully automated SPME and SPDE methods are mainly used in combination with GC/MS-methods, desorbing the analyte from the coating by the heat of the GC-injector (Figures 2d and e). Both methods have been used successfully for the analysis of several volatile substances (drugs of abuse, therapeutics, pesticides, etc.) in different matrices (blood, saliva, hair, food, etc.). SPDE and DI-SPME are only sufficient, when the analytes are volatile. This mainly does not apply to polar, charged and very large substances and proteins in solution, which tend to remain in aqueous samples, like blood. The analysis of highly polar substances or proteins even with recently developed methods is thus still a challenge.

Separation

Using chromatographic methods, the successful separation of substance mixtures prior to identification and quantification strongly depends on the surface modification of the column filling and on the chemical and physical properties of the analytes. Most column fillings used for LC are based on surface-modified porous silica or ceramic microparticles. Octadecyl (C18)- and octyl (C8)-silanes chemically bonded to silica are widely used in LC-columns. Nonpolar surfaces may collapse when a 100% aqueous mobile phase is used. Therefore, polar groups (amino-, cyano-, phenyl-groups, polyethylenglycol) have been introduced to attain a better chromatographic behavior of polar analytes or to get a better performance at extreme pH. For example, sulfonated fluorocarbon polymer coatings are used as strong cationic-exchangers and quaternary amine coatings as strong anionic-exchangers. Additionally, there is a broad variety of covalently bonded chiral surface coatings (e.g. ß-cyclodextrin). Capillary columns with, e.g., dimethyl-, phenyl-, and cyanopropyl-polysiloxane coatings, usually not covalently bonded on the capillary-wall, are used mostly in GC. The stability of these coatings is attained by highly crosslinked polymers.

Detection

Typically applied detection methods, besides the already mentioned UV/VIS-absorption spectrum and molecular mass and mass fragments, are usually fluorescence or electrochemical detection in the case of LC and in the case of GC flame ionization, electron capture, and thermal conductivity. The achieved limit of detection substantially depends on a sufficient foregoing extraction and separation. With advanced extraction methods combined with susceptible chromatographic methods detection limits in the ppt range can be obtained. This, for example, was demonstrated by using SPE in combination with LC/MS detecting aflatoxins [60]. But, in everyday forensic toxicological analysis the LoD is in the ppm or ppb range, strongly depending on the type and state of the sample and the applied extraction methods. Nevertheless, the applied analytical method, the condition of the analytical instrument as well as properties of the analyte also influence the LoD. Thus, the availability of devices allowing a simple and efficient detection of the analyte in the sample, e.g. by surfaces that repel unspecific binding of matrix components, will improve the limits of detection of forensic and clinically relevant analytes, or at least help to simplify the analytical procedure.

3.1.2.2
New Attempts in Analytical Surface Chemistry

New test methods using novel biosensor architectures in combination with very sensitive detection techniques like surface plasmon resonance spectroscopy (SPS, see Chapters 1.2 and 6.7 [61], surface plasmon enhanced fluorescence spectroscopy (SPFS) [62] (see Chapter 5.3), quartz-crystal microbalance (QCM) [63–66], and cantilever sensors [67, 68] can possibly reduce the above-mentioned difficulties in analysis by the detection of analytes in crude samples without a foregoing extraction step. Engineered surfaces that are attributed with matrix-component-repelling properties and specific binding sites for the desired analyte should be able to extract and separate the analyte from the matrix in one step. With such a sensor surface the online extraction, separation and detection of analytes would be possible. First attempts have been made by using analyte-adsorbing materials in SPDE and SPME or using a polydimethylsiloxane (PDMS) coated stir bar (Gerstel-Twister®), stir bar sorptive extraction (SBSE) patented by the company Gerstel (www.gerstel.com). This device is coupled with a thermal desorption unit fabricated in a fashion that this extraction method can be directly coupled with LC, GC or GC/MS methods for detecting volatile, semivolatile or nonvolatile substances, depending on the chromatographic method. With the SBSE extraction and enrichment of the analyte can be done in one step, also gaining more sensitivity than achievable with conventional SPME. But still a further chromatographic separation and detection step of the analytes is necessary to complete the analysis. Molecular-imprinted molecules are introduced as the stationary phase in LC methods to perform enantioseparation of D,L-phenylalanine by using D-phenylalanine imprinted poly(methacrylic acid-co-ethylene glycol dimethylacrylate) (p(MAA-co-EGDMA)) microbeads [69]. Here, the surface coating of the microbeads has been used to separate the enantiomers. Nevertheless, in forensic science as well as for clinical applications novel biosensor-based test devices have not yet been implemented in daily laboratory routine. One reason may be the lack of novel biosensor surfaces suitable for detecting substances in biological samples in trace or ultratrace amounts.

However, many surface architectures have been developed for biosensing purposes in the last decade (see also Chapters 1.2 and 1.4). Most sensor surfaces are based on surface-adsorbed bovine serum albumin (BSA) [70], self-assembled monolayers [71], plasma-assisted surface modifications [72–75] and hydrophilic polymers like hydrogels [41, 76] with different functional groups for bioconjugation (biotin/streptavidin [77], Ni-His tags [78], NHS esters [79], aldehyde chemistry [80], etc.) as mentioned in Chapter 1.4. Most sensor types contain antibodies against the analyte acting as recognition sites. In Figure 3 different biosensor architectures are sketched. Figure 3(a) shows a sensor architecture based on surface-bound streptavidin via biotinylated thiols or silanes as referred to in Chapters 1.1 and 1.4. The self-assembled monolayer (SAM) can be laterally diluted by a short thiol or silane to reduce sterical hinderance of streptavidin binding to the biotin groups. In this case, a biotinylated capture antibody is coupled to streptavidin by its biotin-anchor group. Another possibility to bind a capture antibody to surfaces is by physisorption of protein G or A (see Figure 3b). These proteins are of microbiological origin. The capture antibody will only bind to protein G or A via its Fc-region, which is a protein domain with no sensing properties. This domain is highly conserved in many species. Capture antibodies can also be covalently bound to brush-like oligomers and polymers (see Figure 3c) with a dense distribution of functional groups (–COOH, $–NH_2$). Instead of carboxylic or amine-terminated polymers, oligomers or aliphatic thiols carrying the same functional groups can be used as well. This is made possible by using protein-coupling chemistry (e.g. EDC/NHS) to form amide bonds between the antibody and the functional groups of the oligomers or polymers (see Chapters 1.1 and 1.4). The polymers or oligomers can be anchored to the surface via sulfur bonds or by SiO_x-coupling techniques. Physisorbed (spin-coated) or covalent bound swellable polymer networks (hydrogels) with functional groups might be biofunctionalized with capture antibodies in the same way (see Figure 3d). Molecular-imprinted polymers (MIPs) [47, 81–88] as well as aptamers [89, 90] can also be used as sensing

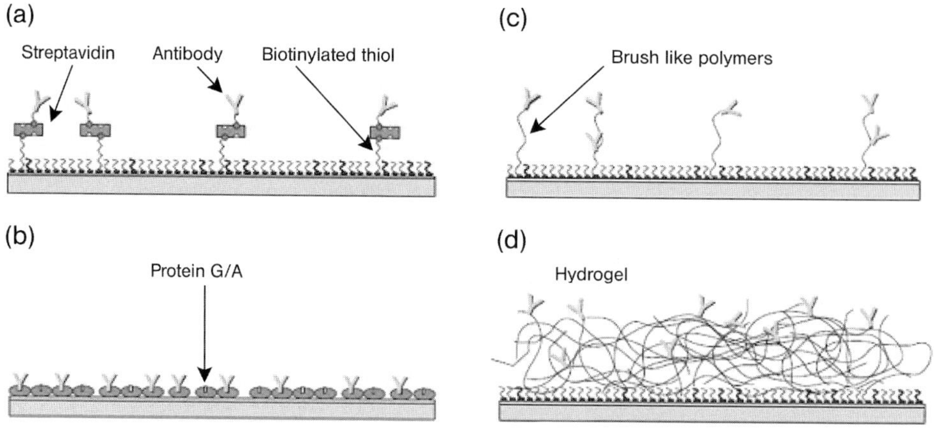

Figure 3 Antibody-based sensor designs. Surface-bound antibodies via (a) streptavidin-biotin coupling, (b) protein G or A complex and covalent bound antibodies onto polymer brushes, (c) polymer networks like hydrogels (d).

molecules. Chemical surface-grafted molecular imprinted homopolymers have been used as artificial adrenergic receptors in enzyme-linked assays [91]. MIPs can also be used instead of an antibody as the recognition element [92]. MIPs have been also introduced to optical biosensor applications [93], for example in infrared evanescent wave spectroscopy [94], SPS [95–98], or surface-acoustic-wave-based biosensor applications (QCM [99, 100]) as recognition sites. Some approaches have been made to implement molecular-imprinted polymers as recognition sites for pharmaceuticals as well as for proteins. For example, glassy carbon electrodes have been modified with molecular imprinted polymers (MIPs) by dissolving the template molecule (here the acidic form of diclofenac), methacrylic acid (MAA) and the crosslinking monomer ethyleneglycol dimethacrylate (EGDMA) in acetonitrile. α,ά-azo-bis-isobuyronitrile (AIBN) was used as a radical initiator. This mixture was spin coated onto the electrode and stored overnight at 60 °C in nitrogen atmosphere to assure polymerization, which results in a p(MAA-co-EGDMA)-covered electrode. Using this method, it was possible to detect diclofenac in aqueous solution electrochemically [81]. p(MAA-co-EGDMA) is frequently used to produce MIPs. More critical is the fabrication of protein-imprinted polymers because proteins can easily denature during polymerization. Such denaturizing of proteins would lead to MIPs with recognition cavities different from non-denaturized proteins as a template because the tertiary structure differs. Mild polymerization conditions have to be chosen in order to prevent protein denaturation. Shi and coworkers have used plasma-assisted polymerization to produce protein-imprinted polymers [86, 87]. In this approach the template protein was adsorbed on freshly cleaved mica followed by preparing a spin-coated trehalose-overlayer. Afterwards, the sugar layer was covered with plasma-polymerized hexafluoropropylene by using a 13.56-MHz radiofrequency glow-discharge reactor to carry out the polymerization. Afterwards, the polymer was glued onto a support (e.g. microscope coverslip). Then the mica was simply peeled off, subsequently exposing the protein partially. Afterwards, the protein was removed with NaOH solution, so finally free cavities as recognition sites were generated.

Moreover, the implementation of such surfaces (like functionalized hydrogels or plasma polymers) as coatings on fibers used in SPME or SPDE methods will open new facilities to reach much lower limits of detection when analyzing forensic and medical relevant substances in samples consisting of complex matrices. Using thermoresponsible polymers, of which the mesh aperture can possibly be expanded simply by changing the temperature, it will be possible to load and release the analyte by varying the temperature. In the literature, it has been shown that not only small molecules can be captured and released by these kinds of polymers, but also chiral recognition could be obtained [101]. By this means, a further increase in recovery may be realized compared to other extraction methods like SPME or SPDE. The chiral recognition could be useful when enantiomers show differences in their pharmacological activity. L-methadone (L-Polamidone, levomethadone), the R(-)-enantiomer of methadone used as a opiate substitute in the therapy of opiate addiction, for example exhibits a 50× higher pharmacological effect than the S(+)-enantiomer. For pharmaceutical and toxicological reasons the stereoselective analysis of such substances would be of advantage to consider side effects and the toxicity of certain substances. MIPs have been already prepared to separate phenylalanine racemates using D-phenylalanine imprinted p(MAA-co-EGDMA) [69, 102]. The adjusting of an aperture that is small enough not to allow larger molecules (e.g. serum albumin) to diffuse into the extraction layer may further improve the recovery of low-abundant substances. If such surfaces bear protein-repelling properties, proteins disturbing the analysis would not get into the analytical device, especially in the case of DI-SPME. With novel polymers like that, also less-volatile substances would be analyzable by DI-SPME or SPE instead of using SPME. The noted surface attributes may also help to improve immunochemical assays (see Chapter 1.1).

3.1.2.3
New Forensic Fields of Application of Novel Surface-Based Biosensors

It has previously been shown that some surface architectures are able to prevent unspecific protein adsorption by using polar and hydrophilic polymers. Unfortunately, most published sensor surface architectures have not been tested at the analysis of crude and/or degraded tissue samples and there still is no advanced sensor architecture commercially available for routine analysis of highly degraded postmortem samples. The analysis of pathologically relevant marker proteins has been and still is a challenge in forensic diagnostic routine. One reason is the fact that only little is known about the postmortem changes of pathologically relevant marker proteins and how far this influences the interpretation of post-mortem-collected data. To date, it is still unclear how precisely these data reflect the status shortly before death or its onset. Postmortem diagnostic would help considerably to elucidate functional causes of death (cardiac or brain infarction, diabetic coma, etc.) [103]. As mentioned above, forensic and pathologically relevant marker proteins underlie postmortem processes like autolysis and putrefaction, which can cause a decomposition of the analyte, especially proteins [31, 104–107]. In Figure 4 the postmortem changes of the proteome are visualized by 2D-gel electrophoretic separation of proteins before (control sample stored at –80 °C to prevent degradation) and after aging of muscle (*M. iliopsoas*) of the same individual at room temperature for 7 days. In particular, most large proteins are underlying a relative fast decomposition. This indicates that an immediate and quick analysis of complex body fluids or tissues is of high interest in order to be able to use pathologically relevant proteins for forensic diagnostic purposes. But not only postmortem decompositions have to be considered, also the effects of simple hemolysis of the erythrocytes, bacterial contamination and fungal decay are a big analytical challenge. Photometric-based analysis can be disturbed by self-absorbing hemoglobin. The combination of novel sensor architectures with surface plasmon enhanced fluorescence spectroscopy (SPFS) [62] or quartz-crystal microbalance (QCM) [63–66] could help to overcome the above-mentioned difficulties. One crucial feature of such sensor architectures would be the prevention of nonspecific adsorption of matrix components that can be realized by using ethylene glycol-oligomers and polymers or surface-attached dextran brushes (see Chapter 1.4). Such materials attributed with specific recognition sites provided by covalently coupled antibodies or MIPs would then restrain matrix effects and allow a specific extraction of the desired analyte. Such surface architectures can be fabricated by combining different materials in a layer structure or by using new engineered polymers. Due to the possibility to attach such surfaces onto a sensor support, biosensoric methods can be applied. All necessary analytical steps could then be done in only one step.

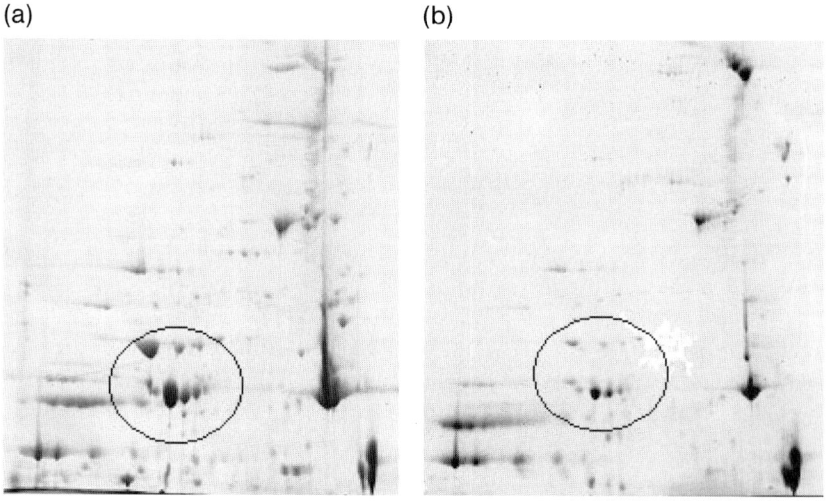

Figure 4 M. iliopsoas of one individual. 400 mg tissue, 2D-GE: first dimension on 18-cm IPG-strips (pH range 3–10), second dimension with SDS-PAGE. (a) Stored at –80 °C; (b) aged over 7 days at room temperature.

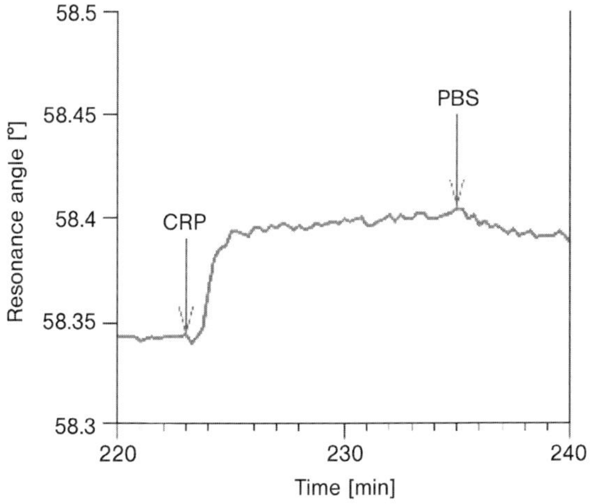

Figure 5 Binding kinetic of CRP onto CRP-antibody coupled onto a mixed thiol–SAM passivated with ethanolamine and BSA.

The C-reactive protein (CRP) in clinical chemistry is used as an indicator of inflammatory processes. An elevated CRP level in blood, for example, may indicate a systemic infection. During autopsy, some infections are not macroscopically detectable. Histological or microbiological examinations have to be applied. The fast and reliable detection of such marker proteins already during autopsy would mean a substantial improvement in forensic diagnostics. We currently are working on a sensor surface capable of detecting CRP. The sensor surface consists of a self-assembled monolayer of 6-mercapto-1-hexanol and 11-mercaptoundecanoic acid (MUA) [108]. The self-assembling was performed by coadsorption of the thiol (2.5 mM/7.5 mM in ethanol) for 3 h. An polyclonal antibody against CRP was immobilized on the thiol surface by using a NHS/EDC-coupling routine (20 mM/10 mM) [108], which is widely used as a coupling technique of proteins [79]. With this routine the carboxylic group of MUA is transformed into an activated ester that then can react with the amine group of an amino acid residue of the antibody (50 μg/mL). The activation was performed in pure water for 2 h and the coupling of the antibody in PBS buffer (pH 7.4). The remaining activated MUA-ester groups were passivated by rinsing the surface with a 1 M ethanolamine aqueous solution (pH 8.5) [109] and with bovine serum albumin (BSA) solution (0.1% in PBS-buffer of pH 7.4) [110, 111]. The binding of CRP to its surface-coupled antibody was recorded with SPS in real time. Figure 5 shows the binding kinetic of this event, which visualizes the fast antigen binding. BSA was used to prevent unspecific binding of the CRP, which is demonstrated in Figure 6(a). Without BSA, CRP shows unspecific binding to the SAM as shown in Figure 6(b). The next step in implementing biosensors using novel surface chemistry is to apply it to real samples and to use other surface architectures like plasma polymers or polyethylene-glycol-like polymers to avoid nonspecific adsorption of the biosensor surface by itself. An additional surface-modification step (here the use of BSA and

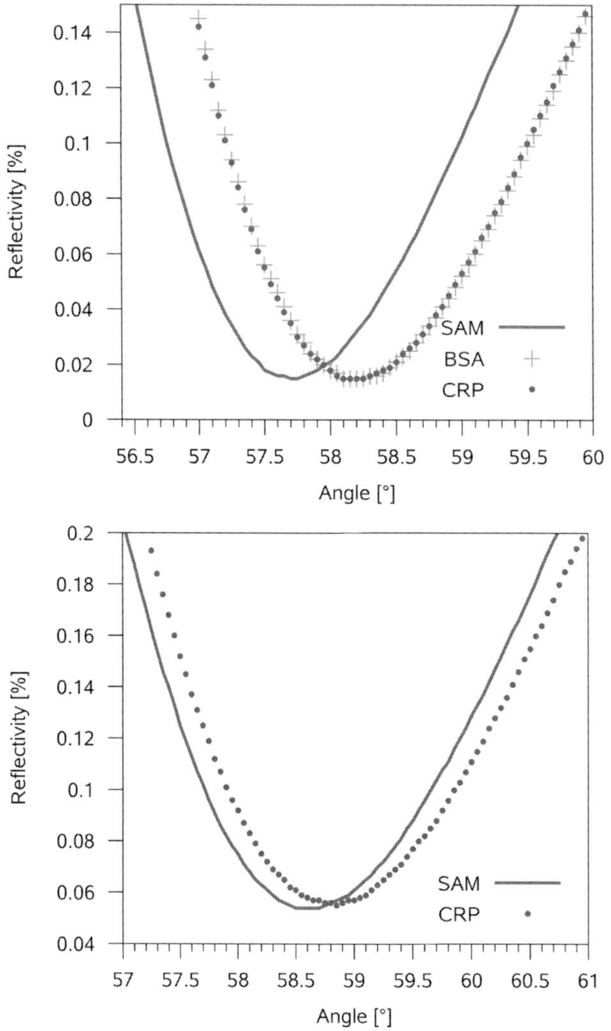

Figure 6 (a) Using BSA as a passivating molecule, no unspecific CRP binding was observable. The SPR scan (minimum of the scan is zoomed out) measured after adding CRP to the BSA-passivated SAM does not show any shift. The BSA and the CRP curves show exactly the same SPR minimum. (b) Without BSA, CRP shows unspecific binding onto the SAM which was passivated with ethanolamine only. This is recognized as a shift of the SPR minimum to larger angles.

ethanolamine) then would become unnecessary to prevent nonspecific adsorption. The same applies for the recognition site. Artificial recognition sites have the advantage that they are more robust against environmental effects (e.g. temperature) and intrinsic effects (e.g. lysozymes).

There is also an increasing interest in the field of biosensor applications in forensic proteomics [112]. The focus is on the development of novel biosensors for the detection of marker proteins that are characteristic of time-dependent postmortem changes. Such a device would allow a much more precise estimation of the time since death than is possible with current methods (e.g. measuring the body temperature). As shown in Figure 4 time-dependent changes in the proteome after death can be visualized using 2-dimensional gel electrophoresis (2D-GE). Given the assumption that changes in the abundance of specific proteins or protein patterns are correlated with the postmortem interval (PMI), it is most likely possible to generate specific antibodies against this protein. Biosensor applications based on these antibodies can probably be used to determine the time since death more accurately than is currently possible with the standard methods. Time-dependent alterations of the proteome in cerebrospinal fluid (CSF) of Alzheimer patients have already been observed [104]. In this case, antemortem and postmortem samples of the same patients were compared. Other publications deal with the estimation of the PMI by using troponin I as a marker protein [31, 105–107], which was detected by Western Blotting. But the application of SPR biosensors will offer much more sensitive and versatile analytical opportunities, especially by the use of single-chain Fv antibody fragments coupled onto mixed thiol–SAM. Such SAMs consist of PEG-terminated thiols for preventing nonspecific protein adsorption and carboxyterminated thiols for amine coupling of the antibody, as described in Chapter 1.2. These surfaces thus bear antifouling properties.

At the Institute of Legal Medicine at the University of Cologne, the postmortem changes of the proteome under realistic conditions are currently under systematic investigation. Potential specific protein patterns or marker proteins indicating the mentioned causes of death have not yet been identified. However, several protein modifications have been detected, which are assumed to be associated with SIDS [113–119]. But to date, the pathomechanism of SIDS still remains unclear. Research activities mainly focused on single proteins and their pathogenic modification and differences in up- and down-regulation compared to a control group of symptom-free children. The research of our group is not only focused on single (marker) proteins but also on SIDS-specific protein patterns. Promising first results achieved by 2D-GE show reproducible differences of the proteome from SIDS victims compared to the control group (Figure 7, boxes). The observed altered set of proteins may include candidates for SIDS biomarkers. Further investigations are necessary to endorse these findings. In the case of the apparent life-threatening event (ALTE), which is thought to be a preform of SIDS, there are hints that novel biosensors could be useful subsidiary tools in the early diagnosis of imperilled children with a high risk to develop ALTE and SIDS, respectively. New biosensor surfaces in combination with SPFS or other sensitive techniques have the potential to produce reliable results by using only a few microlitres of the sample for analysis. This can be assumed because some groups have successfully shown that it is possible to detect human chorionic gonadotropin (hCG) with a LoD of 6×10^{-13} mol dm^{-3} [77] or prostate-specific antigen with a LoD of 8×10^{-14} mol dm^{-3} [120, 121] under laboratory conditions (see Chapter 1.4). With such sensitive methods and optimized sensor architectures like three-dimensional dex-

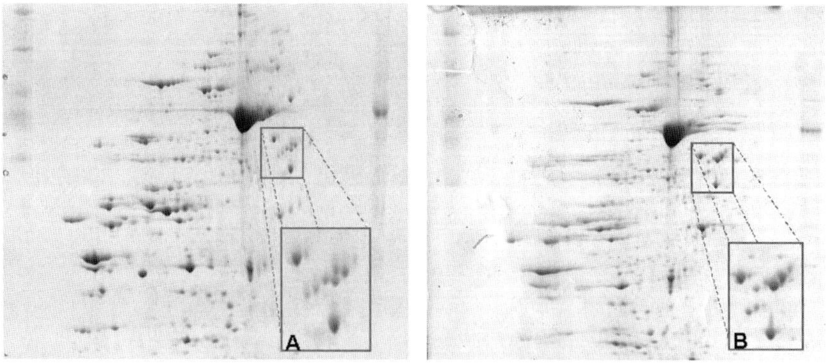

Figure 7 Cardiac muscle of two infants. 500 μg tissue, 2D-GE: first dimension on 18-cm IPG-strips (pH range 3–10), second dimension with SDS-PAGE. (a) Tissue of a SIDS victim; (b) tissue of an infant with a different cause of death. The spots in the boxes are magnified.

trans the need for large sample volumes is obsolete. In general, there are many options for the modification and functionalization of sensing surfaces. A review on this issues and the underlying coupling chemistry is given in Chapters 1.1 and 1.3. The combination of multiple-capture antibodies raised against proteins involved in SIDS, respectively ALTE, and immobilized on a certain sensor surface could be conceivable to develop a suitable sensing device. An overview of possible formats concerning the detection with antibodies in SPR biosensors is given in Figure 13 in Chapter 1.2. This is important because for ethical and medical reasons, it is, however, nearly impossible to take blood samples lager than 500 μL from infants. It has been demonstrated that SPFS in principle is able to detect femtomolar and attomolar concentrations as already mentioned [120, 121]. Moreover, a regeneration step can be applied to the sensor surface for repeated use of the biosensor as is described in Chapter 1.2.

The potential of novel biosensor devices in addition to forensic proteomic may have the same impact in forensic science as DNA analysis had a few years ago. The aim of the described proteomic project is to identify marker proteins indicating SIDS or other causes of death to develop biosensor methods for diagnostic purposes. The identified marker proteins can then be used as templates in the production of artificial (e.g. molecular-imprinted polymers) or biological (e.g. antibodies) recognition sites. The availability of such marker proteins is the prerequisite in order to build up a biosensor platform for diagnosis.

3.1.2.4
Further Perspectives and Applications

The development of surfaces with the above-described properties will not only allow a detection of analytes present at concentrations far below the LoD reached with the methods that are established today. Another main advance is the possibility of the simultaneous detection of more than one analyte even in samples consisting of complex matrices. This could be realized by a surface design that combines several capture antibodies raised against different antigens (here: analytes) and immobilized on different areas of the sensor. An area-specific detection could be performed with surface plasmon microscopy [122–125], cantilever sensors [67, 68] or mass-sensitive QCM, using a multichannel approach, for example.

Even when extraction steps are applied for prepurification, proteins and protein fragments as well as other endogenic substances (e.g. hormones, neurotransmitters, amino acids) are still present in the extract. In such cases, the presence of interferring substances often overlaps the chromatographic behavior of the substances of interest. This is causing a higher limit of detection of standard analytical methods. To overcome this difficulty in forensic analysis larger sample volumes are required. Derivatization of the analyte to achieve different chromatographic behavior and more significant molecule fragments for mass spectroscopy is applied as well. Generally, many preanalytical steps have to be applied to overcome the above-mentioned difficulties that are the everyday challenges in toxicological analysis. This means to apply a lot of man power, consumables, and expensive analytical equipment. Biosensoric techniques combined with new functional surfaces, like plasmapolymers or self-assembled monolayers (see Chapter 1.1), may help to save time and money because the purification step can be done simultaneously during the analysis. With the realization of new surface materials a simultaneous extraction and sensing could be achieved. Biosensors have been already used in tandem with mass spectrometry to identify binding partners of certain proteins. Subpicomole levels of material have been recovered using the 'sandwich elution' method in the case of inhibitors to the HIV-1 protease captured on SPR sensor chip. MS was subsequently used to estimate the relative binding affinities of components competing for the binding site of the immobilized protein. Binding of an inhibitor to a protease bound on the chip can also be evaluated in terms of the enzyme activities in which MS is used to monitor the generation of proteolysis peptides [126, 127]. In this context, mass-spectrometry-assisted immunoassay-based biosensors have been proposed as a new class of analytical methods for clinical reasons [128]. The tandem measurements done by biosensors and mass spectrometry can be used to screen for binding partners of the sensing molecule (e.g. enzymes, antibodies, etc.) bound to the surface. The biosensor usually gives information about the presence and the kinetic parameters of a binding event (e.g. antibody–antigen–reaction), but the detailed structure may remain uncertain, because detailed structure information usually is beyond the resolution. With tandem mass spectroscopy the analyte can be further characterized. Mass spectroscopy is widely used in forensic analysis and proteomic research to identify and to quantify analytes. In forensic analysis immunoassays are mainly used as a screening method. The crossreactivity of analytes belonging to the same pharmacological/toxicological active substance group (like Diazepam to the benzodiazepine group) is used in immunoassay-based screening methods to get evidence of the presence of such substances. For forensic reasons, the presence has to be proven by applying a second more precise analytical method. Biosensor tandem mass spectroscopy would certainly help to make the daily lab routine faster by gaining the same accuracy as the standard analytical routine. As pointed out before, specialized sensor surfaces can be used for a precleaning step (extraction) and to identify analytes in a one-step procedure. Furthermore, the biosensor platform kinetic will give useful information about the binding events. Surface plasmon microscopy has been used already to analyze laterally structured coatings [123]. With this method, it was shown that sensing of interfacial binding events and recognition reactions is possible [122, 124, 125]. With optical biosensors different surface structures can be brought to one sensor surface. Different sensing areas at the sensor surface can be attained by structuring the surfaces with different functional groups [129]. In principle, the simultaneous analysis of different substances with the same biosensor is possible with such techniques. Biosensor arrays are now commercially available. One example is the ProteOn XPR36 array biosensor from Bio-Rad. With this technology, six different chemistries, respectively, surface modifications can be achieved by using six parallel channels. These channels can be simultaneously addressed with the sample. Using this technique, enzyme interactions of different small inhibitors were simultaneously measured at six different concentrations [130]. But many other array platforms from other companies like Biacore, Genoptics, GWC Technology are now available. Rich and Myszka summarized in their review paper the state-of-the-art of optical biosensors [131].

The availability of novel sensor surfaces will allow the development of small devices that can be used at the point of scene (during autopsy and roadside testing at the scene of crime). Nevertheless, only one commercial available QCM-based test device (Biosens®, Biosensor Applications Sweden AB, www.biosensor.se) is currently under practical testing for forensic routine applications. In this case, the desired analyte is extracted from a urine sample by condensation. Subsequently, the analyte is transferred to a quartz crystal sensor modified with an antibody–substrate complex. Any antibody with a higher affinity to the analyte than to the surface-coupled substrate binds to the analyte and loses its contact to the surface. By this means, the antibody gets displaced. This results in a material loss that can be detected by QCM measurements.

The commercial availability of novel biosensors using intelligent surface chemistry would help to gain further progress in forensic toxicology and forensic diagnostic. The combined advantages of such systems as pointed out in the previous chapters and the, in principle, very low manufacturing costs would enable forensic labs to use such analytical tools in addition to the available standard methods. The broad variety of possible surface architectures for sensing purpose is one of the biggest advantages, because daily analytical routine with sensor-based methods would be easily adaptable to the everyday challenges in forensics. In principle, the detection of nearly all analytes of interest is feasible with the currently available standard methods (e.g. coupled MS/MS methods), but the purchase of all devices representing the recent progress in trace analysis most probably will not be affordable for any forensic institute. From the economical point of view it is not possible to have the latest developments in analytical technology installed for exceptional analytes. Because of this dilemma and of analytical reasons there will always be an analytical gap in forensics. Novel biosensor technology would help to close this gap.

3.1.3 Summary and Conclusions

Modern trace analysis and sample pretreatment for (bio)analytical reasons strongly depends on surface chemistry. Surface modifications can be found in nearly all areas of analytical chemistry, regardless if utilized to extract, separate or detect the analyte. One approach to separate analytes from matrix compounds directly is represented by the different variations of solid-phase microextraction, such as HS-SPME, DI-SPME or head-space solid-phase dynamic extraction (HS-SPDE). These methods are working with a fiber or hollow needle with a special coating that absorbs the analyte. SPME and SPDE methods are mainly used in combination with GC/MS-methods, desorbing the analyte from the coating by the heat of the GC injector. They are sufficient methods to analyze many volatile substances (drugs of abuse, therapeutics, pesticides, etc.) from complex matrices, but their detection limit could be improved substantially by the availability of further novel coatings (like functionalized hydrogels or plasma polymers).

A main challenge in trace analysis is analyte interference with matrix effects in complex samples as usually accrue in forensic science, like blood, saliva, hair or postmortem tissue. This is especially crucial if the analyte shows pharmacological or toxicological activities at very low concentrations or is rapidly metabolized or excreted. Furthermore, postmortem decomposition and autolysis, particularly of proteins and hemolysis of blood, bacterial contamination and fungal decay may hinder the analytical process. Therefore, very sensitive, fast and specific analytical methods have to be applied to overcome this problem. Using conventional immunoassay-based methods or conventional surface biosensors may lead to false results or misinterpretations due to crossreactivity of the used antibodies or deficient access of the analytes to the recognition site. This and the broad variety of forensic and clinically relevant substances imply the importance of the development of novel analytical screening devices. A promising approach is the combination of sensor architectures, attributed with antifouling properties, with very sensitive detection methods like surface plasmon enhanced fluorescence spectroscopy (SPFS) [62] or quartz-crystal microbalance (QCM). These sensors are miniaturizable in principle and will allow the development of small devices that can be used at the point of scene or point of care.

Another application of such novel biosensors is located in the expanding field of forensic proteomics. Given the assumption that specific proteins or protein patterns can be correlated with the postmortem interval (PMI) or certain causes of death, the generation of specific antibodies or artificial recognition sites against the concerned proteins would be possible. Such specific recognition sites will allow the development of biosensor applications that can be used to determine the time since death or the cause of death more accurately than is currently possible with the standard methods. There are some hints to specific differences of the proteome from SIDS victims compared to a control group. If that could be confirmed, the combination of an antibody-based biosensor surface with a very sensitive technique like SPFS may allow the development of a useful diagnostic tool in the recognition of children at high risk for ALTE and SIDS.

Thus, the development of novel biosensor surfaces combined with the mentioned sensitive detection techniques also promises some great advances in the analysis of medical or toxicological substances as in the estimation of basic forensic questions like the PMI or certain causes of death.

References

1 Carmona, G.N., Jufer, R.A., Goldberg, S.R., Gorelick, D.A., Greig, N.H., Yu, Q.S., Cone, E.J. and Schindler, C.W. (2000) *Drug Metabolism and Disposition: The Biological Fate of Chemicals*, **28**, 367.

2 Hoffman, R.S., Thompson, T., Henry, G.C., Hatsukami, D.K. and Pentel, P.R. (1998) *Journal of Toxicology – Clinical Toxicology*, **36**, 3.

3 Mehrani, H. (2004) *Process Biochemistry*, **39**, 877.

4 Stewart, D.J., Inaba, T., Tang, B.K. and Kalow, W. (1977) *Life Sciences*, **20**, 1557.

5 Warner, A. and Norman, A.B. (2000) *Therapeutic Drug Monitoring*, **22**, 266.

6 Salmon, A.Y., Goren, Z., Avissar, Y. and Soreq, H. (1999) *Clinical and Experimental Pharmacology and Physiology*, **26**, 596.

7 Schulz, M. and Schmoldt, A. (2003) *Pharmazie*, **58**, 447.

8 Baumgartner, A.M., Jones, P.F., Baumgartner, W.A. and Black, C.T. (1978) *Journal of Nuclear Medicine*, **19**, 696.

9 Baumgartner, A.M., Jones, P.F. and Black, C.T. (1981) *Journal of Forensic Sciences*, **26**, 576.

10 Goldberger, B.A., Caplan, Y.H., Maguire, T. and Cone, E.J. (1991) *Journal of Analytical Toxicology*, **15**, 226.

11 Goldberger, B.A., Darraj, A.G., Caplan, Y.H. and Cone, E.J. (1998) *Journal of Analytical Toxicology*, **22**, 526.

12 Ropero-Miller, J.D., Goldberger, B.A., Cone, E.J. and Joseph, R.E. (2000) *Journal of Analytical Toxicology*, **24**, 496.

13 Sachs, H. and Kintz, P. (1998) *Journal of Chromatography B – Analytical Technologies in, the Biomedical and Life Sciences*, **713**, 147.

14 Sachs, H., Uhl, M., HegeScheuing, G. and Schneider, E. (1996) *International Journal of Legal Medicine*, **109**, 213.

15 Thieme, D., Anielski, P., Grosse, J., Sachs, H. and Mueller, R.K. (2003) *Analytica Chimica Acta*, **483**, 299.

16 Thieme, D., Grosse, J., Sachs, H. and Mueller, R.K. (2000) *Forensic Science International*, **107**, 335.

17 Thieme, D. and Sachs, H. (2007) *Forensic Science International*, **166**, 110.

18 Uhl, M. and Sachs, H. (2004) *Forensic Science International*, **145**, 143.

19 Arnold, W. and Sachs, H. (1994) *Fresenius Journal of Analytical Chemistry*, **348**, 484.

20 Denk, R., Raff, I. and Sachs, H. (1992) *Kriminalistik*, **46**, 253.

21 Moeller, M.R., Fey, P. and Sachs, H. (1993) *Forensic Science International*, **63**, 43.

22 Tsatsakis, A.M., Tzatzarakis, M.N., Psaroulis, D., Levkidis, C. and Michalodimitrakis, M. (2001) *American Journal of Forensic Medicine and Pathology*, **22**, 73.

23 Negrusz, A., Bowen, A.M., Moore, C.M., Dowd, S.M., Strong, M.J. and Janicak, P.G. (2002) *Journal of Analytical Toxicology*, **26**, 471.

24 Negrusz, A., Moore, C., Deitermann, D., Lewis, D., Kalceiak, K., Kronstrand, R., Feeley, B. and Niedbala, R.S. (1999) *Journal of Analytical Toxicology*, **23**, 429.

25 Negrusz, A., Moore, C.M., Hinkel, K.B., Stockham, T.L., Verma, M., Strong, M.J. and Janicak, P.G. (2001) *Journal of Forensic Sciences*, **46**, 1143.

26 Negrusz, A., Moore, C.M., Kern, J.L., Janicak, P.G., Strong, M.J. and Levy, N.A. (2000) *Journal of Analytical Toxicology*, **24**, 614.

27 Chambers, D.R. (1980) *Medicine Science and the Law*, **20**, 146.

28 Fountoulakis, M., Hardmeier, R., Hoger, H. and Lubec, G. (2001) *Experimental Neurology*, **167**, 86.

29 Hufflonergan, E., Parrish, F.C. and Robson, R.M. (1995) *Journal of Animal Science*, **73**, 1064.

30 Ilian, M.A., Bickerstaffe, R. and Greaser, M.L. (2004) *Meat Science*, **66**, 231.

31 Sabucedo, A.J. and Furton, K.G. (2003) *Forensic Science International*, **134**, 11.

32 Schwab, C., Bondada, V., Sparks, D.L., Cahan, L.D. and Geddes, J.W. (1994) *Hippocampus*, **4**, 210.

33 Horbett, T.A. (1995) *Proteins at Interfaces II*, **602**, 1.

34 Horbett, T.A. and Brash, J.L. (1987) *ACS Symposium Series*, **343**, 1.

35 Kaufman, E.D., Belyea, J., Johnson, M.C., Nicholson, Z.M., Ricks, J.L., Shah, P.K., Bayless, M., Pettersson, T., Feldoto, Z., Blomberg, E., Claesson, P. and Franzen, S. (2007) *Langmuir*, **23**, 6053.

36 Czeslik, C. (2006) *Chemie in Unserer Zeit*, **40**, 238.

37 Kingshott, P. and Griesser, H.J. (1999) *Current Opinion in Solid State & Materials Science*, **4**, 403.

38 Lofas, S. and Johnsson, B. (1990) *Journal of the Chemical Society – Chemical Communications*, 1526.

39 Marchant, R.E., Yuan, S. and Szakalas-gratzl, G. (1994) *Journal of Biomaterials Science – Polymer Edition*, **6**, 549.

40 Osterberg, E., Bergstrom, K., Holmberg, K., Riggs, J.A., Vanalstine, J.M., Schuman, T.P., Burns, N.L. and Harris, J.M. (1993) *Colloids and Surfaces A – Physicochemical and Engineering, Aspects*, **77**, 159.

41 Carrigan, S.D. and Tabrizian, M. (2005) *Langmuir*, **21**, 12320.

42 Li, L.Y., Chen, S.F., Zheng, J., Ratner, B.D. and Jiang, S.Y. (2005) *Journal of Physical Chemistry B*, **109**, 2934.

43 Andersson, L.I. (2000) *Analyst*, **125**, 1515.

44 Bereczki, A., Tolokan, A., Horvai, G., Horvath, V., Lanza, F., Hall, A.J. and Sellergren, B. (2001) *Journal of Chromatography A*, **930**, 31.

45 Hu, S.G., Wang, S.W. and He, X.W. (2003) *Analyst*, **128**, 1485.

46 Jodlbauer, J., Maier, N.M. and Lindner, W. (2002) *Journal of Chromatography A*, **945**, 45.

47 Alexander, C., Andersson, H.S., Andersson, L.I., Ansell, R.J., Kirsch, N., Nicholls, I.A., O'Mahony, J. and Whitcombe, M.J. (2006) *Journal of Molecular Recognition*, **19**, 106.

48 Crescenzi, C., Bayoudh, S., Cormack, P.A.G., Klein, T. and Ensing, K. (2001) *Analytical Chemistry*, **73**, 2171.

49 Alves, C., Santos-Neto, A.J., Fernandes, C., Rodrigues, J.C. and Lancas, F.M. (2007) *Journal of Mass Spectrometry*, **42**, 1342.

50 Centini, F., Masti, A. and Comparini, I.B. (1996) *Forensic Science International*, **83**, 161.

51 da Silva, R.C., Zuin, V.G., Yariwake, J.H., Eberlin, M.N. and Augusto, F. (2007) *Journal of Mass Spectrometry*, **42**, 1358.

52 Fucci, N., De Giovanni, N. and Chiarotti, M. (2003) *Forensic Science International*, **134**, 40.

53 Lambropoulou, D.A., Sakkas, V.A. and Albanis, T.A. (2002) *Analytical and Bioanalytical Chemistry*, **374**, 932.

54 Bicchi, C., Cordero, C., Liberto, E., Rubiolo, P. and Sgorbini, B. (1024) *Journal of Chromatography A*, **2004**, 217.

55 Jochmann, M.A., Kmiecik, M.P. and Schmidt, T.C. (2006) *Journal of Chromatography A*, **1115**, 208.

56 Lachenmeier, D.W., Kroener, L., Musshoff, F. and Madea, B. (2003) *Rapid Communications in Mass Spectrometry*, **17**, 472.

57 Lipinski, J. (2001) *Fresenius Journal of Analytical Chemistry*, **369**, 57.

58 Musshoff, F., Lachenmeier, D.W., Kroener, L. and Madea, B. (2002) *Journal of Chromatography A*, **958**, 231.

59 Musshoff, F., Lachenmeier, D.W., Kroener, L. and Madea, B. (2003) *Forensic Science International*, **133**, 32.

60 Wendt, J., Helle, N., Bohlje, M. and Bremer, D. (2007) *AppNote Gerstel*.

61 Knoll, W. (1998) *Annual Review of Physical Chemistry*, **49**, 569.

62 Liebermann, T. and Knoll, W. (2000) *Colloids and Surfaces A – Physicochemical and Engineering, Aspects*, **171**, 115.

63 Rodahl, M., Hook, F., Fredriksson, C., Keller, C.A., Krozer, A., Brzezinski, P., Voinova, M. and Kasemo, B. (1997) *Faraday Discussions*, 229.

64 Rodahl, M., Hook, F. and Kasemo, B. (1996) *Analytical Chemistry*, **68**, 2219.

65 Rodahl, M., Hook, F., Krozer, A., Brzezinski, P. and Kasemo, B. (1995) *Review of Scientific Instruments*, **66**, 3924.

66 Rodahl, M. and Kasemo, B. (1996) *Sensors and Actuators A – Physical*, **54**, 448.

67 Shu, W.M., Laue, E.D. and Seshia, A.A. (2007) *Biosensors & Bioelectronics*, **22**, 2003.

68 Yan, X.D., Ji, H.F. and Thundat, T. (2006) *Current Analytical Chemistry*, **2**, 297.

69 Khan, H., Khan, T. and Park, J.K. (2008) *Separation and Purification Technology*, **62**, 363.

70 Brynda, E., Houska, M., Brandenburg, A. and Wikerstal, A. (2002) *Biosensors & Bioelectronics*, **17**, 665.

71 Chaki, N.K. and Vijayamohanan, K. (2002) *Biosensors & Bioelectronics*, **17**, 1.

72 Bender, K., Fraser, S., Förch, R., Jenkins, A.T.A., Köper, I., Naumann, R., Schiller, S. and Knoll, W. (2005) Plasma Polymer Supported Lipid Bilayers. In: Plasma Polymers and Related Materials, Ankara.

73 Liu, S., Vareiro, M., Fraser, S. and Jenkins, A.T.A. (2005) *Langmuir*, **21**, 8572.

74 Zhang, Z., Knoll, W., Foerch, R., Holcomb, R. and Roitman, D. (2005) *Macromolecules*, **38**, 1271.

75 Zhang, Z.H., Chen, Q., Knoll, W., Foerch, R., Holcomb, R. and Roitman, D. (2003) *Macromolecules*, **36**, 7689.

76 Carrigan, S.D., Scott, G. and Tabrizian, M. (2005) *Langmuir*, **21**, 5966.

77 Vareiro, M.L.M., Liu, J., Knoll, W., Zak, K., Williams, D. and Jenkins, A.T.A. (2005) *Analytical Chemistry*, **77**, 2426.

78 Sigal, G.B., Bamdad, C., Barberis, A., Strominger, J. and Whitesides, G.M. (1996) *Analytical Chemistry*, **68**, 490.

79 Adanyi, N., Varadi, M., Kim, N. and Szendro, I. (2006) *Current Applied Physics*, **6**, 279.

80 Rich, R.L. and Myszka, D.G. (2005) *Journal of Molecular Recognition*, **18**, 1.

81 Blanco-Lopez, M.C., Lobo-Castanon, M.J., Miranda-Ordieres, A.J. and Tunon-Blanco, P. (2003) *Analytical and Bioanalytical Chemistry*, **377**, 257.

82 Boeckl, M.S., Baas, T., Fujita, A., Hwang, K.O., Bramblett, A.L., Ratner, B.D., Rogers, J.W. and Sasaki, T. (1998) *Biopolymers*, **47**, 185.

83 Chianella, I., Piletsky, S.A., Tothill, I.E., Chen, B. and Turner, A.P.F. (2003) *Biosensors & Bioelectronics*, **18**, 119.

84 Dickert, F.L., Lieberzeit, P.A., Hayden, O., Gazda-Miarecka, S., Halikias, K., Mann, K.J. and Palfinger, C. (2003) *Sensors*, **3**, 381.

85 Nicholls, I.A. and Rosengren, J.P. (2001) *Bioseparation*, **10**, 301.

86 Shi, H.Q. and Ratner, B.D. (2000) *Journal of Biomedical Materials Research*, **49**, 1.

87 Shi, H.Q., Tsai, W.B., Garrison, M.D., Ferrari, S. and Ratner, B.D. (1999) *Nature*, **398**, 593.

88 Shoji, R., Takeuchi, T. and Kubo, I. (2003) *Analytical Chemistry*, **75**, 4882.

89 Warsinke, A., Lettau, K., Werner, D., Micheel, B. and Kwak, Y.K. (2003) *Technisches Messen*, **70**, 585.

90 Yamana, K., Ohtani, Y., Nakano, H. and Saito, I. (2003) *Bioorganic & Medicinal Chemistry Letters*, **13**, 3429.

91 Piletsky, S.A., Piletska, E.V., Chen, B.N., Karim, K., Weston, D., Barrett, G., Lowe, P. and Turner, A.P.F. (2000) *Analytical Chemistry*, **72**, 4381.

92 Surugiu, I., Danielsson, B., Ye, L., Mosbach, K. and Haupt, K. (2001) *Analytical Chemistry*, **73**, 487.

93 Svitel, J., Surugiu, I., Dzgoev, A., Ramanathan, K. and Danielsson, B. (2001) *Journal of Materials Science – Materials in Medicine*, **12**, 1075.

94 Jakusch, M., Janotta, M., Mizaikoff, B., Mosbach, K. and Haupt, K. (1999) *Analytical Chemistry*, **71**, 4786.

95 Lai, E.P.C., Fafara, A., VanderNoot, V.A., Kono, M. and Polsky, B. (1998) *Canadian Journal of Chemistry – Revue Canadienne de Chimie*, **76**, 265.

96 Li, P., Huang, Y., Hu, J.Z., Yuan, C.W. and Lin, B.P. (2002) *Sensors*, **2**, 35.

97 Nishimura, S., Yoshidome, T. and Higo, M. (2001) *Bunseki Kagaku*, **50**, 733.

98 Raitman, O.A., Arslanov, V.V., Pogorelova, S.P. and Kharitonov, A.B. (2003) *Doklady Physical Chemistry*, **392**, 256.

99 Dickert, F., Hayden, O., Lieberzeit, P., Palfinger, C., Pickert, D., Wolff, U. and Scholl, G. (2003) *Sensors and Actuators B – Chemical*, **95**, 20.

100 Kanekiyo, Y., Sano, M., Iguchi, R. and Shinkai, S. (2000) *Journal of Polymer Science Part A – Polymer Chemistry*, **38**, 1302.

101 Sugiyama, K., Rikimaru, S., Okada, Y. and Shiraishi, K. (2001) *Journal of Applied Polymer Science*, **82**, 228.

102 Khan, H. and Park, J.K. (2006) *Biotechnology and Bioprocess Engineering*, **11**, 503.

103 Kernbach-Wighton, G. (2006) *Rechtsmedizin*, **27**, 1.

104 Finehout, E.J., Franck, Z., Relkin, N. and Lee, K.H. (2006) *Clinical Chemistry*, **52**, 1906.

105 Zhu, B.-L., Ishikawa, T., Michiue, T., Li, D.-R., Zhao, D., Bessho, Y., Kamikodai, Y., Tsuda, K., Okazaki, S. and Maeda, H. (2007) *Legal Medicine*, **9**, 241.

106 Zhu, B.-L., Ishikawa, T., Michiue, T., Li, D.-R., Zhao, D., Kamikodai, Y., Tsuda, K., Okazaki, S. and Maeda, H. (2006) *Legal Medicine*, **8**, 94.

107 Zhu, B.-L., Ishikawa, T., Michiue, T., Li, D.-R., Zhao, D., Oritani, S., Kamikodai, Y., Tsuda, K., Okazaki, S. and Maeda, H. (2006) *Legal Medicine*, **8**, 86.

108 Briand, E., Salmain, M., Herry, J.-M., Perrot, H., Compere, C. and Pradier, C.-M. (2006) *Biosensors and Bioelectronics*, **22**, 440.

109 Blasco, H., Lalmanach, G., Godat, E., Maurel, M.C., Canepa, S., Belghazi, A., Paintaud, G., Degenne, D., Chatelut, E., Cartron, G. and Le Guellec, C. (2007) *Journal of Immunological Methods*, **325**, 127.

110 Aizawa, H., Tozuka, M., Kurosawa, S., Kobayashi, K., Reddy, S.M. and Higuchi, M. (2007) *Analytica Chimica Acta*, **591**, 191.

111 Balamurugan, S., Mendez, S., Balamurugan, S.S., O'Brien, M.J. and Lopez, G.P. (2003) *Langmuir*, **19**, 2545.

112 Kroener, L., Juebner, M., Bender, K., Huesgen, H. and Rothschild, M.A. (2006) Forensic proteomics: A novel approach in forensic science. In: Proceedings of the XXth Congress of International Academy of Legal Medicine.

113 Kilpatrick, D.C., James, V.S., Blackwell, C.C., Weir, D.M., Hallam, N.F. and Busuttil, A. (1998) *Forensic Science International*, **97**, 135.

114 Prandota, J. (2004) *Pediatric Research*, **55**, 589.

115 Prandota, J. (2004) *American Journal of Therapeutics*, **11**, 517.

116 Rahim, R.A., Boyd, P.A., Patrick, W.J.A. and Burdon, R.H. (1996) *Archives of Disease in Childhood*, **75**, 451.

117 Sawaguchi, T., Patricia, F., Kadhim, H., Groswasser, J., Sottiaux, M., Nishida, H. and Kahn, A. (2003) *Early Human Development*, **75**, S87.

118 Sawaguchi, T., Patricia, F., Kadhim, H., Groswasser, J., Sottiaux, M., Nishida, H. and Kahn, A. (2003) *Early Human Development*, **75**, S75.

119 Sawaguchi, T., Patricia, F., Kadhim, H., Groswasser, J., Sottiaux, M., Nishida, H. and Kahn, A. (2003) *Early Human Development*, **75**, S109.

120 Yu, F., Persson, B., Lofas, S. and Knoll, W. (2004) *Analytical Chemistry*, **76**, 6765.

121 Yu, F., Persson, B., Lofas, S. and Knoll, W. (2004) *Journal of the American Chemical Society*, **126**, 8902.

122 Piscevic, D., Tarlov, M. and Knoll, W. (1994) *Abstracts of Papers of the American Chemical Society*, **208**, 237.

123 Rothenhausler, B. and Knoll, W. (1988) *Nature*, **332**, 615.

124 Schmitt, F.J. and Knoll, W. (1991) *Biophysical Journal*, **60**, 716.

125 Zizlsperger, M. and Knoll, W. (1998) *Horizons 2000 – Aspects of Colloid and Interface Science at, the Turn of the Millennium*, **109**, 244.

126 Downard, K.M. (2006) *Proteomics*, **6**, 5374.

127 Downard, K.M. (2006) *Proteomics*, **6**, 6380.

128 Nedelkov, D. (2006) *Expert Review of Proteomics*, **3**, 631.
129 Chang-Yen, D.A., Myszka, D.G. and Gale, B.K. (2006) *Journal of Microelectromechanical Systems*, **15**, 1145.
130 Bravman, T., Bronner, V., Lavie, K., Notcovich, A., Papalia, G.A. and Myszka, D.G. (2006) *Analytical Biochemistry*, **358**, 281.
131 Rich, R.L. and Myszka, D.G. (2007) *Journal of Molecular Recognition*, **20**, 300.

3.2
Modification of Surfaces by Photosensitive Silanes

Xiaosong Li, Swapna Pradhan-Kadam, Marta Álvarez, and Ulrich Jonas

3.2.1
Introduction

The fabrication of nanosized structures and devices is of fundamental importance in current academic research as well as for state-of-the-art technological applications. The structure formation at the nanoscale can be approached by two complementary strategies, the 'top-down' and 'bottom-up' technologies. Top-down approaches are based on miniaturization of macroscale components down to the nanometer level. In recent decades great improvements have been achieved in this field and techniques like particle or energy-beam lithography, microcontact printing and some techniques based on scanning probe microscopy (SPM) are well developed [1]. Nevertheless, top-down techniques usually have the limitation in the degree of miniaturization, which in most practical cases is around 100 nm. For example, the minimum feature size in photolithography is limited by the irradiation wavelength [2].

The bottom-up technique is based on self-assembly concepts that are ubiquitously present in the biological world. For instance, base pairing in DNA strands, molecular recognition between antibodies and antigens, etc. Such self-assembly concepts mostly rely on the specific interactions at the molecular level between the involved components. In modern technology, the application of self-assembly processes in the bottom-up approach is often based on the formation of self-assembled monolayers (SAMs) by molecules that adsorb to a surface and aggregate into a dense film. Often they are covalently bound to the substrate surfaces. The newly formed functional surfaces provide the possibility to manufacture and attach mesoscale objects onto these surfaces, based on attractive interactions that are strongly influenced by the outmost surface molecules on the substrate and the objects [3]. This approach potentially represents a new fabrication method in the quest for miniaturization, which may lead to smaller and lighter devices as desired in fields such as microelectronics, optics, and sensors. A specific system, self-assembled alkyltrichlorosiloxane monolayers on glass surfaces, have been the subject of intense investigations since their discovery by Sagiv [4] in the 1980s.

Surface Design: Applications in Bioscience and Nanotechnology
Edited by Renate Förch, Holger Schönherr, and A. Tobias A. Jenkins

ISBN: 978-3-527-40789-7

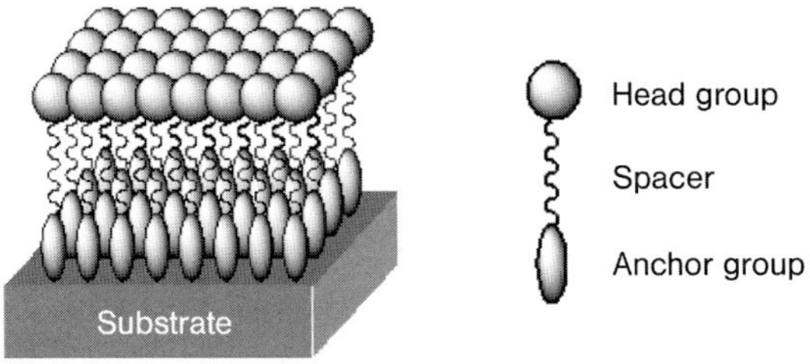

Figure 1 Profile of a self-assembled monolayer fragment and the cartoon of the constituting surface-active molecule.

3.2.2
Application of Patterned Functional Surfaces

Conceptually, self-assembled monolayers can be prepared simply by adding a solution of the desired molecules onto the substrate surface, waiting for the assembly process to complete, and finally washing off any excess. Particularly for alkylthiols on gold, monolayers with a thickness of molecular dimensions (several nanometers or even less) are formed and can range from disordered to highly ordered and oriented layer structures. SAMs can incorporate a wide range of groups both in the spacer region (often alkyl chains) and at the terminal functionality (e.g. amino, succinimidyl, and photoprotecting groups). Depending on their functionalities, these monolayers can substantially alter the properties of the surface such as friction, adhesion, chemical resistance, wettability, etc. [5]. Therefore, a variety of surfaces with specific interactions can be produced with a high degree of chemical control.

An additional level of structural hierarchy can be introduced by lateral patterning of the surface layers by top-down techniques. Two examples from the literature by Hammond [6] and by Heath [7] are presented here. Hammond's group developed a new technique for the self-organization of SiO_2 and polystyrene colloids onto a patterned polyelectrolyte surface. Patterning was achieved in this case by microcontact printing (Figure 2a). Heath reported a method for generating complex, spatially separated patterns of multiple types of semiconducting metallic nanocrystals based on lithographic patterning of photosensitive silane monolayers on SiO_2 substrates (Figure 2b).

Organosilanes with particular functional groups are used extensively for generating very robust organic monolayers on silica and other oxidic surfaces [4]. In our work, we focused on alkyltriethoxylsilanes since they are less reactive than trimethoxy- and trichlorosilanes and therefore much easier to handle under ambient laboratory conditions. The surface-modification process with these alkyltriethoxylsilanes, called silanization, is schematically outlined in Figure 3.

The first step in the silanization process is the hydrolysis of the alkoxysilanes to give hydroxyl silanes that can then interact with the silicon-dioxide surface and further condensate to form covalent bonds. When the anchor group bears more

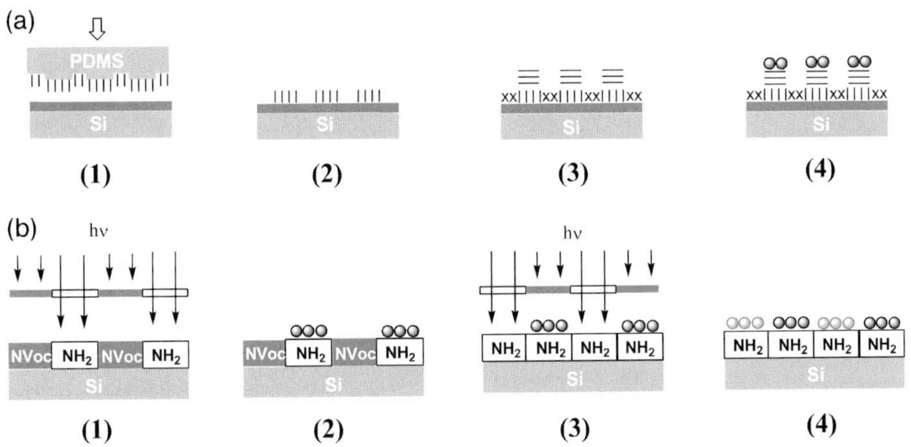

Figure 2 Schematics of the fabrication processes for structural hierarchies at surfaces: (a) Hammond's procedure: (1) pattern transfer by stamping of mercaptohexadecanoic acid (COOH) onto an Au layer on a Si wafer, (2) an acid-terminated SAM pattern was obtained after stamping, (3) an inert hydrophilic resist SAM (marked as crosses) was deposited from solution followed by polyelectrolyte multilayers on the COOH SAM, (4) colloidal particles (SiO_2, PS) were deposited onto polyelectrolyte patterns [6]. (b) Heath's procedure: (1) an amino-functionalized silane layer with photolabile NVoc groups was partly deprotected by irradiation through a mask in the near UV, (2) amine-stabilized nanoparticles assembled on deprotected amino groups, (3) deprotection was repeated in the NVoc region, (4) another type of amine-stabilized nanoparticles was assembled onto the freshly deprotected amino groups, and therefore yielded a binary nanoparticle array [7].

than one functional groups such as di- or tri-functional alkoxysilanes, it is possible that the hydroxysilanes first condense into oligomeric and polymeric species, which then adsorb onto the surface. Condensation of these groups and the formation of aggregates that adsorb onto the surface may lead to the formation of inhomogeneous layers, which has to be prevented so that a homogeneous coating can form via surface diffusion of physisorbed silanes [8].

As mentioned before, when functional SAMs of silanes are combined with common patterning techniques (e.g. photolithography, microcontact printing, etc.), lateral functionalized patterns on planar substrates with micro- to nanometer dimensions can be created. These patterns can define the spatial location of adsorbing objects based on different polarities, ionic charges, or chemical reactivities in different surface areas (Figure 4). The above-mentioned process of Heath and coworkers for generating complex, spatially separated patterns of multiple types of nanoparticles was achieved by attaching photosensitive protecting groups to a preformed amino silane monolayer followed by mask irradiation in a standard photolithography process [7]. We further expanded this concept towards the control of mesoscale assembly by showing regioselective colloid assembly after adsorption of photosensitive alkyltriethoxysilanes and direct monolayer patterning with light, as discussed in detail below [9].

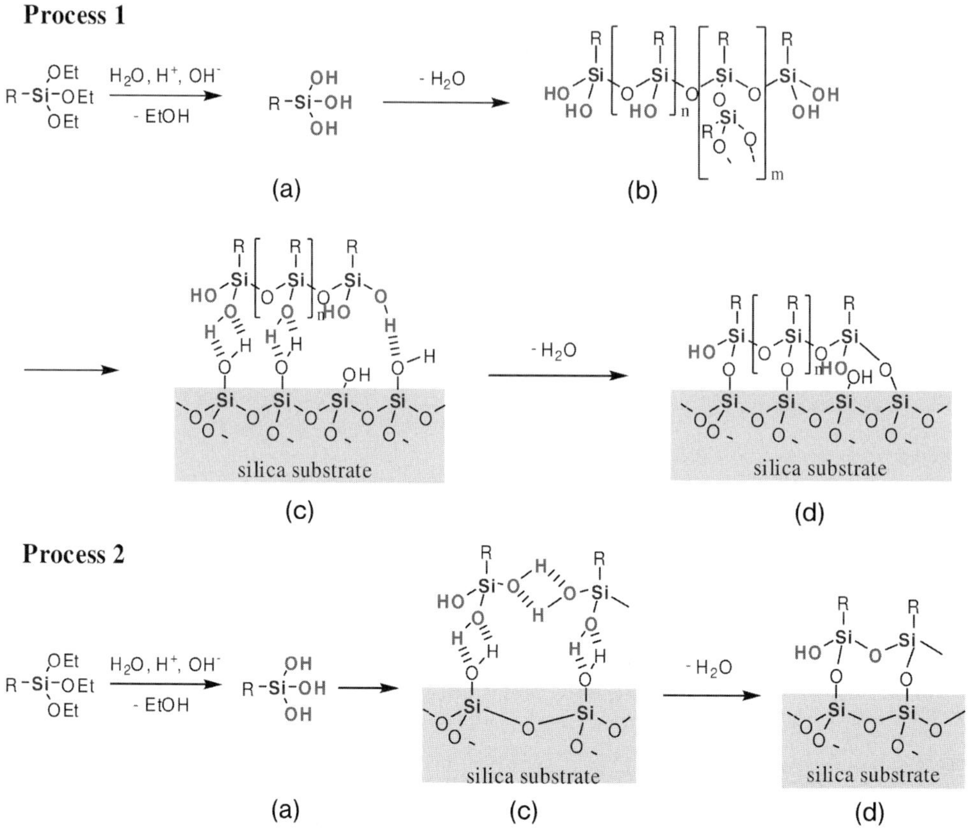

Figure 3 (a) Hydrolysis of the alkyltriethoxylsilanes in solution; (b) condensation in solution; (c) physisorption of hydrolyzed alkyltriethoxylsilanes or polymerized silane aggregates onto the silica substrate surface; (d) condensation with the surface and formation of a silane layer.

Irradiation with UV light is used to either degrade a compound on the surface or activate certain groups [10]. Specifically, photolithography has been widely applied in patterning or light-directed synthesis. Fodor and coworkers demonstrated that photoprotected amine groups could be activated by irradiating the layer with UV light through a mask, which induced the release of the photoprotecting group in the irradiated regions [11]. Calvert and coworkers showed that organosilane layers on silica substrate surfaces could be cleaved or degraded with high-energy, deep-UV irradiation, rendering the surface amenable to further SAM modification or even mammalian cell adhesion and growth [12, 13].

The common procedure to introduce photosensitive moieties onto a substrate involves first the modification of a surface with SAMs to introduce functional groups (e.g. amino, carboxyl, hydroxyl groups), which then react in a second step with photosensitive protecting groups. Since all of the required reaction steps

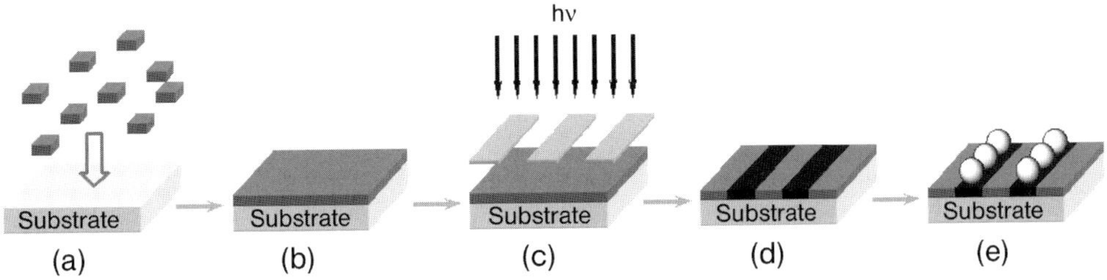

Figure 4 Illustration of the procedure for patterning of surface functional groups by direct monolayer photolithography: (a) chemisorption of photosensitive silanes onto the bare substrate (e.g. quartz, silica wafer); (b) formation of a photoprotected functional silane layer (the light gray surface resembles the photoprotecting group, e.g. NVoc or benzoin derivatives); (c) irradiation of the photoprotected substrate through a mask; (d) resulting chemical patterns on the substrate surface with the deprotected functionality (in black, for instance, amino, hydroxyl, or carboxylic groups) and the nonirradiated protected regions (in gray); (e) site-selective deposition of molecules or mesoscopic objects (grey spheres) onto the irradiated regions.

occur at the substrate surface, the disadvantage of this approach lies in the limited control over the reaction yield, layer composition and analysis of the products. Nevertheless, many scientists have used this technique in the past for postmodification of functionalized gold or silica surfaces [14–16]. A further development is achieved by another approach, involving a surface-modification process with functionalized alkoxysilanes, carrying covalently attached photoprotecting groups. These silanes are synthesized prior to deposition onto the surface. The advantage of synthesizing a silane that directly incorporates a photoprotecting group versus postmodification of a surface lies in the specific preparation and characterization of a well-defined molecular structure and the possibility to prepare mixed monolayers. By this approach, the photoreactive silane can be fully characterized by NMR and mass spectrometry prior to deposition, which is much more difficult after bonding to a surface. Patterning of the adsorbed silane layer can be achieved in the same way as in conventional photolithography by direct irradiation through a mask.

The conceptual requirements for a photoprotected silane are as follows: the photoprotecting group is sensitive to light of a specific wavelength but stable to other wavelengths and all chemical reagents encountered during the whole layer preparation and modification process [17]. The deprotection wavelength is only absorbed by the protecting group and does not affect other parts of the silane molecule or other components in the monolayer and substrate. Moreover, the photodeprotection step does not harm the protected functionality (like reaction of the photoproduct with the remaining monolayer) and the resulting photoproduct is easy to remove. Another factor of great importance is the lifetime of the excited state, which is responsible for the deprotection reaction. The longer the excited state exists before cleavage occurs, the higher are the chances for undesirable quenching processes or side reactions to occur, which in turn reduce the efficiency of the cleavage reaction [18].

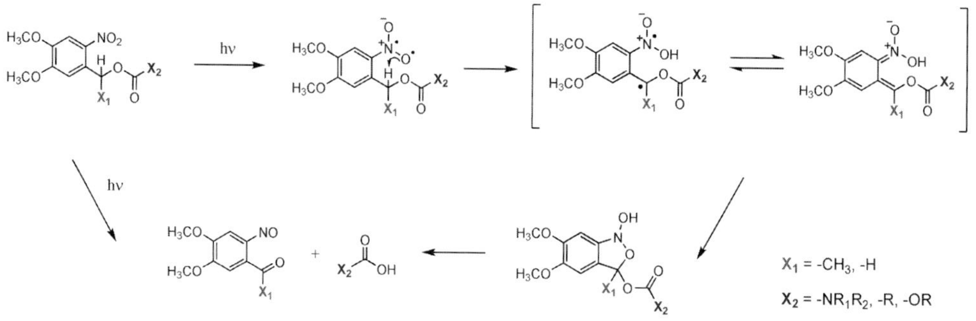

Figure 5 Mechanism of the photolytic cleavage of NVoc-protected carboxyl compounds.

One of the most popular carbamate protecting groups is the nitroveratryloxycarbonyl group (NVoc) used for protection of the amine group, which upon deprotection decarboxylates to give an aldehyde, the free amine and CO_2. However, NVoc protection can also be used for photoprotection of alcohols and carboxylic acids leading to the corresponding NVoc-ester and carbonates [19].

The mechanism of NVoc photodeprotection is shown in Figure 5, where the nitro group is firstly excited by irradiation to produce a diradical, the primary photochemical process is an intramolecular hydrogen-abstraction from the benzyl C–H bond in the *ortho* position to the excited nitro group. An electron redistribution follows to form an azinic acid (aci-nitro group), which rearranges to the nitroso group. In the case of NVoc-protected primary amines the only problem associated with the NVoc group is the potential side reaction of the formed aldehyde photoproduct with the released amine group to give an imine [20].

Benzoin esters (Bzn) have also become very popular as photoprotecting groups due to their high photosensitivity. In 1964 Sheehan and coworkers showed that it was possible to cyclize benzoin acetate into 2-substituted benzofuran and to release acetic acid by irradiation at 366 nm [21]. The carboxylic acid was thereby released with a high quantum yield ($\Phi = 0.64$) thus making this route generally attractive for the protection of carboxylic acids. They also found that substitutions at the benzylic ring can significantly increase the reaction rate and that 3′,5′-dimethoxybenzoin led to a fast and smooth cyclization upon photolysis, whereas 4,4′-dimethoxybenzoin only gives trace amounts of benzofuran. On the other hand, if no substituents are present the mechanism occurs via an α-cleavage of the diradical, resulting from the carbonyl excitation, which does not lead to the formation of benzofuran and liberation of the desired free acid moiety [23, 24]. Dimethoxybenzyl moiety has also been used by other groups [25–27] as a photoprotecting group and they found that the reaction occurred within 10^{-10} s after the absorption of a photon. The benzofuran photoproduct is nonpolar and inert, therefore it can be readily separated from the liberated acid or polar compounds.

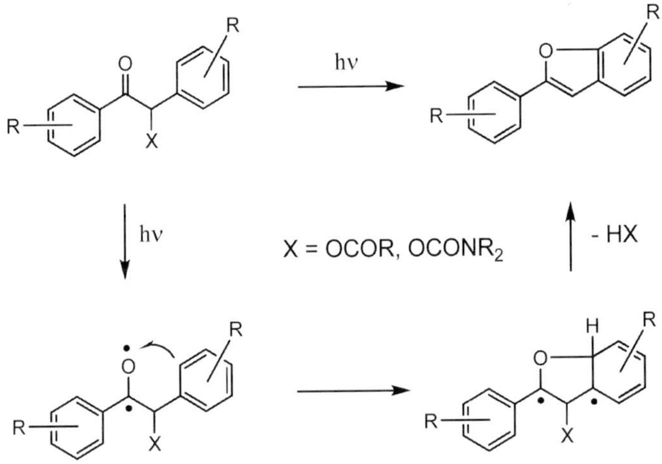

Figure 6 Photolytic cleavage of benzoin ester via a diradical mechanism in the presence of substituents to yield benzofuran and the liberated functional fragment.

The mechanism of photocleavage of a benzoin derivative via a diradical process is shown in Figure 6. In the case of a protected aminoalkyltriethoxsilane presented further below, the group X is triethoxysilanyl semicarbazide to yield the corresponding amino functionality after the photocleavage.

For covering other spectral regions at shorter wavelength than the one for NVoc and due to its high photosensitivity, we have been working with benzoin ester as a photoprotecting group besides the synthesis of Me-NVoc protected alkyltriethoxysilanes with amino-, hydroxyl-, and carboxyl groups. For all photoprotected silanes the possibility to create photopatterned surface structures by direct monolayer lithography was demonstrated. Furthermore, the concept of orthogonal photopatterning was introduced by depositing a mixture of a NVoc- and a Bzn-protected silane onto a substrate. The resulting mixed silane layer was irradiated first at 411 nm to deprotect the NVoc group through a mask with alternating gold stripes and then the same mask was rotated by 90° and irradiated at 254 nm to deprotect the Bzn group. In this way a square pattern with distinct regions of different surface functionalities, (amine, carboxylic acid, mix of both) (Figure 7c and d) was produced [19]. The resulting chemical contrast between exposed and nonirradiated regions was used to direct the assembly process of specific targets onto the activated areas.

Based on the initial results with the original NVoc structures, the synthesis procedures were modified to introduce an additional methyl group in the benzylic position (X_1 group in Figure 5, X_1 = CH_3) of the NVoc protecting group, to yield the so-called Me-NVoc group. As an electron-donating group, the methyl group activates the adjacent benzylic C atom for the formation of the 5-member ring, which reduces the irradiation time for deprotection with respect to the NVoc without the methyl group. Additionally, the corresponding benzyl ketone photopro-

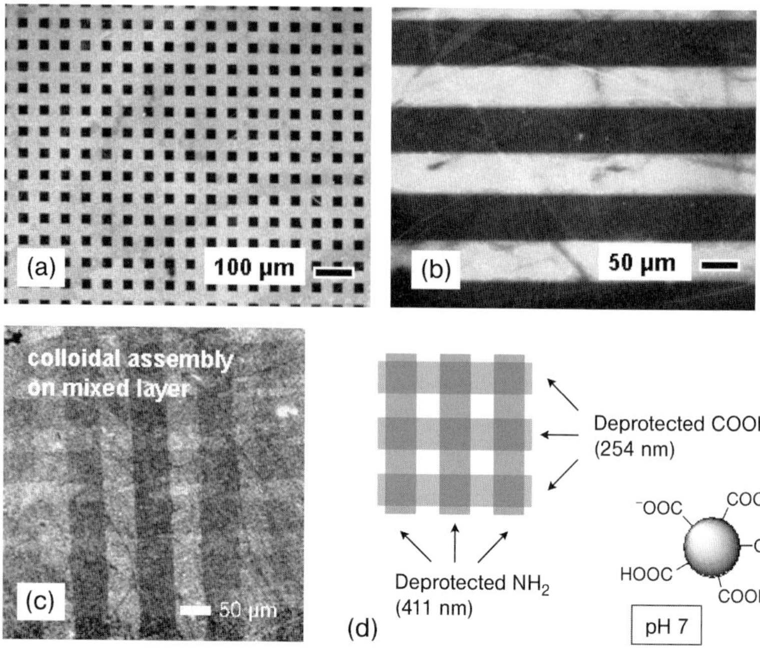

Figure 7 Optical micrographic images (dark field) of assembly patterns with carboxylated PBMA particles (diameter 183 nm) adsorbed from aqueous suspension at pH 7 onto silane layers of (a) (NVoc-protected NH_2 / free NH_2); (b) (Bzn-protected COOH / free COOH) irradiated through a mask; (c) shows a mixed layer of NVoc-protected amino and Bzn-protected carboxyl silane, irradiated at two different wavelengths (254 nm and 411 nm) through a 90°-rotated line mask (according to scheme d) [19].

duct is less reactive than benzaldehyde in imine formation, thus reducing side reactions with the liberated amine group. The deprotection of Me-NVoc alkyltriethoxysilane leading to free carboxylic surface groups has been proved to be achieved by a two-photon process in a monolayer on quartz substrate [28]. The Me-NVoc group was used to protect the functional groups $-NH_2$, –COOH and –OH in alkyltriethoxysilanes, while the Bzn group was applied to –COOH and $-NH_2$ moieties in the silanes.

The newly synthesized Me-NVoc and Bzn silanes were used for photolithography and soft lithography to generate micro- and nanometer patterns of SAMs.

3.2.3
Synthesis of Photosensitive Silane Molecules

Synthesis of Me-NVoc protected alkyltriethoxysilanes with $-NH_2$, –COOH and –OH functional groups was achieved in four-step procedures (Figure 8). Commercially available 3,4-dimethoxy acetophenone (1) was nitrated in the *ortho* position

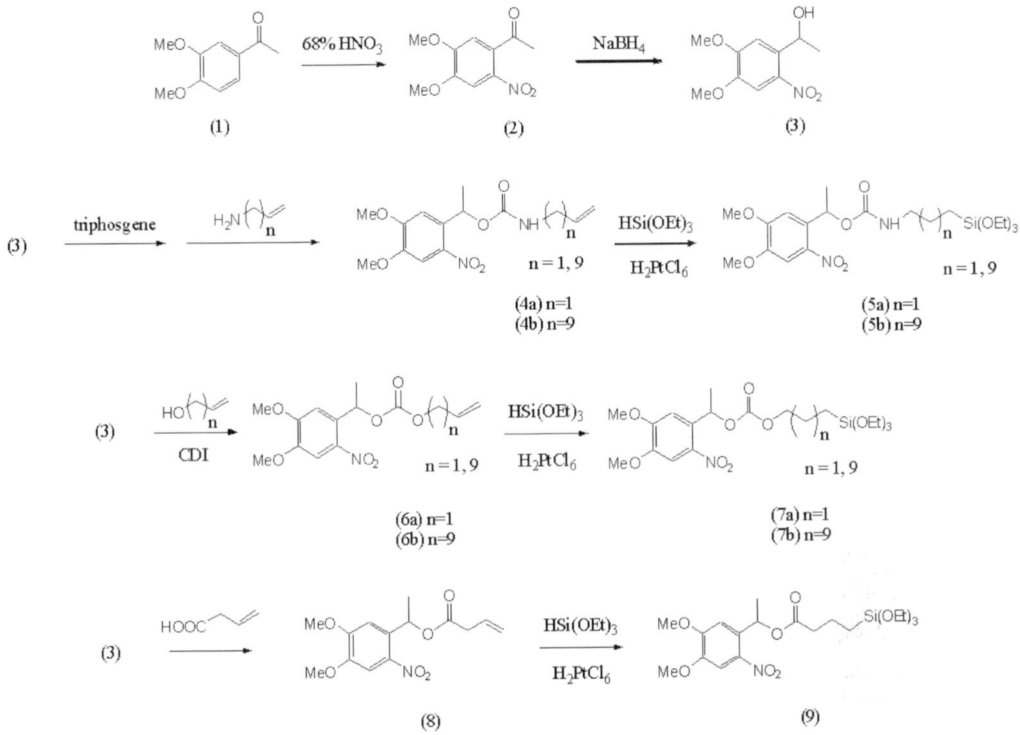

Figure 8 Synthesis of Me-NVoc protected $-NH_2$, –OH, and –COOH functional alkyltriethoxysilanes.

to the acyl group by nitric acid using standard procedures [29] and then converted to the corresponding alcohol by reduction of the ketone with sodium borohydrate in 58% overall yield. The unsaturated silane precursors with different alkyl chain length (n = 1 or 9) were obtained by chloroformylation, carbonate formation, or Mitsunobu esterification. Further hydrosilylation of the terminal double bond in these precursors afforded the Me-NVoc protected $-NH_2$, –COOH, and –OH silanes in around 40% overall yield.

The synthesis of the alkyltriethoxysilane bearing a Bzn-protected amino head group was carried out via a benzotriazole-mediated conversion of benzaldehyde to a benzoin-protected allyl amide with a terminal double bond. The first step employs an umpolung reaction by the activation of the benzaldehyde with benzotriazole to yield an acyl anion precursor (11) that was easily lithiated and trapped with the electrophilic 4,5-dimethoxybenzaldehyde followed by acid work-up [30]. After chloroformylation of the hydroxyl group in the benzoin structure, allylamine was introduced via a urethane linkage. The last step involved the hydrosilylation of the terminal double bond with triethoxysilane to yield the desired Bzn-NH_2 silane.

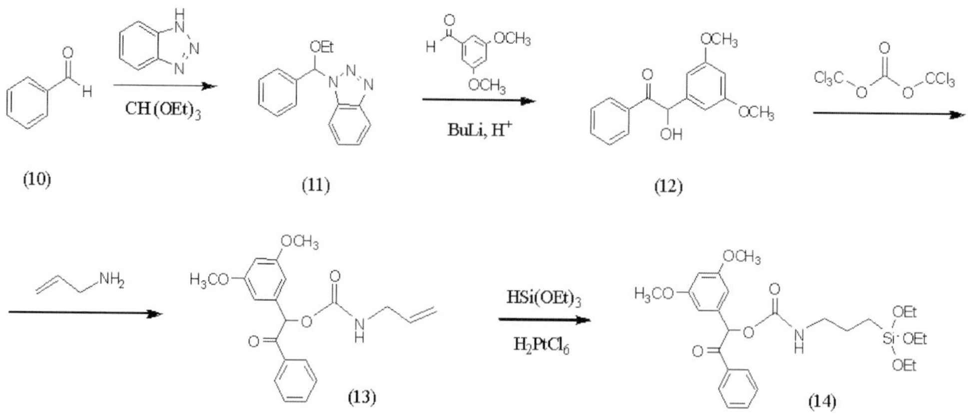

Figure 9 Synthesis of 3′,5′-dimetoxybenzoin-protected $-NH_2$ triethoxysilane.

3.2.4
Photodeprotection of Me-NVoc and Benzoin Silanes in Solution

UV-Vis measurements of silane solutions, as depicted in Figure 10, show that the Me-NVoc group has an absorption maximum at ~350 nm, while the Bzn group presents one at ~250 nm. The appropriate irradiation sources used for their deprotection in solution were UV crosslinker devices operating at 365 nm and 254 nm, respectively, and photodeprotection was monitored by UV-Vis spectroscopy. Before irradiation of the Me-NVoc silane in solution (0.1 wt% in THF), two strong absorbance maxima at 350 nm and 300 nm were observed, while after irradiation at 365 nm (400 mW cm^{-2}) the peak at lower wavelength disappeared and a broader band with maximum at 374 nm resulted (Figure 10a). The redshift indicates the release of the benzyl ketone photoproduct. In the case of the Bzn silane, the decrease of the absorption at 250 nm and the increase at 300 nm upon irradiation at 254 nm indicates the deprotection of Bzn and formation of the benzofuran photoproduct (Figure 10b).

3.2.5
Patterning Self-Assembled Monolayers by Photolithography

Quartz substrates were silanized with Me-NVoc and Bzn silanes from vapor phase and solution phase, respectively. The silane monolayers were characterized by ellipsometry and AFM to confirm uniform monolayer formation (layer thickness about 1.4 nm for Me-NVoc, and 0.45 nm for Bzn).

In analogy to the photodeprotection of the silane in solution, the protecting groups in the monolayers on solid substrates could be cleaved by irradiation to liberate the functional groups. The Me-NVoc photopatterned silane layers obtained by irradiation with a mask aligner at 365 nm were visualized by confocal

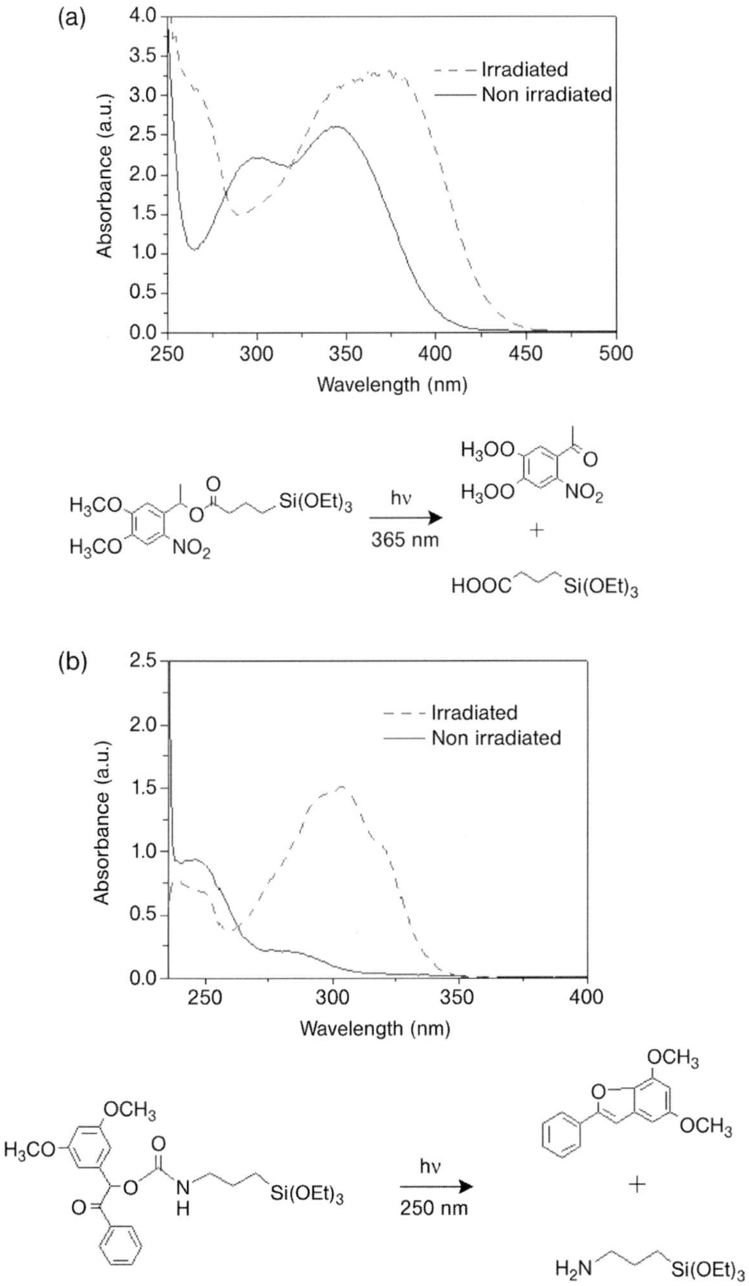

Figure 10 UV-Vis spectra of (a) Me-NVoc protected carboxyl silane in solution (0.01 wt% in THF) before (straight line) and after (dotted line) irradiation at 365 nm (400 mW cm^{-2}); (b) 3′,5′-dimethoxybenzoin protected amino silane in solution (0.1 wt% in THF) before (straight line) and after (dotted line) irradiation at 254 nm (610 mW cm^{-2}).

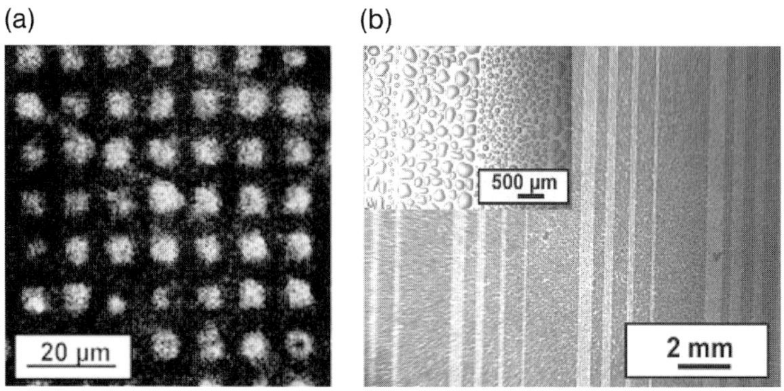

Figure 11 (a) Confocal microscopy image of a pattern obtained by irradiation of a Me-NVoc silane layer with a mask aligner at 365 nm followed by decoration with the dye Streptavidin-Cy5. (b) Optical microscopic picture of water-condensed patterns on a Bzn silane layer, which has been irradiated at 254 nm through a gold mask in a UV crosslinker device.

microscopy after labeling with fluorescent dyes. With the Bzn silane layers deprotection patterns were prepared by irradiation through a mask at 254 nm in a UV crosslinker device. For these irradiated layers water-condensation patterns were obtained after cooling the substrates and visualized by optical microscopy (Figure 11), which is a very simple and rapid method for visualization of differences in surface polarity. The irradiated regions (brighter stripes) are more hydrophilic, which leads to larger water droplets, while the nonirradiated regions are less polar, leading to smaller condensation droplets.

3.2.6
Summary and Conclusion

In summary, functional alkyltriethoxysilanes are particularly useful building blocks for surface modification of solid substrates since they form very robust surface layers under ambient conditions. In order to apply standard photolithography techniques for the lateral patterning of such functional surface layers, photoprotected amino-, hydroxyl-, and carboxylic acid alkyltriethoxysilanes with an NVoc protecting group and a 3′,5′-dimethoxylbenzoin-protected aminoalkyltriethoxysilane were synthesized. Upon irradiation with UV light the photoprotecting groups are cleaved to leave the free functional group in the silane, which was shown in solution and in self-assembled surface layers. By this technique, surface patterns of functional groups could be achieved by a simple adsorption and mask irradiation process, termed direct monolayer lithography.

In future research, the synthesis scheme is planned to be extended to other functional groups (e.g. Bzn-protected alkyltriethoxylsilanes with –OH and –SH functions), as well as to new photoprotecting groups with distinct absorption characteristics different from the NVoc and Bzn groups, to further extend the orthogonal photoprotection strategy to mixtures of even three protecting groups. Current activities are geared towards two-photon deprotection and the combination with metallic nanoobjects, which show particular field enhancements in their vicinity upon irradiation.

Acknowledgment

Financial support by the European Commission through the Nano3D project, NMP4-CT-2005-014006 is acknowledged.

References

1 Gates, B.D., Xu, Q., Stewart, M., Ryan, D., Willson, C.G. and Whitesides, G.M. (2005) New Approaches to Nanofabrication: Molding, Printing, and Other Techniques. *Chemical Reviews*, **105**, 1171.

2 Xia, Y. and Whitesides, G.M. (1998) Soft Lithography. *Angewandte Chemie (International Edition in English)*, **37**, 550.

3 Ulman, A. (1991) *An Introduction to Ultrathin Organic Films*, Academic Press, Boston.

4 Sagiv, J. (1980) Organized monolayers by adsorption. 1. Formation and structure of oleophobic mixed monolayers on solid surfaces. *Journal of the American Chemical Society*, **102**, 92.

5 Xia, Y., Gates, B., Yin, Y. and Lu, Y. (2000) Monodispersed colloidal spheres: old materials with new applications. *Advanced Materials*, **12**, 693.

6 Chen, K.M., Jiang, X., Kimerling, L.C. and Hammond, P.T. (2000) Selective self-organization of colloids on patterned polyelectrolyte templates. *Langmuir*, **16**, 7825.

7 Vossmeyer, T., Jia, S., DeIonno, E., Diehl, M.R., Kim, S.H., Peng, X., Alivisatos, A.P. and Heath, J.R. (1998) Combinatorial approaches toward patterning nanocrystals. *Journal of Applied Physics*, **84**, 3664.

8 Ishida, H. and Miller, J.D. (1984) Substrate effects on the chemisorbed and physisorbed layers of methacryl silane-modified particulate minerals. *Macromolecules*, **17**, 1659.

9 Jonas, U., del Campo, A., Kruger, C., Glasser, G. and Boos, D. (2002) Supramolecular Chemistry And Self-assembly Special Feature: Colloidal assemblies on patterned silane layers. *Proceedings of the National Academy of Sciences of the United States of America*, **99**, 5034.

10 DePuy, C.H. and Chapman, O.L. (1977) *Molekül-Reaktionen und Photochemie*, Wiley-VCH.

11 Fodor, S.P., Read, J.L., Pirrung, M.C., Stryer, L., Lu, A.T. and Solas, D. (1991) Light-directed, spatially addressable parallel chemical synthesis. *Science*, **251**, 767.

12 Dulcey, C.S., Georger, J.H. Jr., Krauthamer, V., Stenger, D.A., Fare, T.L. and Calvert, J.M. (1991) Deep UV photochemistry of chemisorbed monolayers: patterned coplanar molecular assemblies. *Science*, **252**, 551.

13 Brandow, S.L., Chen, M.S., Aggarwal, R., Dulcey, C.S., Calvert, J.M. and Dressick, W.J. (1999) Fabrication of Patterned Amine Reactivity Templates Using 4-Chloromethylphenylsiloxane Self-Assembled Monolayer Films. *Langmuir*, **15**, 5429.

14 Vossmeyer, T., DeIonno, E. and Heath, J.R. (1997) Light-Directed Assembly of Nanoparticles. *Angewandte Chemie (International Edition in English)*, **36**, 1080.

15 Stenger, D.A., Georger, J.H., Dulcey, C.S., Hickman, J.J., Rudolph, A.S., Nielsen, T.B., McCort, S.M. and Calvert, J.M. (1992) Coplanar molecular assemblies of amino- and perfluorinated alkylsilanes: characterization and geometric definition of mammalian cell adhesion and growth. *Journal of the American Chemical Society*, **114**, 8435.

16 Elender, G., Kuhner, M. and Sackmann, E. (1996) Functionalisation of Si/SiO_2 and glass surfaces with ultrathin dextran films and deposition of lipid bilayers. *Biosensors and Bioelectronics*, **11**, 565.

17 Pillai, V.N.R. (1980) Photoremovable Protecting Groups in Organic Synthesis. *Synthesis*, 1.

18 Becker, H.G.O. (1991) *Einführung in die Photochemie*, Deutscher Verlag der Wissenschaften, Berlin.

19 Campo, A.d., Boos, D., Spiess, H.W. and Jonas, U. (2005) Surface Modification with Orthogonal Photosensitive Silanes for Sequential Chemical Lithography and Site-Selective Particle Deposition. *Angewandte Chemie (International Edition in English)*, **44**, 4707.

20 Cameron, J.F. and Frechet, J.M.J. (1991) Photogeneration of organic bases from o-nitrobenzyl-derived carbamates. *Journal of the American Chemical Society*, **113**, 4303.

21 Sheehan, J.C. and Wilson, R.M. (1964) Photolysis of Desyl Compounds. A New Photolytic Cyclization. *Journal of the American Chemical Society*, **86**, 5277.

22 Sheehan, J.C., Wilson, R.M. and Oxford, A.W. (1971) Photolysis of methoxy-substituted benzoin esters. Photosensitive protecting group for carboxylic acids. *Journal of the American Chemical Society*, **93**, 7222.

23 Lewis, F.D., Lauterbach, R.T., Heine, H.G., Hartmann, W. and Rudolph, H. (1975) Photochemical.alpha. cleavage of benzoin derivatives. Polar transition states for free-radical formation. *Journal of the American Chemical Society*, **97**, 1519.

24 Lipson, M. and Turro, N.J. (1996) Picosecond investigation of the effect of solvent on the photochemistry of benzoin. *Journal of Photochemistry and Photobiology A – Chemistry*, **99**, 93.

25 Bochet, C.G. (2001) Orthogonal Photolysis of Protecting Groups. *Angewandte Chemie (International Edition in English)*, **40**, 2071.

26 Chamberlin, J.W. (1966) Use of the 3,5-Dimethoxybenzyloxycarbonyl Group as a Photosensitive N-Protecting Group. *The Journal of Organic Chemistry*, **31**, 1658.

27 Bochet, C.G. (2000) Wavelength-selective cleavage of photolabile protecting groups. *Tetrahedron Letters*, **41**, 6341.

28 Álvarez, M., Best, A., Pradhan-Kadam, S., Koynov, K., Jonas, U. and Kreiter, M. (2008) Single-photon and two-photon induced Photocleavage for Monolayers of an Alkyltriethoxysilane with a Photoprotected Carboxylic Ester. *Advanced Materials*.

29 Baker, J.W., Cooper, K.E. and Ingold, C.K. (1928) LVIII. The nature of the alternating effect in carbon chains. Part XXIV. The directive action in aromatic substitution of certain groups containing triple linkings. *Journal of the Chemical Society*, 426.

30 Katritzky, A.R., Lang, H., Wang, Z., Zhang, Z. and Song, H. (1995) Benzotriazole-mediated conversions of aromatic and heteroaromatic aldehydes to functionalized ketones. *The Journal of Organic Chemistry*, **60**, 7619.

3.3
Solid-Supported Bilayer Lipid Membranes

Ingo Köper and Inga Vockenroth

3.3.1
Introduction

Cell membranes, which form the outer boundary of any cell and of the organelles inside the cells, are highly complex architectures. In principle, they are composed of a lipid double layer. The amphiphilic molecules organize such that their hydrophobic parts form the inner layer of the membrane, while the hydrophilic head groups point towards the outside. These bilayer structures form the platform for the incorporation and attachment of membrane proteins. The lipid bilayer itself is typically composed of a large variety of different lipids, lipopolysaccharides and other species. Systematic investigation of the membrane properties and of incorporated proteins and associated processes are therefore difficult and limited to a few examples only. For example, patch-clamp experiments allow for the study of single ion-channel current fluctuations in a setup where a glass pipette is brought in contact with the membrane [1]. However, these experiments need trained personnel and are impeded by the highly fragile nature of the measurement system. Recently, attempts have been made to develop a high-throughput system, where whole cells are placed automatically on field effect transistors and ion-channel currents are recorded, however, these approaches are still in their early stage [2–4].

For the systematic study of membranes and membrane-related processes, various model systems have been proposed that should allow an easier characterization of the membrane architecture by minimizing the number on components involved [5, 6]. A suitable model system should be as close as possible to the natural archetype. At the same time it should allow an easy fabrication, be accessible to analytical techniques and should ideally form a robust system. The latter criterion is important, especially when membranes are to be used in applications. An upcoming trend in the last years had been the use of membrane proteins as sensing units for all kind of analytes, e.g. toxins [7, 8]. These approaches require a biomimetic platform, ideally a membrane, to host the proteins in a controlled, reproducible and stable fashion.

Vesicles (or liposomes) have been used extensively as model membrane systems (cf. T. L. Williams, Chapter 3.5). Here, lipids form a spherical bilayer system that

Surface Design: Applications in Bioscience and Nanotechnology
Edited by Renate Förch, Holger Schönherr, and A. Tobias A. Jenkins

ISBN: 978-3-527-40789-7

separates an inner compartment from the bulk solution [9]. In model studies, they have been used to study the uptake or release of substances to and from the inside of the vesicle. In a typical experiment, fluorophores have been used to monitor the transport across the bilayer, either by passive diffusion or by transport processes controlled by membrane proteins and pores [10, 11]. For an electrical characterization, which is the most direct approach for sensing applications, the use of vesicles is rather limited. Furthermore, their limited lifetime is an additional drawback for practical applications, since vesicles tend to fuse with other vesicles over time.

Planar bilayer lipid membranes (BLMs) have been established as the standard model membrane system for electrical characterization of membrane properties and for the measurement of protein activity in terms of ion flux through proteins [12, 13]. In principle, these systems consist of a support, typically made of Teflon that contains an aperture. The size of the aperture can vary from hundredth of a micrometer down to the nanometer range. Recently, progress has been made in fabrication of more and more advanced devices. The aperture can be covered by a lipid film, which under the right condition thins out to a lipid bilayer. Thus, the membrane separates two compartments that can be individually addressed. By placing electrodes on both sides of the bilayer, its electrical properties can be determined in terms of resistance and capacitance. Typical membrane resistances are in the $G\Omega$ range, a value that is also known from patch-clamp experiments on whole cells.

The main drawback of BLMs is their inherent instability. They are prone to rupture and have a typical lifetime of the order of hours. However, recent progress has been made for enhancing their stability by using nanoporous supports, either with statistically distributed apertures or by using engineered substrates [14]. Additionally, protective coatings have been used and these protected membranes could be stored and reused after weeks [15–17]. However, long-time measurements and robust sensing devices using BLMs are still not feasible [18, 19].

A more stable model membrane system can be obtained by using solid supports underneath the membrane [19–22]. The first attempts have been made by fusion of vesicles with a solid support, e.g. a silicon wafer, a glass slide or a gold electrode. This approach allows for characterization of the membrane using a large variety of surface-analytical techniques. Small proteins could be incorporated and studied. However, the solid-supported membrane has the drawback that in most cases, no electrical sealing properties can be observed. Furthermore, the membrane is separated from the support by typically a thin layer of water. This relatively weak interaction does not provide any ion reservoir underneath the membrane and does not give enough space for the functional incorporation of complex membrane proteins. The latter typically denature upon contact with the support or do not incorporate into the membrane at all. For practical applications, solid-supported membranes also lack the necessary long-term stability.

A promising continuation of the solid-supported membranes has been the introduction of soft polymeric cushions between membrane and support (Figure 1) [23, 24]. This decouples the membrane from the latter and provides enough

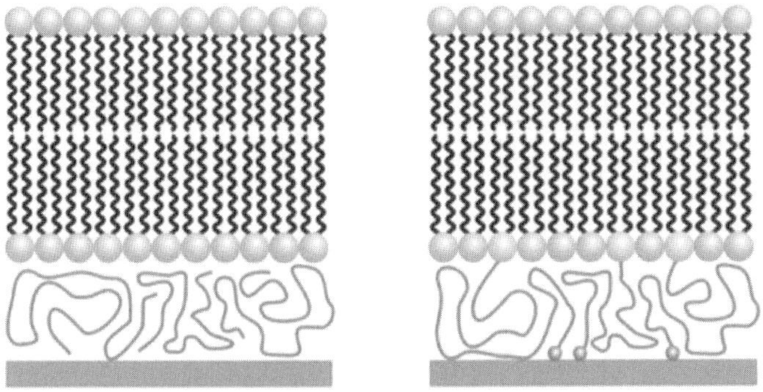

Figure 1 Schematic representation of two polymer cushioned bilayers. In the left case, the polymer acts as a linker between the surface and the bilayer merely by attractive interactions, while in the right case the polymer is covalently anchored to the surface. Furthermore, the polymer exhibits lipid groups that are organized within the lower leaflet of the bilayer during vesicle fusion.

space to host extramembrane parts of a protein. Although a variety of polymers has been used, sufficient electrical sealing properties have always been a challenge.

3.3.2
Tethered Bilayer Lipid Membranes

In tethered bilayer membranes (tBLM), the inner leaflet of a lipid bilayer membrane is covalently coupled via a polymeric spacer group to a solid support (Figure 2) [19, 25–28]. Such systems have been shown to provide excellent stability and electrical properties similar to BLMs or natural membranes, as known from

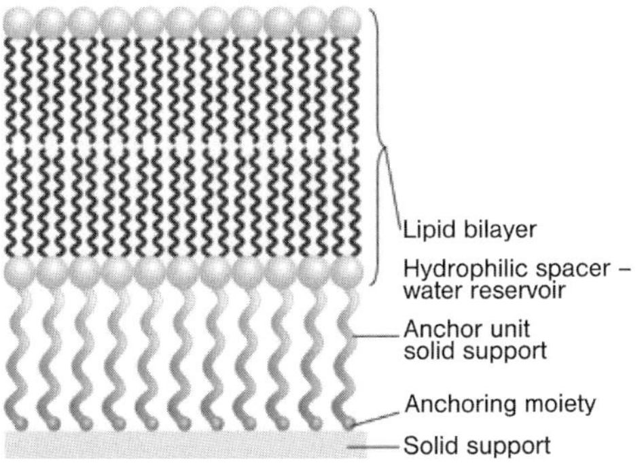

Figure 2 Schematic representation of a tethered bilayer. The first leaflet consists of anchor lipids that are attached to the support via a tethering unit that is bound covalently to the support. In the case of gold surfaces the anchorage is achieved via sulfur chemistry.

patch-clamp experiments. The incorporation of ion-channel proteins is possible and thus the use in biosensing applications has been proposed. By using various spacer groups, the membrane properties can be adapted to a selected protein [29]. Furthermore, the use of different anchor groups allows for the attachment on different surfaces. Typically, gold has been used as substrate, which then can be used directly as an electrode. The lipids are then bound to the metal substrate through thiol chemistry. For other studies, silane chemistry has been used to couple the membrane to glass substrates or oxidic surfaces.

A molecular toolbox of anchor lipids with different spacer and anchor groups has been established that all allow for the formation of functional tethered membranes [29]. Figure 3 shows the synthetic scheme for anchor lipids with different anchor functionalities.

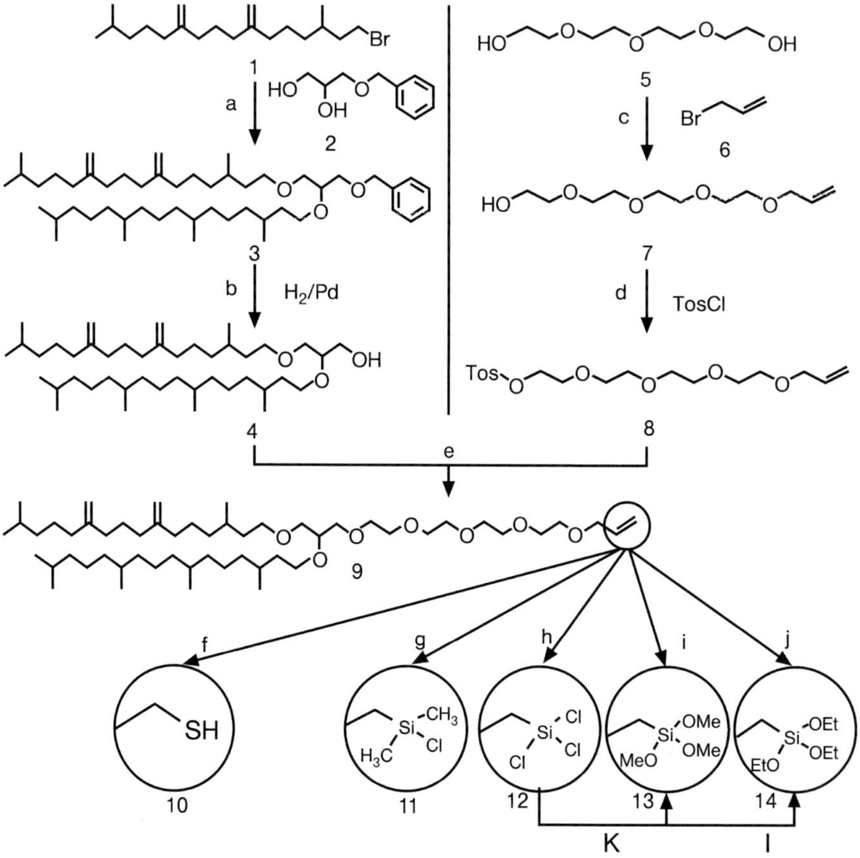

Figure 3 Synthetic scheme for anchor lipids with different anchor functionalities [29]. The lipid part of the molecules consists of two phytanyl chains that are coupled via a glycerol unit to an oligomeric spacer. In principle, different lengths of the spacer unit can be used. The terminal double bond can be converted into various anchor groups.

Using the elegant synthesis via a precursor molecule, the terminal double bond can be easily modified into various silanes or into a thiol group for attachment to a gold substrate. The hydrophobic part of the anchor lipid consists of two phytanyl chains, which are know to form stable and electrically sealing membranes. Furthermore, these lipids are in a fluid phase under ambient conditions. The hydrocarbon chains are linked to an oligo-ethylene glycol chain via a glycerol unit. The length of the polymeric chain has been varied from 4 to 14 units. Other molecules have a lipoic acid group as anchoring moiety. This has the advantage to reduce the oxidation probabilities of the thiol groups. The simple thiol groups tend to dimerize and open upon reaction with the gold surface. The anchor lipids can be grafted to a solid support either by self-assembly or by Langmuir–Blodgett transfer. Self-assembly is typically used for thiol compounds from a dilute ethanolic solution and for reactive trichlorosilane anchors. The anchor lipids from densely packed monolayers when allowed to self-assemble for about 24 h. The alkoxy silanes do not easily self-assemble due to the formation of three-dimensional structures. However, they can be spread on the air/water interface of a Langmuir film balance and transferred to a solid support.

The lipid monolayers can be completed to bilayers either by fusion with small unilamellar vesicles, Langmuir–Schaefer transfer or a solvent-exchange method. The membrane-formation process can be followed using various surface-analytical techniques.

For example, surface plasmon resonance spectroscopy (SPR) allows measurement of thin layers at a gold interface with Ångstrom resolution (see Chapter 1.3). Typically, a setup in the Kretschmann configuration is used. In the scan mode, reflectivity changes are monitored as a function of the angle of incidence of the incoming laser beam. In the kinetic mode, reflectivity changes occurring at a fixed angle are monitored as a function of time.

SPR spectra can be analyzed using Fresnel equations and a layer model including the prism, gold and the thiolipid monolayer. After vesicle fusion, a fourth layer corresponding to the outer leaflet of the bilayer is added (Figure 4). In the case of silane-based membranes, ellipsometry has been shown to be a useful technique to characterize thin layers at a silicon interface [29, 30].

However, the mere optical characterization is often not enough to prove the existence of a functional lipid bilayer. Especially for the characterization of incorporated membrane proteins, e.g. ion channels, the electrical properties of the membrane are important. Typical experiments to study the electrical properties of a membrane and ion-transport phenomena through the membrane and through embedded membrane proteins are patch-clamp experiments. In these experiments, whole cells, membrane patches or giant liposomes are investigated and recordings of single ion-channel fluctuations are possible [31, 32]. However, these techniques require a high level of experimental skill and are limited to laboratory use only. Recently, automated patch-clamp systems have been established, that typically use micromachined substrates [2, 3]. However, these systems are still not useful for systematic investigation. Another approach is the use of bilayer lipid membranes (see above). Here, electrical characterization of the membrane is

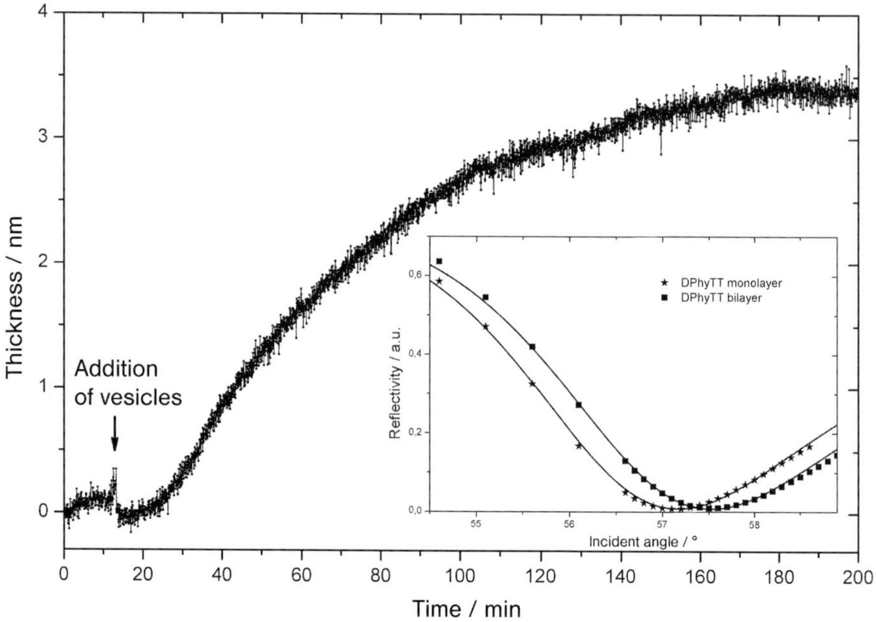

Figure 4 Surface plasmon resonance experiment showing vesicle fusion kinetics on a monolayer of DPhyTT (10 in Figure 3). The inset shows angular scans before and after fusion. By analyzing the scans, the thickness of the layer can be obtained [29].

possible and single ion-channel fluctuations can be measured. Typical values for the membrane resistance are in the GΩ range, measured over typical apertures in the micrometer range [13].

Typical tBLM experiments, however, are performed on electrodes with dimension of about 0.5 cm^2. Here, a path clamp recording is not possible due to the high noise level. Impedance spectroscopy, however, is a useful technique to characterize the electrical parameters of thin films attached to an electrode. The solid support is typically used as the working electrode, while a reference and counter electrode are placed in the buffer solution. A small-amplitude sinusoidal potential is applied at various frequencies and the resulting current is measured. The experimental data can be displayed in Bode plots, where the amplitude (impedance) of the resulting current and the phase shift are plotted as a function of the applied frequency. The data can be analyzed by fitting to simple equivalent circuits. Typically, a membrane can be described by an RC element, a resistor and a capacitor in parallel. In a Bode plot, resistances can be identified by a plateau in the impedance and a low phase angle, whereas capacitances result in a –1 slope of the impedance and a phase angle of ideally 90°.

3.3.3
Protein Incorporation

tBLMs can be used to study functionally incorporated membrane proteins. The high background resistances of the membrane makes them especially suited for the investigation of ion-channel proteins. A functional incorporation of these proteins or of channel-forming peptides leads to a decrease in the membranes resistance, which can be measured using impedance spectroscopy.

Valinomycin is a small ion carrier peptide that selectively transports potassium ions [33]. A tBLM has been created on a highly doped silicon wafer, which can be used as a working electrode. The incorporation of the peptide led to a decrease of the membrane resistance, when potassium ions were present in the electrolyte [30]. The bilayer showed a resistance of about 3.5 $M\Omega\,cm^2$ in 100 mM NaCl and decreased to a value below 0.5 $M\Omega\,cm^2$.

α-hemolysin (α-HL) is an exotoxin and secreted by *Staphylococcus aureus*. The water-soluble 293-residue polypeptides form heptameric pores in lipid bilayers [34–36]. The heptamer has a mushroom form, with the stem penetrating the membrane bilayer and the cap extending into the extracellular space [36]. The complete assembly process is still not fully elucidated. The most prominent pathway describes the assembly of the monomers on the cell membrane. The pore forms upon subsequent collision of the monomers during lateral diffusion in the bilayer [37–39]. The monomers can be easily manipulated and therefore α-HL has

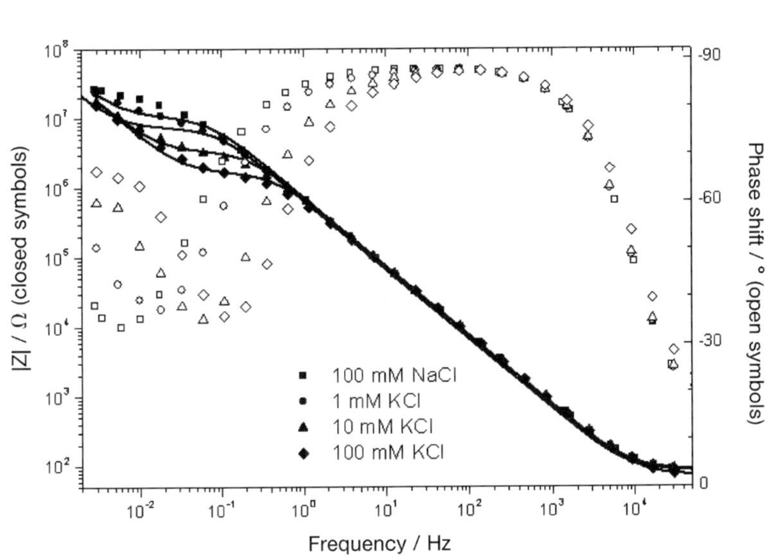

Figure 5 Bode plot of the incorporation of valinomycin in a tBLM based on DPhyTDC (11 in Figure 3) The functional incorporation can be seen by a decrease in the membrane resistance in the presence of potassium ions. Sodium ions are not transported and lead to a higher membrane resistance [30]. Solid lines represent fits to an R(RC)C equivalent circuit, the bilayer is described by an RC element, while the spacer can be modeled by a capacitor. The electrolyte contributes as a resistive element.

been proposed as an ideal biosensing element in novel device architectures. Engineered pores have been used for the detection of metal ions, organic compounds, proteins and DNA [7, 8, 40]. In most assays, a binding site has been placed in the lumen in or near the lumen of the pore. Upon binding of a specific analyte, the conductance of the protein will be modulated in a characteristic way, corresponding to the concentration and type of analyte [8].

Currently, BLMs have been used to incorporate the protein and record fluctuations of the transmembrane currents [8, 41–44]. However, the limited lifetime of the tBLMs make practical applications difficult, even if recent progress has been to stabilize the membranes using crystalline S-layers [45–47] or protective layers such as agarose [48].

In solid-supported membrane, only a partial incorporation has been shown, because the limited space between membrane and support inhibited the complete reconstitution [49].

We used a tBLM in order to provide a membrane with a spacer region for the functional incorporation. DPhyTL has been shown to form dense and insulating membranes [50, 51]. The functional incorporation of the pore can be seen by a decrease in membrane resistance (Figure 6) [52]. The functional incorporation could be simultaneously monitored using optical techniques (data not shown) [52]. SPR showed an increase in membrane thickness.

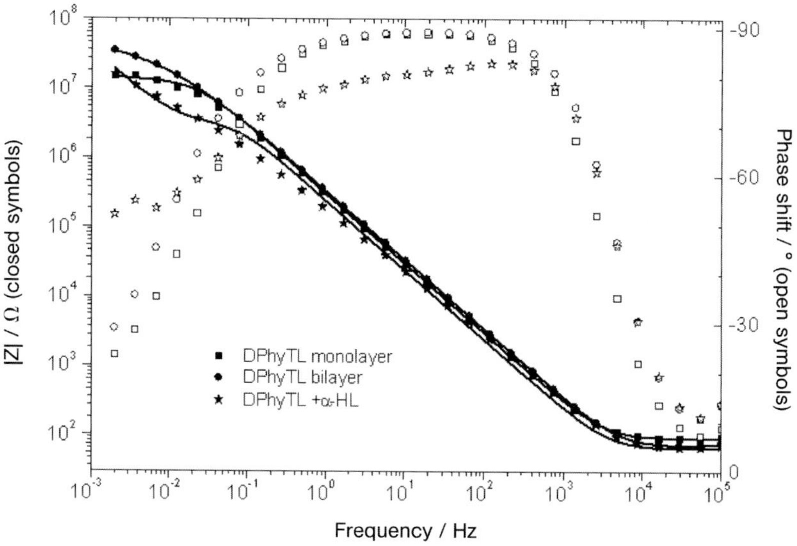

Figure 6 Bode plot of a DPhyTL (**11** in Figure 3, with a lipoic acid moiety as anchor group) monolayer, bilayer and a bilayer with incorporated α-hemolysin pores in NaCl/Mops buffer ($c = 0.1$ M) at 0 V potential; closed symbols show the impedance, open symbols the phase shift; fits are shown as solid lines (only for the impedance). An increase in membrane resistance after bilayer formation and a significant decrease of resistance by a factor of 13 upon pore incorporation can be seen as a shift in the impedance [52].

The architecture of the membrane has a significant effect on the probability of protein incorporation. When the anchor lipid DPhyTL is replaced by modified molecules with larger spacer units or more extended anchor groups. A higher amount of the protein can be incorporated, as seen by a larger decrease in membrane resistance [52]. A similar effect could be seen by using the small channel-forming peptide M2, the transmembrane fragment of the acytylcholine receptor [53]. Using modified anchor lipids, membranes can be created that are optimized to host selected membrane proteins. By variation of the spacer and anchor group, the membrane properties, i.e. electrical properties, packing density and hydration of the spacer layer can be modified.

3.3.4 Conclusion and Outlook

Tethered bilayer lipid membranes are solid-supported model membrane that allow for the systematic investigation of membrane-related processes. Their high electrical resistance makes them especially suited for the study of ion channels or pore-forming peptides. Thus, the architecture opens the possibility to be used in novel biosensing devices, where selected membrane proteins would be used as actual sensing units for specific analytes. Recently, first single-channel recordings in a tBLM have been shown, however, system integration still has to be shown [54, 55]. Nevertheless, such a system would provide a very high selectivity and sensitivity for the detection of analytes using a detection mechanism based on single-molecule interactions. Furthermore, the concept can be easily extended to an array format, which would be an important step towards high-throughput screening assays.

Alternative applications of the tBLM concept include the systematic study of membrane–protein interactions, e.g. to investigate the mechanisms of how antimicrobial peptides interact with membranes of various composition. The stable and open membrane architecture allows here for the use of diverse surface-analytical techniques.

Recently, first investigations have been made that use tBLMs as a model membrane system to study fundamental processes in food science, e.g. the denaturation of proteins at a lipid interface.

Acknowledgments

Several people have contributed to this work. We are especially thankful to P.P. Atanasova, V. Atanasov, N. Knorr and W. Knoll.

References

1 Sackmann, B. and Neher, E. (1983) *Single-Channel Recordings*, Kluwer Academic/Plenum Publishers.

2 Fertig, N., Tilke, A., Blick, R.H., Kotthaus, J.P., Behrends, J.C. and ten Bruggencate, G. (2000) Stable integration of isolated cell membrane patches in a nanomachined aperture. *Applied Physics Letters*, **77** (8), 1218–1220.

3 Pantoja, R., Nagarah, J.M., Starace, D.M., Melosh, N.A., Blunck, R., Bezanilla, F. and Heath, J.R. (2004) Silicon chip-based patch-clamp electrodes integrated with PDMS microfluidics. *Biosensors & Bioelectronics*, **20**, 509–517.

4 Shieh, C.-C. (2004) Automated high-throughput patch-clamp techniques. *Drug Discovery Today*, **9** (13), 551–552.

5 Chan, Y.H.M. and Boxer, S.G. (2007) Model membrane systems and their applications. *Current Opinion in Chemical Biology*, **11** (6), 581–587.

6 Hianik, T. (2006) Structure and physical properties of biomembranes and model membranes. *Acta Physica Slovaca*, **56** (6), 687–805.

7 Bayley, H. and Cremer, P.S. (2001) Stochastic sensors inspired by biology. *Nature*, **413**, 226–230.

8 Bayley, H. (1997) Building doors into cells. *Scientific American*, 42–47.

9 Edwards, D.A., Schneck, F., Zhang, I., Davis, A.M.J., Chen, H. and Langer, R. (1996) Spontaneous vesicle formation at lipid bilayer membranes. *Biophysical Journal*, **71**, 1208–1214.

10 Döbereiner, H.-G. (2000) Properties of giant vesicles. *Current Opinion in Colloid & Interface Science*, **5**, 256–263.

11 Claessens, M.M.A.E., van Ort, B.F., Leermakers, F.A.M., Hoekstra, F.A. and Cohen Stuart, M.A. (2004) Charged lipid vesicles: effects of salts on bending rigidity, stability, and size. *Biophysical Journal*, **87**, 3882–3893.

12 Ottova, A., Tvarozek, V., Racek, J., Sabo, J., Ziegler, W., Hianik, T. and Tien, H. (1997) Self-assembled BLMs: biomembrane models and biosensor applications. *Supramolecular Science*, **4**, 101–112.

13 Winterhalter, M. (2000) Black lipid membranes. *Current Opinion in Colloid & Interface Science*, **5**, 250–255.

14 Eray, M., Dogan, N. and Liu, L. (1994) Highly stable bilayer lipid membranes (BLMs) formed on microfabricated polyimide apertures. *Biosensors & Bioelectronics*, **9**, 343–351.

15 Yuan, H., Leitmannova-Ottova, A. and Tien, H. (1996) An agarose-stabilized BLM: a new method for forming bilayer lipid membranes. *Materials Science and Engineering C*, **4**, 35–38.

16 Schmidt, J. (2005) Stochastic sensors. *Journal of Materials Chemistry*, **15** (8), 831–840.

17 Jeon, T.J., Malmstadt, N. and Schmidt, J.J. (2006) Hydrogel-encapsulated lipid membranes. *Journal of the American Chemical Society*, **128** (1), 42–43.

18 Heyse, S., Stora, T., Schmid, E., Lakey, J.H. and Vogel, H. (1998) Emerging techniques for investigating molecular interactions at lipid membranes. *Biochimica et Biophysica Acta*, **88507**, 319–338.

19 Knoll, W., Frank, C.W., Heibel, C., Naumann, R., Offenhäusser, A., Rühe, J., Schmidt, E.K., Shen, W.W. and Sinner, A. (2000) Functional tethered lipid bilayers. *Reviews in Molecular Biotechnology*, **74**, 137–158.

20 Tien, H.T. and Ottova, A. (1999) From self-assembled bilayer lipid membranes (BLMs) to supported BLMs on metal and gel substrates to practical applications. *Colloids and Surfaces A*, **149**, 217–233.

21 Tien, H.T. and Ottova, A. (1998) Supported planar lipid bilayers (s-BLMs) as electrochemical biosensors. *Electrochimica Acta*, **43** (23), 3587–3610.

22 Sackmann, E. (1996) Supported membranes: scientific and practical applications. *Science*, **271**, 43–48.

23 Tanaka, M. and Sackmann, E. (2005) Polymer-supported membranes as models of the cell surface. *Nature*, **437** (7059), 656–663.

24 Tamm, L.K. and McConnell, H.M. (1985) Supported phospholipid bilayers. *Biophysical Journal*, **47**, 105–113.

25 Raguse, B., Braach-Maksvytis, V., Cornell, B.A., King, L.G., Osman, P.D.J., Pace, R.J. and Wieczorek, L. (1998) Tethered lipid bilayer membranes: formation and ionic reservoir characterization. *Langmuir*, **14**, 648–659.

26 Cornell, B.A., Krishna, G., Osman, P.D., Pace, R.J. and Wieczorek, L. (2001) Tethered-bilayer lipid membranes as a support for membrane-active peptides. *Biochemical Society Transactions*, **29** (4), 613–617.

27 Köper, I. (2007) Insulating tethered bilayer lipid membranes to study membrane proteins. *Molecular BioSystems*, **3** (10), 651–657.

28 Cornell, B.A., Braach-Maksvytis, V.L.B., King, L.G., Osman, P.D.J., Raguse, B., Wieczorek, L. and Pace, R.J. (1997) A biosensor that uses ion-channel switches. *Nature*, **387**, 580–583.

29 Atanasov, V., Atanasova, P., Vockenroth, I., Knorr, N. and Köper, I. (2006) Highly insulating tethered bilayer membranes. A generic approach for various substrates. *Bioconjugated Chemistry*, **17**, 631–637.

30 Atanasov, V., Knorr, N., Duran, R.S., Ingebrandt, S., Offenhäuser, A., Knoll, W. and Köper, I. (2005) Membrane on a Chip. A functional tethered lipid bilayer membrane on silicon oxide surfaces. *Biophysical Journal*, **89** (3), 1780–1788.

31 Suarez-Isla, B.A., Wan, K., Lindstrom, J. and Montal, M. (1983) Single-channel recordings from purified acetylcholine receptors reconstituted in bilayers formed at the tip of patch pipets. *Biochemistry*, **22**, 2319–2323.

32 Riquelme, G., Lopez, E., Garcia-Segura, L.M., Ferragut, J.A. and Gonzalez-Ros, J.M. (1990) Giant liposomes: a model system in which to obtain patch-clamp recordings of ionic channels. *Biochemistry*, **29**, 11215–11222.

33 Naumann, R., Walz, D., Schiller, S.M. and Knoll, W. (2003) Kinetics of valinomycin-mediated K+ ion transport through tethered bilayer lipid membranes. *Journal of Electroanalytical Chemistry*, **550–551**, 241–252.

34 Gray, G.S. and Kehoe, M. (1984) Primary sequence of the a-toxin gene from Staphylococcus aureus Wood 46. *Infection and Immunity*, **46** (2), 615–618.

35 Valeva, A., Hellmann, N., Walev, I., Strand, D., Plate, M., Boukhallouk, F., Brack, A., Hanada, K., Decker, H. and Bhakdi, S. (2006) Evidence that clustered phosphocholine head groups serve as sites for binding and assembly of an oligomeric protein pore. *The Journal of Biological Chemistry*, **281** (36), 26014–26021.

36 Cheley, S., Malghani, M.S., Song, L., Hobaugh, M., Gouaux, J.E., Yang, J. and Bayley, H. (1997) Spontaneous oligomerization of a staphylococcal alpha-hemolysin conformationally constrained by removal of residues that form the transmembrane beta-barrel. *Protein Engineering*, **10** (12), 1433–1443.

37 Reichwein, J., Hugo, F., Roth, M., Sinner, A. and Bhakdi, S. (1987) Quantitative analysis of the binding and oligomerization of staphylococcal alpha-toxin in target erythrocyte membranes. *Infection and Immunity*, **55** (12), 2940–2944.

38 Walker, B., Krishnasastry, M., Zorn, L. and Bayley, H. (1992) Assembly of the oligomeric membrane pore formed by staphylococcal a-hemolysin examined by truncation mutagenesis. *Journal of Biological Chemistry*, **267** (30), 21782–21786.

39 Walker, B., Braha, O., Cheley, S. and Bayley, H. (1995) An intermediate in the assembly of a pore-forming protein trapped with a genetically-engineered switch. *Chemistry and Biology*, **2**, 99–105.

40 Braha, O., Gu, L.-Q., Zhou, L., Lu, X., Cheley, S. and Bayley, H. (2000) Simultaneous stochastic sensing of divalent metal ions. *Nature Biotechnology*, **18**, 1005–1007.

41 Siwy, Z., Apel, P., Baur, D., Dobrev, D.D., Korchev, Y.E., Neumann, R., Spohr, R., Trautmann, C. and Voss, K.-O. (2003) Preparation of synthetic nanopores with transport properties analogous to biological channels. *Surface Science*, **532–535**, 1061–1066.

42 Bayley, H., Braha, O. and Gu, L.-Q. (2000) Stochastic Sensing with Protein Pores. *Advanced Materials*, **12** (2), 139–142.

43 Belmonte, G., Cescatti, L., Ferrari, B., Nicolussi, T., Ropele, M. and Menestrina, G. (1987) Pore formation by Staphylococcus aureus alpha-toxin in lipid bilayers: Dependence upon temperature and toxin concentration. *European Biophysics Journal*, **14**, 349–358.

44 Menestrina, G. (1986) Ionic channels formed by staphylococcus aureus alpha toxin: Voltage-dependent inhibition by divalent and trivalent cations. *Journal Membrane Biology*, **90**, 177–190.

45 Schuster, B. and Sleytr, U.B. (2002) Single channel recordings of alpha-hemolysin reconstituted in S-layer-supported lipid bilayers. *Bioelectrochemistry*, **55**, 5–7.

46 Schuster, B., Pum, D., Braha, O., Bayley, H. and Sleytr, U.B. (1998) Self-assembled a-hemolysin pores in an S-layer-supported lipid bilayer. *Biochimica and Biophysica Acta*, **1370**, 208–288.

47 Schuster, B., Sleytr, U., Diederich, A., Bähr, G. and Winterhalter, M. (1999) Probing the stability of S-layer-supported planar lipid membranes. *European Biophysical Journal*, **28**, 583–590.

48 Kang, X.-F., Cheley, S., Rice-Ficht, A.C. and Bayley, H. (2007) A storable encapsulated bilayer chip containing a single protein nanopore. *Journal of the American Chemical Society*, **129** (15), 4701–4705.

49 Glazier, S.A., Vanderah, D.J., Plant, A.L., Bayley, H., Valincius, G. and Kasianowicz, J.J. (2000) Reconstitution of the pore-forming toxin a-hemolysin in phospholipid/18-octadecyl-1-thiahexa(ethylene oxide) and phospholipid/n-octadecanethiol supported bilayer membranes. *Langmuir*, **16**, 10428–10435.

50 Schiller, S.M., Naumann, R., Lovejoy, K., Kunz, H. and Knoll, W. (2003) Archaea analogue thiolipids for tethered bilayer lipid membranes on ultrasmooth gold surfaces. *Angewandte Chemie*, **42** (2), 208–211.

51 Naumann, R., Schiller, S.M., Giess, F., Grohe, B., Hartman, K.B., Kärcher, I., Köper, I., Lübben, J., Vasilev, K. and Knoll, W. (2003) Tethered lipid bilayers on ultraflat gold surfaces. *Langmuir*, **19**, 5435–5443.

52 Vockenroth, I.K., Atanasova, P.P., Jenkins, A.T.A. and Köper, I. (2008) Incorporation of alpha-hemolysin in different tethered bilayer lipid membrane architectures. *Langmuir*, **24** (2), 496–502.

53 Vockenroth, I.K., Atanasova, P.P., Long, J.R., Jenkins, A.T.A., Knoll, W. and Köper, I. (2007) Functional incorporation of the pore forming segment of AChR M2 into tethered bilayer lipid membranes. *Biochimica et Biophysica Acta (BBA) – Biomembranes*, **1768** (5), 1114–1120.

54 Keizer, Henk M., Dorvel, B.R., Andersson, M., Fine, D., Price, R.B., Long, J.R., Dodabalapur, A., Köper, I., Knoll, W., Anderson, P.A.V., and Duran, R.S. (2007) Functional ion channels in tethered bilayer membranes – implications for biosensors. *ChemBioChem*, **8** (11), 1246–1250.

55 Andersson, M., Keizer, H.M., Zhu, C.Y., Fine, D., Dodabalapur, A. and Duran, R.S. (2007) Detection of single ion channel activity on a chip using tethered bilayer membranes. *Langmuir*, **23** (6), 2924–2927.

3.4 Interaction of Structured and Functionalized Polymers with Cancer Cells

Anika Embrechts, Chuan Liang Feng, Ilona Bredebusch, Christina E. Rommel, Jürgen Schnekenburger, G. Julius Vancso, and Holger Schönherr

3.4.1 Introduction

The interaction of cells with surfaces plays a crucial role for the success or failure of implantable polymers for controlled drug delivery and in applications in the field of tissue engineering, where polymers are used to assist regeneration or three-dimensional tissue growth [1]. Cell/polymer interfaces are formed when a drug-delivery device is implanted or tissue is regenerating inside a body (*in vivo*) or outside the body in a tank reactor (*in vitro*). Since interactions with the local environment have been shown to play a decisive role in stem-cell differentiation and in stem-cell dedifferentiation [2], cell-surface interactions are also an important aspect of cancer research, as outlined in this chapter.

A substantial fraction of reactions and molecular interactions in biology, including molecular recognition processes, occur at interfaces [3]. Under physiological conditions, the low energy barrier for mobility in-plane facilitates complex reactions through clustering and/or conformational changes and high surface area geometries may enhance reaction turnover rates. Unique organic microenvironments can further enhance specific affinities and reactions and surface-energy minimization can orient specific structures at interfaces.

In general, cells interact with the external environment via transmembrane proteins, which transmit both information and molecules from the outside of the cell to the inside. Many of these proteins are receptors, characterized by an extracellular ligand-binding domain and an intracellular signaling domain. When a receptor binds its cognate ligand (typically a protein or peptide), a change in receptor conformation or affinity for other cellular molecules initiates an intracellular cascade of enzyme-mediated reactions resulting in amplification of the signal. This process of signal transduction ultimately affects gene regulation. On the average mammalian cell dozens of different types of receptors are present. Thus, cell functions including survival, proliferation, migration, and differentiation are governed by the integrated signals from many types of molecules.

Surface Design: Applications in Bioscience and Nanotechnology
Edited by Renate Förch, Holger Schönherr, and A. Tobias A. Jenkins

ISBN: 978-3-527-40789-7

Two broad classes of extracellular ligands for biomaterials are growth factors and the extracellular matrix (ECM). Growth factors are typically characterized as relatively soluble peptides or proteins (M_w = 5000–50 000 g/mol). Growth factors acting on a cell may be produced by the same cell, neighboring cells in contact or distant cells. The ECM is a complex network of proteins and carbohydrates, which provides physical scaffolding for cells and tissues, it helps to provide a permeability barrier between tissue compartments, and enables polarization of tissue structures, such as skin. ECM molecules are quite large (M_w > 100 000 g/mol) and contain several functional domains or regions. Within these ECM molecules small adhesion-mediating peptide domains have been identified and characterized (the first was the arginine-glycine-aspartate, or RGD) [4]. These peptides interact with a class of cell-surface adhesion receptors called integrins. Like growth factors, integrins mediate many aspects of cell behavior. Hence, manipulation of integrin ligation by including peptide-adhesion domains in synthetic biomaterials is an area of intense research and development in current biomaterials [5].

As introduced in Chapters 1.1 and 1.4, there are several possibilities to include peptide or protein-adhesion domains on a biomaterial. A given substrate can be functionalized (conjugated) with a particular peptide/protein or more than one type to analyze if the cells react to a specific peptide or protein, or to a pattern of these molecules. The dimensions of these chemical patterns can play an important role in cell life and death, as also elaborated in Chapter 2.4 [6]. In addition to chemical patterns, topographic structures, such as grooves or pillars on a surface can affect, for instance, cell motility [7]. Since cell behavior can be controlled by these factors as well as substrate modulus, the study of cell-surface interactions may also reveal information on cell function itself. This is in particular crucial for diseases such as cancer.

Cancer is a major public-health problem in all developed countries. Currently, one in four deaths in the United States is due to cancer [8]. In Europe, with an estimated 3.2 million new cases and 1.7 million deaths each year, cancer remains an important public-health problem and the aging of the European population will cause these numbers to continue to increase even if age-specific rates remain constant [9]. Thus, there is a great need to learn more about the mechanisms of the disease in order to improve current therapies and to develop new approaches.

The diagnosis 'cancer' is a collective term describing different behaviors and phenotypes of tumors, which are unified and characterized by an uncontrolled growth and spread of abnormal cells. This tumorigenesis is a multistep process involving multiple genetic alteration that push normal cells into highly malignant phenotypes with metastasis being the final state [10]. Metastases that are resistant to conventional therapies in turn, are the major cause of death from cancer [11]. Formation of metastases is again a multistep process and only cells that pass the steps can establish metastases. In particular, the behavior of detachment is quite abnormal compared to ordinary epithelial and endothelial cells, which undergo apoptosis via the anoikis phenomenon (cell death), when they detach from their familiar surrounding material (extracellular matrix) [12]. This implies that there are specific changes in the configuration of cellular proteins, cell-surface receptors, cytoskeleton (shape and elasticity of the cells) [13, 14] and especially in the adhesion molecules profile (e.g. integrins, cadherins) resulting from various genetic alterations [15]. Metastasizing cells start the establishment of a microenvironment and proliferation in a distant organ [14]. Trying to describe these phenomena from the perspective of a single cell, cancer development depends upon changes in the heterotypic interactions between incipient tumor cells and their normal neighbors [10] or – to reduce it even more – cancer development depends upon changes in the interaction of the cells with the surrounding surfaces of the cell microenvironment. Examples of such 'surfaces' are the components of the extracellular matrix (ECM), basal membranes and ligands on neighboring cells. In general, a crucial element for the development of cancer and metastases is the interaction of tumor cells with the host stroma. Already in 1889 Paget hypothesized that the nonrandom pattern of cancer metastases in distant organs is due to specific factors in the host tissue (the 'soil'), which stimulate certain favored tumor cells (the 'seed') to adhere ('seed and soil' hypothesis) [16]. Today, there are several reports suggesting that factors produced by each of the individual tissues in the body may exert differential effects on tumor cell growth [17–19].

However, although we know some of the details in cancer development and metastasis there remain a series of open questions. Moreover, every single step in tumorigenesis and metastasis might be a prospective target for new cancer therapies, assuming that there are models to study the complex interactions. Particularly with regard to the reduction of complexity, but also to reduce animal experiments at the early stage of basic research, there is a great need for adequate *in vitro* models. A substrate that mimics the nanostructured defined and ordered chemical patterns of the tumor environment, is a prerequisite for these models. Tumor-cell biologists currently lack artificial structures for the study of defined environmental surface properties influencing tumor progression and metastasis.

To address these shortcomings, suitable micro- and nanostructured patterns and structures that model the tumor environment must be fabricated. Materials science and chemistry are, as mentioned in Chapters 1.1 and 2.4, advanced to a point that allows studies of cell response to nanometer-scale variations in peptide spacing in 2D. However, in their natural environment cells reside in a complex *three-dimensional* environment. The ECM provides, unlike a surface in a 2D cell culture, the appropriate architecture for 'normal' cell growth and development. Recent reports corroborate the essential need of yet unavailable, well-defined 3D environments, e.g., for the study of cell-matrix interactions [20, 21] or cancer-cell migration, which is intimately linked to cancer progression and the development of new therapeutic concepts [22, 23]. Unfortunately, for the more relevant 3D situation, such investigations are to date beyond the scope of established matrix fabrication [24, 25].

Among the techniques and approaches for the fabrication of 2D chemical patterns microcontact printing (μCP) with peptides or proteins was introduced on, e.g., gold and silicon/glass in conjunction with self-assembled monolayers (SAMs) [26]. In addition, polymers were functionalized with self-assembled structures [27] or first were activated to form a monolayer afterwards [28, 29]. Patterns

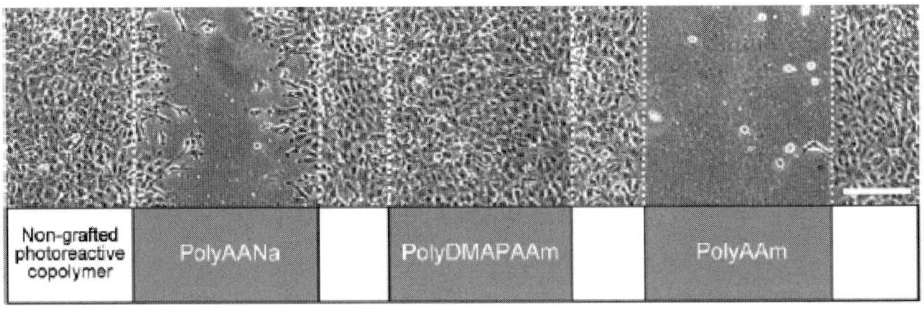

Figure 1 Bovine endothelial cell cultures on photografted micropatterned surfaces. PolyAANa = poly(acrylic acid) sodium salt; PolyDMAPAAm = poly(N-[3-(dimethylamino)propyl]acrylamide) methiodide; PolyAAm = poly(acryl amide). The scale bar is 200 μm. (Reprinted with permission from J. Biomed. Mater. Res., 53, 2000, 584–591, Copyright 2000 by John Wiley & Sons, Inc.).

were also created with peptides or proteins on activated polymers by μCP [30] and by photografting approaches (Figure 1). Here, bovine endothelial cells were cultured on different polymers, grafted on a photoreactive-copolymer-coated PET film. The differential cell response is clearly visible as the photografting allows one to graft the three different polymers next to each other [31]. Since photografting utilized light to initiate the reaction, the pattern sizes that can be obtained are limited by the diffraction limit. In addition, sub-500-nm structures are difficult to fabricate without expensive and very sophisticated lithography equipment.

Controlled variations in topography were introduced, among others, by lithography [7, 32], embossing [32] or photoablation [33]. Besides chemical and topographical patterns, combinations of these patterns have been used to study cell behavior [33, 34]. As shown in Figure 2, hydrophobicity can be a factor of influence to cell migration or spreading. Cells migrating along hydrophilic aminosilane functionalized topographic ridges show anisotropic behavior; it is obvious that the hydrophobic ethylsilane ridges are bridged by the cell (Figure 2a) [35]. By contrast, the long filopodia observed showed few signs of recognizing the adhesive boundaries (Figure 2b). When the adhesive tracks were perpendicular to the gratings, cells followed the adhesive cue, despite the difference in topography with a ridge height of 3.0 μm (Figure 2c).

All previously mentioned techniques and methods were used to date on different polymers. Some of these polymers are biodegradable, such as poly(lactic acid) (PLA) or poly(glycolic acid) (PGA), or can be tailored to have a much wider range of mechanical and chemical properties [36], like Teflon for vascular grafts and high-density polyethylene (HDPE) for use in hip implants.

This early work already demonstrated the interplay of chemical patterns and topographical structures. However, to model the tumor environment the surface design must be refined considerably. This identified need is corroborated in at least two directions:

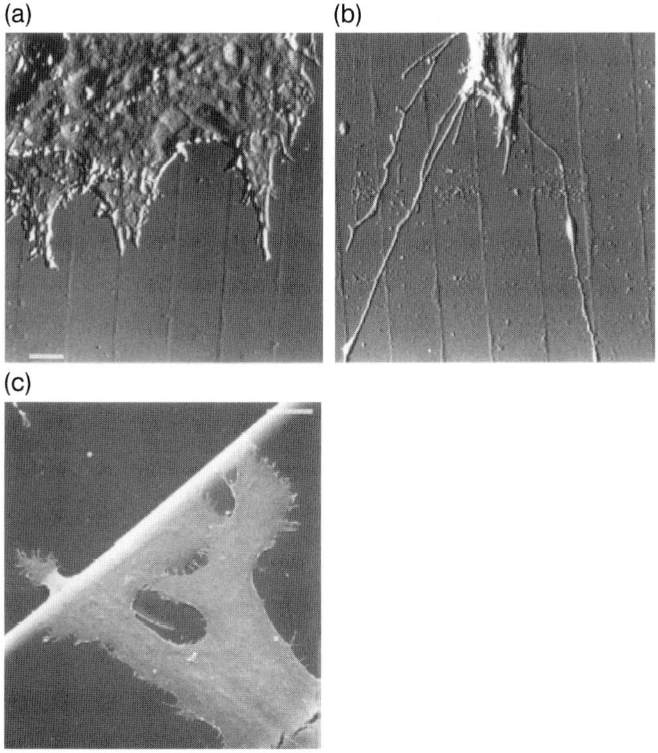

Figure 2 Scanning probe images of BHK cells on 5-μm period amino/methylsilane adhesive gratings on quartz. (a) Many cells spanned successive tracks on 5-μm adhesive gratings. This micrograph shows three processes from the leading edge of such a cell that are clearly extending in parallel along aminosilane tracks. The aminosilane tracks are elevated compared to the methylsilane tracks. Scale bar, 5 μm. (b) More highly aligned cells sometimes extended spindle-like filopodia that showed few signs of recognizing adhesive boundaries. (c) Scanning electron micrographs of BHK cells aligning to a 25-μm wide adhesive track superimposed orthogonally onto a 25-μm pitch, 3.0-μm deep topographic grating. One end of the cell comprised the familiar pincer arrangement, but that seemed to have faltered at the ridge edge. Scale bar 5 μm. (Reprinted from [35]; S. Britland, H. Morgan, B. Wojciak-Stothard, M. Riehle, A. Curtis, C. Wilkinson, Synergistic and Hierarchical Adhesive and Topographic Guidance of BHK Cells, Expt. Cell. Res., 228, 313–325, Copyright 1996, with permission from Academic Press).

1. All different length scales that are relevant for cell-surface interactions must be implemented. This refers in particular to the molecular-scale definition of the micro- and nanostructured patterns and structures, as demonstrated by the group of Spatz (Figure 3) [37]. Surfaces were decorated by gold nanoparticles that were regularly ordered by exploiting block copolymer micelle self-assembly. This self-assembly process yields defined nanoparticle spacings. Via immobilized RGD thiols on these < 8 nm diameter particles the binding of one integrin per particle can be realized. For a separation distance of ≥ 73 nm between the adhesive nanoparticles attachment and spreading of the analyzed

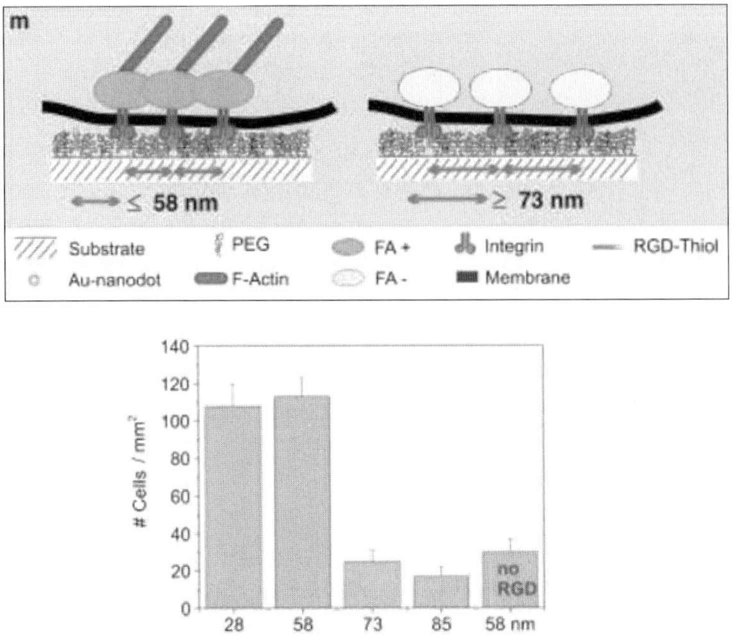

Figure 3 Top: Scheme of biofunctionalized nanopattern to control integrin clustering in cell membranes: Au dots are functionalized by c(RGDfK)-thiols and areas between cell-adhesive Au dots are passivated by PEG against cell adhesion. Therefore, cell adhesion is mediated entirely via c(RGDfK)-covered Au nanodots. A separation of Au/RGD dots by ≥ 73 nm causes limited cell attachment and spreading and actin stress fiber formation, because of restricted integrin clustering. This is indicated by no focal adhesion activation (FA–), whereas distances between dots of ≤ 58 nm caused focal adhesion activation (FA+). Bottom: Number of cells attached per unit area for the different surfaces. (Reproduced with permission from [37]; M. Arnold, E. A. Cavalcanti-Adam, R. Glass, J. Blummel, W. Eck, M. Kantlehner, H. Kessler, J. P. Spatz, ChemPhysChem 2004, 5, 383, Copyright 2004 by Wiley-VCH).

cells was limited. The formation of focal adhesions and actin stress fibers was substantially reduced. The authors attributed this cellular response to restricted integrin clustering and indentified the range between 58 and 73 nm as a universal length scale for integrin clustering and activation.

2. The model systems should evolve from entirely 2D to 3D. First successful experiments on single cells that are incorporated individually into 3D structures show the feasibility to investigate the relation of shape and function of single cells or clusters of cells in a 3-dimensional (3D) microenvironments (Figure 4) [38].

These structures were fabricated in a two-step replication process comprising hot embossing of PS followed by inverted microcontact printing to selectively functionalize the plateau surface between the microwells with a passivating poly(L-lysine)-g-poly(ethylene glycol) (PLL-g-PEG) layer. The interior surface of the

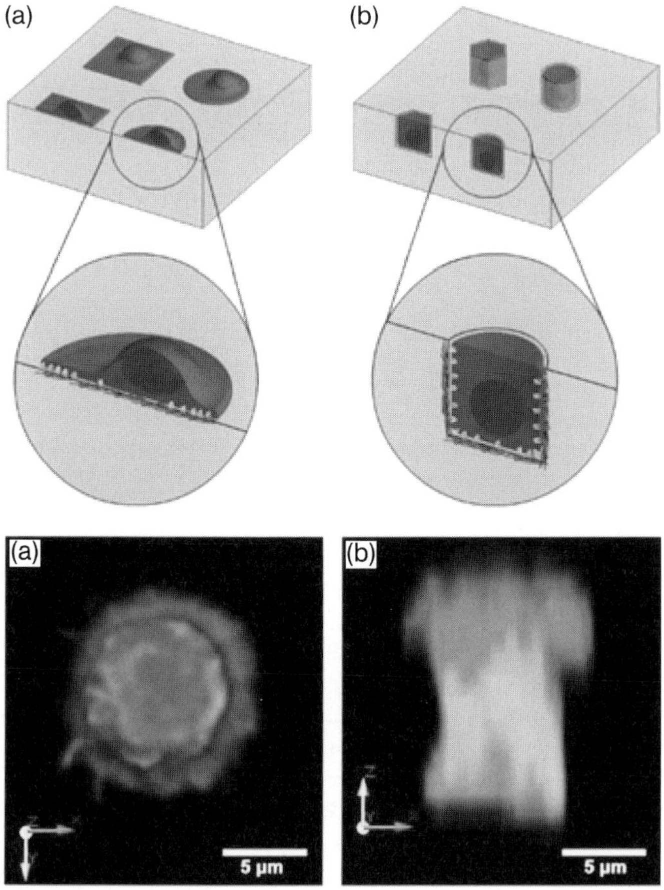

Figure 4 Top: Scheme of the concept of (a) conventional 2D patterning of cells and (b) micro-3D culturing of single cells. The surface of the microwells exhibits cell binding properties, while the plateau surface inhibits adsorption of proteins and attachment of cells. Bottom: 3D reconstruction of a confocal laser scanning microscopy (CLSM z-stack of a single epithelial cell (MDCK II) attached inside a microwell (10 μm diameter, 11 μm depth); 3 h culture time in fetal calf serum (FCS) containing medium; fixed in PFA and filamentous actin stained with FITC-phalloidin: (a) view from below; (b) view from the side. (Reprinted from [38]; M. R. Dusseiller, D. Schlaepfer, M. Koch, R. Kroschewski, M. Textor, An inverted microcontact printing method on topographically structured polystyrene chips for arrayed micro-3D culturing of single cells, Biomaterials, 26, 5917–5925, Copyright 2005, with permission from Elsevier).

microwells was subsequently functionalized by spontaneous adsorption of proteins or functionalized PLL-g-PEG. While this outstanding example illustrates the feasibility of fabricating well-defined microwell structures, it is still beyond the scope of current surface and interface engineering to implement the patterning and structures on all the mentioned length scales. Thus, although advanced micro- and nanofabrication methods have been developed and also appropriate porous scaffolds as ECM mimics for cell growth have been devised in tissue engineering [39–44], these lack the deterministic precision and control required for molecular-scale structure-function studies. Clearly, new approaches are required to provide the basis for fundamental studies in this area. This need for complex, yet well-defined, topographically and biochemically fully controlled 3D matrices represents a grand challenge for materials chemistry and surface design.

As outlined in this chapter, the first steps of implementing 3D structures using intrinsically nanostructured block copolymer materials have been successfully undertaken. The structures fabricated by recently developed and refined soft lithographic methods, such as reactive and inverted microcontact printing, have been tested with pancreatic cancer cells as a relevant model system to be able to systematically study the influence of defined environmental surface properties on tumor progression and metastasis, among others.

3.4.2
Cancer-Cell-Surface Interactions

Interactions of cells with surfaces, matrices, and other cells in general play a crucial role in most fundamental cellular functions, such as motility, growth, differentiation and apoptosis (programmed cell death), as mentioned in the introduction [45]. In this chapter, we focus on the investigation of pancreas adenocarcinoma cells [46] on tailored surfaces that were micro- and/or nanostructured, and functionalized homogeneously or in patterns with PLL, fibronectin and PEG. These cells (PaTu8988T) [46] grow typically as monolayers and are characterized by a high proliferation rate and rapid migration. Resting cells appear rounded and develop an average cell height of ~15 μm. Migrating cells appear flattened with a spread fibroblast-like morphology (Figure 5). In addition, migrating PaTu8988T cells show extended filopodia cell structures involved in the formation of new focal adhesions. These can be discerned both in SEM and AFM images of fixated cells that were allowed to settled on surfaces exposing well-defined surface chemistry.

In the high-resolution AFM image of a cancer cell on a homogeneously fibronectin-covered block copolymer surface, the convergence of the length scales of the filopodia and the microphase-separated blockcopolymer morphology is evident. Clearly the cell–surface interactions are mediated by elements with comparable length scales.

At the investigated block length ratio and molar mass the PS_{690}-*b*-$PtBA_{1210}$ block copolymer exhibits a cylindrical morphology, in which the PS cylinders possess a spacing of ~ 88 nm [47]. As elucidated in detail in Chapter 2.4, the surface exposes the lower surface energy block prior to functionalization, thus the underlying fibronectin layer (Figure 5b and c) is homogeneous. The contrast in the AFM image is related to the local differences in energy dissipation and hence does not reflect any differences in surface chemistry. However, by exploiting advanced surface-modification approaches it is in principle feasible to expose both blocks in

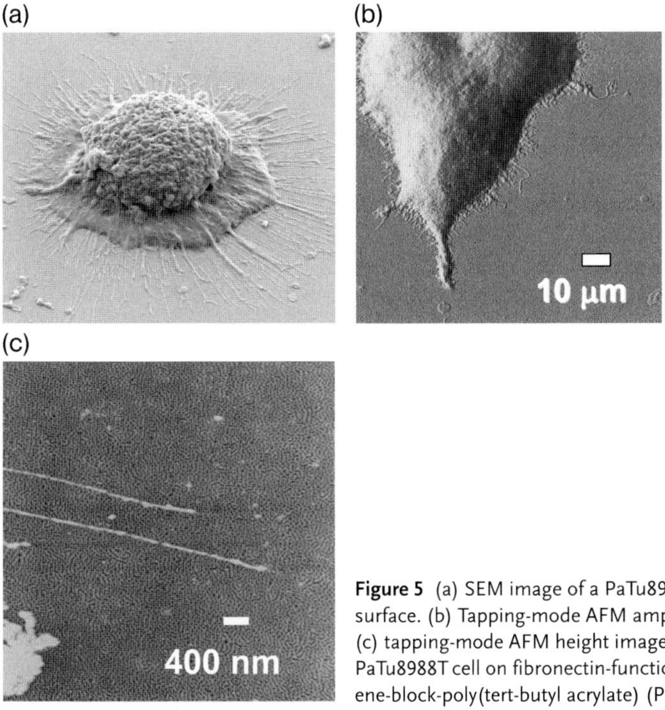

Figure 5 (a) SEM image of a PaTu8988T cell on silicon surface. (b) Tapping-mode AFM amplitude and (c) tapping-mode AFM height images, respectively, of PaTu8988T cell on fibronectin-functionalized poly-styrene-block-poly(tert-butyl acrylate) (PS_{690}-b-$PtBA_{1210}$).

regular nanopatterns at these surfaces. By further exploiting blockcopolymers the corresponding distances can be tuned in the range of ~15 to 100 nm [48]. Thereby, microphase separation further opens avenues towards structuring on multiple length scales via a synergistic combination of top-down patterning techniques with blockcopolymer self-assembly.

In general, blockcopolymers are materials that are increasingly used in micro- and nanotechnology based on their ability to self-assemble into useful structures on the nanometer length scale (see also Chapter 2.1). This self-assembly into nanostructures is caused by interplay between the chemical incompatibility and the covalent connection of the blocks of the blockcopolymer, leading to a thermodynamic microphase separation process [48]. The shape and size of the structures can be defined by controlling the chemical and thermodynamic properties of the polymer system. In addition, the formation and long-range ordering of these structures can also be further influenced by extrinsic factors like the surface energy at interfaces between the blockcopolymer and its environment and the presence of mechanical or electrical forces.

In the first experiments discussed here the reliable conjugation of proteins, as well as PEG for the required surface passivation [49], on activated polystyrene-block-poly(tert-butyl acrylate) (PS_{690}-b-$PtBA_{1210}$) was investigated [50]. In particular the blocking ability of PEG layers with different number average molar mass M_n was addressed. It is well established that the molar mass/chain length and the

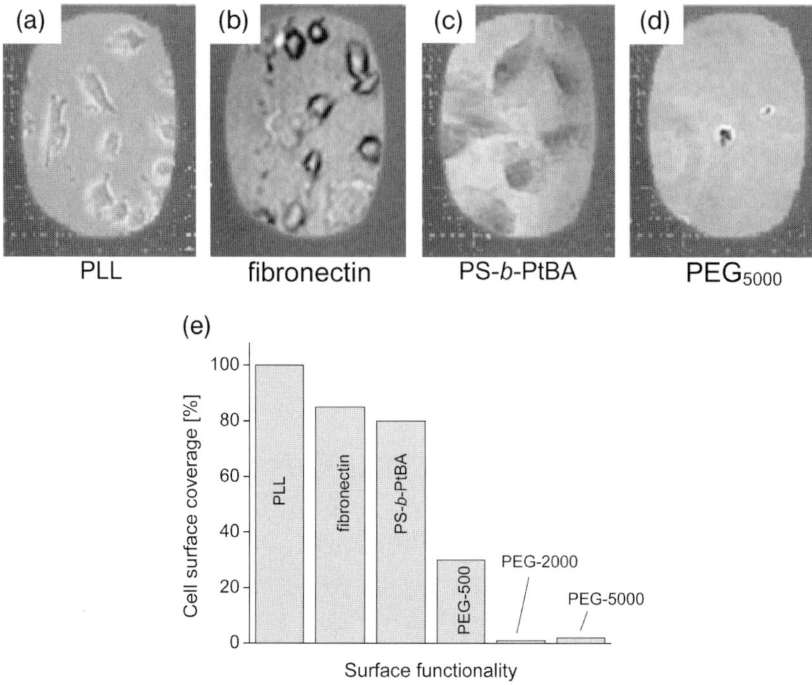

Figure 6 Representative wide-field optical micrographs (90 μm × 160 μm) of (a) PLL-covered PS_{690}-b-$PtBA_{1210}$ film, (b) fibronectin-covered PS_{690}-b-$PtBA_{1210}$ film, (c) a neat PS_{690}-b-$PtBA_{1210}$ film and (d) PEG_{5000}–NH_2-covered PS_{690}-b-$PtBA_{1210}$ film. (e) Histogram of cell-surface coverage (normalized to coverage on PLL) vs. polymer surface composition. (Reprinted from [50]; C. L. Feng, A. Embrechts, I. Bredebusch, A. Bouma, J. Schnekenburger, M. García-Parajó, W. Domschke, G. J. Vancso, H. Schönherr, Tailored interfaces for biosensors and cell-surface interaction studies via activation and derivatization of polystyrene-block-poly(tert-butyl acrylate) thin films, Eur. Polym. J., 43, 2177–2190, Copyright 2007, with permission from Elsevier).

grafting density of PEG molecules determine to a large extent the antifouling properties [51]. Wide-field optical microscopy was applied to monitor the dependence of the cell-surface interaction on surface chemistry (Figure 6). On homogeneous layers of PLL and fibronectin the cells not only adhered, but also stretched out on the film (Figure 6a and b), which is indicative of the full adhesive capacity of PaTu8988T cells and the unrestricted formation of focal adhesions. On the PEG layers with high molar mass only isolated cells were observed that exhibited a rounded appearance. These cells could not attach and died after prolonged contact with the surface. The histogram of cell-surface coverage (Figure 6e) quantifies the observed differences.

Using the optimized surface chemistry the interaction of adherent PaTu8988T pancreas adenocarcinoma cells with submicrometer-sized patterns of passivating PEG and cell-adhesive fibronectin was investigated on samples prepared by reactive μCP (Chapter 2.4) [52]. Transfer printing and covalent coupling of PEG_{5000}-

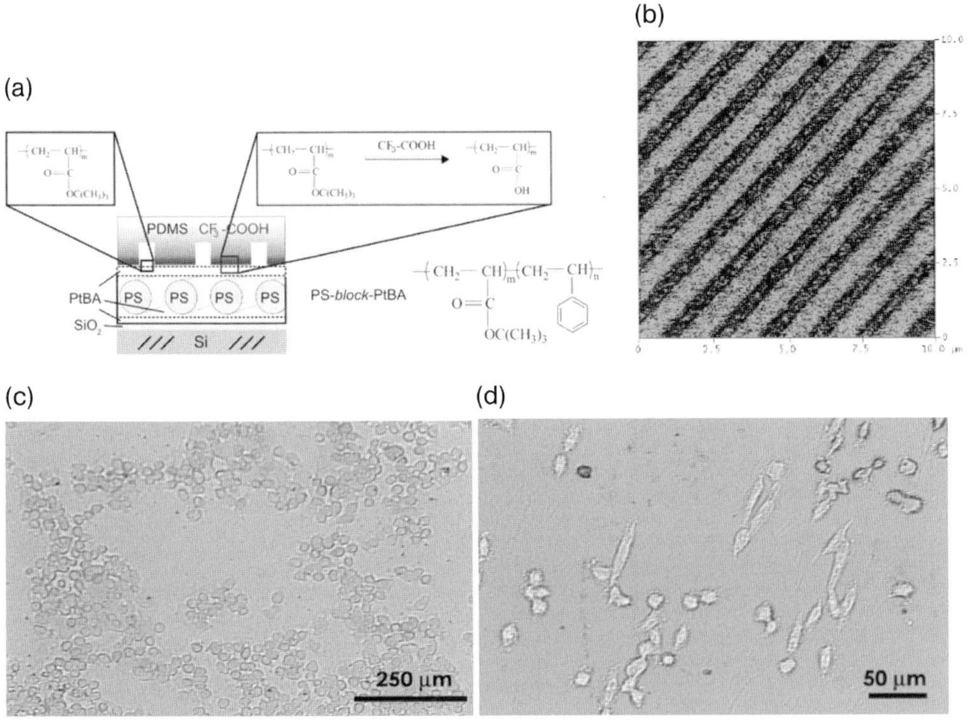

Figure 7 (a) Schematic of a localized deprotection reaction carried out in the contact areas of the stamp and the PtBA skin layer. (b) Friction force AFM image of NHS activated PS_{690}-b-$PtBA_{1210}$ diblock-copolymer film after transfer of PEG_{5000}-NH_2 (friction forces [a.u.] increase from dark to bright contrast) shows the faithful transfer and coupling of the PEG. The transferred PEG layer is characterized by a higher friction force compared to the unfunctionalized areas. (c) Optical microscopy image of PaTu8988T pancreatic cancer cells on unpatterned fibronectin-functionalized films showed few, if any elongated cells. (d) Optical microscopy image of PaTu8988T pancreatic cancer cells on a pattern of fibronectin (300 nm line width) and PEG_{5000}-NH_2 (500 nm line width) on a block-copolymer film prepared by direct reactive μCP of PEG_{5000}-NH_2 on a TFA-prehydrolyzed and an NHS/EDC-activated PS_{690}-b-$PtBA_{1210}$ film, followed by functionalization of the unreacted areas with fibronectin. Cells were cultured for 24 h on the nanopattern. (Reproduced with permission from [52]; C. L. Feng, A. Embrechts, I. Bredebusch, J. Schnekenburger, W. Domschke, G. J. Vancso, H. Schönherr, Adv. Mater. 2007, 19, 286, Copyright 2007 by Wiley-VCH).

NH_2 onto previously hydrolyzed and NHS-ester-activated films of PS_{690}-*b*-$PtBA_{1210}$ using submicrometer-sized stamps led to faithful transfer of the passivating species (Figure 7).

The interactions of PaTu8988T cells with this nanopatterned fibronectin–PEG sample was compared to a homogeneously fibronectin-functionalized film. Wide-field optical microscopy provided clear evidence for an effect of the binary cell-adhesive/cell-repulsive nanopatterns on cancer-cell morphology (Figure 7). On the nanopatterned substrates exposing 300-nm wide lines of fibronectin and 500-nm wide lines of PEG_{5000}, the cells were observed to spread in highly unusual very elongated shapes. 31 ± 9% of the cells were found to be elongated on the patterns (mean aspect ratio 2.9 ± 0.8). By contrast, on the unpatterned fibronectin only 4 ± 2% of the cells showed an elongated morphology (mean aspect ratio 2.2 ± 0.5).

These observations indicate that the pancreas adenocarcinoma cells recognize the differently patterned areas. This represents a more complex process than the mere differentiation of the chemical functionality of the homogeneously functionalized blockcopolymer films (see above). The cells are apparently forced to stretch along the direction defined by the submicrometer-sized pattern of cell adhesive (fibronectin) and nonadhesive (PEG_{5000}) stripes and can distinguish the in-plane directionality of the anisotropic surface chemistry. As a consequence, the cells spread out in the direction of the highest fibronectin coverage (lowest PEG coverage).

The use of even smaller stamps would render the fabrication of smaller feature sizes possible, provided that diffusion of the transferred molecular species is negligible. However, as amply demonstrated in the literature [53–56], conventional PDMS stamps cannot be applied in this respect. Severe problems due to the deformation (or collapse) of these soft elastomeric stamps [53, 54] or the diffusion of ink molecules [55, 56], limit the general applicability for obtaining patterns on the 100-nm size range. However, appropriate stamps with smaller features and optimized mechanical properties have been reported [57–61]. Alternatively, by using a prestructured substrate and flat featureless stamps the problems originating from instabilities of small posts of μCP stamps can be avoided. This methodology of 'inverted microcontact printing' (i-μCP) was first applied, as mentioned above, by the group of Textor to microfabricate micrometer-sized wells of defined chemical character for single-cell studies [38].

This method was successfully extended to the PS-PtBA block copolymer platform introduced in Chapter 2.4 and in this chapter [62]. Following hot embossing the PtBA skin layer is deprotected using trifluoro acetic acid, followed by activation with EDC/NHS. This postembossing wet-chemical treatment activates the entire surface of the topographically structured film for the attachment of primary amines (Figure 8).

Hence, this i-μCP methodology provides access to topographic structures by imprinting approaches, as well as controllable surface chemistry and high coverages of passivating (e.g. PEG) and active species (e.g., proteins and DNA) via established wet-chemical conjugation strategies. These molecules are applied to the surfaces exposed on top of the structures using a featureless, flat stamp that has been inked accordingly. In the absence of ink diffusion, which is the case for (bio)polymers, such as PEG or proteins of considerable molar mass, exclusively the contacted areas are functionalized. The remainder of the surface, including the side walls of the structures remain unreacted, and thus functionalized with NHS-active ester moieties. In a subsequent step these can be backfilled with active molecules of interest in a reaction in buffer. Thereby inverted microcontact printing provides a versatile approach for the fabrication of patterned biointerfaces across the

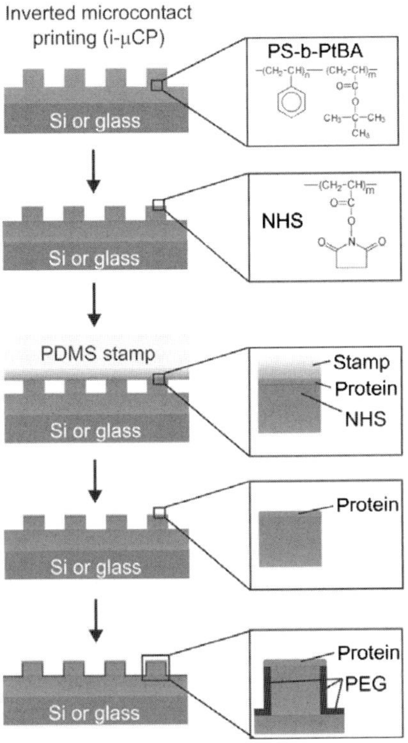

Figure 8 Schematic of the reactive microcontact printing steps of the i-μCP process. The PtBA skin layer exposed at the topographically structured block-copolymer film surface is hydrolyzed using TFA and subsequently activated with EDC/NHS. A hydrated protein can be transferred in the micro- or nanocontact between the oxidized elastomeric PDMS stamp and the structured film resulting in covalent attachment. In a subsequent step the noncontacted areas are passivated by covalently coupling PEG-NH_2 from buffered solution. (Reproduced with permission from [62]; A. Embrechts, C. L. Feng, C. Mills, M. Lee, I. Bredebusch, J. Schnekenburger, W. Domschke, G. J. Vancso, H. Schönherr, Langmuir 2008, 24, 8841, Copyright 2008 by American Chemical Society).

length scales, from tens of micrometers down to 150 nanometers and likely below. Due to the glassy PS blocks and the robust covalent amide linkages, the derivatized films showed excellent stability under a broad range of processing conditions.

Figure 9 shows a fluorescence micrograph of a line pattern of bovine serum albumin (BSA) and PEG that has been exposed to pancreas adenocarcinoma cells. Before the analysis the immobilized protein and the cells were labeled with lissamine rhodamine B sulfonyl chloride. From the micrograph the high definition of the pattern obtained by i-μCP becomes obvious. In addition, it can be observed that the cells adhere to and bridge across the BSA-functionalized lines.

More detailed studies were conducted with PaTu8988S pancreas adenocarcinoma cells. Pancreatic adenocarcinomas are highly invasive tumors forming metastases early in tumor development. Tumor growth and invasion is promoted by a tumor microenvironment mainly composed of host epithelial cells, stromal fibroblast and extracellular matrix proteins [63]. Fibronectin is known to be upregulated in pancreatic tumors, to increase tumor-cell survival, and to promote tumor proliferation and metastasis by cytokine-mediated pathways [64]. Hence, the patterns investigated exposed topographically elevated lines functionalized on top with fibronectin. The PaTu8988S cells used in this study are characterized by a nonpolar organization of the cell cytoplasma and tumor growth in a nude mouse model [46]. PaTu8988S cells express fibronectin binding integrin subunits

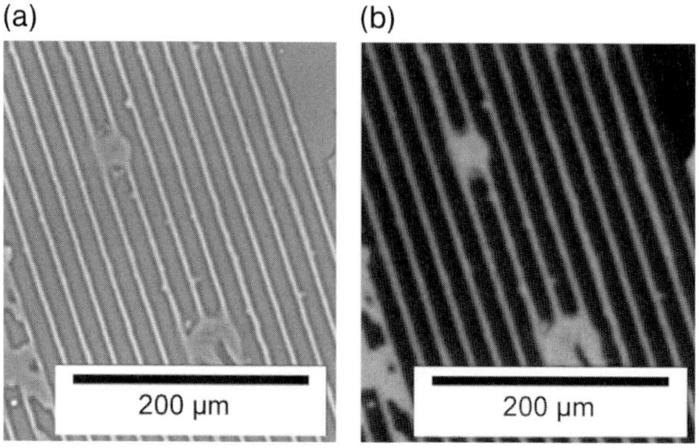

Figure 9 (a) Bright-field optical and (b) fluorescence microscopy images of cancer cells on BSA functionalized ridges that were passivated in between with PEG.

and bind to fibronectin-coated surfaces [65]. As shown in Figure 10, the cancer cell attached exclusively to the topographically elevated adhesive ridges [62], and in some cases cells were observed to bridge the distance between two ridges. High-resolution AFM analysis of fixated cells showed how lamellopodia and filopodia grew along the direction of the ridges.

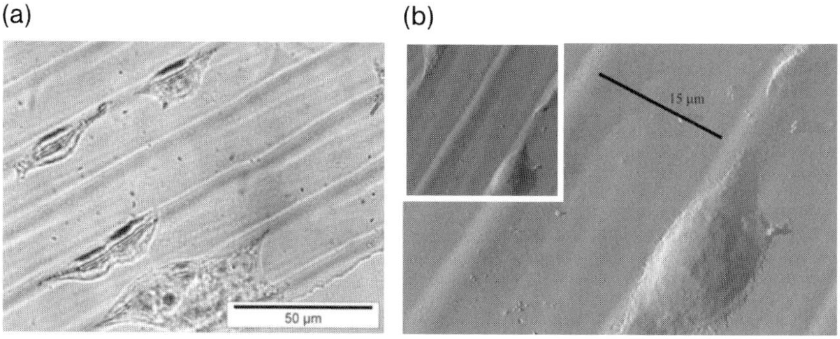

Figure 10 (a) Optical microscopy and (b) TM-AFM phase images of fixated PaTu8988S cells on topographically structured and chemically patterned substrate reveal the preferential attachment on the 5-μm wide fibronectin-functionalized mesas (line spacing for all images: 15 μm). (Reproduced with permission from [62]; A. Embrechts, C. L. Feng, C. Mills, M. Lee, I. Bredebusch, J. Schnekenburger, W. Domschke, G. J. Vancso, H. Schönherr, Langmuir 2008, 24, 8841, Copyright 2008 by American Chemical Society).

3.4.3
Outlook

The blockcopolymer platform and multilength scale patterning methodologies introduced in this chapter comprised a first step in the direction of the required micro- and nanostructured patterns and structures that model the tumor environment. Even with these 'simple' model systems systematic studies of cell adhesion and migratory behavior, as well as protein up- or downregulation as a result of cell-surface interactions can be addressed now. In the on-going work variations in the type of protein and vertical and lateral distances are investigated. In addition, microwell-type experiments, such as those carried out by the Textor lab (Figure 4), with nanostructured polymer surfaces are within reach. Once the block copolymer nanostructure has been exploited for obtaining self-assembled nanometer scale patterns, the pathway to the required well-defined 3D structures and matrices will be opened. These developments promise to provide ultimately new incentives for the development of new therapeutic concepts for the treatment of cancer.

Acknowledgments

The support of the EU (NoE Nano2Life) is gratefully acknowledged. This work has been financially supported by the MESA$^+$ Institute for Nanotechnology of the University of Twente, the Council for Chemical Sciences of the Netherlands Organization for Scientific Research (CW-NWO) in the framework of the *vernieuwingsimpuls* program (grant awarded to HS) and a BMBF grant (Dr. J. Schnekenburger).

References

1 Kumar, C.S.S.R. (ed.) (2006) *Tissue, Cell and Organ Engineering*, Wiley-VCH, Weinheim.

2 (a) Mitsiadis, T.A. *et al.* (2007) *Experimental Cell Research*, **313**, 3377; (b) Jones, D.L. and Wagers, A.J. (2008) *Nature Reviews. Molecular Cell Biology*, **9**, 11; (c) Moore, K.A. and Lemischka, I.R. (2006) *Science*, **311**, 1880.

3 Castner, D.D.G. and Ratner, B.D. (2002) *Surface Science*, **500**, 28.

4 Ruoslahti, E. (1996) *Annual Review of Cell and Developmental Biology*, **12**, 697.

5 Griffith, L.G. (2000) *Acta Materialia*, **48**, 263.

6 Chen, C.S., Mrksich, M., Huang, S., Whitesides, G.M. and Ingber, D.E. (1997) *Science*, **276**, 1425.

7 Curtis, A. and Wilkinson, C. (1997) *Biomaterials*, **18**, 1573.

8 Jemal, A. *et al.* (2007) *CA–A Cancer Journal for Clinicians*, **57**, 43.

9 Ferlay, J. *et al.* (2007) *Annals of Oncology*, **18**, 581.

10 Hanahan, D. and Weinberg, R.A. (2000) *Cell*, **100**, 57.

11 Fidler, I.J. (2002) *Seminars in Cancer Biology*, **12**, 89.

12 Rennebeck, G., Martelli, M. and Kyprianou, N. (2005) *Cancer Research*, **65**, 11230.

13 Hugo, H. *et al.* (2007) *Journal of Cellular Physiology*, **213**, 374.

14 Wittekind, C. and Neid, M. (2005) *Oncology*, **69** (Suppl 1), 14.

15 Nguyen, D.X. and Massague, J. (2007) *Nature Reviews – Genetics*, **8**, 341.

16 Paget, S. (1989) *Cancer Metastasis Reviews*, **8**, 98.

17 Fidler, I.J., Kim, S.J. and Langley, R.R. (2007) *Journal of Cellular Biochemistry*, **101**, 927.

18 Langley, R.R. and Fidler, I.J. (2007) *Endocrine Reviews*, **28**, 297.

19 Morrissey, C. and Vessella, R.L. (2007) *Journal of Cellular Biochemistry*, **101**, 873.

20 (a) Hwang, N.S., Kim, M.S., Sampattavanich, S., Baek, J.H., Zhang, Z. and Elisseeff, J. (2006) *Stem Cells*, **24**, 284; (b) Even-Ram, S., Artym, V. and Yamada, K.M. (2006) *Cell*, **126**, 645; (c) Ramunas, J., Illman, M., Kam, A., Farn, K., Kelly, L., Morshead, C.M. and Jervis, E.J. (2006) *Cytometry A*, **69A**, 1202.

21 (a) Cukierman, E., Pankov, R., Stevens, D.R. and Yamada, K.M. (2001) *Science*, **294**, 1708; (b) Zaman, M.H., Trapani, L.M., Sieminski, A.L., MacKellar, D., Gong, H., Kamm, R.D., Wells, A., Lauffenburger, D.A. and Matsudaira, P. (2006) *Proceedings of the National Academy of Sciences of the United States of America*, **103**, 10889.

22 Smalley, K.S.M., Lioni, M. and Herlyn, M. (2006) *In Vitro Cellular & Developmental Biology – Animal*, **42**, 242.

23 Hegerfeldt, Y., Tusch, M., Brocker, E.B. and Friedl, P. (2002) *Cancer Research*, **62**, 2125.

24 (a) Brandl, F., Sommer, F. and Goepferich, A. (2007) *Biomaterials*, **28**, 134; (b) Desai, T.A. (2000) *Medical Engineering & Physics*, **22**, 595; (c) Andersson, H. and van den Berg, A. (2004) *Lab on a Chip*, **4**, 98; (d) Hutmacher, D.W. (2001) *Journal of Biomaterials Science – Polymer Edition*, **12**, 107; (e) Hutmacher, D.W., Sittinger, M. and Risbud, M.V. (2004) *Trends in Biotechnology*, **22**, 354; (f) Zhang, H., Hutmacher, D.W., Chollet, F., Poo, A.N. and Burdet, E. (2005) *Macromolecular Bioscience*, **5**, 477; (g) Leong, T., Gu, Z., Koh, T. and Gracias, D.H. (2006) *Journal of the American Chemical Society*, **128**, 11336; (h) Peppas, N.A., Hilt, J.Z., Khademhosseini, A. and Langer, R. (2006) *Advanced Materials*, **18**, 1345; (i) Langer, R. and Tirrell, D.A. (2004) *Nature*, **428**, 487.

25 (a) Tourniaire, G., Collins, J., Campbell, S., Mizomoto, H., Ogawa, S., Thaburet, J.-F. and Bradley, M. (2006) *Chemical Communications*, 2118; (b) Bretagnol, F., Valsesia, A., Sasaki, T., Ceccone, G., Colpo, P. and Rossi, F. (2007) *Advanced Materials*, **19**, 1947; (c) Wang, Z., Hu, H., Wang, Y., Wang, Y., Wu, Q., Liu, L. and Chen, G. (2006) *Biomaterials*, **27**, 2550; (d) Falconnet, D., Koenig, A., Assi, F. and Textor, M. (2004) *Advanced Functional Materials*, **14**, 749.

26 (a) Kumar, A., Biebuyck, H.A. and Whitesides, G.M. (1994) *Langmuir*, **10**, 1498; (b) Mrkisch, M. and Whitesides, G.M. (1995) *Tibtech*, **13**, 228.

27 (a) Kwok, C.S., Mourad, P.D., Crum, L.A. and Ratner, B.D. (2000) *Biomacromolecules*, **1**, 139; (b) De Silva, M.N., Desai, R. and Odde, D.J. (2004) *Biomedical Microdevices*, **6**, 219.

28 Genzer, J. and Efimenko, K. (2000) *Science*, **290**, 2130.

29 (a) Khademhosseini, A., Jon, S., Suh, K.Y., Tran, T.-N.T., Eng, G., Yeh, J., Seong, J. and Langer, R. (2003) *Advanced Materials*, **15**, 1995; (b) Lahann, J., Balcells, M., Rodon, T., Lee, J., Choi, I.S., Jensen, K.F. and Langer, R. (2002) *Langmuir*, **18**, 3632.

30 (a) Hyun, J., Zhu, Y., Liebmann-Vinson, A., Beebe, T.P., Jr., and Chilkoti, A. (2001) *Langmuir*, **17**, 6358; (b) Lee, K., Kim, D.J., Lee, Z., Woo, S.I. and Choi, I.S. (2004) *Langmuir*, **20**, 2531; (c) Feng, C.L., Embrechts, A., Vancso, G.J. and Schönherr, H. (2006) *European Polymer Journal*, **42**, 1954; (d) Feng, C.L., Vancso, G.J. and Schönherr, H. (2006) *Advanced Functional Materials*, **16**, 1306; (e) Feng, J., Gao, C. and Shen, J. (2004) *Chemistry of Materials*, **16**, 1319.

31 Nakayama, Y., Anderson, J.M. and Matsuda, T. (2000) *Journal of Biomedical Materials Research*, **53**, 584.

32 Xia, Y., Rogers, J.A., Paul, K.E. and Whitesides, G.M. (1999) *Chemical Reviews*, **99**, 1823.

33 Schwarz, A., Rossier, J.S., Roulet, E., Mermod, N., Roberts, M.A. and Grault, H.H. (1998) *Langmuir*, **14**, 5526.

34 (a) Mrkisch, M., Chen, C.S., Xia, Y., Dike, L.E., Ingber, D.E. and Whitesides, G.M. (1996) *Proceedings of the National Academy of Sciences of the United States of America*, **93**, 10775; (b) Khademhosseini, A., Jon, S., Suh, K.Y., Tran, T.T., Eng, G., Yeh, J., Seong, J. and Langer, R. (2003) *Advanced Materials*, **15**, 1995; (c) Lussi, J.W., Michel, R., Reviakine, I., Falconnet, D., Goessl, A., Csucs, G., Hubbell, J.A. and Textor, M. (2004) *Progress in Surface Science*, **76**, 55.

35 Britland, S., Morgan, H., Wojciak-Stothard, B., Riehle, M., Curtis, A. and Wilkinson, C. (1996) *Experimental Cell Research*, **228**, 313.

36 Lavik, E. and Langer, R. (2004) *Applied Microbiology and Biotechnology*, **65**, 1.

37 Arnold, M., Cavalcanti-Adam, E.A., Glass, R., Blummel, J., Eck, W., Kantlehner, M., Kessler, H. and Spatz, J.P. (2004) *ChemPhysChem*, **5**, 383.

38 Dusseiller, M.R., Schlaepfer, D., Koch, M., Kroschewski, R. and Textor, M. (2005) *Biomaterials*, **26**, 5917.

39 (a) Langer, R. and Vacanti, J.P. (1993) *Science*, **260**, 920; (b) Levenberg, S., Rouwkema, J., Macdonald, M., Garfein, E.S., Kohane, D.S., Darland, D.C., Marini, R., van Blitterswijk, C.A., Mulligan, R.C., D'Amore, P.A. and Langer, R. (2005) *Nature Biotechnology*, **23**, 879.

40 (a) Hartgerink, J.D., Beniash, E. and Stupp, S.I. (2001) *Science*, **294**, 1684; (b) Kim, M.S., Yeon, J.H. and Park, J.K. (2007) *Biomedical Microdevices*, **9**, 25.

41 (a) Stetlerstevenson, W.G., Liotta, L.A. and Kleiner, D.E. (1993) *FASEB Journal*, **7**, 1434; (b) Stetlerstevenson, W.G., Aznavoorian, S. and Liotta, L.A. (1993) *Annual Review of Cell Biology*, **9**, 541; (c) Kleinman, H.K. and Martin, G.R. (2005) *Seminars in Cancer Biology*, **15**, 378.

42 Murphy, W.L., Dennis, R.G., Kileny, J.L. and Mooney, D.J. (2002) *Tissue Engineering*, **8**, 43.

43 Hahn, M.S., Miller, J.S. and West, J.L. (2006) *Advanced Materials*, **18**, 2679.

44 Bratton, D., Yang, D., Dai, J. and Ober, C.K. (2006) *Polymers for Advanced Technologies*, **17**, 94.

45 (a) Blau, H.M. and Baltimore, D.J. (1991) *The Journal of Cell Biology*, **112**, 781; (b) Ruoslahti, E. and Obrink, B. (1996) Experimental Cell Research, **227**, 1; (c) Cameron, J., Wilson, B.E., Clegg, R.E., Leavesley, D.I. and Pearcy, M.J. (2005) *Tissue Engineering*, **11**, 1.

46 Elsasser, H.P., Lehr, U., Agricola, B. and Kern, H.F. (1992) *Virchows Archive B: Cell Pathology Including Molecular Pathology*, **61**, 295.

47 Feng, C.L., Vancso, G.J. and Schönherr, H. (2005) *Langmuir*, **21**, 2356.

48 (a) Park, C., Yoon, J. and Thomas, E.L. (2003) *Polymer*, **44**, 6725; (b) Hamley, I.W. (2003) *Nanotechnology*, **14**, R39.

49 (a) Groll, J., Fiedler, J., Engelhard, E., Ameringer, T., Tugulu, S., Klok, H.-A., Brenner, R.E. and Möller, M. (2005) *Journal of Biomedical Materials Research Part A*, **74A**, 607; (b) Harbers, G.M., Emoto, K., Greef, C., Metzger, S.W., Woodward, H.N., Mascali, J.J., Grainger, D.W. and Lochhead, M.J. (2007) *Chemistry of Materials*, **19**, 4405.

50 Feng, C.L., Embrechts, A., Bredebusch, I., Bouma, A., Schnekenburger, J., García-Parajó, M., Domschke, W., Vancso, G.J. and Schönherr, H. (2007) *European Polymer Journal*, **43**, 2177.

51 Deible, C.R., Petrosko, P., Johnson, P.C., Beckman, E.J., Russell, A.J. and Wagner, W.R. (1998) *Biomaterials*, **19**, 1885.

52 Feng, C.L., Embrechts, A., Bredebusch, I., Schnekenburger, J., Domschke, W., Vancso, G.J. and Schönherr, H. (2007) *Advanced Materials*, **19**, 286.

53 (a) Lauer, L., Klein, C. and Offenhausser, A. (2001) *Biomaterials*, **22**, 1925; (b) Roca-Cusachs, P., Rico, F., Martínez, E., Toset, J., Farré, R. and Navajas, D. (2005) *Langmuir*, **21**, 5542; (c) Bietsch, A. and Michel, B. (2000) *Journal of Applied Physiology*, **88**, 4310; (d) Delamarche, E., Schmid, H., Michel, B. and Biebuyck, H. (1997) *Advanced Materials*, **9**, 741.

54 Bessueille, F., Pla-Roca, M., Mills, C.A., Martinez, E., Samitier, J. and Errachid, A. (2005) *Langmuir*, **21**, 12060.

55 Libioulle, L., Bietsch, A., Schmid, H., Michel, B. and Delamarche, E. (1999) *Langmuir*, **15**, 3000.

56 Delamarche, E., Schmid, H., Bietsch, A., Larsen, N.B., Rothuizen, H., Michel, B. and Biebuyck, H. (1998) *The Journal of Physical Chemistry. B*, **102**, 3324.
57 Pla-Roca, M., Fernandez, J.G., Mills, C.A., Martinez, E. and Samitier, J. (2007) *Langmuir*, **23**, 8614.
58 Michel, B., Bernard, A., Bietsch, A., Delamarche, E., Geissler, M., Juncker, D., Kind, H., Renault, J.P., Rothuizen, H., Schmid, H., Schmidt-Winkel, P., Stutz, R. and Wolf, H. (2001) *IBM Journal of Research and Development*, **45**, 870.
59 Coyer, S.R., Garcia, A.J. and Delamarche, E. (2007) *Angewandte Chemie – International Edition*, **46**, 6837.
60 Li, H.-W., Muir, B.V.O., Fichet, G. and Huck, W.T.S. (2003) *Langmuir*, **19**, 1963.
61 Yu, A.A., Savas, T., Cabrini, S., diFabrizio, E., Smith, H.I. and Stellacci, F. (2005) *Journal of the American Chemical Society*, **127**, 16774.
62 Embrechts, A., Feng, C.L., Mills, C., Lee, M., Bredebusch, I., Schnekenburger, J., Domschke, W., Vancso, G.J. and Schönherr, H. (2008) *Langmuir*, **24**, 8841.
63 Mahadevan, D. and von Hoff, D.D. (2007) *Molecular Cancer Therapeutics*, **6**, 1186.
64 (a) Vaquero, E.C., Edderkaoui, M., Nam, K.J., Gukovsky, I., Pandol, S.J. and Gukovskaya, A.S. (2003) *Gastroenterology*, **125**, 1188; (b) Lowrie, A.G., Salter, D.M. and Ross, J.A. (2004) *British Journal of Cancer*, **91**, 1327.
65 Vogelmann, R., Kreuser, E.D., Adler, G. and Lutz, M.P. (1999) *International Journal of Cancer*, **80** (5), 791.

3.5
Fabrication and Application of Surface-Tethered Vesicles

Thomas Lloyd Williams and A. Tobias A. Jenkins

3.5.1
Introduction

3.5.1.1
Lipid Vesicles

Lipid bilayer vesicles (LBVs), or liposomes, refers to the production of closed spheres of phospholipids encapsulating an aqueous environment (Figure 1). LBVs are composed of concentric opposing layers of amphipathic molecules self-assembled to ensure that the hydrophobic regions face one another, and the polar hydrophilic regions orientate towards the aqueous environment. Bangham first described these LBVs in 1965 [1]. After the publication of this article detailing the observations made, much interest went into their characterization, production, and use.

The fabrication of a reliable and reproducible cell-membrane model to study membrane dynamics and specific bimolecular interactions has been desirable. This is because the study of isolated membrane constituents removed from their native environment gives rise to the possibility of events that would not necessarily occur in their natural membrane associated state. Conversely, whole-cell systems can be too complex to dissect the precise role of a given component.

Several model systems have been devised to correlate the properties of natural membranes with the behavior of pure lipids, and in 1968 the first paper reviewed the use of LBVs as a model for biological membranes [2]. The diffusion of solutes entrapped within the aqueous space of LBVs was studied, and it was found that the lipid vesicles were capable of ion discrimination, osmotic swelling, and responding to various physiologic and pharmacologic agents. Proteins and other biological molecules were incorporated into the lipid bilayers or encapsulated in the interior of lipid vesicles.

The adsorption of lipid vesicles to solid surfaces occurs through nonspecific attractive forces, including van der Waals, ionic, hydrophobic, and hydrogen bonding. These forces act indiscriminately between all parts of the self-assembled system and the solid surface and result in the elastic deformation of the soft colloidal

Surface Design: Applications in Bioscience and Nanotechnology
Edited by Renate Förch, Holger Schönherr, and A. Tobias A. Jenkins

ISBN: 978-3-527-40789-7

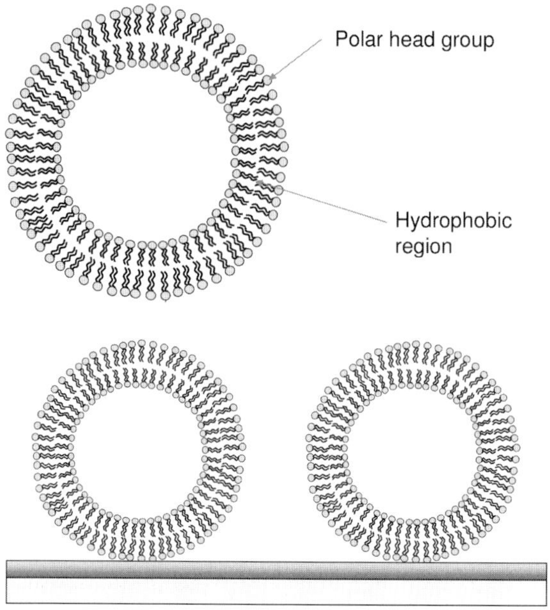

Figure 1 Schematic of a lipid bilayer vesicles (LBV), whereby the polar head groups face the aqueous environment to effectively shield the hydrophobic hydrocarbon chains, and an illustration of surface-tethered vesicles.

vesicles [3]. The stress exerted on the lipid vesicles as they adhere increases their fragility, leakage, rupture and fusion. The decoupling of lipid vesicles from a solid surface using tethers or cushions can prevent the deformation of LBVs by providing specific binding sites. Such decoupling ensures that the whole vesicle surface is not in direct contact with the surface, and only a discrete number of specific ligand–receptor interactions occur. The application of site-specific interactions between the lipid vesicles and the solid surface results in adhesion forces dependent on the strength and number of the specific ligand–receptor interactions. This review focuses on the construction of such tethered lipid vesicles, the various support systems such as polymer and self-assembled monolayers, the specific receptor–analyte tether system, and composition of the lipid vesicles. The review will also discuss the uses of these systems for which they have been used to study of bimolecular interactions and the permeability of lipid membranes *in vitro*.

3.5.1.2
Fabrication of Surface-Tethered Lipid Vesicles

Lipid vesicles are formed by agitating amphipathic molecules in solution, causing the self-assembly of closed lipid bilayers entrapping the surrounding aqueous environment. Amphipathic molecules form a heterogeneous population of vesicles consisting of one or more bilayers. The size and lamellar number of the bilayers varies depending upon the method of fabrication. Small unilamellar vesicles (SUVs) contain a single lipid bilayer and have a diameter less than 100 nm, large unilamellar vesicles (LUVs) contain a single lipid bilayer and have a diameter between 100 nm to several micrometers, multilamellar vesicles (MLVs) contain between five and twenty concentric lipid bilayers and have a diameter below 1000 nm, and multivesicular vesicles (MVVs) contain multicompartment structures and a diameter in the micrometer range [4].

3.5.1.3 Polymer-, SAM- and Streptavidin-Tethered LUVs

The direct physisorption of bilayer lipid vesicles to a surface causes inherent problems such as lack of orientational control, change in biological activity, and distortion of the vesicle bilayer. It has been shown that vesicles may nonspecifically adsorb or may form a stable vesicular layer, or may rupture and fuse to form supported lipid bilayers [5, 6]. The outcome is highly dependent on the surface energy of the substrate and factors such as the size and properties of the lipids, ionic strength and substrate morphology.

Decoupling of the lipid vesicles from the solid surface has previously been employed to maintain the structural integrity of the model membrane system. The use of soft polymeric supports has been investigated by a number of researchers, including the authors of this chapter. Rossetti *et al.* used poly(lysine)-poly(ethylene glycol)-copolymer-coated substrates to physisorb bilayer vesicles via electrostatic interactions, while planar lipid bilayers become decomposed and lifted away from the copolymer support [7]. The covalent immobilization of LBVs to a support ensures the tethering of intact vesicles, and has been used in various applications including in the design of optochemical sensors. Nguyen *et al.* used the sol-gel method to create thin films prepared from hydrolyzed tetramethylorthosilicate. These gels were coupled to DMPC, cholesterol, dihexadecyl phosphate (5:4:1 molar ratio) lipid vesicles encapsulating a carboxyfluorescein marker [8]. The sol-gel gelatinizes and covalently tethers the vesicles to the solid support and ensures high mechanical and chemical stability, minimum dye quenching, and a reproducible fabrication method.

Other covalent attachment strategies include the functionalization of a gold surface with a plasma-polymerized maleic anhydride/SiO_2 multilayer. This method was used to tether intact DMPC/DMPE lipid vesicles to the hydrolyzed maleic anhydride surface using ethyl-3-(3-dimethylaminopropyl)-carbodiimide-N-hydroxysuccinimide (EDC-NHS) peptide-coupling chemistry, and was visualized using reflection interference contrast microscopy (see Figure 3c). The EDC catalyzes the formation of an amide bond from a carboxylic acid and is used in conjunction with NHS to increase coupling efficiency and to create a stable amine-reactive product. The amino-functionalized phosphatidylethanolamine lipid was conjugated to the carboxylic acid groups on the swollen maleic anhydride film [9].

Additional strategies employed to tether intact bilayer vesicles include the use of supported monolayers and bilayers to decouple the vesicles and the surface.

Self-assembled mono- and bilayers are extensively reviewed by Ulman [10]. Functionalized oligonucleotides have been covalently attached to the surface of preformed lipid vesicles and tethered to complimentary oligonucleiotide presenting supported bilayers. Boxer prepared lipid–DNA conjugates by removing the dimethoxytrityl group of the oligonucleotides and reacted it with an iodination reagent, such as $(PhO)_3PCH_3I$, to render the 5′-end electrophilic. The product was reacted with a lipid-thiolate followed by a series of deprotection, cleavage and purification steps to yield the desired conjugate [11]. A complementary set of oligonucleotides were synthesized and modified in the same fashion to insert into the supported bilayer.

The biospecific recognition between biotin and streptavidin is well established, and has become a common model system for creating more complex surface architectures because each streptavidin is capable of binding up to four biotin molecules with very high affinity ($K_a = 1 \times 10^{-15}$ mol^{-1}) [12]. The high-affinity interactions are the result of multiple van der Waals and hydrogen bonds, which causes the ordering of surface polypeptide loops to bury the biotin into the streptavidin interior [13]. The carboxy-terminus of the biotin molecule, which is not involved in the binding process to streptavidin, can be covalently linked to structural elements. The covalent attachment of biotin to lipids and thiols allows for the preparation of molecular layered architectures at the water/air interface to form 2D crystalline structures to tether vesicles. This strategy to tether lipid vesicles has been extensively exploited for various applications including to fabricate self-assembled microarrays [14].

3.5.1.4 Fabrication of LBVs

Various strategies exist to produce LBVs and typically rely on the physical manipulation of solubilized amphiphiles to produce a homogeneous population of vesicles. The first publications concerning LBVs used ultrasonic waves to disrupt the heterogeneous multilamellar vesicles, either using a tip or water-bath sonicator [15, 16]. This method was shown to produce a homogeneous vesicle population of 300 ± 30 Å. However, prolonged sonication causes substantial chemical degradation even in the presence of an N_2 environment [17].

Methods relying on solvent replacement remove the need to use sonication and thereby eliminate the potential degradation of the amphiphilic molecules and the need for long sonication times. An ethanolic solution of lipid is rapidly injected through a glass gas-tight syringe in an ethanolic solution (maximum 7.5% ethanol), and then concentrated on a centrifugal ultrafiltration device. Ensuring that the temperature is kept above the solvent boiling point causes the evaporation of the organic solvent. This produces single bilayer vesicles indistinguishable from sonicated lipid vesicles, with an average diameter of 26.5 nm and an internal volume of 5.2×10^{-4} ml μmol dm^{-3} [18].

Extrusion is one of the most accessible methods for producing a monodispersed population of unilamellar lipid vesicles with controlled size. Extruded vesicles are produced by forcing a lipid suspension through a polycarbonate membrane with a defined pore size. This method produces a highly homogeneous population of vesicles at high lipid concentrations, and provides a nondestructive method in terms of amphiphile hydrolysis, oxidation and degradation. The lipid dispersion is typically subjected to 19 passes through the polycarbonate filter to avoid contamination of the sample by large vesicles that might not have passed through the filter [19]. The extrusion method is a preferred method because they are free from organic solvents and detergents; therefore, the arrangement of their constituent lipids is at equilibrium and causes no destabilization constraints on the lipids. The lipid vesicles are relatively homogeneous in size, 112 ± 6 nm using a 100-nm pore diameter filter [20].

3.5.1.5
Composition of Solid-Supported LBVs

Various factors must be taken into consideration when selecting the lipid composition of the vesicles. The precise lipid composition of a cell membrane affects its structural and physiological properties. The phase transition temperature (T_c) is the temperature at which a lipid changes from the ordered gel phase where the hydrocarbon chains are fully extended and closely packed, to a disordered liquid-crystalline phase where the hydrocarbon chains are randomly orientated and fluid. Above the phase transition temperature of a lipid, the polar head groups increase in mobility and the hydrocarbon chains change to an increasingly unstable conformation [21]. The T_c is directly affected by the hydrocarbon chain length, unsaturation, and head-group species [22]:

- With increasing hydrocarbon chain length: the phase transition temperature increases as van der Waals interactions become stronger.
- With increasing unsaturation of the hydrocarbon chains: the phase transition temperature decreases because a lower temperature is required to induce an ordered packing arrangement.
- Position of the unsaturated bond: lower phase transition temperature observed with double bonds situated in the middle of the hydrocarbon chain.
- Nature of the head group: phosphatidylethanolamine causes the highest phase transition, followed by phosphatidylserine, then phosphatidylcholine, and then phosphatidylglycerol.
- Increasing pH of the medium lowers the T_c.
- Controlling the transition temperature of the lipid vesicle system provides a useful means of regulating the physical properties of the system because choosing a lipid with a high T_c provides a nonleaky packing system.

3.5.2
Results and Discussion

3.5.2.1
The Effect of Lipid Chain Length

Lipid vesicles composing either 1,2-Dimyristoyl-sn-Glycero-3-Phosphocholine (DMPC), 1,2-Dipalmitoyl-sn-Glycero-3-Phosphocholine (DPPC), or 1,2-Distearoyl-sn-Glycero-3-Phosphocholine (DSPC) and encapsulating a water-soluble BODIPY 650/665 dye were tethered to a solid support using a biotinylated thiol monolayer. Using SPFS, the relative change in dye concentration (measured by the change in fluorescence) over time due to the natural porosity of the vesicle bilayer was monitored (Figure 2). The results were fitted to equation (3), below.

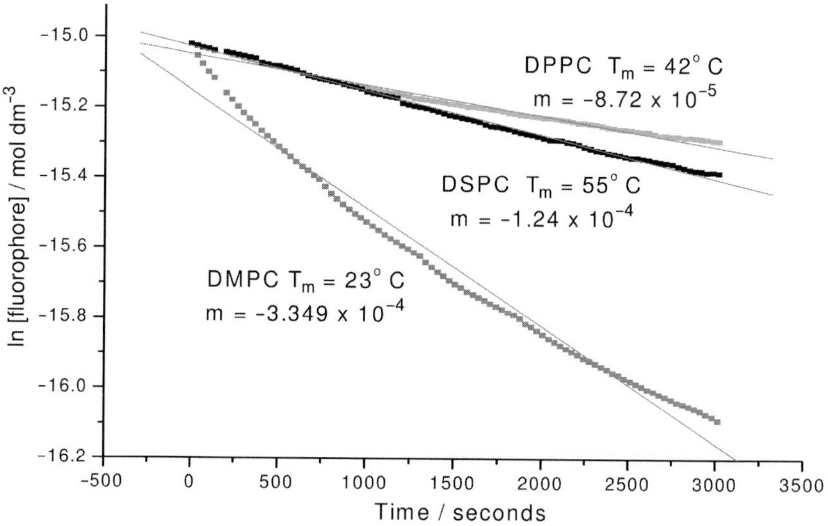

Figure 2 Normalized ln fluorophore concentration (mol dm^{-3}) as a function of time (seconds) for the three vesicle systems containing either DMPC (14 carbon hydrocarbon chains), DPPC (16 carbon hydrocarbon chains), or DSPC (18 carbon hydrocarbon chains) as determined using SPFS.

3.5.2.2
Modeling Passive Diffusion Through Membrane

A simple theoretical model has been developed to fit the decrease in fluorescence–time curves of SPFS data [23]. The assumption that all vesicles are of a uniform size with a total internal volume *V*, and a total area *A* and membrane thickness *d*. Concentration of fluorescent dye within vesicles is C_{in}. A straightforward diffusion model based on Fick's 1st law, for the diffusion flux, J of dye through the membrane is given by:

$$J = \frac{KDc_{\text{in}}}{d} \tag{1}$$

where K relates to the partitioning of dye into the membrane, $K = C_{\text{mem}} / C_{\text{in}}$ and D is the diffusion coefficient of the dye through the membrane. The change in internal concentration of dye with time can thus be approximated:

$$\frac{\mathrm{d}C_{\text{in}}}{\mathrm{d}t} = \frac{AJ}{V} = \frac{AKDC_{\text{in}}}{Vd} \tag{2}$$

This is a first-order kinetics equation, which can be further simplified and integrated to give:

$$\ln C_{\text{in}} = \ln C_{\text{in,o}} - kt \tag{3}$$

where k is the first-order rate constant:

$$k = \frac{AKD}{Vd} \tag{4}$$

This gives dimensionally correct units of rate constant k in s^{-1}. Since the membrane partition coefficient K is not known, the term D^* is termed the apparent diffusion coefficient, and is given by KD ($D^* = KD$). Hence, dye leakage, measured as a decrease in fluorescence was fitted to (3) to quantify the effect of passive dye leakage through the vesicle bilayer as a function of lipid composition and T_c. As the external volume is many times larger than the internal volume of vesicles (ca. 0.5 ml) and the fluorophore experiences enhanced fluorescence only within a ca. 200 nm zone close to the gold slide, the assumption was made that the effective concentration of dye outside of the vesicles was zero.

Using the passive diffusion model based upon Fick's first law of diffusion it was shown that DMPC-tethered vesicles offered the lowest resistance to dye permeation (apparent dye diffusion coefficient = 1.76×10^{-14} cm^2 s^{-1}). This was because the DMPC lipids ($T_c = 23$ °C) were in the liquid phase at room temperature; therefore, the hydrocarbon chains were disordered and offered a weak barrier to dye diffusion. Moreover, lipids with the longer hydrocarbon chains, DPPC ($T_c = 42$ °C) and DSPC ($T_c = 55$ °C), which were in the gel phase at room temperature, ensured that the hydrocarbon chains were ordered and fully extended, and offered a greater barrier to the diffusion of the dye through the bilayer. The DSPC-tethered vesicle system had an apparent dye diffusion coefficient of 8.9×10^{-15} cm^2 s^{-1}, while the DPPC-tethered vesicle system had an apparent dye diffusion coefficient of 5.76×10^{-15} cm^2 s^{-1}. Therefore, lipids in the gel phase offer a higher barrier to dye diffusion than phospholipids in the disordered liquid phase. The increased barrier against membrane porosity is attributed to increased interchain bonding as increased van der Waals forces between neighboring chains is observed between longer hydrocarbon chains. The passive diffusion model allows the quantitative determination of the stability of LBVs and can also be used to determine the permeation of the model membranes as a result of toxin permeation.

The choice of bilayer components facilitates the long-term stability of the vesicles. The inclusion of lipids containing unsaturated bonds increases the lipid's risk of oxidation, and thus a shorter shelf life. The inclusion of lipids from biological sources, such as from egg, bovine and soybean, typically contain significant levels of unsaturated fatty acids and are therefore inherently less stable than their synthetic counterparts. Natural biological membranes tend to carry a charge on their surface, and this charge depends on the nature of the membrane and its cellular location. Therefore, the chemical composition of the model bilayer vesicles is reflected in the choice of phospholipid species and the overall net charge. The charge on biological membranes has been shown to be important and has been shown to partially control the activity of membrane-bound enzymes [24].

The production of a single LBV system may not yield the necessary parameters desired for a particular system, and therefore, may not be an appropriate model. The inclusion of complex lipid mixtures of two or more individual lipid species has been used to reproduce specific charges, unsaturation, T_c and biological functions. Various composition mixtures have been used to mimic certain membranes including a 5:3:2 (%wt) of PE:PS:PC composition to reproduce the function of native brain tissue to investigate the depth of water penetration in lipid bilayers [25].

3.5.2.3
The Effect of Cholesterol

The inclusion of cholesterol in the bilayer of lipid vesicles significantly affects bilayer fluidity, permeability and integrity. Cholesterol typically comprises between 5 and 30% of eukaryotic membranes and is found at high proportions in densely packed membranes such as in liver, spinal cord and brain-cell membranes. Cholesterol acts as a molecular 'cement' by filling in gaps and defects found in lipid bilayers. It ensures membrane fluidity over a wider range of temperatures and eliminates the sharp thermal transition between the gel and the liquid phases [26], giving intermediate properties to the membrane between the two phases. Cholesterol has been demonstrated to inhibit the flow of water through membrane bilayers, and reduces the passive water permeability in direct proportion to the concentration of cholesterol in the membrane [27]. The thermodynamics observed for a large proportion of cholesterol in DMPC vesicles promotes the formation of a liquid-ordered phase, while low proportions of cholesterol promotes the formation of a gel phase [28]

The inclusion of cholesterol into the bilayer of tethered lipid vesicles has been shown to significantly affect the diffusion of an encapsulated fluorescent marker through the lipid bilayer. Varying the mole per cent of cholesterol from 0–70% in tethered DMPC vesicles allowed for the determination of the apparent dye diffusion coefficients for each of the cholesterol/DMPC systems from the leakage of the marker through the lipid bilayer (Table 1).

Table 1 Summary of the apparent dye diffusion coefficients as a function of cholesterol content in DMPC-tethered vesicles loaded with a fluorescent marker.

Mole percentage of cholesterol included in vesicles	Apparent dye diffusion coefficient ($cm^2\ s^{-1}$)
0	8.90×10^{-14}
10	2.63×10^{-14}
20	3.47×10^{-14}
30	1.47×10^{-14}
40	1.74×10^{-14}
50	1.72×10^{-14}
60	7.78×10^{-15}
70	7.93×10^{-15}

The inclusion of cholesterol in the vesicle bilayer caused a dose-dependant reduction in the porosity of the lipid bilayer to the fluorescent marker, this was because the cholesterol altered the phase-transition temperature of the lipid bilayer, and caused the hydrocarbons to adopt an intermediate condensed ordered gel phase and reduced the mobility of the fatty-acid acyl chains [29]. The inclusion of a small percentage (10%) of cholesterol caused a >3-fold improvement in the ability of the biomimetic vesicles to retain their fluorescent content compared to the cholesterol-deficient tethered vesicles. At medium cholesterol concentrations (20–50%), there was around a 5-fold improvement in fluorescent probe retention compared to the system devoid of sterol, and there was an approximately 11-fold improvement in dye retention when high levels (60–70%) of cholesterol were included in the tethered lipid bilayer.

3.5.2.4 Biophysical and Microscopy Techniques to Study Tethered Lipid Vesicles

Lipid vesicles in suspension emerged as a practical tool for drug delivery in the 1970s; however, they encountered inherent difficulties because of their *in-vivo* stability. Through systematic research, the stability and encapsulation efficiency of bilayer vesicles has been greatly improved. Various analytical and biophysical

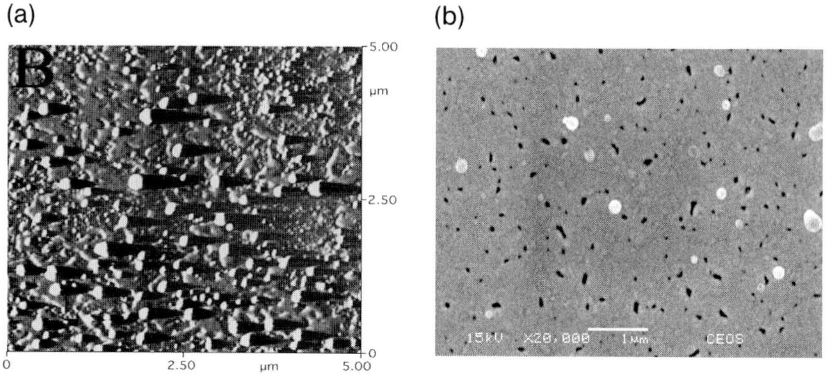

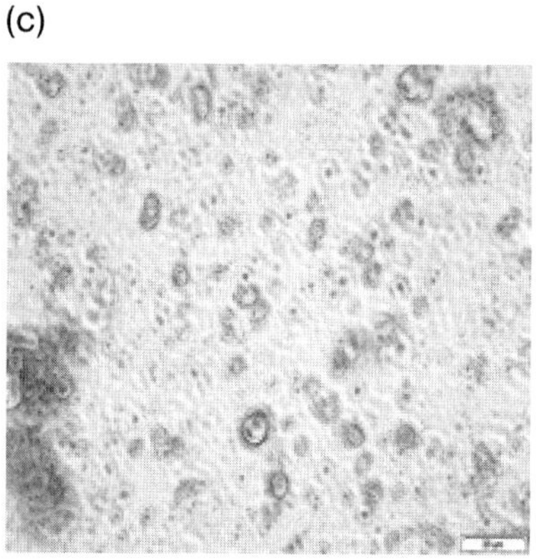

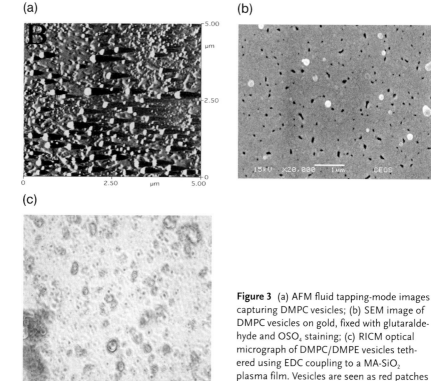

Figure 3 (a) AFM fluid tapping-mode images capturing DMPC vesicles; (b) SEM image of DMPC vesicles on gold, fixed with glutaraldehyde and OSO_4 staining; (c) RICM optical micrograph of DMPC/DMPE vesicles tethered using EDC coupling to a MA-SiO_2 plasma film. Vesicles are seen as red patches surrounded by bright halos.

techniques have been employed to characterize the bulk solution vesicles, such as patch-clamp [30] and electrochemistry [31]. However, it was not until the development of immobilization strategies that surface-sensitive techniques were employed. Visualization of tethered vesicles has been possible using a range of microscopic techniques including atomic force microscopy [32], scanning electron microscopy and reflection interference contrast microscopy (Figure 3) [9].

Visualization of lipid vesicles using AFM allows for the determination of vesicular dimensions and ensures tethering of the vesicles does not cause the rupture of the intact vesicles. Palmer *et al.* used AFM to model the cytoskeleton and plasma membrane of eukaryotic cells by encapsulating monomeric actin (G-actin) within egg-yolk-derived phosphatidylcholine small unilamellar vesicles [32]. This model was devised to provide a minimalist model of the cell that mimicked the cell membrane and the underlying cytoskeletal matrix. SEM and TEM provide a quick method to quantify the mean diameter of tethered vesicles and to ensure the decoupling results in intact vesicles adsorbing to the surface. Differential interference contrast microscopy has been used to confirm that the encapsulated BODIPY 650/665 dye was retained upon tethering of DMPC/DMPE vesicles to plasma-polymerized maleic anhydride/SiO_2 multilayers [9].

3.5.2.5
The Application of Surface-Tethered Lipid Vesicles

The tethering of bilayer vesicles to thin metal films allows for the application of surface-sensitive techniques such as surface plasmon resonance spectroscopy (SPS) and surface plasmon field-enhanced fluorescence spectroscopy (SPFS) to be utilized to monitor real-time bimolecular interactions, including the density and orientation of capture ligands [33]. Tethered LBVs were successfully applied to the study of protein toxin–membrane interactions. DMPC vesicles encapsulating Alexa Fluor 647 were tethered to a SAM via streptavidin–biotin coupling to determine the membrane permeation events that follow the introduction of a secretary phospholipase A_2 enzyme [34]. Using combined SPR/SPFS, the direct binding of the enzyme to the model membrane was visualized in real time, followed by the subsequent lysis of the lipid vesicles (Figure 4a). The addition of the lytic enzyme to the tethered vesicle surface was shown to bind, with approximately 0.7 ng mm^{-2} PLA_2 binding. Lysis was not observed for ~ 7 min, which was the lag period during which the enzyme binds and begins to hydrolyze the *sn-2* acyl bonds of the phospholipids. This was in good agreement with other groups reporting the latency period of membrane hydrolysis by PLA_2 [35].

The inclusion of dimethyl eicosodienoic acid (DEDA) in the bilayer of the tethered lipid vesicles prevented the hydrolysis of the phospholipids (Figure 4b). The binding of the PLA_2 to the tethered lipid vesicles without subsequent hydrolysis provided a means of determining the equilibrium dissociation constant (the avidity of an analyte:ligand interaction in terms of the ease of separating the binding), which was determined to be 1×10^{-5} mol dm^{-3}.

The immobilization of small unilamellar DMPC vesicles containing the GM1 ganglioside and encapsulating the BODIPY650/665 fluorescent marker was used to measure the binding of cholera toxin (*Ctx*) and was shown as a novel method for the determination of potential toxin inhibitors [23]. The formation of pores through the tethered lipid bilayer as a consequence of cholera toxin insertion, and the resulting flux of the encapsulated marker were monitored in real time using combined SPR/SPFS. The diffusion of the BODIPY dye was accurately quantified using the passive diffusion model devised by Williams and Jenkins [23]. The equilibrium dissociation constant (K_D) for the interaction between the tethered biomimetic membrane and the cholera toxin was $(1.16 \pm 0.2) \times 10^{-6}$ mol dm^{-3}, and was shown to be in agreement with other reported K_D values [36]. The change in fluorescence upon membrane permeation by *Ctx* showed an unexpected, initial increase in measured fluorescence, followed by a decrease (Figure 5).

The enhancement in fluorescence upon *Ctx* binding was shown not to result from tethering strategy or vesicular distortion, and was not a result of dye quenching. It was believed to be the result of a dehydration effect on the dye as it traverses the *Ctx* pore in the lipid membrane. A working model was devised to explain the possible enhancement observed in measured fluorescence (Figure 6), and was believed to be the result of a temporary dehydration of the dye molecule and as it interacts and passes through the *Ctx* pore. The model also accounts for

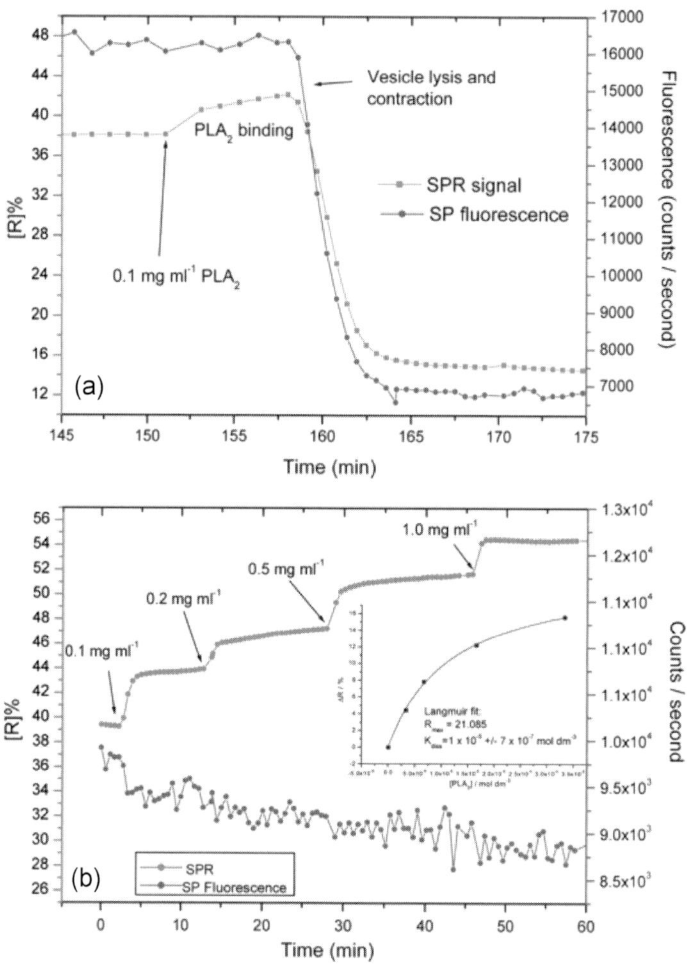

Figure 4 (a) SPR measurements (left y-axis) and SPFS measurements (right y-axis) showing the initial attachment of PLA_2 followed by the subsequent lysis of vesicles. (b) SPR and SPFS measurements of the titration of PLA_2 at concentrations between 3.3×10^{-6} mol dm^{-3} to 3.3×10^{-5}mol dm^{-3} into the DMPC tethered vesicle system containing a PLA_2 inhibitor, dimethyl eicosodienoic acid to prevent vesicle lysis but allow enzyme–membrane interactions.

the greater fluorescence enhancement observed when the β-subunit was used compared with whole *Ctx*, since the α-subunit may partially impede the passage of the marker through the charged toxin pore.

Blocking the GM1 receptor with lanthanide III salts inhibited the binding of cholera toxin to the tethered lipid vesicles. Europium III (Eu^{3+}) chloride was shown to be an effective means of fluorescently labeling sialic acid containing sugars such as GM1 [37]. Bound Eu^{3+} effectively shields the GM1 receptors from the binding of the β-subunits of the cholera toxin, and in doing so prevents the

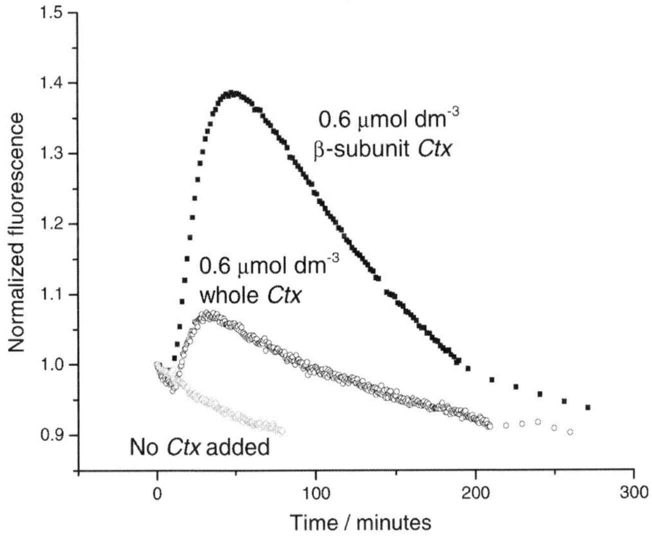

Figure 5 SPFS measurement of change in fluorescence of surface-tethered vesicles on addition of whole cholera toxin and β-subunit cholera toxin. The passive leakage of marker from the vesicles not exposed to toxin is also shown.

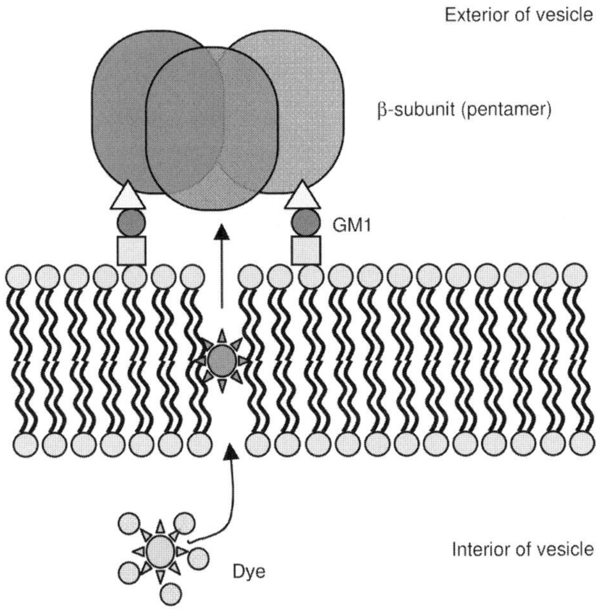

Figure 6 Schematic of proposed mechanism of fluorescent enhancement on binding of *Ctx* to GM1 modified lipid. Fluorescent dye, partially quenched by solvent water is dequenched in lipid membrane made leaky by attachment of β-subunit of *Ctx* to GM1 receptors on membrane surface. To aid clarity, the α subunit is not drawn.

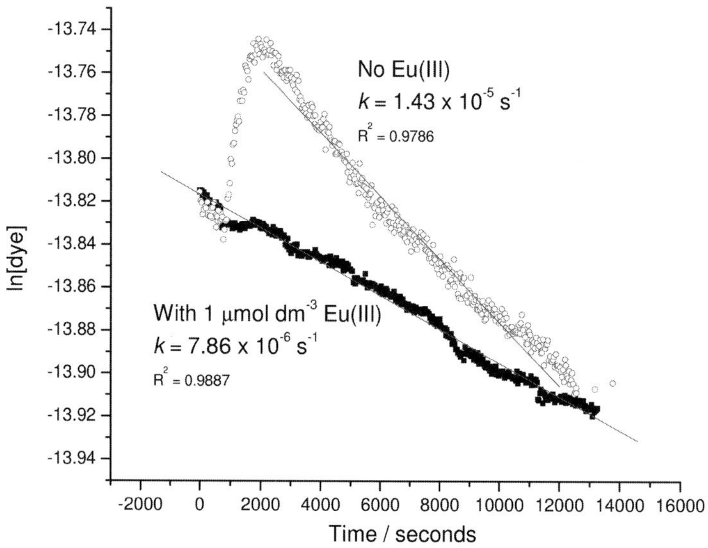

Figure 7 Plot of dye leakage data for vesicles following addition of 0.6 μmol dm^{-3} whole *Ctx*. Gradient from linear fit gives the first-order diffusion coefficient, used to determine the apparent diffusion constant of the dye through the membrane following toxin binding. Note the increase in dye concentration in measurement without Eu^{3+} is an apparent increase, due to the fluorescence change observed in Figure 6, as discussed above.

toxin–membrane interaction and the subsequent increased porosity of the tethered vesicles (Figure 7). The calculated apparent dye diffusion coefficient for the tethered vesicles coupled to Eu^{3+} was significantly lower (6.5×10^{-16} cm^2 s^{-1}) compared to the coefficient determined for control vesicles that were not coupled with Eu^{3+} (1.3×10^{-15} cm^2 s^{-1}). Other lanthanide salts, including lanthanum, erbium, ytterbium, gadolinium and terbium have all shown similar inhibitory affects towards blocking the GM1 receptor from cholera toxin.

A similar support system was used to tether mixed lipid vesicles (1% biotin, 2% GM1, 30% cholesterol, 50% DMPC, 13.5%DMPE, 13.5% DMPS, 1.75% DMPG) to study the effects of amyloid forming peptides towards tethered LBVs. The deposition of insoluble amyloid fibrils is a major pathological hallmark of the amyloidosis disorders, which include Alzheimer's disease, Parkinson's disease and the prion diseases. They are an unusual phenomenon, and are made up of different, typically soluble proteins that undergo a profound conformational change and assemble to form very stable, insoluble fibrils rich in β sheets, that accumulate in the extracellular spaces [38]. The mechanism by which the Aβ protein, the peptide that has been implicated in Alzheimer's disease, causes neurodegeneration has not been definitively identified. One hypothesis concerning Aβ toxicity is that the oligomeric form of the peptide induces channels or pores in neuronal membranes that lead to cell death [39]. The adsorption of Aβ to the surface of the mixed-lipid tethered vesicles was monitored by SPR, resulting in 2.68 ng mm^{-2} of Aβ binding to the surface. An increase in encapsulated marker diffusion

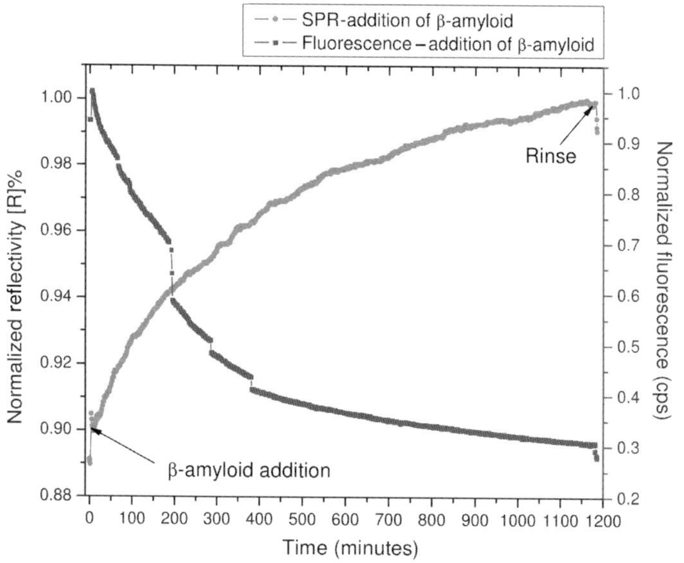

Figure 8 Kinetics plot following the adsorption of 10 μmol dm^{-3} Aβ to mixed phospholipid-GM1-cholesterol biomimetic membrane vesicles.

accompanied the adsorption of the peptide to the membrane surface (Figure 8). The peptide has been shown to preferentially adsorb to negatively charged phospholipids compared to neutral lipids, and the apparent dye diffusion coefficient for the Aβ binding to zwitterionic tethered vesicles was calculated to be 1.08×10^{-17} cm^2 s^{-1}, whereas Aβ binding to negatively charged tethered vesicles showed increased membrane permeation (apparent dye diffusion coefficient = 2.44×10^{-16} cm^2 s^{-1}). Therefore, a realistic mechanism of Aβ toxicity would appear to be the formation of channels, pores or membrane thinning, leading to the flux of solute molecules.

Single-molecule measurements have been made using tethered lipid vesicles to distinguish the static and dynamic properties between identical molecules. The hairpin ribozyme was shown to exhibit two to three orders of magnitude variation in folding kinetics between molecules [40]. Other single-molecule work using tethered lipid vesicles includes a novel tool for single biomolecule spectroscopy, which allowed the immobilization of single protein molecules within lipid vesicles to a surface. The technique was developed to overcome the problems of molecule–surface interactions and was shown to be a promising technique for studying single-molecule large-scale protein dynamics [41].

Oligonucleiotide-functionalized tethered lipid vesicles have been used as a novel method for the amplification of analyte-DNA sensing using electrochemical techniques such as Faradaic impedance spectroscopy [42]. Using this method, the electronic DNA sensor showed marked improvements in specificity and sensitivity, and exhibited a 1×10^5–1×10^6-fold enhancement in oligonucleotide-DNA biorecognition compared to previous electrochemical DNA sensors that exhibited a sensitivity of 1 μA $nmol^{-1}$ [43]. This method has also been extensively used to characterize docking reactions between two populations of tethered vesicles that display complementary DNA. Boxer used epifluorescence microscopy to observe individual docking events and population kinetics between Oregon green labeled egg-yolk phosphatidylcholine vesicles as they mixed via two-dimensional diffusion with Texan red labeled egg-yolk phosphatidylcholine vesicles that were displaying an average of 10 oligomers [44]. It was shown that the average diffusion constant for DNA–lipid diffusion was estimated to be 1–4 $\mu m^2\ s^{-1}$. Other applications of oligomer-tethered lipid vesicles include observing the lateral mobility and diffusivity of DNA tethered vesicles with regard to the development of sensors and separation devices, because it can be used to monitor discrete recognition events or other biological processes. Höök showed that the diffusivity of the tethered vesicle was not dependent on vesicle size or on the length of the DNA tether, but was dependent on anchor size. 1-palmitoyl-2-oleoyl-sn-glycero-phosphocholine (POPC) vesicles tethered via a 30-oligomer DNA-cholesterol anchor showed a 3-fold decrease in the diffusion coefficient compared to POPC vesicles tethered via a 15-oligomer DNA-cholesterol anchor [45]. It was also determined that lipid vesicles anchored via a single cholesterol anchor diffuse at a rate of 1.6 $m^2\ s^{-1}$, whereas vesicles anchored via two cholesterol anchors diffused 1.5 times slower at a rate of 1.16 $m^2\ s^{-1}$.

3.5.3 Conclusions

The expanding uses of surface-tethered lipid vesicles since their first development has provided many novel tools for the characterization of membrane binding events, and the monitoring of membrane, lipid and macromolecule dynamics. The ability to ensure macromolecule functionality and prevention of possible adverse surface–biomolecular interactions lends the strategy to a host of analyte–ligand modeling systems, such as between the amyloid forming peptides and synthetic neuronal lipid vesicles. Another potential strategy is the integration of the tethered lipid vesicles as an element in microfluidic systems as a possible novel strategy for lab-on-chip design by coupling the membrane-associated reactions. The immobilization of the bilayer vesicles provides an attractive model of the cell membrane that can be precisely controlled to study a range of analyte–ligand interactions using surface-sensitive techniques previously unavailable to study these membrane events. Tethered lipid vesicles have shown versatility in the type of strategy employed to ensure an intact and stable model system. Covalent immobilization ensures decoupling of the LBVs from the solid surface, preventing deformation of the lipid bilayer. Tethered lipid vesicles are emerging as a means of initially determining the effectiveness of ligand–analyte interactions, such as DEDA inhibition of PLA_2 and lanthanide ion inhibition of cholera toxin–membrane interactions. The immobilization strategy is also immerging as a novel tool to monitor the interactions between the neurodegenerative amyloid forming peptides and biomimetic model systems. Tethered lipid vesicles provide a simple and convenient model system to gain valuable insight in analyte–ligand interactions and a strategy to specifically pattern surfaces in a controlled manner.

Acknowledgments

Tom Williams appreciates the help and support of Dr Louise Serpell and the Alzheimer's Research Trust. I also wish to thank Natural Environment Research Council for previous funding.

References

1 Bangham, A.D., Standish, M.M. and Watkins, J.C. (1965) Diffusion of univalent ions across the lamellae of swollen phospholipids. *Journal of Molecular Biology*, **13** (1), 238.

2 Sessa, G. and Weissmann, G. (1968) Phospholipid spherules (liposomes) as a model for biological membranes. *Journal of Lipid Research*, **9**, 310–318.

3 Chiruvolu, S., Walker, S., Israelachvili, J., Schmitt, F.-J., Leckband, D. and Zasadzinski, J.A. (1994) Higher order self-assembly of vesicles by site-specific binding. *Science*, **264**, 1753.

4 Rongen, H.A.H., Bult, A. and van Bennekom, W.P. (1997) Liposomes and immunoassays. *Journal of Immunological Methods*, **204** (2), 105–133.

5 Richter, R., Mukhopadhyay, A. and Brisson, A. (2003) Pathways of lipid vesicle deposition on solid surfaces: a combined QCM-D and AFM study. *Biophysical Journal*, **85** (5), 3035–3047.

6 Sackmann, E. and Tanaka, M. (2000) Supported membranes on soft polymer cushions: fabrication, characterization and applications. *Trends in Biotechnology*, **18**, 58–64.

7 Rossetti, F.F., Reviakine, I., Csúcs, G., Assi, F., Vörös, J. and Textor, M. (2004) Interaction of poly(L-lysine)-g-poly(ethylene glycol) with supported phospholipid bilayers. *Biophysical Journal*, **87** (3), 1711–1721.

8 Nguyen, T., McNamara, K.P. and Rosenzweig, Z. (1999) Optochemical sensing by immobilizing fluorophore-encapsulating liposomes in sol–gel thin films. *Analytica Chimica Acta*, **400**, 45–54.

9 Chifen, A.N., Förch, R., Knoll, W., Cameron, P.J., Khor, H.L., Williams, T.L. and Jenkins, A.T.A. (2007) Attachment and phospholipase A(2)-induced lysis of phospholipid bilayer vesicles to plasma-polymerized maleic anhydride/SiO_2 multilayers. *Langmuir*, **23** (11), 6294–6298.

10 Ulman, A. (1991) *An Introduction to Ultrathin Organic Films: From Langmuir–Blodgett to Self-Assembly*, Academic Press, New York, 10: 0127082301.

11 Yoshina-Ishii, C., Miller, G.P., Kraft, M.L., Kool, E.T. and Boxer, S.G. (2005) General method for modification of liposomes for encoded assembly on supported bilayers. *Journal of the American Chemical Society*, **127**, 1356–1357.

12 Chaiet, L. and Wolf, F.J. (1964) The properties of streptavidin, a biotin-binding protein produced by streptomycetes. *Archives of Biochemistry and Biophysics*, **106**, 1–5.

13 Weber, P.C., Ohlendorf, D.H., Wendoloski, J.J. and Salemme, F.R. (1989) Structural origins of high-affinity biotin binding to streptavidin. *Science*, **243** (4887), 85–88.

14 Stamou, D., Duschl, C., Delamarche, E. and Vogel, H. (2003) Self-assembled microarrays of attoliter molecular vessels. *Angewandte Chemie – International Edition*, **42**, 5580–5583.

15 Huang, C.-H. (1969) Studies on phosphatidylcholine vesicles. Formation and physical characteristics. *Biochemistry*, **8** (1), 344–352.

16 Johnson, S.M. and Bangham, A.D. (1969) Potassium permeability of single compartment liposomes with and without valinomycin. *Biochimica et Biophysica Acta – Biomembranes*, **193** (1), 82–91.

17 Hauser, H.O. (1971) The effect of ultrasonic irradiation on the chemical structure of egg lecithin. *Biochemical and Biophysical Research Communications*, **45** (4), 1049–1055.

18 Batri, S. and Korn, E.D. (1973) Single bilayer liposomes prepared without sonication. *Biochimica et Biophysica Acta*, **298**, 1015–1019.

19 MacDonald, R.C., MacDonald, R.I., Menco, B.Ph.M., Takeshita, K., Subbarao, N.K. and Hu, L.-R. (1061) Small-volume extrusion apparatus for preparation of large, unilamellar vesicles. *Biochimica et Biophysica Acta*, **1991**, 297–303.

20 Lapinski, M.M., Castro-Forero, A., Greiner, A.J., Ofoli, R.Y. and Blanchard, G.J. (2007) Comparison of liposomes formed by sonication and extrusion: rotational and translational diffusion of an embedded chromophore. *Langmuir*, **23**, 11677–11683.

21 Linden, M.V., Wiedmer, S.K., Hakala, S.R.M. and Riekkola, M.-L. (2004) Stabilization of phosphatidylcholine coatings in capillary electrophoresis by increase in membrane rigidity. *Journal of Chromatography A*, **105** (1–2), 61–68.

22 Boon, J.M. and Smith, B.D. (2002) Chemical control of phospholipid distribution across bilayer membranes. *Medicinal Research Reviews*, **22** (3), 251–281.

23 Williams, T.L. and Jenkins, A.T.A. (2008) Measurement of the binding of cholera toxin to GM1 gangliosides on solid supported lipid bilayer vesicles and inhibition by europium (III) chloride. *Journal of the American Chemical Society*, **130** (20), 6438–6443.

24 Wojtczak, L. and Nalecz, M.J. (1979) Surface charge of biological membranes as a possible regulator of membrane-bound enzymes. *European Journal of Biochemistry*, **94**, 99–107.

25 Simon, S.A. and McIntosh, T.J. (1986) Depth of water penetration into lipid bilayers. *Methods in Enzymology*, **127**, 511–521.

26 Brown, D.A. and London, E. (2000) Structure and function of sphingolipid- and cholesterol-rich membrane rafts. *The Journal of Biological Chemistry*, **275** (23), 17221–17224.

27 Finkelstein, A. and Cass, A. (1967) Effect of cholesterol on the water permeability of thin lipid membranes. *Nature*, **216** (5116), 717–718.

28 Ipsen, J.H., Karlström, G., Mouritsen, O.G., Wennerström, H. and Zuckermann, M.J. (1987) Phase equilibria in the phosphatidylcholine-cholesterol system. *Biochimica et Biophysica Acta*, **905** (1), 162–172.

29 Demel, R.A. and De Kruyff, B. (1976) The function of sterols in membranes. *Biochimica et Biophysica Acta*, **457** (2), 109–132.

30 Riquelme, G., Lopez, E., Garciasegura, L.M., Ferragut, J.A. and Gonzalezros, J.M. (1990) Giant liposomes: a model system in which to obtain patch-clamp recordings of ionic channels. *Biochemistry*, **29** (51), 11215–11222.

31 Jenkins, A.T.A. and Olds, J.A. (2004) Electrochemical measurement of the interaction of Crotalus adamanteus venom with DMPC vesicles. *Chemical Communications*, **18**, 2106–2107.

32 Palmer, A.F., Wingert, P. and Nickels, J. (2003) Atomic force microscopy and light scattering of small unilamellar actin-containing liposomes. *Biophysical Journal*, **85**, 1233–1247.

33 Vareiro, M.L.M., Liu, J., Knoll, W., Zak, K., Williams, D. and Jenkins, A.T.A. (2005) Surface plasmon fluorescence measurements of human chorionic gonadotrophin: role of antibody orientation in obtaining enhanced sensitivity and limit of detection. *Analytical Chemistry*, **77** (8), 2426–2431.

34 Williams, T.L., Vareiro, M.M.L.M. and Jenkins, A.T.A. (2006) Fluorophore-encapsulated solid-supported bilayer vesicles: a method for studying mem-

brane permeation processes. *Langmuir*, **22** (15), 6473–6476.

35 Sanchez, S.A., Bagatolli, L.A., Gratton, E.-D. and Hazlett, T.L. (2002) A two-photon view of an enzyme at work: crotalus atrox venom PLA2 interaction with single-lipid and mixed-lipid giant unilamellar vesicles. *Biophysical Journal*, **82** (4), 2232–2243.

36 Lauer, S., Goldstein, B., Nolan, R.L. and Nolan, J.P. (2002) Analysis of cholera toxin-ganglioside interactions by flow cytometry. *Biochemistry*, **41** (6), 1742–1752.

37 Alptürk, O., Rusin, O., Fakayode, S.O., Wang, W., Escobedo, J.O., Warner, I.M., Crowe, W.E., Král, V., Pruet, J.M. and Strongin, R.M. (2006) Lanthanide complexes as fluorescent indicators for neutral sugars and cancer biomarkers. *Proceedings of the National Academy of Sciences of the United States of America*, **103** (26), 9756–9760.

38 Serpell, L.C. and Smith, J.M. (2000) Direct visualisation of the beta-sheet structure of synthetic Alzheimer's amyloid. *Journal of Molecular Biology*, **299**, 225–235.

39 Arispe, N., Pollard, H.B. and Rojas, E. (1994) beta-Amyloid Ca(2+)-channel hypothesis for neuronal death in Alzheimer disease. *Molecular and Cellular Biochemistry*, **140** (2), 119–125.

40 Okumus, B., Wilson, T.J., Lilley, D.M.J. and Ha, T. (2004) Vesicle encapsulation studies reveal that single molecule ribozyme heterogeneities are intrinsic. *Biophysical Journal*, **87**, 2798–2806.

41 Boukobza, E., Sonnenfeld, A. and Haran, G. (2001) Immobilization in surface-tethered lipid vesicles as a new tool for single biomolecule spectroscopy. *The Journal of Physical Chemistry B*, **105**, 12165–12170.

42 Patolsky, F., Lichtenstein, A. and Willner, I. (2000) Electrochemical transduction of liposome-amplified DNA sensing. *Angewandte Chemie – International Edition*, **39** (5), 940–943.

43 Korri-Youssoufi, H., Garnier, F., Srivastava, P., Godillot, P. and Yassar, A. (1997) Toward bioelectronics: specific DNA recognition based on an oligonucleotide-functionalized polypyrrole. *Journal of the American Chemical Society*, **119**, 7388–7389.

44 Chan, Y.H.M., Lenz, P. and Boxer, S.G. (2007) Kinetics of DNA-mediated docking reactions between vesicles tethered to supported lipid bilayers. *Proceedings of the National Academy of Sciences of the United States of America*, **104**, 18913–18918.

45 Benkoski, J.J. and Höök, F. (2005) Lateral mobility of tethered vesicle-DNA assemblies. *The Journal of Physical Chemistry. B*, **109** (19), 9773–9779.

3.6
Plasma-Polymerized Allylamine Thin Films for DNA Sensing

Li-Qiang Chu, Wolfgang Knoll, and Renate Förch

3.6.1
Introduction

Over the past two decades plasma-polymerization techniques have been proved to be a simple and effective method for both the preparation of a large variety of functional thin films and the surface modification of different materials [1–5]. Plasma polymerization can be defined as 'formation of polymeric materials under the influence of plasma' [2]. By the careful choice of precursors and deposition parameters, the physical and chemical properties of plasma polymers, as well as their surface properties, can be well tailored, which make them very attractive for biological or biomedical applications. Plasma polymerization of allylamine (ppAA) has attracted considerable interest over the past two decades because of the numerous applications of aminated surfaces in biotechnology. It has previously been reported that ppAA films are suitable as substrates for cell cultures [6, 7] and for the immobilization of biomolecules such as polysaccharides [8] and DNA [9, 10].

The plasma-polymerization process offers some unique advantages over other surface-modification techniques, these include:

1. Only the surface properties of the materials will be altered by the deposition of plasma-polymerized films, while the bulk properties of the substrate materials remain unchanged.
2. It is possible to produce thin films with a great diversity in chemical structure and properties by careful selection of the precursors and the process parameters.
3. Plasma deposition is conformal and hence can be applied to complex substrates with very good adhesion. The resulting films are pinhole free.
4. The deposition process is rapid, adjustable and sterile. Exact control over the film thickness can easily be achieved by adjusting the deposition time.

Detection of specific DNA sequences has become increasingly important in the diagnosis of genetic and infectious diseases, particularly after the entire sequence of the human genome was determined within the human-genome project. Moreover, the identification of gene sequences in bacteria, viruses or food is crucial for

Surface Design: Applications in Bioscience and Nanotechnology
Edited by Renate Förch, Holger Schönherr, and A. Tobias A. Jenkins

ISBN: 978-3-527-40789-7

environmental monitoring, as well as, for food control. A variety of DNA sensors have thus been proposed based on different transduction principles such as the detection of changes in mass, optical or electrochemical properties, while others methods depend on a labeling step for the detection of the hybridized stands [11–14]. Most methods available today rely on the immobilization of single-stranded oligonucleotides (so-called DNA probes or catchers) onto the sensor surfaces that are used as the recognition elements. This is followed by subsequent hybridization of the surface-attached probes with the complementary DNA target from solution. On the other hand, homogeneous DNA-detection methods, which avoid the difficulties associated with probe immobilization, have also been demonstrated by some researchers [15–18].

Surface plasmon enhanced fluorescence spectroscopy (SPFS) was successfully employed for DNA sequence mismatch detection, as well as for the investigation of hybridization kinetics between the surface-attached oligonucleotide and a target DNA from the solution [19–25]. SPFS is a combination of surface plasmon resonance (SPR) and fluorescence spectroscopy [19, 21]. In brief, the excitation of a surface plasmon mode results in an enhanced optical field, which can excite fluorophores located within the evanescent tail of the surface plasmon mode, resulting in a very strong fluorescence signal. Requirements for the technique are a thin metal film (generally gold or silver) typically deposited on a glass slide and an optimum distance of about 50 nm between the metal/dielectric interface and the surface at which immobilization and hybridization occur. The SPR evanescent field decays exponentially into the dielectric medium with a penetration depth of approximately $L = 150$ nm, which provides the surface sensitivity of SPFS. That is, only fluorophores adsorbed, adhered, or bound to the surface will be excited, while the fluorophores in the bulk solution will not be excited.

Peptide nucleic acids (PNA) are DNA analogs in which an uncharged pseudo-peptide chain replaces the negatively charged sugar–phosphate backbone of natural DNA, resulting in an achiral and neutral DNA mimic [26–28]. PNA is capable of sequence-specific binding to DNA or RNA sequences obeying the Watson–Crick hydrogen-bonding rules [27]. In this study we demonstrate how thin ppAA films can be used as positively charged surfaces in a new DNA-detection method. This method takes advantage of the surface sensitivity of SPFS, as well as, the different electrostatic properties of PNA and DNA and the plasma-polymer surface.

3.6.2
Experimental Section

3.6.2.1
Materials and Substrates

The allylamine monomer (i.e. 2-propen-1-ylamine, 99%) was purchased from Sigma-Aldrich (Germany). The monomer was outgassed three times before use, but was not purified further. The substrates for SPFS measurements were

LaSFN9 glass slides (Hellma Optik, Jena, Germany) coated with 2 nm of Cr and 50 nm of gold by thermal evaporation. To ensure optimum adhesion of the plasma polymer to the gold, a monolayer of 1-octadecanethiol (5 mM in ethanol, 10 min immersion) was self-assembled on the gold. The ppAA films for FT-IR and XPS measurements were deposited onto 30 × 25 mm^2 glass slides coated with 80 nm of gold.

Fluorescently labeled PNA (P1: ACATGCAGTGTTGAT-Cy5) was purchased from Applied Biosystems, Foster City, CA. The ssDNA used in this work, including the complementary target DNA (T1: 3′-ATC AAC ACT GCA TGT-5′) and the single-mismatch target DNA (T2: 3′-ATC AAC ACT ACA TGT-5′) were purchased from MWG Biotech, Ebersberg, Germany. PBS buffer (0.01 M phosphate buffer, 0.0027 M potassium chloride and 0.137 M sodium chloride, pH 7.4, at 25 °C) (Sigma-Aldrich) was used to prepare the PNA and DNA solutions. PNA/DNA hybridization was carried out by adding 100 nmol L^{-1} of P1 into a solution of 100 nmol L^{-1} of either T1 or T2. The mixture was shaken at room temperature for 0.5 h allowing for the hybridization of PNA/DNA.

3.6.2.2 Plasma Deposition of ppAA Films

Plasma polymerization was carried out in a home-built capacitively coupled cylindrical radio frequency (13.56 MHz) plasma reactor (Figure 1). The reaction chamber, enclosed in a Faraday cage, consists of a cylindrical Pyrex tube, 30 cm in length and 10 cm in diameter. The plasma power was generated by a RF generator (Coaxial, RFG150) passed through a matching network and delivered to the reactor via a coil placed around the exterior of the reactor tube. The reaction chamber was evacuated using a rotary pump (Leybold Trivac, D16B) and a liquid-nitrogen cold trap was used to collect excess organic vapor. A baratron module (MKS, Type 627B) was connected near the gas inlet to monitor the pressure inside the chamber. Side arms at the reactor inlet allow for the introduction of different precursor vapors and standard gases. The flow rate of allylamine vapor was controlled by a Kobold floating-ball flowmeter.

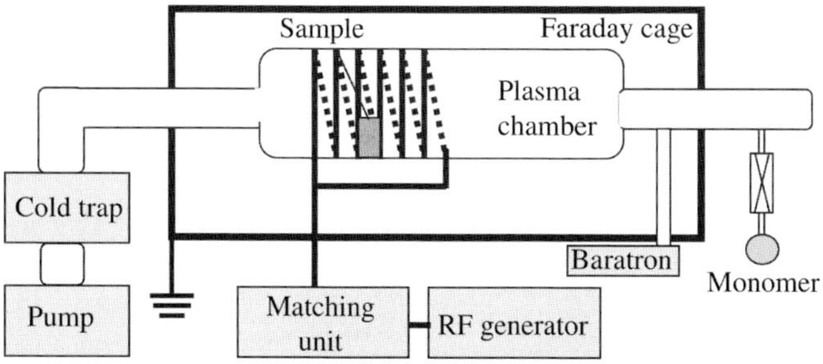

Figure 1 Schematics of the plasma-polymerization system.

The plasma input power (P_{peak}) employed was 5 or 100 W under a continuous-wave (CW) mode. The process pressure was 0.1 mbar for all ppAA deposition. Plasma polymers are known to contain some unbonded low molecular weight material. In order to remove this soluble part from the ppAA films, the samples were submersed in pure ethanol and shaken for different lengths of time [29]. After extraction, the films were washed with excess ethanol to remove small molecules adsorbed on the surface, and then dried at $T = 50\,°C$ for 2 h. In order to achieve an optimized fluorescence signal in SPFS and not to lose too much intensity due to energy transfer to the metallic substrate, the chromophores have to be sufficiently separated from the substrate surface. In the present work the thickness of the ppAA film had to be 40 nm to achieve an optimum signal for SPFS experiments.

3.6.2.3 Film Analysis

Fourier transform infrared spectroscopy (FT-IR) was carried out using a Nicolet 850 spectrometer operated in the reflection mode. FT-IR spectra were recorded at a resolution of $4\,cm^{-1}$. X-ray photoelectron spectroscopy (XPS) was carried out using a Physical Electronics 5600 A equipped with a MgK_{α} (1253.6 eV) X-ray source operated at 300 W. A pass energy of 117.40 eV was used for the survey spectrum. Surface charging was controlled using an electron gun. The spectra were recorded using a 45° take-off angle relative to the surface normal. The C1s XPS spectra were analyzed using MultiPak 5.0 and XPSPeak 4.1.

3.6.2.4 SPR and OWS Measurements

SPR and OWS measurements were carried out with a home-built setup based on the Kretschmann configuration as described before [30–32]. The SPR and OWS spectra were recorded against air immediately after plasma polymerization for freshly deposited ppAA films and after ethanol extraction. The SPR curve can be fitted using the Fresnel equations obtaining the optical thickness (nd) of the dielectric medium. If the refractive index of the thin film is available, one can determine the geometrical thickness of the film. In contrast to SPR, OWS can distinguish between the refractive index n and the thickness d of the film provided that at least two optical waveguide modes can be excited.

3.6.2.5 SPFS Measurements

The SPFS measurements were carried out at room temperature with a home-built setup as described before [19–21]. The emitted fluorescence light from the sample surface was collected by a photomultiplier tube (Hamamatsu), coupled to a photon counter (Hewlett-Packard). A lens (f = 40 mm; Owis) focused the light through an interference filter ($\lambda = 670\,nm$, $\Delta\lambda = 10\,nm$) used to block the scattered and outcoupled excitation light. The whole fluorescence detection unit was mounted on a goniometer such that its position was fixed relative to the sample. The ppAA-coated slides were attached to a flow cell made of stainless steel for liquid exchange, which is covered by a quartz glass window with a low intrinsic fluorescence background. A reference experiment was carried out by injecting 1 nmol L^{-1} fluorescent PNA into the flow cell of SPFS. Then, the mixture of PNA and DNA targets were sequentially introduced into the flow cell. The fluorescent signal was recorded in real time. NaOH solution of 50 mM was added to regenerate the sensor surface.

3.6.3 Results and Discussion

3.6.3.1 Film Analysis

FT-IR was used to determine the chemical structure of the ppAA films before and after ethanol extraction. Figure 2 shows typical FT-IR spectra for ppAA films deposited in a CW plasma mode with increasing input powers. The band at 3360 cm^{-1} indicates the presence of primary and secondary amines within the ppAA films. With increasing power from 5 to 100 W, the bands for CH_3 and CH_2 at 2960–2850 cm^{-1} increased. A similar trend could also be observed for the bands at 2240–2185 cm^{-1}, which are ascribed to the absorption of C≡C or C≡N structures. These results are ascribed to the higher fragmentation of the monomer structure taking place under high plasma power conditions. Analysis of the elemental composition at the surface of the ppAA films using XPS (Table 1), showed that the N/C ratio of ppAA deposited at low P_{peak} (5W) was higher than that deposited at higher P_{peak} (100 W). This is in agreement with previously published work.

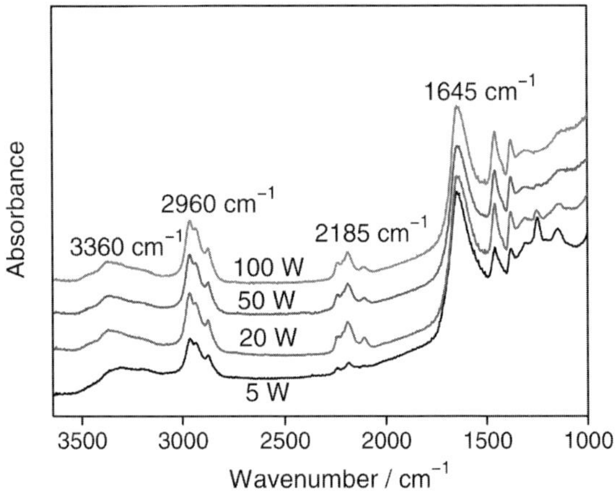

Figure 2 FT-IR spectra of the ppAA films deposited at different plasma input powers.

Table 1 Surface composition from XPS measurements for the ppAA films deposited at different input powers.

Plasma power	%C	%N	%O	N/C
5 W	77.2	20.0	2.8	0.26
100 W	79.6	17.9	2.5	0.22

Figure 3 shows two OWS spectra of a ppAA film (deposited at 100 W for 7 min) before and after ethanol extraction. Because there are more than two optical waveguide modes in each spectrum, both the refractive index n and thickness d of the film can be determined from the simulation. For the fresh ppAA, the thickness was found to be d = 601 nm and refractive index was n = 1.594, while for the extracted ppAA, d and n were 470 nm and 1.585, respectively. It is evident that after ethanol extraction the ppAA film shows a large decrease in thickness because of the loss of material. The decrease in refractive index n suggests a decrease in film density, which could be a result of voids within the film after extraction of low molecular weight material and/or because of residual solvent within the film.

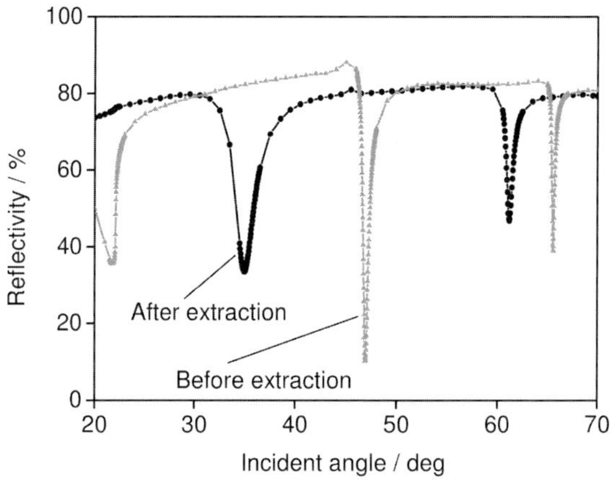

Figure 3 Typical s-polarized OWS spectra of ppAA films (d = 601 nm) before and after ethanol extraction. Plasma conditions: 100 W, CW, 0.1 mbar.

Figure 4 shows the change in thickness of a 601-nm thick ppAA film as a function of exposure time in ethanol. Clearly, 8 h ethanol extraction results in a distinct decrease in film thickness, probably because of the loss of unbonded, or noncrosslinked material within the freshly deposited ppAA. After an additional 7 h of extraction, only small additional changes in thickness were observed. After another 15 h of extraction, no more changes were observed. The decrease is com-

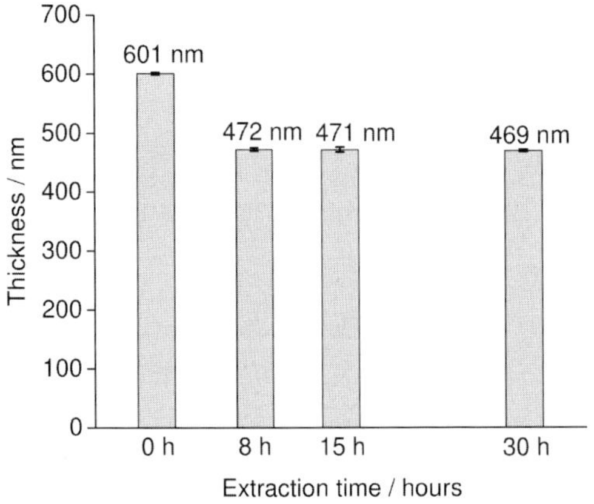

Figure 4 Thickness evolution of a 601-nm ppAA film upon ethanol extraction with increasing extraction time. Plasma conditions: 100 W, CW, 0.1 mbar.

parable to the error in the measurements. Therefore, ethanol extraction overnight was found to be sufficient to stabilize the ppAA films at the present conditions.

While these measurements suggest the loss of material from the deposited ppAA film, FT-IR analysis showed no significant changes in the chemical structure of the films before and after overnight ethanol extraction, as shown in Figure 5. The overall decrease of the band intensity is due to the decrease of the film's thickness. This could be confirmed by XPS (data not shown here), which showed no measurable change in the composition of the ppAA surface.

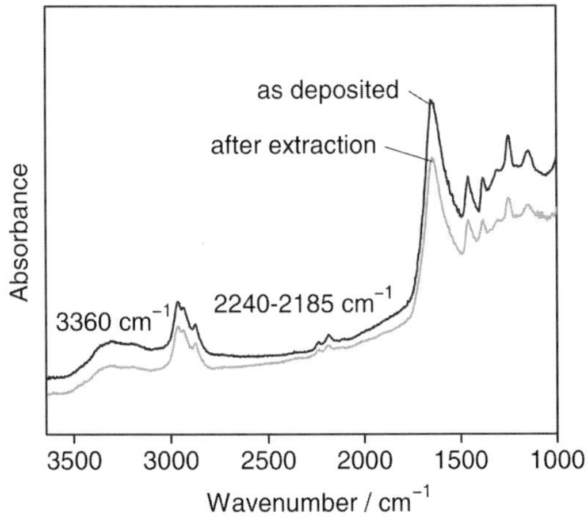

Figure 5 FT-IR spectra of a ppAA film before (upper line) and after 15 h ethanol extraction (lower line). Plasma conditions: 100 W, CW, 0.1 mbar.

The removal of unbound material by ethanol extraction clearly improved the stability of a 100-W ppAA film in PBS buffer solution, as could be observed in real-time SPR measurements (Figure 6a). For the freshly deposited ppAA (d = 46 nm), a rapid decrease in SPR reflectivity could be observed for the first 100 min. This decrease in reflectivity can be directly related to a decrease in the film thickness. After the initial fast decrease, a slow process appeared to take over and the film seemed stable after approximately 12 h. This is consistent with the data described above. A similar experiment performed on ppAA previously extracted in ethanol overnight showed no change in the SPR reflectivity measured over the same time period.

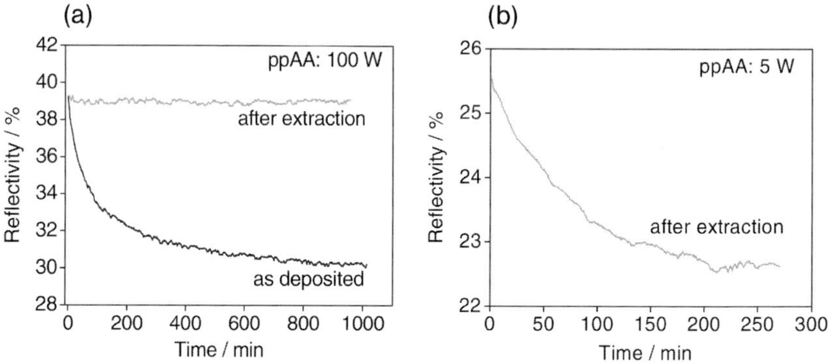

Figure 6 SPR kinetic measurement showing the swelling behavior of different ppAA films in PBS buffer solution. (a) a 100-W ppAA film (d = 46 nm) before and after extraction, respectively; (b) a 5-W ppAA film (d = 12 nm) after extraction.

PpAA films deposited at a P_{peak} of 5 W (d = 12 nm) showed a more pronounced effect upon extraction in ethanol. Following the same procedure of a 15-h immersion in ethanol showed that this time was insufficient to stabilize the film. As shown in Figure 6(b) the SPR reflectivity continued to decrease for at least 3–4 h more. Considering the fact that the extraction process already removed the soluble materials for ppAA, the decrease in Figure 6(b) seems to suggest that the 5-W ppAA film undergoes structural change, e.g., rearrangement of polymer chains or reorientation of some functional groups, in buffer solution. In contrast, a 100-W ppAA film is believed to have a higher crosslinking density and thus have no significant structural change in PBS buffer solution.

3.6.3.2
Comparison of PNA and DNA Adsorption on a ppAA Film

Due to the difference in the electrostatic features between PNA and DNA (see Figure 7(a), they will show a different affinity towards a positively charged surface. Figure 7(b) shows the SPR and SPFS data after DNA immobilization and the SPFS data after PNA immobilization on a ppAA surface (P_{peak} = 100 W). PBS buff-

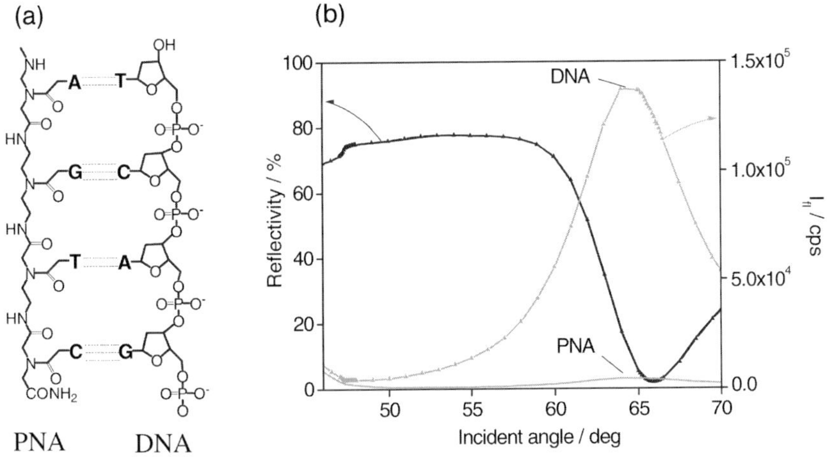

Figure 7 (a) Chemical structures of PNA and DNA, respectively. A, T, G and C represent the nucleobases adenine, thymine, guanine and cytosine, respectively. (b) SPFS spectra after Cy5-labeled PNA (i.e. P1) and DNA adsorbing on the ppAA surface. PNA and DNA have the same sequences (i.e. ACA TGC AGT GTT GAT-Cy5). The concentration of PNA and DNA solution was 1 nm in PBS buffer solution (pH = 7.4).

er solution (pH = 7.4) was used throughout the experiments. The high fluorescence emission of 1.5×10^5 cps observed for the DNA immobilization shows that the affinity for DNA is much higher than that for PNA. This could be explained by the strong electrostatic interaction between the DNA and the ppAA surface.

3.6.3.3 Detection of Different DNA Targets

By taking advantage of the different electrostatic properties of PNA and DNA, as well as the surface sensitivity of SPFS, a new DNA-detection method was developed [33]. Scheme 1 illustrates the concept of the present DNA-detection method. The labeled PNA is neutral and hence it does not adsorb onto the positively charged surface (the upper part in Scheme 1). The fluorophores in the bulk solution cannot be detected by SPFS. As a consequence, no fluorescence signal can be detected by SPFS. On the other hand, as shown in the lower part of Scheme 1, the specific hybridization of target DNA to the labeled PNA results in a negatively charged hybrid, which will adsorb onto the positively charged surface (100-W ppAA films in this case) by electrostatic interaction. The dye associated with PNA will thus be excited and can be detected using SPFS. In this way, the presence of a particular target DNA can be detected without the necessity to label DNA. Here, the labeled PNA serves not only as the DNA catcher recognizing a particular target DNA, but also as a fluorescent indicator.

In a typical experiment for detecting different DNA sequences, the fluorescently labeled PNA (i.e. P1) was firstly added into different target DNA solutions, i.e.

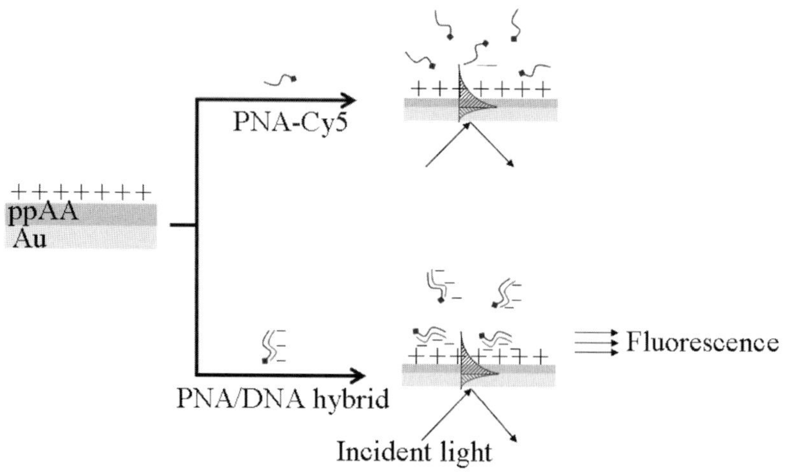

Scheme 1 The concept of DNA detection using SPFS and fluorescent PNA as DNA probes.

T1 (fully complementary) and T2 (single base-pair mismatch), respectively. The PNA/DNA mixture was shaken for 0.5 h, allowing for the hybridization of PNA and DNA and was then introduced into the flow cell of the SPFS. Figure 8(a) shows the real-time SPFS detection of the fluorescent signal upon injection of the two different solutions. The P1 solution was used here as a reference. One could observe a small increase of the fluorescence intensity after the P1 injection. In the case of single-mismatch DNA (i.e. T2/P1), the fluorescence signal showed a small increase, almost comparable to that for the pure P1 injection, most likely because the hybridization between T2 and P1 was unsuccessful, leading to the same response as P1 only. However, for fully complementary DNA target (i.e. T1/P1), a strong increase of the SPFS signal can be observed. This clearly shows that the T1/P1 hybrid, now a negatively charged entity, is able to bind to the ppAA surface. Figure 8(b) shows the SPFS data after injection of various sample solutions. The hybrid of PNA and the fully complementary DNA target (i.e. T1/P1) results in a strong fluorescent signal. In contrast, the hybrid of PNA and a DNA strand with one single mismatch (T2/P1) shows a weak fluorescent signal, which is only slightly higher than the reference PNA solution. It is apparent that this DNA-detection method can discriminate effectively between different DNA sequences with one base-pair mismatch.

As shown in Figure 8(a), the fluorescence signal returns to the baseline after rinsing the sensor surface with 50 mM NaOH, indicating that all of the hybridized PNA strands were removed from the ppAA surface and that the ppAA surfaces are reuseable after regeneration.

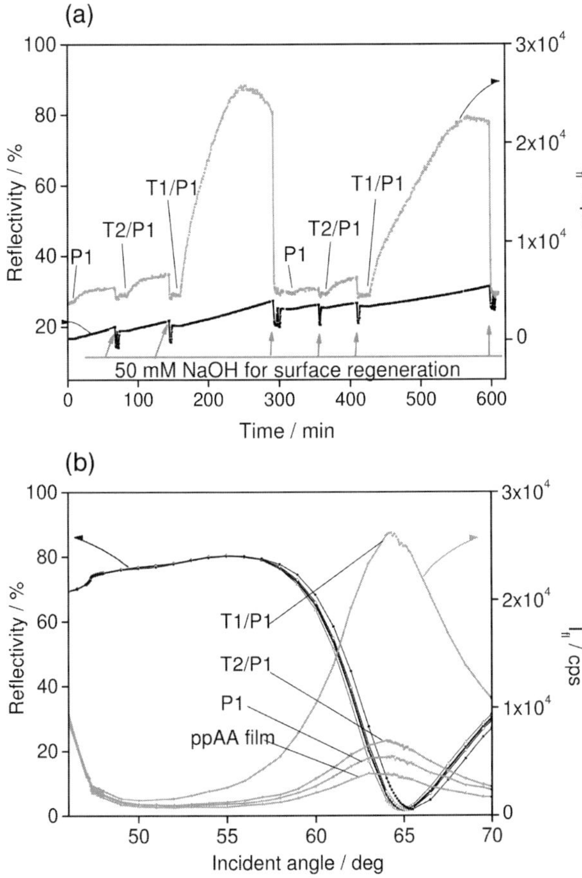

Figure 8 (a) SPFS kinetic measurements during injecting 1 nmol L^{-1} of P1, T1/P1 and T2/P1, respectively. The surface was regenerated with 50 mM NaOH solution for 2 min, followed by PBS buffer rinse. (b) SPFS data after adsorption of various species. The SPFS data of the 100-W ppAA film in PBS buffer is given as a reference.

3.6.4 Conclusions

We demonstrated the use of labeled PNA as probes for homogeneous DNA detection. Several advantages are offered by this method: (i) It avoids the step of immobilization of DNA probes on sensor surfaces, and thus reduces the complexity and the cost for sensor preparation. (ii) Since commercially available PNA are used not only as DNA probes, but also as a fluorescent indicator, no labeling to the target DNA is required. (iii) DNA/PNA hybridization occurs in solution, and thus is faster than that occurring on the sensor surface. Consequently, the present method promises a rapid identification of a specific gene. (iv) The method will make use of all the advantages associated with using PNA as DNA probe, including the high affinity, specificity and easy requirements for the buffer solution.

Plasma-polymerized allylamine thin films were employed as the positively charged coatings in this method. Plasma polymerization exhibited some particular advantages, e.g., good control of film thickness, film chemistry, as well as good adhesion on the substrates. It is apparent that ethanol extraction is an easy approach to stabilize the ppAA film with a high retention of functionalities.

References

1 Biederman, H. and Osada, Y. (1992) *Plasma Polymerization Processes*, Elsevier, New York.

2 Yasuda, H. (1985) *Plasma Polymerization*, Academic, Orlando.

3 Inagaki, N. (1996) *Plasma Surface Modification and Plasma Polymerization*, Technomic, Lancaster, PA.

4 Ratner, B.D. and Castner, D.G. (1997) *Surface Modification of Polymeric Biomaterials*, Plenum, New York.

5 d'Agostino, R., Favia, P. and Wertheimer, M.R. (eds) (2005) *Plasma Processes and Polymers*, Wiley-VCH, Weinheim.

6 Harsch, A., Calderon, J., Timmons, R.B. and Gross, G.W. (2000) Pulsed plasma deposition of allylamine on polysiloxane: a stable surface for neuronal cell adhesion. *Journal of Neuroscience Methods*, **98**, 135–144.

7 Griesser, H.J., Chatelier, R.C., Gengenbach, T.R., Johnson, G. and Steele, J.G. (1994) Growth of human-cells on plasma polymers – putative role of amine and amide groups. *Journal of Biomaterials Science – Polymer Edition*, **5** (6), 531–554.

8 Dai, L.M., St. John, H.A.W., Bi, J.J., Zientek, P., Chatelier, R.C. and Griesser, H.J. (2000) Biomedical coatings by the covalent immobilization of polysaccharides onto gas-plasma-activated polymer surfaces. *Surface and Interface Analysis*, **29** (1), 46–55.

9 Chen, Q., Förch, R. and Knoll, W. (2004) Characterization of pulsed plasma polymerization allylamine as an adhesion layer for DNA adsorption/hybridization. *Chemistry of Materials*, **16**, 614–620.

10 Zhang, Z., Chen, Q., Knoll, W., Förch, R., Holcomb, R. and Roitman, D. (2003) Plasma polymer film structure and DNA probe immobilization. *Macromolecules*, **36**, 7689–7694.

11 Kostrikis, L.G., Tyagi, S., Mhlanga, M.M., Ho, D.D. and Kramer, F.R. (1998) Molecular beacons – spectral genotyping of human alleles. *Science*, **279** (5354), 1228–1229.

12 Taton, T.A., Mirkin, C.A. and Letsinger, R.L. (2000) Scanometric DNA array detection with nanoparticle probes. *Science*, **289** (5485), 1757–1760.

13 Gerion, D., Chen, F.Q., Kannan, B., Fu, A.H., Parak, W.J., Chen, D.J., Majumdar, A. and Alivisatos, A.P. (2003) Room-temperature single-nucleotide polymorphism and multiallele DNA detection using fluorescent nanocrystals and microarrays. *Analytical Chemistry*, **75** (18), 4766–4772.

14 Hansen, J.A., Mukhopadhyay, R., Hansen, J.O. and Gothelf, K.V. (2006) Femtomolar electrochemical detection of DNA targets using metal sulfide nanoparticles. *Journal of the American Chemical Society*, **128** (12), 3860–3861.

15 Ho, H.A., Boissinot, M., Bergeron, M.G., Corbeil, G., Dore, K., Boudreau, D. and Leclerc, M. (2002) Colorimetric and fluorometric detection of nucleic acids using cationic polythiophene derivatives. *Angewandte Chemie – International Edition*, **41** (9), 1548–1551.

16 Gaylord, B.S., Heeger, A.J. and Bazan, G.C. (2002) DNA detection using water-soluble conjugated polymers and peptide nucleic acid probes. *Proceedings of*

the National Academy of Sciences of the United States of America, **99** (17), 10954–10957.

17 Storhoff, J.J., Lucas, A.D., Garimella, V., Bao, Y.P. and Muller, U.R. (2004) Homogeneous detection of unamplified genomic DNA sequences based on colorimetric scatter of gold nanoparticle probes. *Nature Biotechnology*, **22** (7), 883–887.

18 Stoeva, S.I., Lee, J.S., Thaxton, C.S. and Mirkin, C.A. (2006) Multiplexed DNA detection with biobarcoded nanoparticle probes. *Angewandte Chemie – International Edition*, **45** (20), 3303–3306.

19 Liebermann, T. and Knoll, W. (2000) Surface-plasmon field-enhanced fluorescence spectroscopy. *Colloids and Surfaces A – Physicochemical and Engineering Aspects*, **171** (1–3), 115–130.

20 Liebermann, T., Knoll, W., Sluka, P. and Herrmann, R. (2000) Complement hybridization from solution to surface-attached probe-oligonucleotides observed by surface-plasmon-field-eahanced fluorescence spectroscopy. *Colloids and Surfaces A – Physicochemical and Engineering Aspects*, **169** (1–3), 337–350.

21 Neumann, T., Johansson, M.L., Kambhampati, D. and Knoll, W. (2002) Surface-plasmon fluorescence spectroscopy. *Advanced Functional Materials*, **12** (9), 575–586.

22 Kambhampati, D., Nielsen, P.E. and Knoll, W. (2001) Investigating the kinetics of DNA-DNA and PNA-DNA interactions using surface plasmon resonance-enhanced fluorescence spectroscopy. *Biosensors & Bioelectronics*, **16** (9–12), 1109–1118.

23 Yao, D.F., Yu, F., Kim, J.Y., Scholz, J., Nielsen, P.E., Sinner, E.K. and Knoll, W. (2004) Surface plasmon field-enhanced fluorescence spectroscopy in PCR product analysis by peptide nucleic acid probes. *Nucleic Acids Research*, **32** (22), 177–192.

24 Yao, D.F., Kim, J., Yu, F., Nielsen, P.E., Sinner, E.K. and Knoll, W. (2005) Surface density dependence of PCR amplicon hybridization on PNA/DNA probe layers. *Biophysical Journal*, **88** (4), 2745–2751.

25 Robelek, R., Niu, L.F., Schmid, E.L. and Knoll, W. (2004) Multiplexed hybridization detection of quantum dot-conjugated DNA sequences using surface plasmon enhanced fluorescence microscopy and spectrometry. *Analytical Chemistry*, **76** (20), 6160–6165.

26 Nielsen, P.E., Egholm, M., Berg, R.H. and Buchardt, O. (1991) Sequence-selective recognition of DNA by strand displacement with a thymine-substituted polyamide. *Science*, **254** (5037), 1497–1500.

27 Egholm, M., Buchardt, O., Christensen, L., Behrens, C., Freier, S.M., Driver, D.A., Berg, R.H., Kim, S.K., Norden, B. and Nielsen, P.E. (1993) PNA hybridizes to complementary oligonucleotides obeying the Watson–Crick hydrogen-bonding rules. *Nature*, **365** (6446), 566–568.

28 Egholm, M., Buchardt, O., Nielsen, P.E. and Berg, R.H. (1992) Peptide nucleic-acids (PNA) – oligonucleotide analogs with an achiral peptide backbone. *Journal of the American Chemical Society*, **114** (5), 1895–1897.

29 Chu, L.Q., Knoll, W. and Förch, R. (2006) Stabilization of plasma-polymerized allylamine films by ethanol extraction. *Langmuir*, **22** (13), 5548–5551.

30 Knoll, W. (1998) Interfaces and thin films as seen by bound electromagnetic waves. *Annual Review of Physical Chemistry*, **49**, 569–638.

31 Knoll, W. (1997) in *Handbook of Optical Properties II: Optics of Small Particles, Interfaces, and Surfaces* (eds R.E. Hummel and P. Wißmann), CRC Press, Boca Raton, pp. 373–400.

32 Knoll, W. (1991) Optical characterization of organic thin films and interfaces with evanescent waves. *MRS Bulletin*, **16**, 29–39.

33 Chu, L.Q., Förch, R. and Knoll, W. (2007) Surface-plasmon-enhanced fluorescence spectroscopy for DNA detection using fluorescently labeled PNA as 'DNA indicator'. *Angewandte Chemie – International Edition*, **46** (26), 4944–4947.

4
Nanoparticles and Nanocontainers

Surface Design: Applications in Bioscience and Nanotechnology
Edited by Renate Förch, Holger Schönherr, and A. Tobias A. Jenkins

ISBN: 978-3-527-40789-7

4.1
Defined Colloidal 3D Architectures

Nina V. Dziomkina, Mark A. Hempenius, and G. Julius Vancso

4.1.1
Introduction

Organized colloidal structures have intriguing properties that make them useful in a wide range of applications in optics [1], as chemical sensors [2], in data-storage devices [3], etc. Colloidal crystals [4] have also been of tremendous interest in relation to a wide range of scientific phenomena. Colloidal particles suspended in solution provide fascinating models for studying basic physics problems including phase transitions [5], fundamental problems of kinetics of crystallization [6, 7], and the physics of nucleation and growth [8, 9]. Due to the great scientific interest and wide range of applications, the fabrication of colloidal structures has been intensively reviewed [10–14].

Colloidal crystals formed by a stack of colloidal layers of one or different particle sizes are of increasing interest for photonic crystal applications [15]. Photonic devices have important potential applications in optical-communication systems, optical chips, and all optical computers [16, 17]. Particularly important are crystals possessing a photonic bandgap (PBG), which are periodic structures that are able to block the propagation of light through the crystal in one, or more, directions [18, 19]. In this case, the periodicity of such PBG structures in a specific direction has to be in the order of the forbidden wavelength. The most vivid examples of photonic-crystal materials can be found in nature in opals, iridescent butterfly wings, and in threads and spines of the sea mouse [20, 21]. The opal itself consists of closely packed silica spheres that are often used, in addition to polymeric microspheres, for making artificial colloidal crystals. The efficiency of the light-propagation blockage is highly dependent on the colloidal crystal symmetry and refractive-index contrast between the building blocks [22–25]. It was found that a close-packed structure such as a face-centered cubic (FCC) structure with a packing efficiency of 0.74 that is favorable in spontaneous colloid self-assembly [26], should have at least a refractive-index contrast of 2.8 in order to open up a PBG [22]. On the contrary for diamond-like symmetries with a packing efficiency of 0.34, a minimum refractive-index contrast of 1.9 [25] is enough to fulfil the PBG requirements [23–25]. The lower required refractive-index contrast allows a broader

Surface Design: Applications in Bioscience and Nanotechnology
Edited by Renate Förch, Holger Schönherr, and A. Tobias A. Jenkins

ISBN: 978-3-527-40789-7

choice of dielectric materials. In addition, the PBG for the diamond-like structure is less sensitive to structural disorder such as size dispersion, displacement and roughness of the crystal building blocks than close-packed structures [27]. Although the diamond structure with tetrahedral packing is very well suited to fabricate PBG crystals, its formation is not favored by spontaneous colloidal self-assembly. In order to fabricate colloidal crystals with a diamond-like structure, one can think of mimicking the packing of ionic crystal lattices based on scaling up the ions with the corresponding ionic radii and charges to colloidal dimensions. For example, the zincblende type of packing has a diamond lattice structure but consists of oppositely charged ions of different sizes. Possessing a complete PBG, these structures could be promising candidates for photonic-crystal applications. If colloidal crystals with a zincblende structure are stable, then a prerequisite for successful fabrication of such structures is a full control over size and surface charge of the polymer microparticles. In addition, if PBG structures of practical relevance were to be built using colloidal particles as building blocks, one must find a colloidal assembly approach that would allow one to control packing symmetry, packing order and packing efficiency of such colloidal crystals.

In addition to colloidal crystals, two-dimensional colloidal assemblies have found applications as microlenses in imaging [28] and as physical masks in nanosphere lithography [29]. Hollow colloidal particles placed individually on a surface provide an opportunity to locally study reactions that take place in hollow (nano)-capsules. Such capsules can be used as delivery vehicles for a controlled release of substances such as drugs, cosmetics, dyes and inks [30].

Commonly used methods for assembly of colloidal crystals are based on gravitational [31–34], electrostatic [35–40], and capillary forces [41–51], or on the use of physical confinement in combination with pressure and flow [52–55]. Although sedimentation in a gravitational field is conceptually straightforward and simple, very precise control of parameters, such as the size and the density of the colloidal spheres, is essential for successful self-assembly. Controlled sedimentation of colloidal particles in a gravitational field leads to the formation of 3D polycrystalline structures, usually exhibiting a close-packed FCC lattice [26, 33, 56, 57]. Silica colloidal particles are commonly used in this type of fabrication due to their high density. Unfortunately, the process of sedimentation is rather slow and successful fabrication of good-quality colloidal crystals can take up to several weeks [31–34]. In order to speed up the sedimentation process, filtration and centrifugation have been applied [58]. Although the use of filtration and centrifugation shorten the time of crystal manufacturing, these techniques do not yield better quality crystals. Subsequent growth of crystalline domains in various directions [59] leads to random orientation of the crystalline grains formed. A significant improvement of ordering of colloidal crystals compared with crystals grown on a bare substrate can be done by sedimentation of colloids in a physical confinement [55]. The crystals obtained demonstrate much better ordering and orientation. Further improvement of ordering of colloidal crystals in physical confinement can be induced by ultrasonic agitation [52, 60], or by mechanical shear in addition to the above-mentioned sonication [53], or by sedimentation under an oscillatory shear [54].

Another inconvenience of the sedimentation method is that it is difficult to control the number of colloidal layers, while for some applications, such as integration of colloidal crystals into microchips, the crystal thickness plays an essential role. The use of a vertical deposition technique allows one to control the thickness of the crystalline layers formed [41–50].

On flat substrates, control of colloidal crystallization becomes more effective when capillary forces are exploited to drive the colloidal assembly. In practice, this is usually realized by controlled solvent evaporation [41, 42, 44, 61, 62]. In the control of ordering, domain size and thickness of 2D and 3D colloidal crystals, various parameters must be simultaneously considered including the choice of substrate, tilt angle of the substrate with respect to the colloidal suspension (change in the meniscus shape), choice of solvent, evaporation rate, temperature, concentration of colloidal particles and relative humidity. Despite all efforts to date, colloidal monolayers and crystals formed on bare substrates without topological or chemical patterns by these methods possess a close-packed hexagonal structure. Taking a hexagonally formed monolayer as a template substrate, binary colloidal multilayers (AB, AB_2, AB_3) can be grown using a layer-by-layer deposition process of two building blocks of different sizes [63]. Crystals with two layers consisting of either silica, or PS colloidal particles were subsequently deposited. As layer-by-layer growth fabrication is a time-consuming approach, a stepwise spin-coating strategy [64] was applied to fabricate AB_2 and AB_3 binary colloidal crystals from suspension. The diameter ratio between the two constituents was shown to influence the packing symmetry. Essential for this type of growth (epitaxy on a prefabricated colloidal layer or on a periodically corrugated potential) is a tight fit of the particles into the interstitial sites of the templating substrate.

It is clearly seen from the examples above that it is difficult to control colloidal ordering and crystal orientation without the use of adequate templates that direct crystallization. Templates used for 'nucleation' of colloidal crystal layers are usually engineered by the modification of their surfaces including physical (lithographical) [65–68] or chemical [69–73] patterning. Control of surface chemistry and the corresponding patterns of chemical functional units can be achieved by using self-assembled monolayers (SAMs) [69] of organic molecules that have good adhesion to the substrates (for example, thiols to gold [35–38, 69–71], silanes to oxide substrates [39, 40, 69, 71]) or polyelectrolytes [36, 37, 72] in layer-by-layer supramolecular assembly either on homogeneous, or on prepatterned substrates. The layer-by-layer method allows one to tune the surface charge (positive, negative, or neutral), as well. Soft lithography [71] and nanoimprint lithography [73] are widely applied techniques for the 'printing' of the molecules used for surface modification. The topological pattern of the stamp determines the chemical pattern on the substrate surface. Ordering of colloidal particles on such chemically modified surfaces is governed by the combined effects of electrostatic and capillary forces. Charged microspheres self-assemble on the oppositely charged areas of substrates when the substrate is slowly taken out from the colloidal suspension [35–40]. Depending on the ratio between colloidal particle size and the size of the patterned area, colloidal aggregates of very complex forms can be obtained [37].

Colloidal multilayers on chemically modified substrates can be selectively grown by the vertical deposition technique, when a substrate is withdrawn from the suspension with a certain speed [39]. The crystals grown were reported to exhibit mainly nonclose-packed hexagonal structures.

Gravitational and capillary forces are always present under normal conditions, however, other interactions such as electrostatic forces can also be utilized for colloidal crystal engineering. In these approaches, colloidal particles with an electric charge are electrostatically deposited from their aqueous dispersions onto surface domains bearing opposite charges [35–40]. The interactions of colloidal particles with direct and alternating electric fields on electrode surfaces were first intensively studied by investigating the formation of 2D colloidal ordering [74–77]. The electric field serves to move the charged colloidal particles toward the electrode with opposite charges and to generate lateral attraction between the colloidal beads [76]. The application of an electric field to the colloidal suspension also promotes the growth of 3D colloidal crystals on planar electrode surfaces [78]. Careful adjustment of colloidal concentration and electric-field strength allows one to selectively form various crystal structures on the electrode surface [7, 79]. A drawback of colloidal crystal-structure engineering using electrodeposition is the dependence of the structure stability on the presence of the electrostatic field. When the field used for assembly is switched off, the crystals formed can redisperse back in the suspension or be transformed back to FCC or RHCP structures by capillary forces when the sample is withdrawn from the suspension. Therefore, the colloidal crystal must be fixated before its withdrawal, for example, by subsequent polymerization of the dispersion medium [80–82]. The ordering and orientation of 2D close-packed colloidal structures (colloidal monolayers) and colloidal crystals could be simultaneously realized when colloidal particles were electrophoretically deposited on a topologically patterned electrode [83–85]. It was shown that the dimensions of the topological pattern play a crucial role in the formation of 2D colloidal structures.

Emerging micro- and nanofabrication technologies, including physical template structuring and controlled assembly of colloidal particles, have a tremendous potential for use in colloidal crystal engineering [85–95]. It was demonstrated that the formation and orientation of colloidal mono- and multilayer structures depends on the dimensions and geometry of the surface-patterned features, pattern directionality to the drying front and the particle diameter [86, 89, 90, 96–99]. Formation of nonclose-packed colloidal multilayer structures can be realized by a precise control of the periodicity of a 2D surface grating [91–93]. Template-directed crystallization and growth also allow one to tailor, to a certain degree, the packing structure, lattice orientation, and the size of the colloidal crystals. It was shown that by slow sedimentation of colloidal particles into the 'holes' of topologically patterned templates, large and regular FCC colloidal crystals with a crystal orientation of (100) planes parallel to the patterned surface that is different from the usual (111) plane orientation with respect to the surface were directly formed [100]. This work, using various forms and sizes of patterned substrates was later extended by different research groups [89, 101–111]. It was also demonstrated that variation in the shape of the pattern, controlled via silicon wafers possessing (100) and (110) crystallographic orientations, results in the formation of colloidal crystals of FCC (100) and (110) lattices, respectively [112]. The basic idea of this approach was successfully borrowed from semiconductor technology. In semiconductor structure manufacturing, controlled growth of inorganic thin films with a certain orientation on topologically patterned surfaces, known as graphoepitaxy or 'artificial epitaxy' [113] has been used with success. Growth of crystals by graphoepitaxy, as its name states, is controlled by substrate topology as opposed to atomic (lattice) epitaxy in which a material is crystallized onto an existing crystal of another material such that the crystal supergrowth results in a continuation of the crystal structure of the substrate [114].

The example of a graphoepitaxial approach underlines the power of using topological templates, or other patterns, such as chemical [37] or optical [115], in directed colloidal crystal assembly to remedy the shortcomings of spontaneous colloidal self-assembly for the manufacturing of tailored colloidal crystals for realistic photonics applications. Therefore, the development of directed, robust methods is essential for obtaining high-quality, uniform, and mechanically stable structures with a controlled packing symmetry and with a predetermined crystal-structure orientation with respect to the substrates (templates) used. One of such methods is the method of 'electrophoretic deposition of charged polymer colloids on patterned substrates' [93]. This method combines the advantages of various deposition techniques described above. The pattern at the electrode surface determines the colloidal crystal structure, orientation and directionality. An applied electric field guides charged colloidal particles towards the patterned electrode as well as shortens the deposition time. A defined deposition time controls the number of deposited colloidal layers. The withdrawal speed of the electrodes determines the quality of the colloidal crystals which defines crystal structure. Application of this method in colloidal crystal growth is presented in this chapter.

4.1.2 Results and Discussion

4.1.2.1 Electrophoretic Deposition of Polymer Colloidal Particles

For the deposition of colloidal particles, an electrophoretic cell was constructed that is schematically illustrated in Figure 1(a). Two glass/ITO electrodes were separated from each other with a 1-mm thick Teflon spacer. One of the electrodes was patterned in the way described in the previous section. The cell was immersed into a colloidal suspension, perpendicular to the suspension surface. The colloidal suspension consisted of polymer colloidal particles with volume fractions of 0.2 vol.% or 0.5 vol.% redistributed in a mixture of ethanol and water (80:20 vol/vol). Negatively charged polystyrene colloidal particles with a diameter of 330 nm, polydispersity index of less than 3% and surface charge densities of

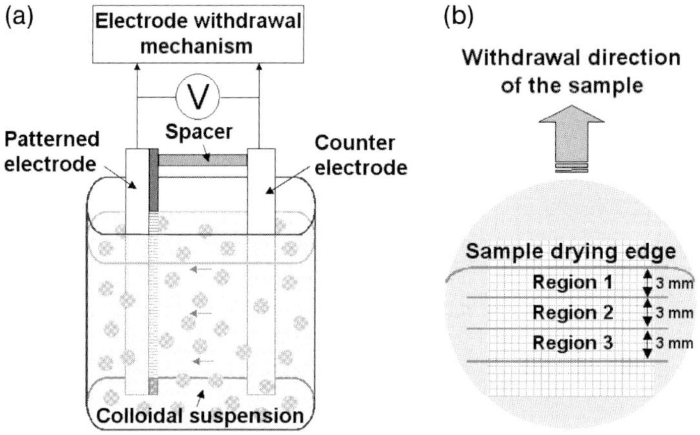

Figure 1 (a) Schematic illustration of the electrophoretic deposition setup; (b) schematic representation of the patterned electrode showing the various regions used in subsequent analyses.

16.2 $\mu C/cm^2$ and 26.5 $\mu C/cm^2$ were used. When a dc electric field was applied, the charged polymer colloidal particles moved to the oppositely charged patterned electrode and were deposited on its surface. After a preset deposition time the cell was withdrawn at a preset speed from the colloidal suspension using a speed control device as presented in Figure 1(b). The patterned electrode was left for several minutes to dry completely under a continuously applied voltage. Electrophoretic depositions were performed in a glove box under a nitrogen atmosphere in order to maintain constant humidity conditions. Patterned electrodes with deposited colloidal particles were marked in three regions starting from the drying edge, Figure 1(b). Effective surface coverage as an equivalent value of the average number of colloidal layers was calculated from SEM images over three regions. In region 1, the surface coverage never exceeded 2 colloidal layers (independent of deposition parameters) likely due to the variable speed in the initial phase of withdrawal. Surface coverage in this region 1 was inhomogeneous and contained areas of colloidal layers that did not correspond to the electrode surface pattern. Homogeneous surface coverage was observed in regions 2 and 3, where the withdrawal speed of the electrodes had stabilized. Below region 3, colloidal crystals were thick, polycrystalline and did not correspond to the electrode pattern. Colloidal crystals in this region were formed upon drying of a droplet of colloidal suspension left at the electrode surface when the electrode broke contact with the colloidal suspension. Therefore, regions below region 3 were not taken into account during the analysis of the surface coverage of grown colloidal crystals. Transitions from one region to another were not sharp and varied from one sample to another depending on deposition parameters. The width of each region was taken as 3 mm.

4.1.2.2
Deposition of Colloidal Particles on Flat (Nonpatterned) Electrode Surfaces

During electrophoretic deposition of colloidal particles on electrode surfaces, the formation of a 'gaseous' structure is usually observed when a weak field is applied (below 25 $V\,cm^{-1}$), Figure 2(a) [75]. Colloidal particles with a diameter of 330 nm and a surface charge density of 16.2 $\mu C/cm^2$ were deposited on the electrode surface at applied voltages of 2.5, 3.0 and 3.5 V during 1 min. These deposition conditions were also applied in electrophoretic colloidal depositions on patterned surfaces, as discussed in the following sections of this chapter. An increase in applied voltage led to an increase in colloidal concentration on the electrode surface and the formation of colloidal crystals with a glassy structure (Figure 2b and c). In principle, colloidal crystal ordering on nonpatterned electrode surfaces can be improved by adjusting a set of deposition parameters, such as electric-field strength, colloidal size and concentration, etc. A consequence of such adjustments is that hexagonally close-packed colloidal monolayers and colloidal crystals are formed [75, 78]. In order to induce a nonclose-packed colloidal crystal structure on the electrode surface, structuring of the electrode surfaces is necessary [83].

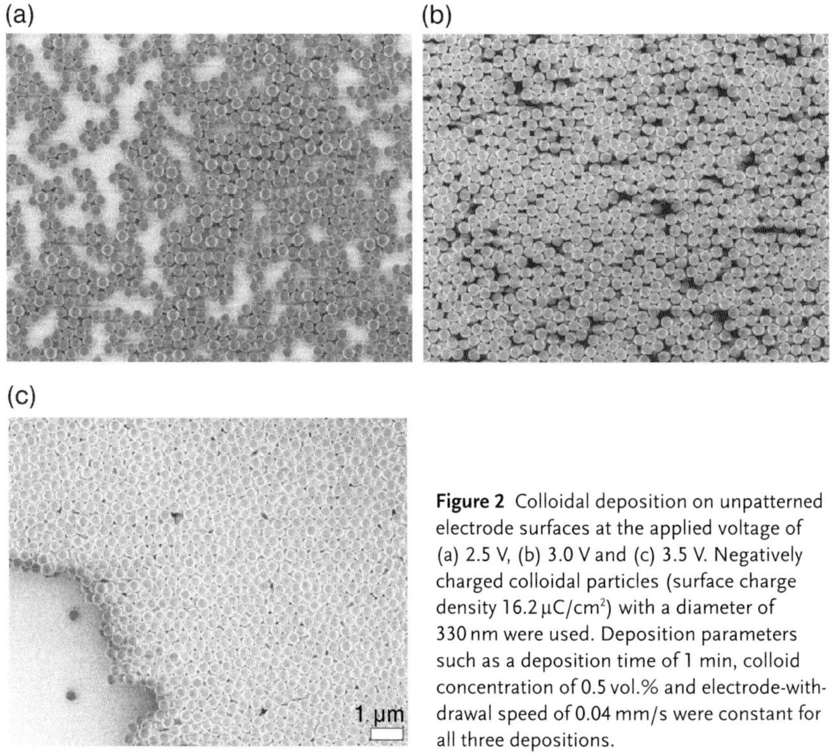

Figure 2 Colloidal deposition on unpatterned electrode surfaces at the applied voltage of (a) 2.5 V, (b) 3.0 V and (c) 3.5 V. Negatively charged colloidal particles (surface charge density 16.2 $\mu C/cm^2$) with a diameter of 330 nm were used. Deposition parameters such as a deposition time of 1 min, colloid concentration of 0.5 vol.% and electrode-withdrawal speed of 0.04 mm/s were constant for all three depositions.

4.1.2.3 Deposition Control of Colloidal Particles on Patterned Surfaces

Ordering of colloidal particles on patterned electrodes can be accomplished by controlling a set of deposition parameters. These include deposition voltage, concentration and surface charge density of polymer colloidal particles, withdrawal speed of electrodes from the suspension, pattern periodicity and the distance between separated electrodes. In the section below, as a representative case study we show the influence of some selected processing and structural parameters on the deposition process and the packing characteristics of the resulting colloidal layers. Effective surface coverage is defined as an average equivalent value of the number of deposited colloidal layers over three regions marked in Figure 1(b).

Surface coverage results for a template periodicity of 350 nm are shown as a function of applied voltage and deposition time in Figure 3(a). The other deposition parameters such as colloidal concentration (0.5 vol.%), withdrawal speed of the electrodes (0.04 mm/s) and colloid surface charge density of 16.2 $\mu C/cm^2$ were kept constant. The surface coverage is presented for region 2 of the sample as shown in Figure 1(b). The three data sets depicted (filled circles, triangles and squares) show the influence of increasing cell voltage, which increases the surface coverage at a given deposition time. In Figure 3(b), surface-coverage data obtained at a deposition time of 60 s are shown for four different electrode pattern periodicities (330, 350, 380 and 450 nm). It is obvious from this figure that for a given pattern periodicity and deposition time, surface coverage begins to dramatically increase if the cell voltage is larger than a given threshold value, indicated by arrows in Figure 3(b). The value of the threshold voltage decreases with increasing pattern periodicity, see Figure 3(c). Eventually, the value of the threshold voltage drops to zero for the pattern periodicity of 450 nm. The decrease in threshold voltage is determined by the electrode geometry, however, pattern quality and colloidal particle polydispersity also play an important role. For a pattern periodicity

Figure 3 Graphs representing the dependence of the number of colloidal layers deposited at various deposition times on different parameters (constant parameters for all graphs are the size of the colloids (330 nm) and electrode-withdrawal speed (0.04 mm/s). Applied voltage dependence: (a) Time dependence (—●— 2 V; —▲— 3 V; —■— 3.5 V). (b) Applied voltage dependence for the deposition time 60 sec and various periodicities (V_T is the threshold voltage) (Pattern periodicities: —■— 330 nm; —◆— 350 nm; —▲— 380 nm; —▼— 450 nm). (c) Threshold voltage versus pattern periodicity: additional constant parameters are colloid concentration (0.5 vol.%) and surface charge density of colloids = 16.2 $\mu C/cm^2$. (d) Spacer thickness dependence: additional constant parameters are applied voltage (3.5 V), colloid concentration (0.5 vol%) and surface charge density of colloids (16.2 $\mu C/cm^2$), deposition time is 60 s. (e) Surface charge density dependence: additional constant parameters are applied voltage (3.5 V) and colloid concentration (0.5 vol%) (—■— Colloid diameter = 330 nm and surface charge density = 16.2 $\mu C/cm^2$; —□— Colloid diameter = 320 nm and surface charge density = 26.5 $\mu C/cm^2$). (f) Concentration dependence: additional constant parameters are applied voltage (3.5 V), colloid concentration (0.5 vol%) and surface charge density of colloids (16.2 $\mu C/cm^2$) (—■— Colloid concentration = 0.5 vol%; —○— Colloid concentration = 0.2 vol%)

higher than 330 nm (equal to the colloid size), deposited colloidal layers became less sensitive to the particle polydispersity, and therefore more colloidal particles were detected at low applied voltages and a decrease in the threshold voltage was found.

In subsequent experiments, applied (deposition) voltages of 3.5 V and 3.9 V were chosen, that are above the threshold voltage for all pattern periodicities. At

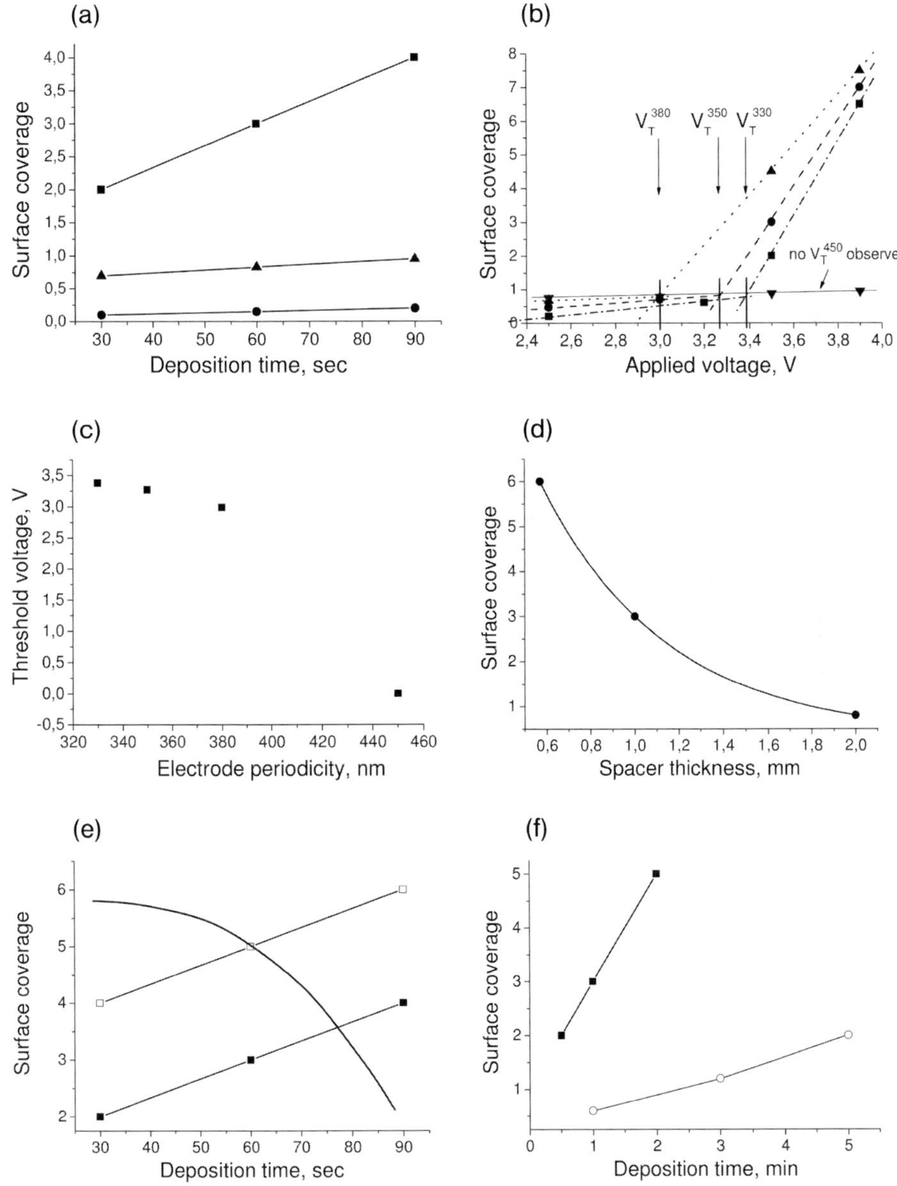

Figure 3 (legend see p. 294)

applied voltages above 3.9 V, control over the colloidal crystallization in colloidal crystals predetermined by the electrode pattern becomes difficult.

The surface coverage versus the distance between electrodes is presented in Figure 3(d) for region 2 (Figure 1b). Colloidal particles were deposited on the electrode surface with a pattern periodicity of 350 nm. A deposition voltage of 3.5 V was applied to the colloidal suspension of 0.5 vol.% and the electrode-withdrawal speed was 0.04 mm/s. The nonlinear dependence of colloidal thickness observed was due to a decrease in electric-field strength. During the deposition, a decrease in electric-field strength leads to a decrease in particle mobility and, therefore, to a lower number of deposited layers. In addition, when electrodes were placed in close proximity to each other (0.57 mm) and immersed in the colloidal suspension, then capillary pressure raised the level of the colloidal suspension between the electrodes over the colloidal suspension surrounding the electrodes, see Figure 1(a). At this small spacing, colloid suspensions remained longer between the electrodes during electrode withdrawal, prolonging the deposition time. At large distances between electrodes (2 mm), the part of the colloidal suspension between electrodes during the deposition was on the same level as the outside suspension, therefore, electrodes were withdrawn from the suspension without trapping the colloidal solution between the electrodes.

Surface charge densities of colloidal particles play an important role in colloidal deposition. Deposition results for colloids with two different surface charges are presented in Figure 3(e). Colloidal particles were deposited on an electrode surface with a pattern periodicity of 350 nm at an applied voltage of 3.5 V. The colloidal concentration was 0.5 vol.% and the electrode-withdrawal speed was 0.04 mm/s. An increase in the surface charge density increased the surface coverage. For both surface charge densities, an increase in deposition time increased the number of layers. We found by SEM observation that good-quality colloidal monolayers were obtained from lower charge density colloids. An increase in surface charge density increases the mobility of colloidal particles and, therefore, colloidal particles arriving at the electrode surface do not have sufficient time to rearrange according to the electrode surface pattern.

Variations in the colloidal volume fraction have a pronounced influence on the colloidal surface coverage, Figure 3(f). Here, we depicted surface coverage versus deposition time for two volume fractions – 0.5 vol.% (filled squares) and 0.2 vol.% (empty squares). Pattern periodicity (350 nm), applied voltage (3.5 V) and withdrawal speed (0.04 mm/s) were kept constant. As expected, with increasing concentration the coverage dramatically increased (for a given deposition time). However, we also observed a deterioration of the packing quality of the crystals for high concentrations.

Colloidal depositions for template electrodes with pattern periodicities of 350 nm and 450 nm and deposition voltages of 3.5 V and 3.9 V are presented in Figure 4. Colloidal particles with a diameter of 330 nm and surface charge density of 16.2 $\mu C/cm^2$ were used. The colloidal concentration was 0.5 vol.%. Surface coverage results for two periodicities and an electrode-withdrawal speed of 0.04 mm/s are shown as a function of deposition time in Figure 4(a). For small packing distances (350 nm), the surface coverage linearly increased with deposition time. However, for a pattern periodicity of 450 nm, the surface coverage was close to one layer even for long deposition times.

Colloidal crystal growth with controlled structure and orientation is highly dependent of the withdrawal speed of electrodes from the colloidal suspension and dependent on the angle at which electrodes are withdrawn. In our first experiments, we did not use a controlled withdrawal of electrodes [93]. Prior to electrode withdrawal from the colloidal suspension, the applied electric voltage was increased to above 10 V, in order to get 'permanent' adhesion of colloidal particles to the electrode surface [75]. At this relatively high voltage, thick colloidal crystals were formed. In order to get colloidal monolayers, the upper layers were washed away in ethanol without turning off the electric field. In this method, it was difficult to predict when washing had to be stopped. The thickness of colloidal crystals could not be controlled by simply increasing the deposition time. Therefore, in order to control the thickness of colloidal crystals, the influence of the withdrawal speed of electrodes from colloidal suspensions needed to be determined. The relationship between surface coverage and withdrawal speed of electrodes from the colloidal suspension is shown in Figure 4(b). Results for applied voltages of 3.5 V and 3.9 V and sample regions 2 and 3 (Figure 1b) are presented. The surface coverage dramatically decreased on increasing the withdrawal speed of the electrodes with a pattern periodicity of 350 nm. The thickness 'gradient' was established over two sample regions. However, almost no influence of the withdrawal speed on crystal thickness was detected over two sample regions for a pattern periodicity of 450 nm. The difference between depositions on electrode surfaces with different periodicities is related to the surface topology of the deposited colloidal layers. For the pattern periodicity of 450 nm, it was possible to grow colloidal crystals only with a maximum thickness of two colloidal layers at any applied voltage used. This is probably due to smoothening of the surface topology of two deposited layers, where the top layer approximates a flat surface. We speculate that colloids deposited on top of these two layers glide back into the bulk colloidal suspension during electrode withdrawal.

In Figure 4(c), surface-coverage data for the three electrode regions are shown, that were obtained for withdrawal speeds of 0.04, 0.1 and 0.17 mm/s. The deposition voltage of 3.5 V, deposition time of 1 min and colloidal concentration of 0.5 vol.% were kept constant. For the pattern periodicity of 350 nm and at withdrawal speeds of 0.04 mm/s, the surface coverage was not homogeneous and increased from the top to the bottom of the sample.

Upon increasing the withdrawal speed (0.1 mm/s), the surface coverage became homogeneous over the whole sample area. Above this value, the number of layers decreased from the top to the bottom of the sample.

However, for periodicities of 450 nm, a slight increase in the number of layers was detected at all withdrawal speeds, but the difference between three depositions at different withdrawal speeds is small. For the two template periodicities (350 and 450 nm) at a withdrawal speed of 0.17 mm/s, different trends in the surface coverage were observed. For the periodicity of 450 nm, the surface coverage

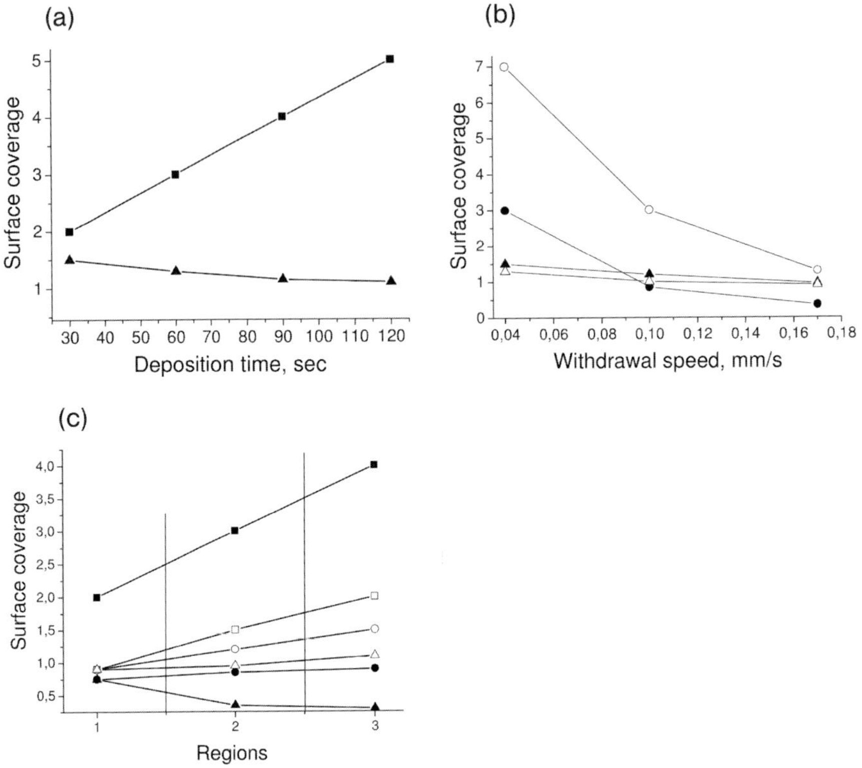

Figure 4 Graphs representing a comparison of colloidal deposition on patterns with different periodicities (constant parameters for all graphs are colloid size (330 nm), colloid concentration (0.5 vol%) and colloid surface charge density (16.2 $\mu C/cm^2$). (a) Pattern periodicity dependence: additional constant parameters are applied voltage = 3.5 V, withdrawal speed of electrodes = 0.04 mm/s (Pattern periodicity: —■— 350 nm; —▲— 450 nm). (b) Applied voltage and withdrawal speed dependence: additional constant parameter is deposition time = 1 min (Pattern periodicity of 350 nm: —●— 3.5 V; —○— 3.9 V; Pattern periodicity of 450 nm: —▲— 3.5 V; —△— 3.9 V). (c) Withdrawal speed and region on the sample (see Figure 1b) dependence: additional constant parameters are applied voltage = 3.5 V and deposition time = 1 min (Pattern periodicity of 350 nm: —■— 0.04 mm/s; —●— 0.1 mm/s; —▲— 0.17 mm/s; Pattern periodicity of 450 nm: —□— 0.04 mm/s; —○— 0.1 mm/s; —△— 0.17 mm/s).

was homogeneous over the sample and for the periodicity of 350 nm, a decrease in surface coverage was detected. We postulate that for the periodicity of 450 nm, deposited colloids interact less with each other via electrostatic forces due to their large separation, and strong electrostatic forces between colloids and electrodes keep the particles on the surface. On the contrary, at the periodicity of 350 nm, colloids are in close proximity to each other and they are not close packed (colloid size is 330 nm), therefore, there is a strong competition between electrostatic forces that act between electrode and colloids and capillary forces acting between colloids. Therefore, for the periodicity of 350 nm, at high withdrawal speeds (0.17 mm/s) colloidal particles were likely driven back into the colloidal suspension.

In separate experiments, the effect of capillary forces on colloidal crystal growth was examined. Colloidal particles were first deposited on a patterned electrode surface with a periodicity of 350 nm by applying an electric field, followed by electrode withdrawal without an applied electric field. In the second experiment, electrodes were kept in colloidal suspension for a certain time without any applied electric field and then pulled out of the colloidal suspension. Deposition parameters used were a colloidal concentration of 0.5 vol.%, applied voltage of 3.5 V for 1 min and a withdrawal speed of 0.04 mm/s. In the two cases, practically no colloidal deposition was observed. Therefore, we conclude that capillary forces at withdrawal speeds used in this study did not have any significant effect on the crystal buildup, as shown by control experiments. This finding also shows the importance of leaving the voltage 'on' to hold the crystals on the electrodes during withdrawal.

4.1.2.4
Formation of Colloidal Monolayers

Electrophoretic colloidal deposition on patterned electrode surfaces allows one to control the symmetry of colloidal monolayers and crystals. Colloidal monolayers with hexagonal and square types of patterns and various periodicities are presented in Figure 5. Pattern periodicities were 350 nm, 380 nm, 400 nm, 450 nm and 500 nm, respectively. Colloidal particles used in this deposition were 330 nm in diameter. An increase in the pattern periodicity from one electrode sample to another improved the surface coverage of colloidal monolayers. The smaller the difference between the particle diameter and the pattern periodicity, the stronger is the influence of the imperfection of the colloidal particle size and the stronger are the interactions between particles and capillary forces acting between the particles during the electrode withdrawal. The examples exhibiting two layers, shown in Figure 5(C) could be obtained by increasing, e.g., the applied voltage or deposition time, as discussed previously. In general, the lower the colloidal concentration (here: 0.2 vol.%) and surface charge density (here: 16.2 $\mu C/cm^2$), the slower the deposition process and therefore, the better the quality of deposited colloidal layers. An increase in the applied voltage from 3.5 V to 3.9 V led to a subsequent increase in the number of layers, but the deposited colloidal layers did not correspond to the colloidal crystals that were expected according to the employed surface pattern. Colloidal multilayers (crystals) formed at an applied voltage of 3.9 V were polycrystalline with a mixed crystal orientation and orientation with respect to the electrode pattern orientation. This result was likely caused by the large number of particles arriving simultaneously at the patterned substrate, therefore, the particles lacked the time required to arrange themselves according to the surface pattern. At an applied voltage of 3.5 V, the quality of deposited colloidal layers depended on the withdrawal speed of the electrodes. We found that withdrawal speeds must be tuned to the pattern periodicity. For a pattern periodicity of the

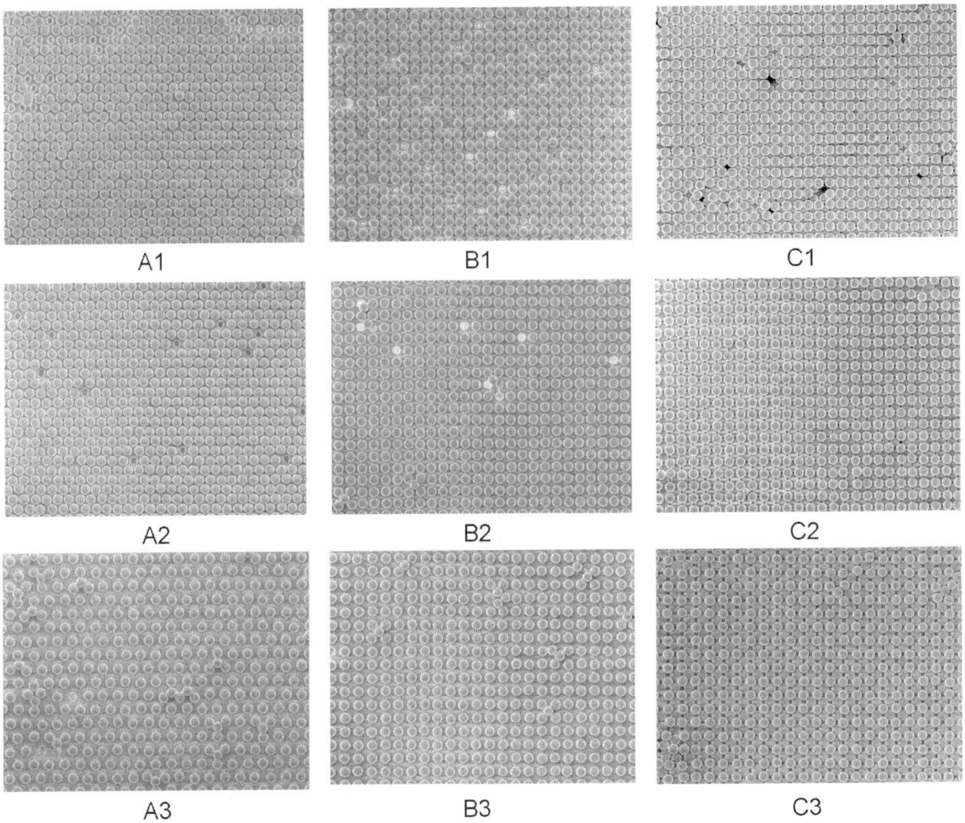

Figure 5 Colloidal deposition on electrodes of different periodicities: (A) hexagonal pattern with periodicities of (A1) 350 nm, (A2) 400 nm and (A3) 500 nm; (B) and (C) square pattern of one and two layers with periodicities of (B1) and (C1) 350 nm, (B2) and (C2) 380 nm and (B3) and (C3) 450 nm.

order of the colloid size, a low speed (in our case 0.04 mm/s) was preferred, since this enhanced the surface coverage as well as the quality of deposited layers. An increase in surface-pattern dimensions should be accompanied by an increase in withdrawal speed. From our experience, for pattern periodicities of 350 nm and 380 nm, a withdrawal speed of 0.07 mm/s was found to be optimal, and for periodicities of 450 nm and 500 nm, withdrawal speeds of 0.1 mm/s and 0.17 mm/s were preferred. By adjusting these deposition parameters, we were able to control the quality and thickness of deposited colloidal layers on the electrode surfaces with various (nonclose-packed) periodicities over large electrode surface areas (several square millimeters). All these results show that the electrophoretic colloidal deposition on patterned electrodes is a versatile method that allows a good control of the location of deposited colloidal particles and orientation and quality of colloidal layers.

4.1.2.5
Colloidal Crystals with FCC and BCC Crystal Structure

The structure that is usually obtained by self-assembly of colloidal particles on a flat substrate is a close-packed structure such as face-centered cubic (FCC) with a packing density of 0.74, stacked in the direction of the close-packed (111) plane [41, 42]. The body-centered cubic (BCC) structure possesses a packing density of 0.68, only slightly lower than that of the FCC, but it is already a problem to fabricate such a structure in a controlled way. Realization of the controlled growth of a BCC colloidal crystal structure will allow one to make a step towards the nonclose-packed structures.

The orientation of colloidal crystal growth can be induced by the patterned surface. The orientation of FCC colloidal crystals was induced by the patterning of the electrode surface in relation to the different planes of the FCC crystal. In Figure 6, three different patterns are presented: a *hexagonal* pattern (Figure 6a) that corresponds to the FCC (111) plane (the most close-packed plane of the FCC structure); a *square* pattern (Figure 6b) that corresponds to the FCC (100) plane (a less close-packed plane than (111)); and a *rectangular* pattern (Figure 6c) that corresponds to the FCC (110) plane (a nonclose-packed plane). Pattern periodicity was maintained to the order of the colloidal particle size for close-packed planes. Negatively charged polystyrene colloidal particles with a diameter of 330 nm, polydispersity index of 1.03 and surface charge densities of 16.2 μC/cm^2 were used. In

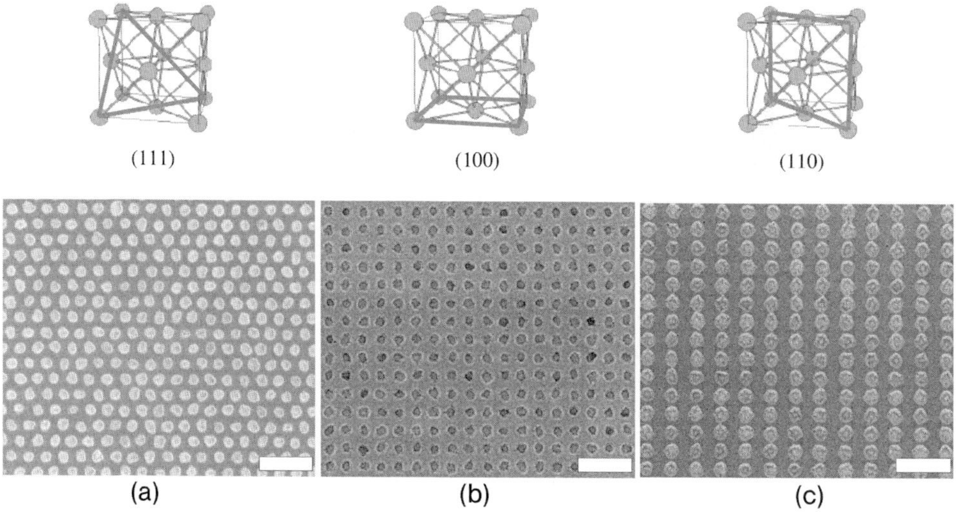

Figure 6 SEM images of electrode surfaces with (a) hexagonal, (b) square, and (c) rectangular patterns. Pattern structures correspond to FCC (111), (100) and (110) crystal planes, shown above the SEM images. The scale bar represents 1 μm. (Reprinted with permission from N.V. Dziomkina, M.A. Hempenius, G.J. Vancso, *Proc. SPIE*, 2005, *5931*, 59310Z, Copyright 2005 by SPIE).

the colloidal suspension, polymer colloidal particles with volume fractions of 0.2 vol.% or 0.5 vol.% were redistributed in a mixture of ethanol and water (80:20 vol/vol). Applied voltages were varied between 2.5, 3.0, 3.5 and 3.9 V. Withdrawal speeds of the electrodes from the colloidal suspension were 0.04, 0.08 and 0.1 mm/s.

Surface coverage and surface structure of colloidal monolayers on the electrode surfaces with a periodicity on the order of the colloid particle diameter are presented in Figure 7 (A1), (B1) and (C1). The hexagonal structure is the most closely packed structure and, therefore, is the most favorable for colloids to be formed on flat electrodes. The formation of almost fully covered colloidal monolayers with a hexagonal structure was easily realized. In this case the electrode pattern influences only the orientation of hexagonal colloid monolayers on the electrode surface, as is shown in Figure 7 (A1). Colloidal layers with square and rectangular structures cannot be fabricated without using patterned electrodes exhibiting the required packing. The growth of colloidal monolayers with square and rectangular structures confirms the influence of a patterned surface on colloidal crystallization, Figure 7 (B1) and (C1).

The easiest direction for the colloidal crystals to grow is the direction of the most close-packed plane, the (111). Therefore, colloidal crystals that are naturally formed onto flat, unpatterned substrates tend to self-assemble into the FCC structure with the (111) plane parallel to the substrate. An FCC (100) crystal could not be formed by self-assembly on flat substrates in a controlled way; therefore, its crystal growth had to be induced by surface structuring. Growth of an FCC crystal from the nonclose-packed plane (110) was rather difficult. We found that one or two layers of FCC (110) structure could be easily formed over large patterned areas at low deposition voltages (3 V or 3.5 V), long deposition times (over 3 min) and high withdrawal speeds (0.1 mm/s), Figure 7 (C1) and (C2). Fabrication of thick, homogeneous in thickness FCC (110) crystals was almost impossible. With an increasing number of colloidal layers and due to the tendency of colloidal particles to pack densely, colloidal crystallization on nonclose-packed (110) patterns led to the formation of domains with a different orientation, Figure 7 (C3), but in most of the areas, the formation of colloidal crystals with either the (111) or the (100) crystal orientation was observed.

The thickness of colloidal crystals could be controlled in two ways. Firstly, a thickness gradient could be maintained across the sample at low withdrawal speeds (e.g. 0.04 mm/s) of the electrodes and at applied voltages (3.5 V and 3.9 V) and deposition times tuned to the desired thickness gradient, as in Figure 4(c). Secondly, a uniform thickness over the whole sample was achieved by applying a high withdrawal speed (0.08 mm/s or even 0.1 mm/s) by varying the applied voltage (3.5 V or 3.9 V) and by tuning the deposition time, as in Figure 4(b). In the last case, the crystal thickness could be increased by increasing the deposition time. Examples of the thickness control of colloidal crystals with different crystal orientations are presented in Figure 7 (A) and (B).

A high-quality FCC colloidal crystal with a very low defect concentration could be grown if the periodicity of the patterned electrode was precisely controlled in

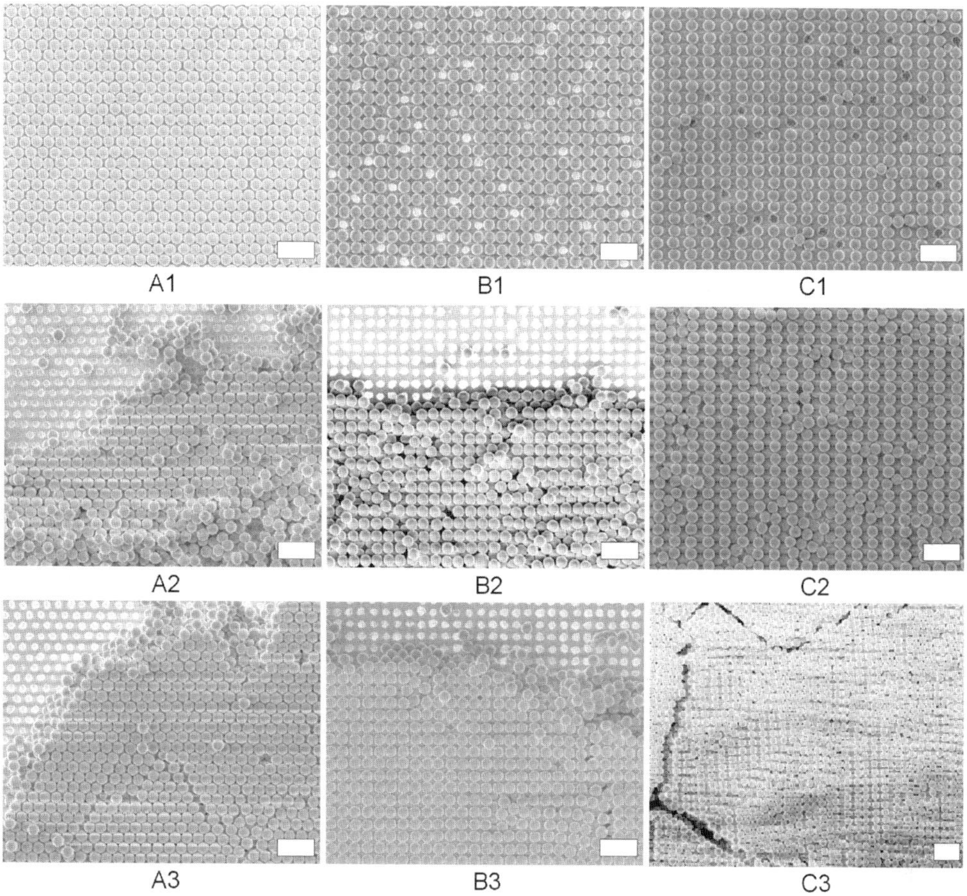

Figure 7 FCC crystals grown on different crystal facets: (A1)–(A3) one, two and five layers of FCC (111); (B1)–(B3) one, two and four layers of FCC (100); (C1)–(C3) one, two and seven layers of FCC (110). The scale bars represent 1 μm. (Reprinted with permission from: N.V. Dziomkina, M.A. Hempenius, G.J. Vancso, *Proc. SPIE*, 2005, *5931*, 59310Z, Copyright 2005 by SPIE).

relation to the dimensions of the FCC colloidal crystal. If there is a small mismatch between the substrate periodicity and colloidal size, line defects would grow through the colloidal crystal as is shown in Figure 7 (A3) for a hexagonal pattern. Defect structures formed in colloidal crystals resemble those formed during the deposition of inorganic materials [114]. In general, defects in colloidal crystals are formed due to the drying process following colloidal deposition when colloids, while trying to pack densely, destroy the hexagonal packing in some parts of the sample, forming line defects. The monodispersity of colloidal particles plays an important role in determining the quality of formed colloidal crystals.

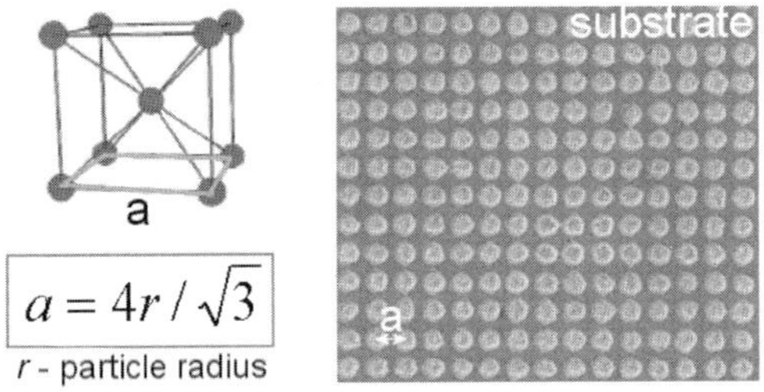

Figure 8 Schematic illustration of the BCC structure (left); SEM image of the patterned electrode surface with a periodicity (a) equal to the distance between colloids in the (100) plane (right).

In the case of BCC colloidal crystal growth, colloidal deposition can be performed on a patterned electrode surface with a square type of structure (Figure 8), that corresponds to BCC (100) crystal planes. The pattern periodicity was chosen equal to the distance between neighboring colloids in the (100) plane. Negatively charged polystyrene colloidal particles with a diameter of 330 nm and a polydispersity index of 1.03 were used. Optimal deposition parameters for the BCC colloidal crystal growth were found as a correlation between the obtained crystals and the used deposition parameters, discussed in the previous sections. These optimal deposition parameters are: a surface charge density of the colloidal particles of 16.2 $\mu C/cm^2$, a concentration of colloidal suspension of 0.2 vol.%, deposition voltage of 3.5 V and a withdrawal speed of 0.08 mm/s. The growth of colloidal crystals as a function of deposition time at constant deposition parameters are presented in Figure 9(A). An increase in the deposition time resulted in a linear increase in colloidal surface coverage. Examples of colloidal crystals with different thicknesses are presented in the SEM images, Figure 9(A), (a–e). Such colloidal crystals covered large electrode areas of several millimeters square. The thickness of the colloidal crystals could be accurately controlled at fixed deposition parameters and was reproducible.

Large, almost defect-free colloidal crystals were grown at deposition times of up to 15 min. At a deposition time of approximately 15 min, crystal defects became significant that eventually destroyed the colloidal crystal homogeneity. Upon increasing the deposition time further to 20 min, polycrystalline close-packed domains and no BCC colloidal crystal domains were observed. Therefore, using this method, colloidal crystals possessing a BCC crystal structure oriented in the direction of the (100) plane could be grown with a thickness of up to five colloidal layers.

In order to increase the number of layers of colloidal crystals further, we varied the meniscus shape of the colloidal suspension between the electrodes. The meniscus can be increased by several methods: (1) by changing the tilt angle of the electrodes with respect to the colloidal suspension by a few degrees or (2) by introducing obstacles for the colloidal suspension of different thickness. We tried both methods and we obtained the same results. The disadvantage of the second method is that not the whole electrode surface takes part in the colloidal deposi-

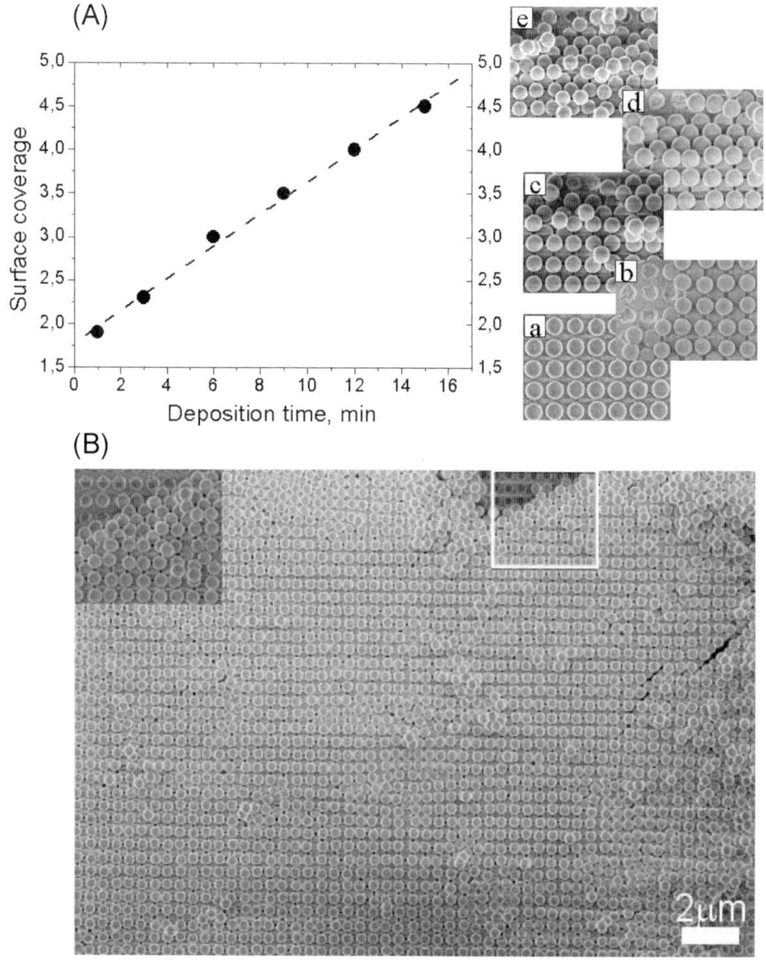

Figure 9 (A) BCC crystal growth by increase in the deposition time: (a) to (e) examples of 1 to 5 colloidal layers, respectively. (B) SEM image of colloidal crystals with a thickness of six layers grown with an electrode tilt of a few degrees. Deposition voltage = 3.5 V, deposition time = 15 min, colloidal concentration = 0.175 vol.%, colloidal surface charge density = 16.2 μC/cm^2 and withdrawal speed = 0.08 mm/s.

tion and thickness control of colloidal crystal growth must be realized by the precise level of immersion of electrodes in the colloidal suspension. The first method is much more practical, as only a tilt angle must be fixed during the colloidal deposition. As a result, by changing the tilt angle by approximately 1°, colloidal crystals possessing a BCC crystal structure and a thickness of up to seven colloidal layers were obtained. A colloidal crystal with a BCC crystal structure and with a thickness of six colloidal layers is presented in Figure 9(B).

4.1.2.6
Layer-by-Layer Colloidal Deposition: Nonclose-Packed Colloidal Crystals with Hexagonal Symmetry

Growth of colloidal crystals on patterned surfaces with pattern periodicities larger than the colloid diameter (Figure 10) or with a pattern with two different periodicities (e.g. the rectangular pattern, Figure 10) is rather difficult. Colloidal particles tend to pack into close-packed polycrystalline structures of different orientation that do not correspond to the surface pattern. In order to improve colloidal crystallization on such types of patterns, layer-by-layer deposition of colloidal particles was performed. After each colloidal layer deposition, the sample was dried in order to assure a better adhesion of the colloidal particles to their deposition site. Deposition parameters for monolayer formation were defined using data presented earlier. Negatively charged colloidal particles with a diameter of 545 nm were deposited onto a hexagonally patterned electrode surface (colloid diameter (D) to pattern periodicity (Λ) ratio $D/\Lambda = 0.75$) at an applied voltage of 3.5 V during 90 s. Electrodes were withdrawn from the colloidal suspension with a colloid volume fraction of 0.2 vol.% with a speed of 0.1 mm/s. After colloidal deposition, electrodes were left to dry for several minutes under the applied voltage. Under these conditions, the first deposited colloidal layer covered an electrode surface over several millimeters, see Figure 11(a). The first colloidal layer already shows some disorder caused either by defects in the substrate pattern or by a local increase in colloid concentration. Then, three more layers were deposited using the same deposition parameters, see Figure 11(b–d).

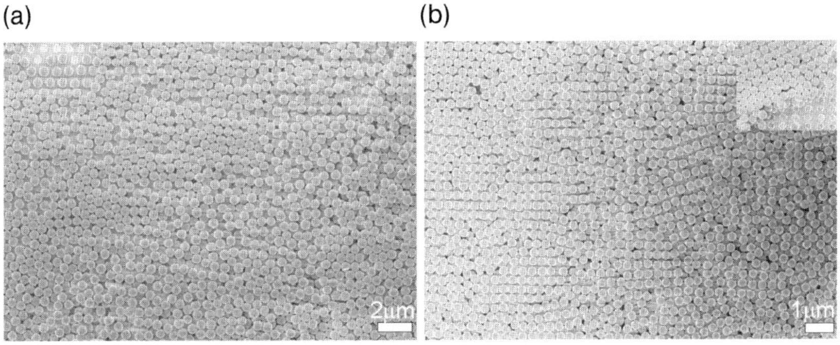

Figure 10 Colloidal particles deposited on (a) a *hexagonally patterned surface* with a ratio of colloidal diameter to surface periodicity of 0.75 (negatively charged colloidal particles with a diameter of 545 nm and surface charge density of 25.3 μC/cm^2, redistributed in an ethanol/water (80:20 vol/vol) mixture (colloidal volume fraction of 0.2 vol.%) were deposited for 3 min at an applied voltage of 3.5 V, withdrawal speed = 0.08 mm/s); (b) *a rectangular patterned electrode* surface with a pattern periodicity of 330 nm by 450 nm in perpendicular directions (negatively charged colloidal particles with a diameter of 330 nm and a surface charge density of 16.2 μC/cm^2, redistributed in an ethanol/water (80:20 vol/vol) mixture (colloidal volume fraction of 0.5 vol.%) were deposited for 2 min at an applied voltage of 3.5 V, withdrawal speed = 0.08 mm/s).

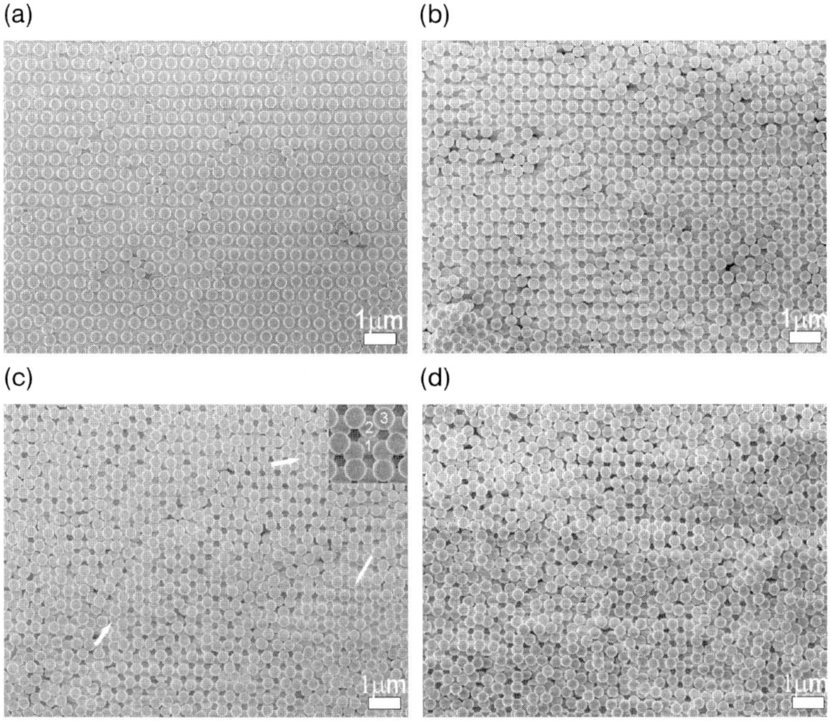

Figure 11 Colloidal multilayers grown on hexagonally patterned substrates (colloid diameter was 545 nm and pattern periodicity was 722 nm): (a) one layer, (b) two, (c) three and (d) four layers. Arrows indicate areas with A–B–A sequences.

Deposition of colloidal particles resembles atomic-layer vacuum deposition of inorganic materials [114]. In both cases a nucleation and growth mechanism operates. Nucleation on a flat surface is usually characterized by a critical nuclear diameter at which, depending on deposition conditions, further crystal growth may continue. Any kind of defect usually acts as a nucleation center. Electrode surfaces featuring a periodic pit structure in a dielectric layer are ideal candidates for colloidal particle nucleation, Figure 11(a). The structure of the first colloidal layer is determined by the pattern of the electrode. Then, the deposited colloidal monolayer on the patterned electrode surface acts itself as a nucleation layer (periodic defect structure) for the second colloidal layer. Deposition of the second layer is more complex. When the colloidal diameter is slightly different from the value of the pattern periodicity ($D/\Lambda = 0.75$), colloidal particles cannot occupy each site in depressions between neighboring colloidal particles of the first layer. Nucleation on the surface starts randomly on the first colloidal layer, and further colloidal layer growth around a nucleation center leads to mismatches between neighboring colloidal crystal grains, Figure 11(b). With deposition of third and fourth layers, Figure 11(c) and (d), the colloidal grain size decreased. Colloidal crystals with three and four colloidal layers possess predominantly an A–B–A layer sequence as found in the hexagonal close-packed (HCP) crystal structure, while small areas of several micrometers with an A–B–C layer sequence (an FCC analog) were also present, Figure 11(c). Growth in a dominating A–B–A sequence instead of an A–B–C is induced by a heightened counter ion concentration in the first two deposited layers in interparticle spaces (formed in the 'tail' of colloidal particles, when an electric field is applied). Approaching colloidal particles are electrostatically attracted by counter ions of deposited colloidal particles and thus take up their positions on top of preliminary deposited colloidal layers. Therefore, a decrease in grain size of colloidal crystals with three and four colloidal layers is caused only by growing defect structures formed in the first colloidal layer. Colloidal crystals presented in Figure 11 possess a low packing density compared to, for example, the close-packed HCP colloidal crystal.

4.1.2.7
Layer-by-Layer Colloidal Deposition: Binary Colloidal Monolayers and Crystals

The crystallization of a mixture of large (L) and small (S) colloidal particles can result in 2D and 3D binary colloidal crystals. Binary colloidal crystals with stoichiometries of LS_2 (atomic analog – AlB_2) and LS_{13} (atomic analog – $NaZn_{13}$, UBe_{13}, etc.) were first discovered in Brazilian opals [116]. Theoretically and experimentally these structures have been intensively studied in hard-sphere colloidal suspensions [117–123]. Recently, it was demonstrated by means of confocal microscopy that colloidal particles of different sizes and opposite charges can yield a variety of colloidal structures, such as LS (atomic analog – NaCl and NiAs), LS_2, LS_6, and CsCl ($R_S/R_L \approx 1$) [79]. The formation of binary colloidal crystals from nanoparticles of metals, semiconductors and metal oxides has recently also received tremendous interest [124–127]. Structures of binary colloidal crystals are determined by the radii ratios of colloidal particles R_S/R_L, the total packing fraction $\phi = \phi_L + \phi_S$ and the relative numbers of small and large colloidal particles n_S/n_L [121–123]. Theoretically, it was calculated that binary colloidal crystals can be obtained for R_S/R_L between 0.15 and 0.5. Below 0.15 the small colloids could freely flow, while above 0.5 (smaller than 0.58) [116] the small colloids would no longer fit inside colloidal crystals formed from large particles [122, 128]. In the binary suspensions, the entropic depletion effect induced by the existence of the small spheres forces large spheres to closely pack into an ordered structure, and then small colloids are spontaneously trapped within voids in the ordered assembly of large colloids [118, 119]. Formation of large binary colloidal crystals with large R_S/R_L ratios was recently demonstrated utilizing accelerated solvent evaporation [46, 129]. A combination of 'colloidal epitaxy' and controlled drying in a layer-by-layer deposition of colloidal particles with $R_S/R_L = 0.5$ allows one to fabricate 2D colloidal crystals with stoichiometries LS, LS_2 and LS_3 (Figure 12) [63]. The self-assembly of small colloidal particles between hexagonally packed larger colloids is the result of the interplay of the geometrical packing arrangement, minimization of the surface tension of the drying liquid film and capillary forces induced at the

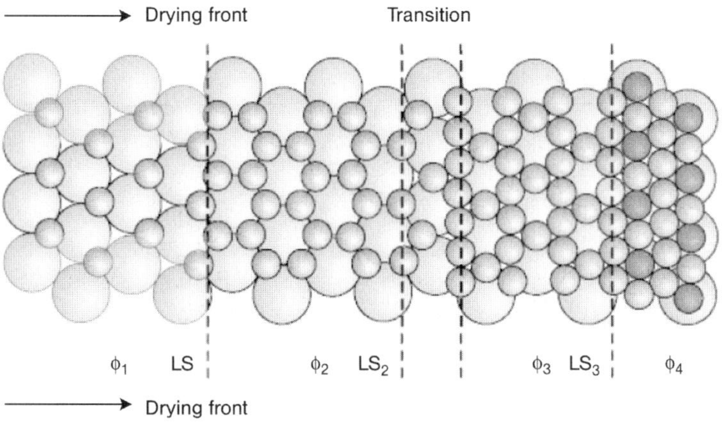

Figure 12 Schematic representation of the structure of small colloidal particles 'nested' onto a hexagonal layer of large colloidal particles $R_S/R_L = 0.5$ as a function of volume fraction ϕ. (Reprinted with permission from [63], Copyright 2002 by American Association for the Advancement of Science).

curved meniscus. In addition to the controlled drying method, LS_2 and LS_3 colloidal crystals were also formed by a stepwise spin-coating method [64]. A disadvantage of layer-by-layer growth of binary colloidal crystals on flat surfaces is the limited possibility for variation of colloidal crystal structures. The utilization of patterned surfaces for the guided deposition of large colloidal particles will lead to the formation of various structures of such particles and, therefore, allow the formation of more complex binary colloidal crystal structures. Moreover, when such binary colloidal crystals consist of colloidal particles of different nature such as polymers and inorganic material, then a selective removal of one type of colloidal particles can lead to the formation of colloidal crystals of the remaining colloidal particles possessing low crystal symmetry [130–132]. In addition to binary colloidal crystals, layer-by-layer deposition of colloidal particles possessing the same charge sign on patterned surfaces with periodicities larger than the colloidal size could open up possibilities to directly grow colloidal crystals with low packing densities. Layer-by-layer deposition of colloidal particles also allows one to introduce planar defects in the form of colloidal layers of different particle size into colloidal crystals.

Binary colloidal monolayers were fabricated on electrode surfaces with hexagonal and square structures under an applied electric field (Figure 13). Layer-by-layer deposition of charged colloidal particles was performed electrophoretically. Optimal deposition parameters were determined separately for each type of colloidal particle and patterned substrate. Large colloidal particles with a diameter of 545 nm and a surface charge density of 25.3 $\mu C/cm^2$ were deposited on a square pattern at 3.9 V during 45 s and withdrawn from the colloidal suspension with a colloidal volume fraction of 0.2 vol.% at a speed of 0.1 mm/s. On hexagonally patterned surfaces, large colloidal particles were deposited at 3.5 V during 90 s, followed by withdrawal (speed = 0.1 mm/s). After deposition, electrodes covered with monolayers of colloidal particles were left to dry for several minutes. Then, electrodes with deposited large colloidal particles were immersed in a colloidal suspension of small colloidal particles at a speed of 0.97 mm/s (the highest speed of the speed-controlled device used) with a preset applied potential on them. Small colloidal particles with a diameter of 225 nm and a surface charge density of 17.7 $\mu C/cm^2$ were deposited on the square pattern at 3.0 V and 3.5 V during 30 s and the substrate was withdrawn from the colloidal suspension (colloidal volume fraction of 0.2 vol.%) at a speed of 0.1 mm/s. On hexagonally patterned surfaces, small colloidal particles with a diameter of 265 nm and a surface charge density of 17.0 $\mu C/cm^2$ were deposited at 3.0 V during 30 and 45 s and substrates were subsequently withdrawn at a speed of 0.1 mm/s. The deposition time was measured from the time electrodes touched the colloidal suspension during electrode immersion. As was mentioned in the previous section, deposition of colloidal particles is very similar to the deposition of inorganic materials. Similar to inorganic materials, colloidal particles deposited onto an electrode surface tend to first nucleate at surface irregularities (or defects). Surface irregularities can artificially be introduced by periodically patterned surfaces. Small colloidal particles arriving at the electrode surface covered with a periodic structure of large particles take their positions in depressions between large colloidal particles, forming binary colloidal layers. On electrodes with the square pattern, two binary colloidal structures were formed with stoichiometries LS and LS_5[1)].

The number of small colloidal particles and, therefore, colloidal crystal stoichiometry, was varied by tuning the applied voltages, but it also could be increased by increasing the deposition time. Colloidal crystals grown on the square type of surface pattern with LS stoichiometry have a binary atomic analog, i.e. NaCl. Structures of binary colloidal layers on hexagonally patterned surfaces can be varied either by tuning colloidal particle diameter or by changing pattern periodicity. Examples of LS_2 and LS_4 structures formed with colloidal particles of radii ratios R_S/R_L equal to 0.49 and 0.41, respectively, are shown in Figure 13(c) and (d). Deposition parameters for both binary colloidal structures were the same. An atomic analog for LS_2 colloidal crystals was found to be AlB_2. To our knowledge, LS_4 does not have a reported isostructural binary atomic analog. At shorter deposition times (20–30 s), small colloidal particles with $R_S/R_L = 0.41$ order in colloidal layers with LS_2 stoichiometry, as well. In the bottom part of the sample with $R_S/R_L = 0.49$, presented in Figure 13(c), at higher applied voltages and longer deposition times, binary colloidal layers with LS_3 stoichiometry (see Figure 12) as well as mixed areas (transition regions) of LS_2 and LS_3 were observed. Decreasing the ratios of colloidal diameters to pattern periodicities for a hexagonally patterned electrode from $D_L/\Lambda = 1$ to 0.83 and 0.78 allows the formation of nonclose-packed binary colloidal monolayers that possess an LS_2 stoichiometry (Figure 13e and f).

1) LS_5 stoichiometry was determined for a colloidal monolayer. In the case of colloidal crystals the crystal structure and stoichiometry are not known, with $D_L/\Lambda = 1$ (Figure 13a and b). A LS_5 structure was also found in the bottom parts of the sample with a LS structure (region 3, Figure 1b).

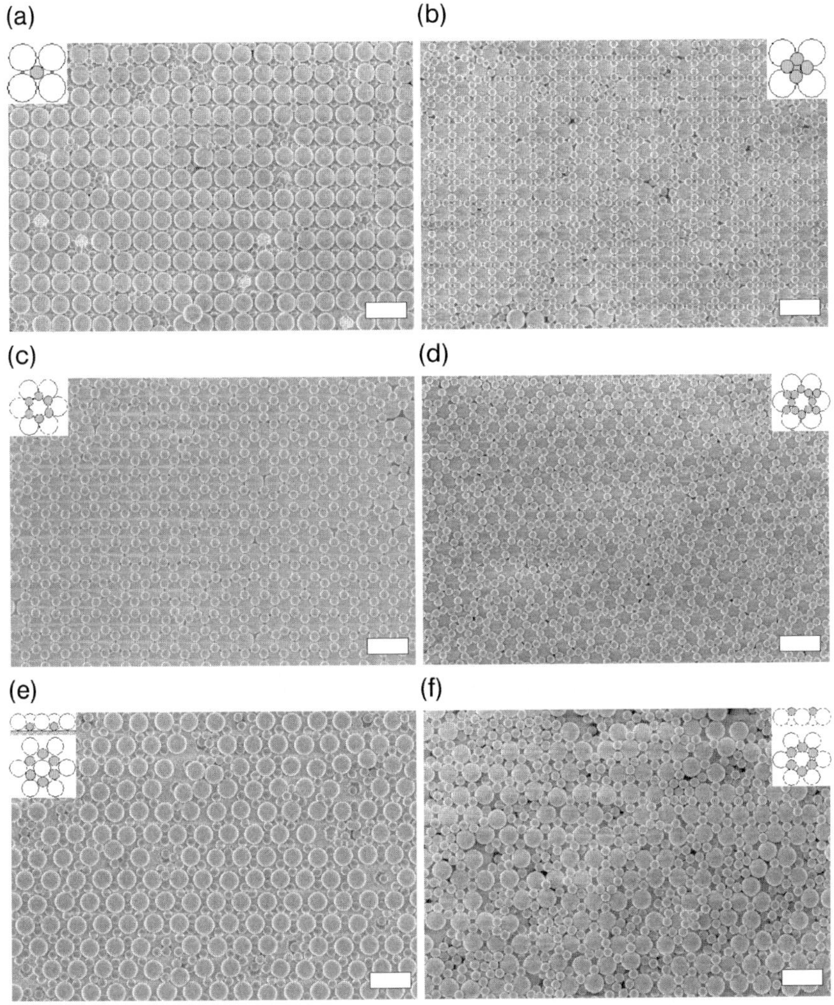

Figure 13 Binary colloidal monolayers with stoichiometries: (a) LS, (b) LS_5, (c), (e) and (f) LS_2 and (d) LS_4. Ratios of colloidal radii R_S/R_L: (a), (b) and (d) 0.41; (c), (e) and (f) 0.49; and ratios of diameters of large particles to pattern periodicities D_L/Λ: (a), (b), (c) and (d) 1; (e) 0.78 and (f) 0.83. Large colloidal particles with a diameter of 545 nm were deposited (a) and (b) at 3.9 V during 45 sec, electrode-withdrawal speeds = 0.1 mm/s; (c)–(f) at 3.5 V during 90 s, electrode-withdrawal speeds = 0.1 mm/s. Small colloidal particles with a diameter of 225 nm were deposited (a) at 3.0 V and (b) at 3.5 V during 30 s, electrode-withdrawal speed = 0.1 mm/s. Small colloidal particles with a diameter of 265 nm were deposited at 3.0 V during (c) and (d) 45 s, and (e) and (f) 30 s, electrode-withdrawal speed = 0.1 mm/s. In the corner of each image, a schematic representation of corresponding monolayer structures is depicted. Scale bar = 1 μm.

Deposition parameters were the same in both cases. Although the colloidal monolayers formed possess defect structures, it is clear that small colloidal particles are lying on the electrode surface (Figure 13e), and small colloidal particles are located on top of large colloidal particles, forming hexagons around them (Figure 13f). The binary colloidal monolayers cover large electrode areas of several square millimeters. An example of the nonclose-packed LS_2 structure is presented in Figure 13(e). The ability to form such nonclose-packed monolayer structures on electrode areas of several micrometers indicates that with some improvements, formation of large-area, nonclose-packed colloidal crystals could be possible.

Electrophoretic deposition of charged colloidal particles on patterned surfaces allows one to realize various complex binary colloidal monolayer structures, including nonclose-packed ones, by changing electrode pattern structure and periodicity.

4.1.2.8
Layer-by-Layer Colloidal Deposition: Formation of Colloidal Crystals with Planar Defects

Control over the formation of defect structures in colloidal crystals is important in their application as photonic crystals. Light propagation in photonic crystals can be guided through artificially created defects. Planar defects in colloidal crystals can be introduced in the form of a colloidal layer or layers formed with colloidal particles of a size different from the main crystal particle size. Electrophoretic deposition of charged colloidal particles allows one to control colloidal crystal structure, orientation and thickness. Planar defects can also be introduced in a controlled way by performing a layer-by-layer deposition method and adjusting deposition parameters.

Negatively charged large colloidal particles with a diameter of 545 nm were deposited on a hexagonally patterned electrode surface with $D_L/\Lambda = 1$ at 3.5 V during 3 min and withdrawn from the colloidal suspension with a colloidal volume fraction of 0.2 vol.% at a speed of 0.04 mm/s. At this withdrawal speed, a thickness gradient in the colloidal crystal was formed over the electrode surface. The thickness and surface coverage of colloidal crystals formed with large colloidal particles could be easily varied by adjusting deposition parameters such as deposition voltage and time, colloidal concentration and electrode-withdrawal speed. Then, the sample was left to dry for several minutes and small colloidal particles were deposited on top of the large ones.

The structure produced using small colloidal particles can be adjusted to one with LS, LS_2, LS_3 and LS_4 stoichiometries depending on deposition parameters and colloidal radii ratios. Examples of colloidal layers with LS_2 stoichiometry and colloidal radii ratios of $R_S/R_L = 0.49$ and 0.41 are shown in Figure 14(a) and (b). Negatively charged small colloidal particles were deposited at 3 V during 45 s ($R_S/R_L = 0.49$) and 30 s ($R_S/R_L = 0.41$), and the electrode was withdrawn at 0.1 mm/s from the colloidal suspension with a colloid volume fraction of 0.2 vol.%. In depositions following the first deposition, electrodes were immersed in colloidal suspension at a speed of 0.97 mm/s and the deposition time was measured from

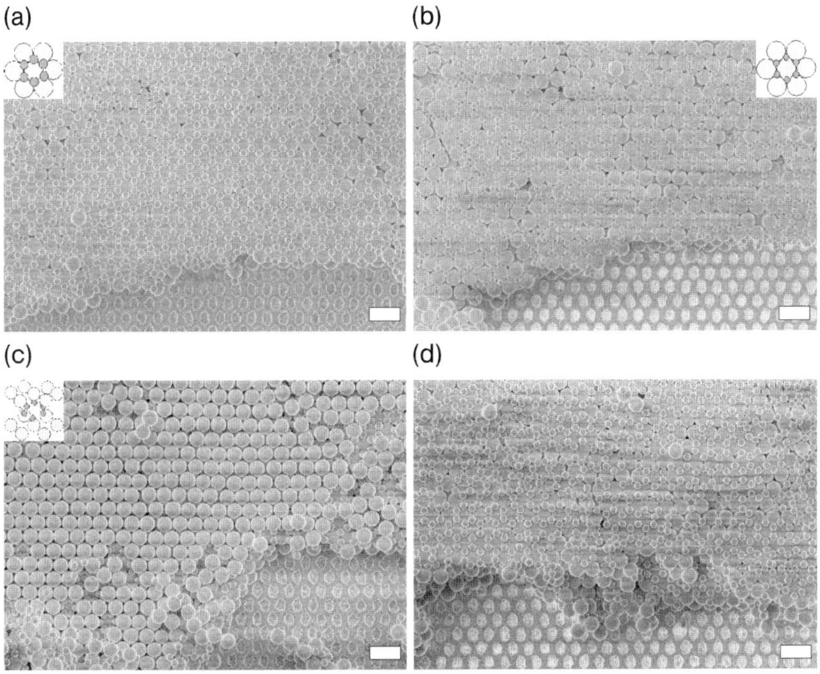

Figure 14 Layer-by-layer deposited binary colloidal layers on a hexagonally patterned electrode surface: (a) two layers of large colloids (R_L) and one layer of small colloids (R_S), $R_S/R_L = 0.49$; (b) three layers of large colloids (R_L) and one layer of small colloids (R_S), $R_S/R_L = 0.41$; (c) sandwich (alternating) structure of colloidal layers with 2 large, 1 small, 2 large and 1 small colloidal particles, $R_S/R_L = 0.49$; (d) sandwich structure of colloidal layers with 2 large, 1 small and 1 large colloidal particles, $R_S/R_L = 0.49$. In the corner of images (a), (b) and (c), a schematic representation of corresponding layer structures is shown (for (c) the view is 'in perspective'). Scale bar = 1 μm.

the time electrodes touched the surface of the colloidal suspension. Using the layer-by-layer method, a layer of small colloidal particles could be deposited among layers of large colloidal particles, breaking the usual sequence of a colloidal crystal, as is schematically presented in the corner of Figure 14(c). In Figure 14(c) only one top layer of large particles was deposited in order to clearly see the changes induced in layers with large colloidal particles, caused by a layer of small particles. Large colloidal particles were deposited at 3.5 V during 6 min and electrodes were withdrawn at speed of 0.1 mm/s.

In order to demonstrate the flexibility of this method, a sandwich structure of alternating colloidal layers with small and large colloidal particles was grown on the hexagonally patterned electrode surface shown in Figure 14(d). Deposition parameters for the first two layers were the same as for those presented in Figure 14(a–c). The next layers of large colloidal particles were deposited at 3.5 V during 2 min and using a withdrawal speed of 0.04 mm/s. The top layer of small colloids was deposited at 3 V during 60 s, the electrode-withdrawal speed was 0.1 mm/s.

Planar defect structures can also be introduced in colloidal crystals grown on electrode surfaces with a square pattern. Structures composed of large and small colloidal particles in this case can possess LS or LS_5 stoichiometries, as presented earlier in Figure 13(a) and (b). Further growth of large colloidal particles on top of the layer of small particles is more complex. The quality of colloidal layers with large colloidal particles is highly dependent on the homogeneity in surface coverage of underlying small colloidal particles.

4.1.2.9
Layer-by-Layer Colloidal Deposition: Colloidal Crystals with NaCl Structure

Colloidal crystals possessing a NaCl structure (LS stoichiometry) can be fabricated by layer-by-layer deposition of large and small colloidal particles on surfaces with a square pattern, where small colloidal particles take their places in between the large ones, as shown in Figure 15. Large colloidal particles were deposited on an electrode surface in close-packed fashion ($D_L/\Lambda = 1$). The ratio of colloidal radii was $R_S/R_L = 0.41$, corresponding to the smallest possible value for the NaCl-type structure, where colloidal particles are arranged in a close-packed manner. Deposition parameters were determined for each type of colloidal particle according to the approach presented earlier. Growth of binary colloidal crystals can be started from one or two colloidal layers of large colloidal particles. Starting the growth of such colloidal crystals with two layers of large colloidal particles has an advantage. Small colloidal particles position themselves precisely among the particles, which decreases the ability of large colloidal particles to rearrange themselves during subsequent depositions. Small colloidal particles deposited onto a monolayer of large colloidal particles did not have any influence in our case on subsequent depositions. Negatively charged large colloidal particles with a diameter of 545 nm were deposited at 3.9 V during 45 s and withdrawn from the colloidal suspension with a colloidal volume fraction of 0.2 vol.% at a speed of 0.1 mm/s (in the case of monolayer formation, Figure 15a) and at 3.5 V during 3 min and a withdrawal speed of 0.04 mm/s in the case of two colloidal layers (Figure 15b). Electrodes were left to dry for several minutes prior to deposition of small colloidal particles. Then, electrodes with deposited large colloidal particles were immersed in a colloidal suspension of small colloidal particles at a speed of 0.97 mm/s under a preset applied potential. Negatively charged small colloidal particles were deposited at 3 V during 30 s (in the case of one layer of large colloidal particles) and 45 s (in the case of two layers), and withdrawn with an electrode speed of 0.1 mm/s from the colloidal suspension with a colloid volume fraction of 0.2 vol.%. The next layer of large colloidal particles was deposited at 3.5 V during 2 min and employing a withdrawal speed of 0.04 mm/s (Figure 15c).

The growth of good-quality binary colloidal crystals using the layer-by-layer deposition method under an applied electric field is very sensitive to the quality of the colloidal monolayers formed in each deposition step. Various defect structures in deposited colloidal layers are indicated by white arrows in Figure 15. Defects in initial layers of large colloidal particles are usually introduced by missing colloidal

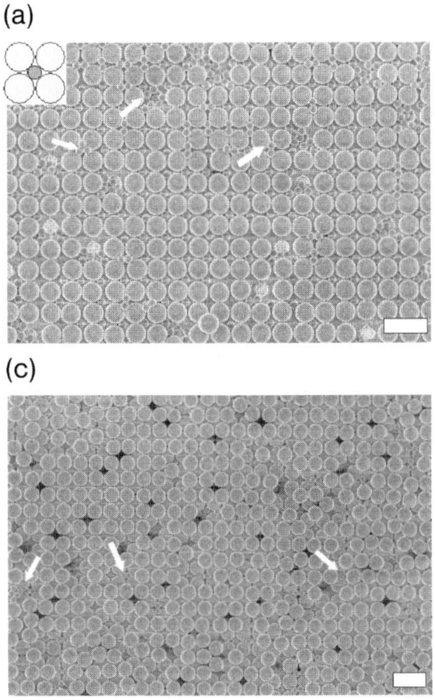

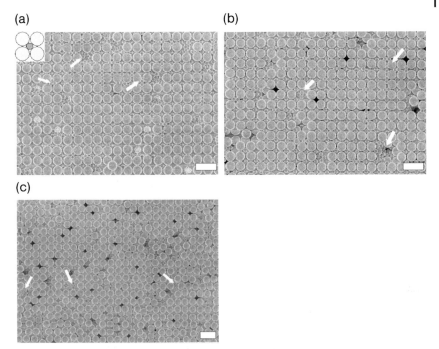

Figure 15 Layer-by-layer colloidal deposition on a square patterned substrate with a radii ratio of $R_S/R_L = 0.41$: (a) colloidal monolayer, (b) and (c) two layers. In the corner of the first image, a schematic representation of corresponding layer structures is shown. Scale bar = 1 μm. Arrows indicate defects in colloidal structures.

particles. Partial coverage by small colloidal particles of such defects occurs, destroying the ordering of subsequently deposited colloidal layers. The homogeneity of deposited small colloidal particles plays an important role in determining the quality of subsequently deposited colloidal layers. Although the colloidal suspensions used were homogeneous, the local concentration of colloidal particles in suspension may vary. The local colloidal concentration of small colloidal particles close to an electrode surface influences their distribution in the layer of large colloidal particles. In addition to single small colloidal particles occupying vacancies between large colloidal particles, double, triple and quadruple aggregates of small colloidal particles as well as no particles at all were observed on colloidal layers. As a consequence of the existence of such diverse colloidal aggregates, displacement of large colloidal particles from their original positions takes place, Figure 15(c). When such defect structures are eliminated, large-area binary colloidal crystals with a NaCl structure can be grown.

4.1.3
Conclusions

The method of 'electrophoretic deposition of charged polymer colloids on patterned substrates' allows an independent control of colloidal crystal structure, orientation and crystal thickness. Patterned surfaces predetermine the location of colloidal particles on the surface and an electric field acts as a driving force in the colloidal crystallization process. The technique was successfully applied to grow single and binary colloidal monolayers, colloidal crystals with FCC crystal structures and (111), (100) and (110) plane orientations, BCC colloidal crystals grown from a (100) crystal plane, nonclose-packed colloidal multilayers with a HCP crystal sequence and binary colloidal crystals with NaCl and AlB_2 analogs. Planar defects were easily introduced and controlled by the set of deposition parameters.

The method of electrophoretic deposition of colloidal particles onto patterned electrode substrates was first compared with the deposition on flat electrode surfaces. Deposition on flat electrode surfaces led to the formation of disordered colloidal monolayers and crystals, while patterned surfaces predetermined the location of colloidal particles on electrode surfaces by physical confinement. Deposition parameters such as electric-field strength, colloid concentration, surface charge density, withdrawal speed of electrodes, etc., were shown to play an important role in colloidal crystal growth. It was found that colloidal particles with low surface charge densities allowed a better control over colloidal crystallization. An optimal deposition voltage of 3.5 V was found for colloidal crystal growth. At higher deposition voltages, more disorder in colloidal crystals occurred and at lower voltages the deposition process was ineffective. The withdrawal speed of electrodes from the colloidal suspension plays an important role in determining the thickness and growth of colloidal crystals with nonclose-packed structures. For different crystal structures, the withdrawal speed was optimized for the systems considered.

It was possible to control the growth of colloidal crystals with FCC and BCC structures with this technique. Colloidal crystals were deposited on electrode surfaces with various symmetries in order to induce colloidal crystal growth from different crystal planes. Colloidal crystals with FCC crystal structure grown in [111], [100] and [110] crystal directions and BCC in the [100] direction were presented. It was found that growth of FCC (111) and (100) could be easily controlled at low withdrawal speeds, while growth of FCC (110) colloidal crystals was limited to two colloidal layers. Thick colloidal crystals grown in the [110] crystal direction finally possessed small polycrystalline regions with (111), (100) and (110) crystal planes. Disorder and polycrystallinity were caused by the geometrical factor of this plane, which is not close-packed in one of the main (perpendicular) directions. Growth of high-quality close-packed FCC colloidal crystals was determined by the precise match of crystal lattice constant and electrode pattern periodicity. The quality of the FCC crystals can be further improved by the exact match of the pattern periodicity and the colloidal particle size. Fabrication of colloidal crystals with a BCC structure from the (100) crystal plane was shown to be possible by applying this deposition technique. To grow a BCC structure was easier than to grow the closed-packed FCC structures due to the higher ratio between the pattern periodicity and the colloidal particle size, where the effect of the electrode surface imperfections and the colloidal-size polydispersity does not play a determining role. Optimal deposition parameters were found for the systems investigated, allowing us to grow up to seven layers of the BCC crystal structure.

Layer-by-layer electrophoretic deposition of colloidal particles with mono- and bimodal distributions on electrode surfaces with periodic patterns was successfully applied. Colloidal binary monolayers with LS, LS_2, LS_3, LS_4 and LS_5 stoichiometries were formed by this method on electrode surfaces with square and hexagonal patterns. A decrease in D/Λ ratio from 1 to 0.75 led to the formation of nonclose-packed binary colloidal monolayers. Layer-by-layer deposition of large and small colloidal particles with radii ratios of 0.41 and 0.49 on square and hexagonal patterns, respectively, led to the formation of binary colloidal crystals with atomic analogs as NaCl and AlB_2, respectively. Deposition of binary colloidal particles of the same charge sign (for example, only negatively charged colloidal particles) was easier to control than deposition of oppositely charged colloidal particles. The number of deposited positively charged colloidal particles was always higher than expected based on colloidal concentration, applied voltage and deposition time. Planar defects could be introduced in colloidal crystals in the form of deposited mono- (multi-) layers of colloidal particles with a colloid diameter different from the colloidal particles forming the main crystal. Growth of nonclose-packed colloidal crystals formed by colloidal particles of the same size and charge was successfully performed. Formation of nonclose-packed colloidal crystals with an A–B–A sequence (analogous to the HCP crystal structure) on hexagonally patterned surfaces with a D/Λ ratio of 0.75 was demonstrated. In addition, the growth of FCC colloidal crystals with the (110) plane parallel to the electrode surface was improved due to a stepwise construction of the crystal. Layer-by-layer electrophoretic deposition of colloidal particles on patterned electrode surfaces is a powerful method for the fabrication of nonclose-packed and binary colloidal crystals with various crystal structures.

The method of the electrophoretic deposition of charged colloidal particles on patterned electrode surfaces proves to be a versatile technique for controlling monolayer formation and colloidal crystal growth with desired structure and orientation in comparison with other available deposition techniques. In order to get better results (thicker crystals, crystal coverage) one has to improve the quality of the electrode surface patterns. The used technique (LIL) for surface patterning introduced imperfections into the pattern shape that, during the colloidal deposition, could give a small displacement of the deposited colloids on the electrode surface. Although the pattern quality of the current electrode surfaces was not very good, excellent deposition results were obtained.

Acknowledgments

The authors gratefully thank Henk A. G. M. van Wolferen for technical assistance with the operation of the LIL setup, Mark Smithers for his help with SEM imaging, and Prof. Laurens (Kobus) Kuipers for the many useful discussions. Funding by the MESA$^+$ Institute for Nanotechnology by the University of Twente in the Photonic Materials Program is acknowledged.

References

1 Colvin, V.L. (2001) *MRS Bulletin*, **26**, 637.
2 Holtz, J.H. and Asher, S.A. (1997) *Nature*, **389**, 829.
3 Gourevich, I., Pham, H., Jonkman, J.E.N. and Kumacheva, E. (2004) *Chemistry of Materials*, **16**, 1472.
4 Hunter, R.J. (2001) *Foundations of Colloid Science*, Oxford University Press.
5 Gast, A.P. and Russel, W.B. (1998) *Physics Today*, **51**, 24.
6 Auer, S. and Frenkel, D. (2001) *Nature*, **409**, 1020.
7 Yethiraj, A. and van Blaaderen, A. (2003) *Nature*, **421**, 513.
8 Gasser, U., Weeks, E.R., Schofield, A., Pusey, P.N. and Weitz, D.A. (2001) *Science*, **292**, 258.
9 Habdas, P. and Weeks, E.R. (2002) *Current Opinion in Colloid & Interface Science*, , **7**, 196.
10 Xia, Y., Gates, B., Yin, Y. and Lu, Y. (2000) *Advanced Materials*, **12**, 693.
11 Wang, D. and Möhwald, H. (2004) *Journal of Materials Chemistry*, **14**, 459.
12 Norris, D.J., Arlinghaus, E.G., Meng, L., Heiny, R. and Scriven, L.E. (2004) *Advanced Materials*, **16**, 1393.
13 Lopez, C. (2003) *Advanced Materials*, **15**, 1679.
14 Dziomkina, N.V. and Vancso, G.J. (2005) *Soft Matter*, **1**, 265.
15 Joannopoulos, J.D., Meade, R.D. and Winn, J.N. (1995) *Photonic Crystals: Molding the Flow of Light*, Princeton University Press, Princeton, NJ.
16 Liu, J.-M. (2005) *Photonic Devices*, Cambridge University Press, Cambridge.
17 Advances in Polymer Science, 2002. *Polymers for Photonic Applications*, **158**. Springer-Verlag, Berlin, Heidelberg.
18 Yablonovitch, E. (1987) *Physical Review Letters*, **58**, 2059.
19 John, S. (1987) *Physical Review Letters*, **58**, 2486.
20 Tayeb, G., Gralak, B. and Enoch, S. (2003) *Optic & Photonics News*, **14**, 38.
21 Biro, L.P., Balint, Zs., Kertesz, K., Vertesy, Z., Mark, G.I., Horvath, Z.E., Balazs, J., Mehn, D., Kiricsi, I., Lousse, V. and Vigneron, J.-P. (2003) *Physical Review E*, **67**, 021907.
22 Busch, K. and John, S. (1998) *Physical Review E*, **58**, 3896.
23 Ho, K.M., Chan, C.T. and Soukoulis, C.M. (1990) *Physical Review Letters*, **65**, 3152.
24 Yablonovitch, E., Gmitter, T.J. and Leung, K.M. (1991) *Physical Review Letters*, **67**, 2295.
25 Ho, K.M., Chan, C.T., Soukoulis, C.M., Biswas, R. and Sigalas, M. (1994) *Solid State Communications*, **89**, 413.
26 Woodcock, L.V. (1997) *Nature*, **385**, 141.
27 Koenderink, A.F., Lagendijk, A. and Vos, W.L. (2005) *Physical Review B – Condensed Matter*, **72**, 153102.
28 Hayashi, S., Kobayashi, M., Kumamoto, Y., Suzuki, K., Suzuki, T. and Hirai, T. (1992) *Journal of Colloid and Interface Science*, **153**, 509.
29 Hulteen, J.C. and Van Duyne, R.P. (1995) *The Journal of Vacuum Science and Technology A*, **13**, 1553.
30 Caruso, F. (2000) *Chemistry – A European Journal*, **6**, 413.

31 Vos, W.L., Megens, M., van Kats, C.M. and Bösecke, P. (1996) *Journal of Physics – Condensed Matter*, **8**, 9503.
32 Mayoral, R., Requena, J., Moya, J.S., Lopez, C., Cintas, A., Miguez, H., Meseguer, F., Vazquez, L., Holgado, M. and Blanco, A. (1997) *Advanced Materials*, **9**, 257.
33 Miguez, H., Meseguer, F., Lopez, C., Mifsud, A., Moya, J.S. and Vazquez, L. (1997) *Langmuir*, **13**, 6009.
34 Bevan, M.A., Lewis, J.A., Braun, P.V. and Wiltzius, P. (2004) *Langmuir*, **20**, 7045.
35 Tien, J., Terfort, A. and Whitesides, G.M. (1997) *Langmuir*, **13**, 5349.
36 Chen, K.M., Jiang, X., Kimerling, L.C. and Hammond, P.T. (2000) *Langmuir*, **16**, 7825.
37 Lee, I., Zheng, H., Rubner, M.F. and Hammond, P.T. (2002) *Advanced Materials*, **14**, 572.
38 Aizenberg, J., Braun, P.V. and Wiltzius, P. (2000) *Physical Review Letters*, **84**, 2997.
39 Fustin, C.-A., Glasser, G., Spiess, H.W. and Jonas, U. (2004) *Langmuir*, **20**, 9114.
40 Masuda, Y., Itoh, T., Itoh, M. and Koumoto, K. (2004) *Langmuir*, **20**, 5588.
41 Dimitrov, A.S. and Nagayama, K. (1996) *Langmuir*, **12**, 1303.
42 Jiang, P., Bertone, J.F., Hwang, K.S. and Colvin, V.L. (1999) *Chemistry of Materials*, **11**, 2132.
43 Kralchevsky, P.A. and Denkov, N.D. (2001) *Current Opinion in Colloid & Interface Science*, **6**, 383.
44 Im, S.H., Kim, M.H. and Park, O.O. (2003) *Chemistry of Materials*, **15**, 1797.
45 McLachlan, M.A., Johnson, N.P., De La Rue, R.M. and McComb, D.W. (2004) *Journal of Materials Chemistry*, **14**, 144.
46 Kitaev, V. and Ozin, G.A. (2003) *Advanced Materials*, **15**, 75.
47 Denkov, N.D., Velev, O.D., Kralchevsky, P.A., Ivanov, I.B., Yoshimura, H. and Nagayama, K. (1992) *Langmuir*, **8**, 3183.
48 van Duffel, B., Ras, R.H.A., de Schryver, F.C. and Schoonheydt, R.A. (2001) *Journal of Materials Chemistry*, **11**, 3333.
49 Reculusa, S. and Ravaine, S. (2003) *Chemistry of Materials*, **15**, 598.
50 Im, S.H., Lim, Y.T., Suh, D.J. and Park, O.O. (2002) *Advanced Materials*, **14**, 1367.
51 Jiang, P. and McFarland, M.J. (2004) *Journal of the American Chemical Society*, **126**, 13778.
52 Park, S.-H., Gates, B. and Xia, Y. (1999) *Advanced Materials*, **11**, 462.
53 Amos, R.M., Rarity, J.G., Tapster, P.R., Shepherd, T.J. and Kitson, S.C. (2000) *Physical Review E*, **61**, 2929.
54 Vickreva, O., Kalinina, O. and Kumacheva, E. (2000) *Advanced Materials*, **12**, 110.
55 Park, S.H., Qin, D. and Xia, Y. (1998) *Advanced Materials*, **10**, 1028.
56 Vos, W.L., Megens, M., van Kats, C.M. and Bösecke, P. (1997) *Langmuir*, **13**, 6004.
57 Megens, M., van Kats, C.M., Bösecke, P. and Vos, W.L. (1997) *Langmuir*, **13**, 6120.
58 Velev, O.D. and Lenhoff, A.M. (2000) *Current Opinion in Colloid & Interface Science*, **5**, 56.
59 Pusey, P.N., van Wegen, W., Barlett, P., Ackerson, B.J., Rarity, J.G. and Underwood, S.M. (1989) *Physical Review Letters*, **63**, 2753.
60 Sasaki, M. and Hane, K. (1996) *Journal of Applied Physics*, **80**, 5427.
61 Denkov, N.D., Velev, O.D., Kralchevsky, P.A., Ivanov, I.B., Yoshimura, H. and Nagayama, K. (1993) *Nature*, **361**, 26.
62 Hartsuiker, A. and Vos, W.L. (2008) *Langmuir*, **24**, 4675.
63 Velikov, K.P., Christova, C.G., Dullens, R.P.A. and van Blaaderen, A. (2002) *Science*, **296**, 106.
64 Wang, D. and Möhwald, H. (2004) *Advanced Materials*, **16**, 244.
65 Rai-Choudhury, P. (ed.) (1997) *Handbook of Microlithography, Micromachining, and Microfabrication*, Vol. 1, Microlithography, SPIE Press, Washington.
66 Kuipers, S., van Wolferen, H.A.G.M., van Rijn, C., Nijdam, W., Krijnen, G. and Elwenspoek, M. (2001) *Journal of Micromechanics and Microengineering*, **11**, 33.
67 van Rijn, C.J.M., Veldhuis, G.J. and Kuipers, S. (1998) *Nanotechnology*, **9**, 343.

68 Berger, V., Gauthier-Lafayer, O. and Costard, E. (1997) *Electronics Letters*, **33**, 425.

69 Ulman, A. (1996) *Chemical Reviews*, **96**, 1533.

70 Delamarche, E., Schmid, H., Bietsch, A., Larsen, N.B., Rothuizen, H., Michel, B. and Biebuyck, H. (1998) *The Journal of Physical Chemistry. B*, **102**, 3324.

71 Xia, Y. and Whitesides, G.M. (1998) *Angewandte Chemie – International Edition*, **37**, 550.

72 Bertrand, P., Jonas, A., Laschewsky, A. and Legras, R. (2000) *Macromolecular Rapid Communications*, **21**, 319.

73 Sotomayor Torres, C. M. (ed.) (2003) *Alternative Lithography: Unleashing the Potentials of Nanotechnology*, Kluwer Academic, New York.

74 Yeh, S.-R., Seul, M. and Shraiman, B.I. (1997) *Nature*, **386**, 57.

75 Trau, M., Saville, D.A. and Aksay, I.A. (1997) *Langmuir*, **13**, 6375.

76 Zhang, K.-Q. and Liu, X.Y. (2004) *Nature*, **429**, 739.

77 Velev, O.D. and Bhatt, K.H. (2006) *Soft Matter*, **2**, 738.

78 Holgado, M., Garcia-Santamaria, F., Blanco, A., Ibisate, M., Cintas, A., Miguiz, H., Serna, C.J., Molpeceres, C., Requena, J., Mifsud, A., Meseguer, F. and Lopez, C. (1999) *Langmuir*, **15**, 4701.

79 Leunissen, M.E., Christova, C.G., Hynninen, A.-P., Royall, C.P., Campbell, A.I., Imhof, A., Dijkstra, M., van Roij, R. and van Blaaderen, A. (2005) *Nature*, **437**, 235.

80 Johnson, S.A., Ollivier, P.J. and Mallouk, T.E. (1999) *Science*, **283**, 963.

81 Gates, B., Yin, Y. and Xia, Y. (1999) *Chemistry of Materials*, **11**, 2827.

82 Bertone, J.F., Jiang, P., Hwang, K.S., Mittleman, D.M. and Colvin, V.L. (1999) *Physical Review Letters*, **83**, 300.

83 Kumacheva, E., Golding, R.K., Allard, M. and Sargent, E.H. (2002) *Advanced Materials*, **14**, 221.

84 Golding, R.G., Lewis, P.C., Allard, M., Sargent, E.H. and Kumacheva, E. (2004) *Langmuir*, **20**, 1414.

85 Allard, M., Sargent, E.H., Lewis, P.C. and Kumacheva, E. (2004) *Advanced Materials*, **16**, 1360.

86 Kumacheva, E., Garstecki, P., Wu, H. and Whitesides, G.M. (2003) *Physical Review Letters*, **91**, 128301.

87 Lin, K.-H., Crocker, J.C., Prasad, V., Schofield, A., Weitz, D.A., Lubensky, T.C. and Yodh, A.G. (2000) *Physical Review Letters*, **85**, 1770.

88 Ye, Y.-H., Badilescu, S., Truong, Vo-Van, Rochon, P. and Natansohn, A. (2001) *Applied Physics Letters*, **79**, 872.

89 Ozin, G.A. and Yang, S.M. (2001) *Advanced Functional Materials*, **11**, 95.

90 Xia, Y., Yin, Y., Lu, Y. and McLellan, J. (2003) *Advanced Functional Materials*, **13**, 907.

91 Yi, D.K., Seo, E.-M. and Kim, D.-Y. (2002) *Applied Physics Letters*, **80**, 225.

92 Hoogenboom, J.P., Rétif, C., de Bres, E., van de Boer, M., van Langen-Suurling, A.K., Romijn, J. and van Blaaderen, A. (2004) *Nano Letters*, **4**, 205.

93 Dziomkina, N.V., Hempenius, M.A. and Vancso, G.J. (2005) *Advanced Materials*, **17**, 237.

94 Venkatesh, S. and Jiang, P. (2007) *Langmuir*, **23**, 8231.

95 Zhao, W., Zheng, Y. and Low, H.Y. (2006) *Microelectronic Engineering*, **83**, 404.

96 Yin, Y. and Xia, Y. (2001) *Advanced Materials*, **13**, 267.

97 Yin, Y., Lu, Y., Gates, B. and Xia, Y. (2001) *Journal of the American Chemical Society*, **123**, 8718.

98 Sun, J., Li, Y., Dong, H., Zhan, P., Tang, C., Zhu, M. and Wang, Z. (2008) *Advanced Materials*, **20**, 123.

99 Varghese, B., Cheong, F.C., Sindhu, S., Yu, T., Lim, C.-T., Valiyaveettil, S. and Sow, C.-H. (2006) *Langmuir*, **22**, 8248.

100 van Bladeren, A., Ruel, R. and Wiltzius, P. (1997) *Nature*, **385**, 321.

101 Yang, S.M. and Ozin, G.A. (2000) *Chemical Communications*, 2507.

102 Kim, E., Xia, Y. and Whitesides, G.M. (1995) *Nature*, **376**, 581.

103 Kim, E., Xia, Y. and Whitesides, G.M. (1996) *Advanced Materials*, **8**, 245.

104 Yang, S.M., Miguez, H. and Ozin, G.A. (2002) *Advanced Functional Materials*, **12**, 425.

105 Miguez, H., Yang, S.M., Tetreault, N. and Ozin, G.A. (2002) *Advanced Materials*, **14**, 1805.

106 Ferrand, P., Egen, M., Griesebock, B., Ahopelto, J., Müller, M., Zentel, R., Romanov, S.G. and Sotomayor Torres, C.M. (2002) *Applied Physics Letters*, **81**, 2689.

107 Ferrand, P., Minty, M.J., Egen, M., Ahopelto, J., Zentel, R., Romanov, S.G. and Sotomayor Torres, C.M. (2003) *Nanotechnology*, **14**, 323.

108 Schaak, R.E., Cable, R.E., Leonard, B.M. and Norris, B.C. (2004) *Langmuir*, **20**, 7293.

109 Zhang, J., Alsayed, A., Lin, K.H., Sanyal, S., Zhang, F., Pao, W.-J., Balagurusamy, V.S.K., Heiney, P.A. and Yodh, A.G. (2002) *Applied Physics Letters*, **81**, 3176.

110 Yin, Y. and Xia, Y. (2002) *Advanced Materials*, **14**, 605.

111 Yin, Y., Li, Z.-Y. and Xia, Y. (2003) *Langmuir*, **19**, 622.

112 Matsuo, S., Fujine, T., Fukuda, K., Juodkazis, S. and Misawa, H. (2003) *Applied Physics Letters*, **82**, 4283.

113 Smith, H.I., Geis, M.W., Thompson, C.V. and Atwater, H.A. (1983) *Journal of Crystal Growth*, **63**, 527.

114 See for example: Freund, L.B. and Suresh, S. (2004) *Thin Film Materials: Stress, Defect Formation and Surface Evolution*, Cambridge University Press, Cambridge.

115 Hayward, R.C., Saville, D.A. and Aksay, I.A. (2000) *Nature*, **404**, 56.

116 Murray, M.J. and Sanders, J.V. (1978) *Nature*, **275**, 201.

117 Dinsmore, A.D., Yodh, A.G. and Pine, D.J. (1996) *Nature*, **383**, 243.

118 Hachisu, S. and Yoshimura, S. (1980) *Nature*, **283**, 188.

119 Eldridge, M.D., Madden, P.A. and Frenkel, D. (1993) *Nature*, **365**, 35.

120 Imhof, A. and Dhnot, J.K.G. (1995) *Physical Review Letters*, **75**, 1662.

121 Bartlett, P., Ottewill, R.H. and Pusey, P.N. (1992) *Physical Review Letters*, **68**, 3801.

122 Bartlett, P. and Pusey, P.N. (1993) *Physica A*, **194**, 415.

123 Hunt, N., Jardine, R. and Bartlett, P. (2000) *Physical Review E*, **62**, 900.

124 Kiely, C.J., Fink, J., Brust, M., Bethell, D. and Schiffrin, D.J. (1998) *Nature*, **396**, 444.

125 Redl, F.X., Cho, K.-S., Murray, C.B. and O'Brien, S. (2003) *Nature*, **423**, 968.

126 Zanchet, D., Moreno, M.S. and Ugarte, D. (1999) *Physical Review Letters*, **82**, 5277.

127 Shevchenko, E.V., Talapin, D.V., Rogach, A.L., Kornowski, A., Haase, M. and Weller, H. (2002) *Journal of the American Chemical Society*, **124**, 11480.

128 Dinsmore, A.D., Yodh, A.G. and Pine, D.J. (1995) *Physical Review E*, **52**, 4045.

129 Wang, J., Ahl, S., Li, Q., Kreiter, M., Neumann, T., Burkert, K., Knoll, W. and Jonas, U. (2008) *Journal of Materials Chemistry*, **18**, 981.

130 Garcia-Santamaria, F., Lopez, C., Meseguer, F., Lopez-Tejeira, F., Sanchez-Dehesa, J. and Miyazaki, H.T. (2001) *Applied Physics Letters*, **79**, 2309.

131 Garcia-Santamaria, F., Miyazaki, H.T., Urquia, A., Ibisate, M., Belmonte, M., Shinya, M., Meseguer, F. and Lopez, C. (2002) *Advanced Materials*, **14**, 1144.

132 Zhou, Z., Yan, Q., Li, Q. and Zhao, X.S. (2007) *Langmuir*, **23**, 1473.

4.2
Nanoparticles at the Interface: The Electrochemical and Optical Properties of Nanoparticles Assembled into 2D and 3D Structures at Planar Electrode Surfaces

Petra J. Cameron

4.2.1
Introduction

Quantum dots are colloidal nanocrystals made from semiconducting materials. Electron waves are confined within the small crystallites, leading to strong quantization of the energy levels. A QD does not contain conduction and valence bands as bulk semiconductors do, instead there are discrete levels well separated in energy (in the case of CdSe energy levels can be >100 meV apart) [5]. There is a bandgap that separates the highest occupied molecular orbital (HOMO) and the lowest unoccupied molecular orbital (LUMO); the bandgap separation is highly dependent on the size and shape of the QD. As a result the particle size and shape controls the wavelength of fluorescence emission, in general the smaller the nanocrystals the higher the emission energy. Figure 1 outlines the general trend exhibited by the majority of QDs, generally speaking the larger the diameter of the QD the longer the wavelength of the fluorescence emission. TEM images of CdSe QDs with a diameter of ~5.5 nm are shown in Figure 2. QDs made by a bottom-up approach have an

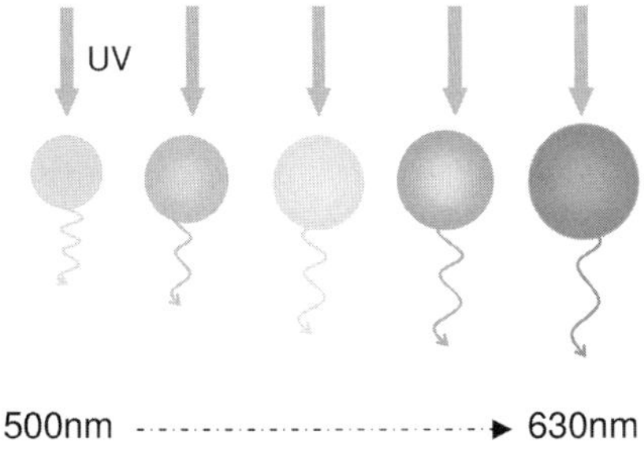

Figure 1 Depiction of the change in wavelength of fluorescence emission with change in size of quantum dots made from the same semiconductor material. The specific size required to emit at a given wavelength depends on the material the dot is made from (i.e. core-shell q-dot or alloyed q-dot), the ligands around the dot and the shape of the dot.

Surface Design: Applications in Bioscience and Nanotechnology
Edited by Renate Förch, Holger Schönherr, and A. Tobias A. Jenkins

ISBN: 978-3-527-40789-7

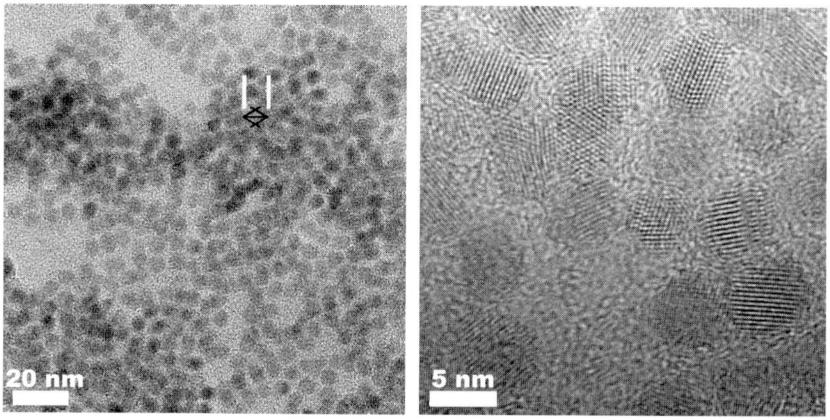

Figure 2 Transmission electron micrograph (TEM) images of CdSe quantum dots with mercaptopropionic acid capping group, single quantum dots are highlighted by the white lines.

outer shell of capping ligands that stabilize the colloidal nanoparticles and play an important role in controlling the size and shape of the particles during synthesis. The TEM images in Figure 2 show CdSe QDs with mercaptopropionic acid capping ligands, the role of the capping ligands is discussed in much more detail in Chapter 4.3.

Several different approaches have been used to model energy levels and optical transitions in quantum dots. The effective-mass approximation (EMA), the tight-binding approach, the empirical pseudopotential method, density functional theory (DFT) and the linear combination of atomic orbital approach have all been used to calculate energy levels for small crystalline spheres [6–12]. In one of the simplest techniques the electrons are assumed to be confined to a spherical, infinite, potential well and energy levels can be calculated based on solutions to the wave equation for a 'particle in a box' model [13]

The models agree that the electron orbitals in IIb-IV (e.g. cadmium chalcogenide) QDs show S, P, D (and so on) symmetry analogous to that found in atoms. As a result, QDs are frequently labeled 'artificial atoms' as in many ways their properties resemble those of organic macromolecules more closely than those bulk semiconductor materials.

The small size of a QD means that its dimensions approach the exciton Bohr radius. An exciton is a bound electron–hole pair formed after optical excitation of the QD. The Bohr radius is a measure of the separation between the electron and hole

$$a = \frac{\varepsilon \hbar^2}{\mu e^2} \tag{1}$$

where ε is the dielectric constant for the crystal, μ is the reduced mass of the electron and hole, e is the charge on an electron and $\hbar$ is Plank's constant (h) divided by 2π. Several of the models outlined above have been used to predict electronic and optical transitions in quantum dots with different sizes and different compositions [14]. The dielectric constant of the QD is an important parameter for accurate modeling and will be discussed in more detail below.

4.2.2
Choice of Substrate – 2D and 3D Arrays

QD synthesis is discussed in detail in Chapter 4.3; suffice to say that QDs can be synthesized in a controlled manner and a variety of functional groups can be introduced at the surface to aid specific attachment of the QDs to a variety of surfaces [14–27]. QDs have been successfully deposited on many different electrode materials, including gold, silver, platinum and tin-doped indium oxide (ITO). The electrode surface is commonly derivatized to increase QD adhesion. In the case of gold, silver and platinum electrodes, self-assembled monolayers of linear bifunctional molecules are formed at the surface. At one end of the linear molecule there is a thiol group that has a strong affinity to the metal surface, at the other end is a functional group that can interact with the QD. Di-thiols are often used as they also form bonds to the QD surface. Carboxylic-acid- or amine-terminated thiols are used to introduce electrostatic forces between the electrode surface and QD. In the case of ITO electrodes, silanization reactions, either in the solution or gas phase, can be used to form a thin silane layer on the ITO surface. In this way, functional groups that interact with the capping ligands on the QDs can be introduced to the ITO surface.

4.2.3
The Dielectric Environment

The surface area to volume ratio for QDs is much greater than in the bulk material; as a result the QD properties vary substantially from those of bulk semiconductors. Both the tight-binding approximation and the effective-mass approximation are frequently used to predict the optical properties of semiconductor nanocrystals [6, 12, 27, 28]. There is still controversy over what value of the dielectric constant should be input into these models to best predict the energy levels and optical transitions in QDs and that give the results that most closely mirror the experimental data. Some models use values for the static dielectric constant, $\varepsilon(0)$ (low-frequency limit of ε), and other models use values for the optical dielectric constant, $\varepsilon(\infty)$ (high-frequency limit of ε). In addition, a lack of experimental data for the dielectric constants of nanocrystals leads to values for the bulk crystalline material being used to describe the nanocrystalline material and hence the introduction of significant errors into the calculations. Semiempirical and *ab-initio* calculations of the dielectric constant of silicon quantum dots have suggested that the nanocrystals should display lower effective dielectric constants than the bulk material due to the influence of the particle surface on the dielectric response [29–31]. One model, developed by Haken, uses a combination of $\varepsilon(0)$ and $\varepsilon(\infty)$ to calculate a size-dependent dielectric constant, ε_{SD}, for semiconductor nanoparticles [6]. This model predicts a decrease in ε_{SD} with decreasing particle size.

In our work [32] we have used surface plasmon resonance measurements to extract an experimental value of the dielectric constant of mercaptopropionic-acid-capped CdSe QDs, albeit at a single wavelength ($\lambda = 632.8$ nm). Submonolayers of CdSe QDs were covalently attached to a self-assembled monolayer on a gold surface. The change in refractive index was monitored as the CdSe layer formed and effective medium theory was used to extract the dielectric constants for individual CdSe QDs. The calculated value, $\varepsilon = 3.4$, encompasses both the nanocrystalline centre and the thin layer of capping ligands. The dielectric constant was found to be considerably lower than literature values for bulk CdSe at this wavelength, which is consistent with the theoretical models outlined in the preceding paragraph.

The properties of QDs are also extremely sensitive to the surrounding dielectric environment. Modifying the nanoparticle by exchanging capping ligands or by surrounding the particle with a second material causes changes in the optical and electronic properties. It has been shown that replacing allylamine capping ligands with thiol capping ligands produced a significant redshift in the exciton emission band [33]. The photoluminescence spectrum of CdSe QDs has been shown to redshift when they are embedded in a ferroelectric $BaTiO_3$ matrix, an effect that the authors contributed to changes in the dielectric environment of the QDs [34].

4.2.4
Electrochemical Measurements on Quantum Dots

The separation between the highest occupied molecular orbital (HOMO) and the lowest unoccupied molecular orbital (LUMO) in a QD is generally much larger than the thermal energy (kT) so the as-synthesized QD exists in the ground state. Electrons can be added to or removed from the particle either electrochemically, photochemically or using strong oxidizing and reducing agents. In this section electrochemical and photochemical charging of QDs is discussed.

Cyclic voltammetry (in the dark) has been used to measure the conduction- and valence-band energies for quantum dots both in the solution phase and attached to an electrode surface [35–42]. The energy of the Fermi level (E_F) in the conducting electrode is controlled potentiostatically (Figure 4). As a negative voltage is applied (relative to the reference electrode) the Fermi level in the electrode increases in energy with respect to the energy levels in the QDs. If the Fermi level is raised above empty energy levels in the QD (e.g. the LUMO level) then electrons can be transferred from the electrode to the QD and a reduction current is measured. The voltage at which electron transfer occurs gives information about the relative position of the LUMO. If a positive voltage is applied then the Fermi level in the electrode is lowered with respect to the energy levels in the QD. When the

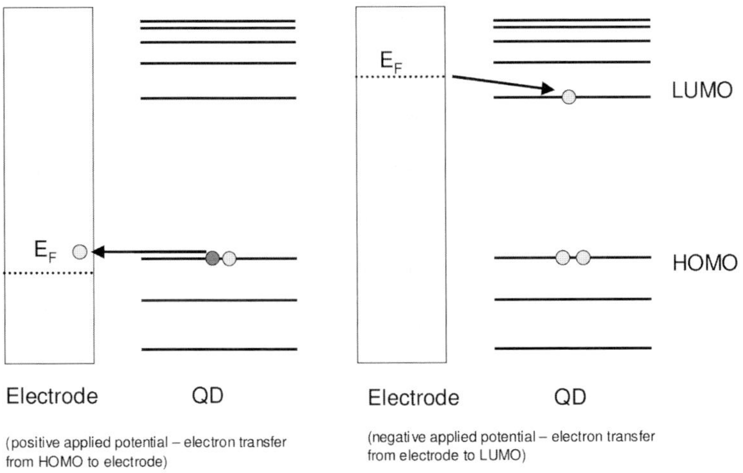

Figure 3 Representation of electrochemical oxidation (left) and reduction (right) of a QD at a conducting electrode surface.

Fermi level is lower in energy than the HOMO, electrons can be transferred from the QD to the electrode and an oxidation current is measured. In the same way, the voltage at which the oxidation current is measured gives information about the HOMO. Electron transfer between QDs and electrodes is represented in Figure 3. It has been shown that the redox peaks measured in the voltammogram vary systematically with nanoparticle size and the oxidation peak to reduction peak separation can be related to the HOMO–LUMO bandgap [38]; the picture is often complicated by extra peaks that are due to chemical degradation of the QDs or to electron injection into bandgap states, also called 'trap states'. Bandgap states are an important and often controversial feature of QDs, it is thought that surface defects can introduce additional electron states between the HOMO and LUMO.

Haram *et al.* [38] measured the cyclic voltammetry of thioglycerol-capped CdS nanoparticles dispersed in anhydrous di-methyl formamide (DMF) with tetrahexylammonium perchlorate (THAP) supporting electrolyte on gold and platinum electrodes. Clear oxidation and reduction peaks were seen that could be related to the optical bandgap and assigned to electron transfer from the HOMO and to the LUMO, respectively (peaks A1 and C1; Figure 4). It was found that the oxidation and reduction peak separation varied systematically with the QD size. The CdS QDs were reactive; multiple electrons could be injected from the electrode that led to particle degradation.

The electrochemical response of surface-attached PbS QDs was measured by Ogawa *et al.* [35] PbS QDs with diameters of 1.9, 2.5 and 4.2 nm and dioctyl sulfosuccinate capping ligands were attached to hexanedithiol self-assembled monolayers on gold electrodes. The background electrolyte was 0.5M Na_2SO_4 in water. Application of a positive potential led to oxidation and anodic dissolution of the

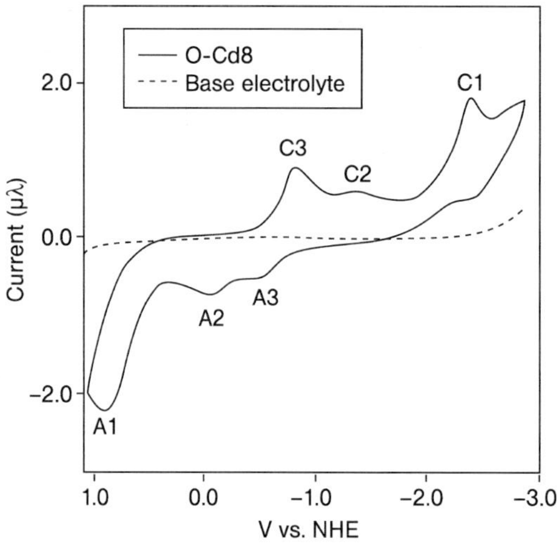

Figure 4 Cyclic voltammogram of thioglycolcapped CdS Q-particles (1 mg/mL of fraction IV) at a Pt electrode. Sweep rate 50 mV s^{-1} and [THAP] 0.05 M. The A1–C1 peak-to-peak separation gives information about the HOMO–LUMO separation. (Reproduced from [38]).

QDs with an oxidation peak centered on ±0.4 V. The QDs were reduced at potentials greater than −1.1 V versus SCE. The smaller the QD the more negative the reduction potential that was measured, suggesting that the reduction peak was related to injection of electrons from the electrode to the HOMO and hence the HOMO–LUMO separation could be estimated. Again, the QDs were found to be very reactive, cyclic voltammetry was not reproducible on second or subsequent scans due to the occurrence of irreversible redox processes.

Poznyak *et al.* [36] measured the aqueous electrochemical response of eight different sizes of CdTe QDs; half of which had 3-mercaptopropionic capping ligands and half had thioglycolicacid ligands. The QDs were attached to gold and ITO surfaces and cyclic voltammograms taken in aqueous buffer solution. The position of certain oxidation and reduction peaks were found to be specific to the size of the CdTe q-dots, and could be related to the optical bandgap. Evidence was found of interband surface states that shifted to more negative potentials with increasing nanoparticle size. Bandgap energies and the presence of trap states have also been measured by cyclic voltammetry of nanoparticle monolayers of PbS and CdTe at the air/water interface in a Langmuir trough [43].

Photoelectrochemical measurements are used to probe the electrical properties of optically excited QDs. Illumination with light of energy greater than the bandgap energy of the QD leads to the excitation of an electron from a ground state to an excited state (with rate k_1, see Figure 5). The excited electron can recombine directly with the hole, causing QD fluorescence (with rate k_2). It can recombine nonradiatively via bandgap states (not shown). When the QD is in proximity to an electrode, the excited electron can be transferred to the electrode surface (rate k_3).

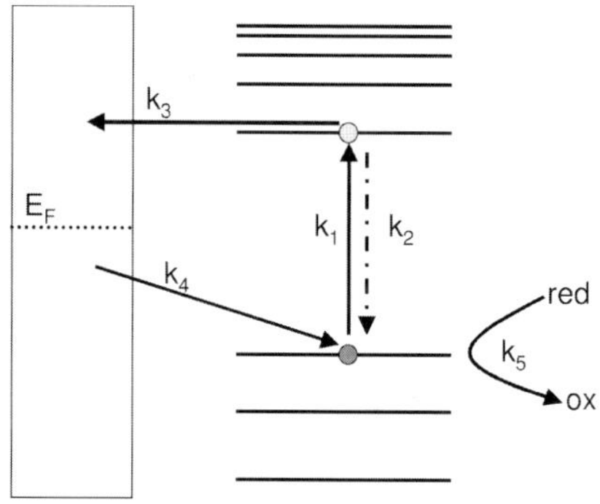

Figure 5 The different processes that can occur in a surface-attached quantum dot under illumination. Photoexcitation (rate k_1), direct radiative recombination of electron and hole (rate k_2), transfer of excited electron to the electrode (rate k_3), transfer of an electron from the electrode to the hole in the QD (if E_F is higher in energy than the HOMO, rate k_4), donation of an electron from an electron donor ('hole scavenger') in the electrolyte to the quantum dot (rate k_5).

If the Fermi level is at an energy above the HOMO then the donated electron can recombine directly with the hole (gold-mediated recombination, rate k_4).

The addition of a 'sacrificial hole scavenger' to the electrolyte increases the probability that permanent charge separation will occur. A hole scavenger is an electron donor that can rapidly transfer an electron to the hole in the QD (with rate k_5), if electron transfer from the hole scavenger occurs on a shorter timescale than the direct recombination of electron and hole, then the excited electron is free to move to the adjacent electrode and will be measured as a photocurrent in the external circuit (Figure 6).

Several electrochemical techniques, including intensity modulated photocurrent spectroscopy (IMPS) and scanning tunnelling microscopy (STM), have been used to measure tunnelling coefficients between QDs and metal surfaces [47–50]. In a series of elegant experiments Vanmaekelberg attached QDs to a variety of dithiol SAMs on gold. The group measured photoinducted electron transfer from CdSe Q-dots across unsaturated and saturated di-thiols ranging in length from 3–12 Å. During IMPS, a light source with sinusoidally varying intensity is used to excite a photocurrent in surface-attached QDs. The delay time between light of a given frequency exciting the QDs and a photocurrent at the same frequency being measured in the external circuit gives information about tunnelling coefficients. The rates of transfer between excited QDs and electrodes are of particular importance in solar cells that use QDs as light-harvesting particles [44–46]. After the adsorption of a photon by the QD and charge separation, the hole is quickly filled by electron donation from a liquid electrolyte or a hole conducting polymer in intimate contact with the QD. The electron can then be transferred to an electrode and used to do work in the external circuit.

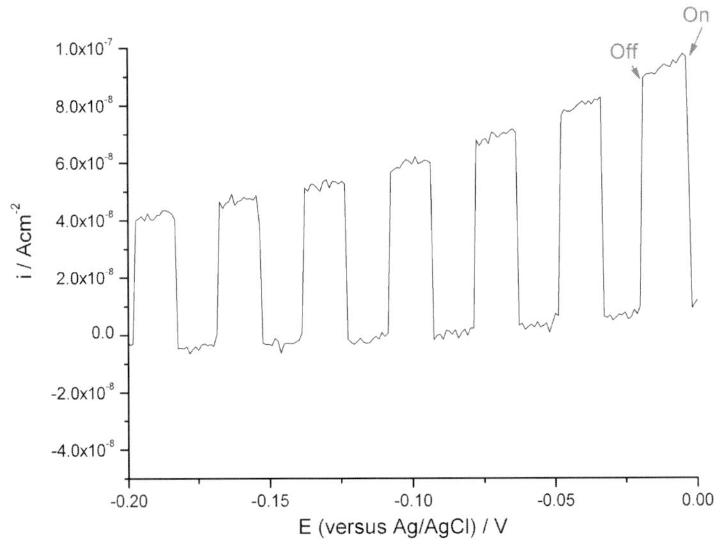

Figure 6 Response of CdZnSe alloy QDs with mercaptoundecanoic-acid capping ligands, covalently bound to a 11-amino-1-undecane-thiol self-assembled monolayer on the surface of a gold electrode, to chopped illumination from an LED at 470 nm. The increase in current under illumination is due to the injection of photoexcited electrons from the surface-bound quantum dots to the gold electrode. The measurement was carried out in a three-electrode mode (Pt counter electrode and Dri-Ref™ Ag/AgCl reference) in 0.1 M Na_2SO_3 electrolyte.

4.2.5 Optical Properties of Charged Quantum Dots

There is great interest in charged QDs as they can have substantially different optical properties compared to ground-state QDs. Electrochemistry has been successfully combined with optical spectroscopy to investigate optical changes in the QD properties upon injecting or removing electrons [51–60]. Changes in QD optical properties also occur when the QD is placed in a strong electric field or when the dielectric environment around the particle is changed [34, 61, 62]. Figure 7 shows a simplified band diagram for a quantum dot. Injecting an extra electron into the 1Se state, for example, can lead to suppression in the optical transitions in which that state is involved (e.g. the $1S_h$ to $1S_e$ transition). The injection of a second electron into the $1S_e$ state can suppress the bandgap transition even further.

Suppression of transitions upon injection of excess electrons can be observed experimentally by the disappearance of peaks in the absorption spectra of the QD. QDs show a very typical adsorption spectrum, with onset of absorption followed by a steady increase in extinction coefficient at shorter wavelengths (see Figure 8a). The second derivative of the absorption spectrum can show fine structure that has

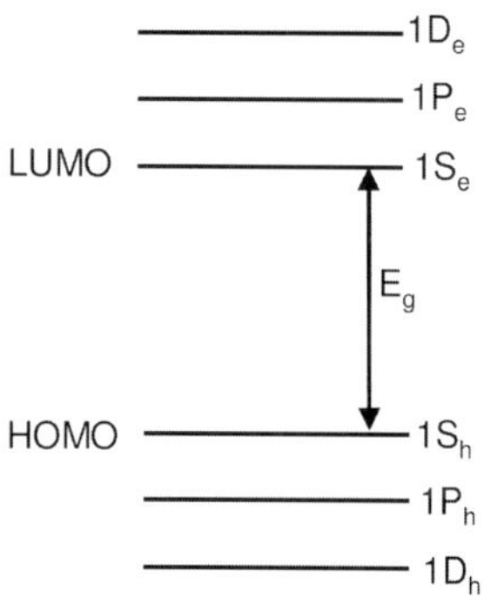

Figure 7 Simplified energy-level diagram for a QD.

been used to identify the individual optical transitions for QDs. Fluorescence emission is also commonly seen at wavelengths slightly longer than the wavelength of adsorption onset due to the radiative recombination of electrons in excited state with hole in 1Sh. This is called 'band-edge fluorescence'. The intensity of the fluorescence emission spectrum can also be changed by charging the QDs, injection of charge into the LUMO has been shown to suppress the fluorescence emission [55].

Wehrenberg *et al.* [57] measured the electrochemical charging of 7.2-nm diameter oleic-acid-capped PbSe quantum dots arranged as a 2D film on a platinum electrode. The platinum was first derivatized with 1,6-hexanedithiol, the QDs were drop cast onto the surface and 1,7-diaminoheptane was added to crosslink the QDs and make a more robust film. After heating and drying under vacuum, cyclic voltammograms of the modified electrode were measured in anhydrous acetonitrile with 0.1 M $LiClO_4$ background electrolyte. Electrochemistry and FT-IR spectroscopy were measured simultaneously. It was found that electrons could be electrochemically injected into the first excited state at negative potentials (> –0.4 V versus Ag wire) leading to a bleach in the interband optical transitions involving this state. At positive potentials electrons could be removed from the ground state and a resulting bleach in the IR absorption intensity of processes involving this state was seen.

Houtepen *et al.* [52] measured the differential capacitance of quantum-dot films on ITO and gold (Figure 8b and c). Pyridine capped CdSe QDs were drop coated onto ITO and 1,6-hexanedithiol-derivatized gold electrodes; the nanoparticles were crosslinked with 1,6-heptanediamine and the samples were annealed at 70 °C and dried under vacuum to remove solvent. Electrochemical measurements were carried out in a solution of 0.1 M $LiClO_4$ in anhydrous acetonitrile. Small steps in potential difference were applied and the resulting current was monitored. The integrated current, after correcting for the background charging current, gave the charge injected into the CdSe QD film after each 25-mV potential step. Negative potentials were applied that corresponded to injection of electrons into the CdSe network. Charge could be injected into the CdSe film below –0.45 V

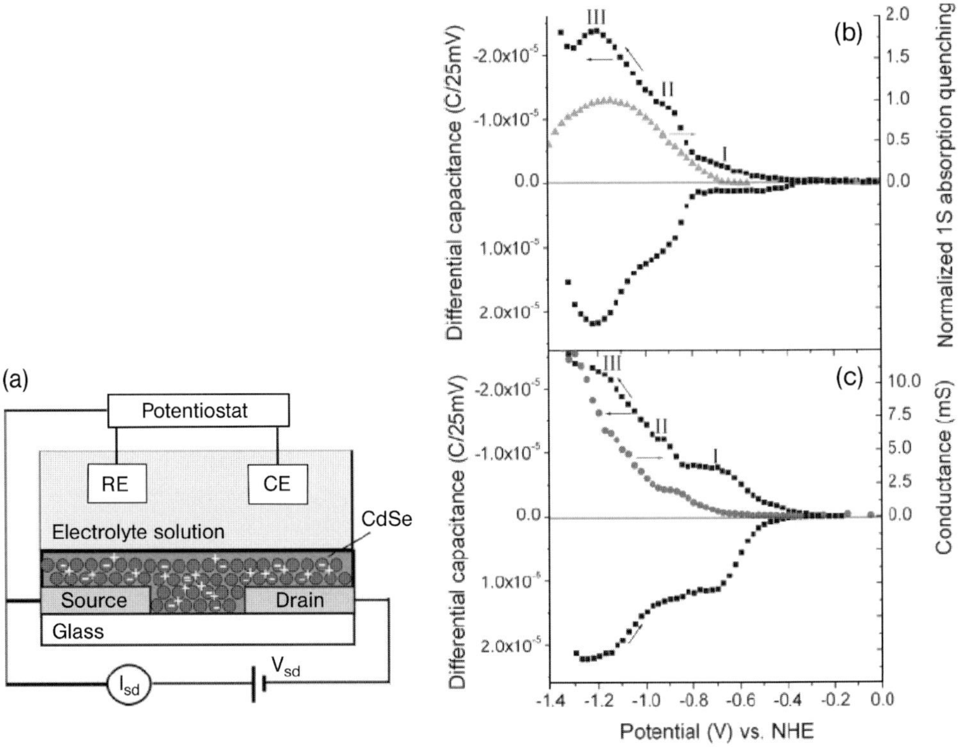

Figure 8 (a) Scheme of the experimental setup for the differential capacitance measurements shown in (b). (b and c) Differential capacitance measurements of a CdSe film. Three charging waves can be seen in (b), two of which lead to 1S absorption quenching. The first wave is thought to be charging of surface states in the CdSe QDs. (Reproduced from [51]).

versus NHE and three charging waves could be clearly identified. Interestingly, the 1S absorption quenching and increase in conductivity did not occur until the second charging wave; this strongly suggested that the first wave was due to electron injection into localized bandgap states/trap states and the second and third waves due to injection into the quantum states of the QD. Potential-dependent absorption spectra showed that injection of electrons into the assembly led to the quenching of some of the optical transitions and redshifts in transition energies. It was suggested that the redshift was caused by the Coulomb interaction between the injected electron and the exciton pair. Quenching of optical transitions occurred only when electrons were injected into conduction-band orbitals and blocked transitions involving these levels.

Several groups have shown that there is both visible absorption quenching and photoluminescence quenching from QDs when they are charged from an elec-

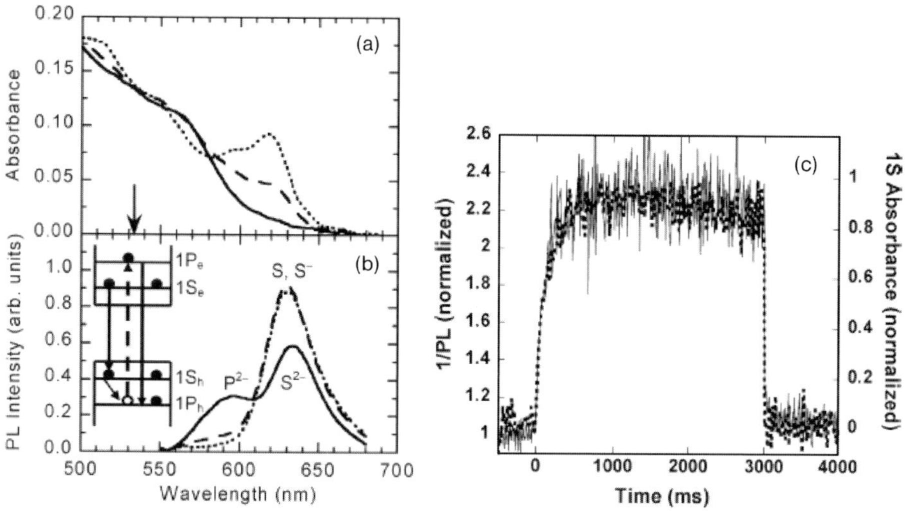

Figure 9 (a and b) Optical bleach and photoluminescence intensity of CdSe QD film at −70 °C, neutral QDs (dotted line), singly charged QDs (dashed line), doubly charged quantum dots (solid line). (Reproduced from [55]). (c) Optical bleach (dashed line) and photoluminescence quenching (solid line) of a CdSe/CdS core-shell nanoparticle film as the potential was stepped from 0 V to −0.6 V and back. (Reproduced from [58]).

trode [55, 56, 58]. In one set of experiments Wang *et al.* [55] deposited a film of ~6.8-nm CdSe QDs with TOPO caps on ITO and measured them in anhydrous N,N-dimethylformamide with 0.1 M tetrabutylammonium tetrafluoroborate background electrolyte. Changes in the photoluminescence intensity were measured as electrons were injected into and removed from the QDs electrochemically under negative bias. Charging the 1Se state led to changes in the visible absorption spectrum with an absorption bleach due to the presence of the injected electrons (Figure 9). Injecting an average of 1 electron per quantum dot did not lead to fluorescence quenching, injecting two electrons per quantum dot led to a strong quenching of the S exciton emission and the appearance of a new peak that was assigned to recombination of an electron in the 1Pe state and a hole in a lower level. It was found that there was a clear difference in kinetics between charging/discharging the 1Se state and fluorescence quenching.

In subsequent experiments Jha *et al.* [58] carried out similar measurements on CdSe/CdS core-shell nanoparticle films. As before, the films were drop cast onto ITO electrodes and crosslinked with 1,7-heptanediamine. The measurements were carried out in anhydrous DMF with 0.1 M tetrabutylammonium perchlorate. In the experiments with CdSe films the kinetics of photoluminescence recovery were found to be slow and did not follow the removal of electrons from the 1Se quantum state. It was postulated that injecting electrons into the nanoparticles created 'quenching centers', e.g. surface redox sites that dominated fluorescence quenching and had to be oxidized by solution species before the photoluminescence could recover. In contrast, the photoluminescence response of the CdSe/CdS core-shell nanoparticles followed the optical absorption bleach closely and the photoluminescence could be switched on and off rapidly by charging and decharging the QDs (Figure 9c). The result was attributed to the removal of selenium from the surface that when reduced was thought to be the main cause of quenching sites.

Fluorescence emission of QDs can also be switched on and off by the application of an electric field [61, 62]. In one set of experiments the single-molecule fluorescence response of CdSe QDs was investigated. The QDs were positioned midway between the fingers of two interdigitated microelectrodes so that they would experience a uniform external electric field applied between the two electrodes. A series of single-dot emission spectra (with laser excitation at 514 nm, 10 K) were taken under an alternating electric field. The emission spectrum varied reversibly in both energy and intensity as a function of the applied field. Delocalized exciton states in the quantum levels of the QD should be highly polarizable under an electric field and hence contribute to a polarizable character of the excited state. In contrast, electrons localized in trap or surface states should have a strong dipole character. The study showed that the first excited state of CdSe had both polar and polarizable characteristics, indicating the presence of both localized and quantum-confined electrons.

In our work we have investigated photoluminescence quenching and recovery of CdSe quantum dots in an aqueous environment. CdSe quantum dots were tethered to a gold electrode and excited using the evanescent tail of a surface plasmon wave propagating in the gold electrode. The application of negative potentials led to quenching of the surface-enhanced fluorescence, the fluorescence recovered to its original level when the cell was returned to 0 V or open circuit.

4.2.6
The Interaction of QDs and Surface Plasmons

Surface plasmon resonance spectroscopy (SPR) can be used to follow the real-time assembly of QDs at metal surfaces [63–66]; an example of SPR curves taken during the formation of a monolayer of CdSe QDs on a gold electrode is shown in Figure 10. SPR is also a useful tool for measuring values of the dielectric constants of QDs as the method directly monitors changes in the dielectric constants of layers as they form. Surface plasmons and surface plasmon resonance spectroscopy are described in detail in elsewhere in this book. Briefly, a laser is used to excite surface plasmon modes at the interface between a metal and a dielectric medium. The resulting optical field decays exponentially both into the metal and the dielectric. A spectrum measures reflectivity as a function of coupling angle of the laser light; a minimum occurs when the incident light excites surface plasmons in the metal. The position of this minimum is very sensitive to changes in dielectric constant close to the metal surface, which means that it can be used to sense adsorbates. In our own work the covalent binding of mercapto-propionic-

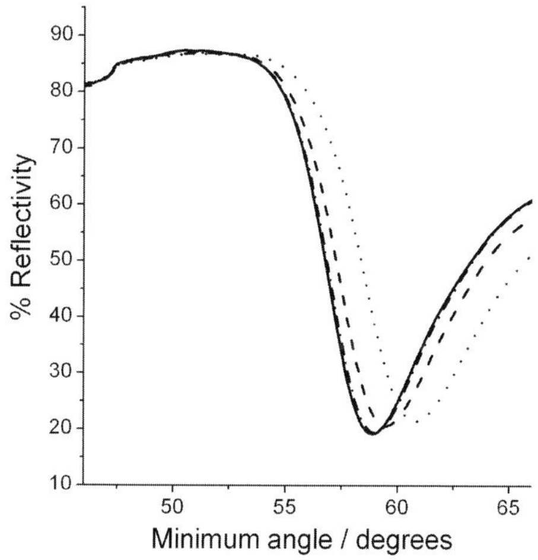

Figure 10 Shift in SPR minimum angle as CdSe is covalently bound to a gold electrode. Background (black line), 1% of a monolayer surface coverage (dash-dot line), 24% coverage (dashed line), 72% coverage (dotted line). The % coverage was calculated using a calibration curve that related coverage from AFM measurements to minimum angle shift.

capped CdSe quantum dots to amine-terminated gold surfaces was followed by surface plasmon resonance spectroscopy (SPR), surface coverage was monitored by measuring changes in the SPR curve at a fixed angle as a function of time. Binding kinetics were extracted by titrating in increasingly concentrated solutions of CdSe and fitting the adsorption curves using a model for simple bimolecular association at a surface [32]. Hutter *et al.* measured the electrostatic binding of CdS to self-assembled monolayers on gold and silver films by SPR [64]. The authors investigated the film structure by scanning electron microscopy and found that the QDs bound to gold surfaces modified with 2-aminoethanedithiol and 1,6-hexanedithiol retained their identity as discrete nanoparticles, whereas on silver the QDs amalgamated and large CdSe crystallite were formed.

In surface plasmon enhanced fluorescence spectroscopy (SPFS) chromophores are placed within the evanescent field close to the metal surface [67–70]. The chromophores can be excited by the surface plasmon field and the de-excitation pathway depends on the chromophore–metal distance. If the chromophore is close to the metal it can decay by Förster energy transfer to the metal. At intermediate distances the decay pathway is complex and can include back coupling of energy to the surface plasmon wave. At slightly larger distances (but still within the tail of the evanescent wave) the chromophore fluoresces. In SPFS the emitted photons are collected and counted, most commonly the signal is used to detect the presence of fluorescently labeled molecules binding at the metal surface. Quantum dots behave in a similar way to molecular chromophores [71–73]. If the absorption band edge of the quantum dot occurs at a longer wavelength than the surface plasmon excitation wavelength, then the surface plasmon field excites QD fluorescence. It has also been shown that the QD fluorescence can couple back to the surface plasmon in the metal layer. In the case of quantum dots close (few tens of nanometers) to an atomically flat gold surface the fluorescence is entirely quenched. If the gold is rough, either due to deliberate nanostructuring or as a function of the deposition method, then a photoluminescence enhancement of the QDs has been shown. Okamoto *et al.* [69] measured a 23-fold enhancement in photoluminescence efficiency from CdSe quantum dots directly deposited on evaporated gold films (as compared to CdSe on quartz). The as-deposited surface roughness was of the order of a few tens of nanometers and was sufficient to induce substantial photoluminescence enhancement without complex nanostructuring of the gold surface. Song *et al.* [72] measured a ~50-fold enhancement in fluorescence efficiency when core-shell CdSe/ZnS QDs were attached on the surface of a nanostructured array of silver columns. We have measured surface-enhanced fluorescence from CdZnSe alloyed QDs ~1.5 nm away from an evaporated gold film, a ten-fold increase in fluorescence relative to the background was measured after the QDs were attached to the surface.

In the author's own work SPFS has been used to probe electrochemically induced changes in the fluorescence emission of a film of CdZnSe-alloyed QDs covalently attached to a gold layer [73]. The gold substrate with the supported QD film acted as the working electrode, a Pt wire acted as the counter electrode and the cell was completed by an Ag/AgCl reference electrode. The electrolyte was 0.1 M Na_2SO_3. The aim of the experiments was to investigate the properties of QDs relevant to their use in biological systems, as a result all the experiments were done in an aqueous environment and no attempt was made to exclude oxygen. When a small electrochemical potential was applied to the underlying gold substrate, a quenching of the plasmon-enhanced fluorescence response from the QDs was observed. The background fluorescence, in contrast, was totally insensitive to the applied potential. The application of negative potentials led to a decrease in the surface plasmon enhanced fluorescence signal (Figure 11). When the cell was returned to 0 V (versus Ag/AgCl) or open circuit the fluorescence recovered over several hundred seconds. The changes in fluorescence intensity occurred at relatively small applied potentials, certainly at potentials far below the 1Se band edge. This suggests that the fluorescence changes are due to the creation of reduced quenching sites at the QD surface at relatively low applied potentials. Fluorescence recovery is likely to be due to reoxidation of the sites by electron-accepting species in the electrolyte.

Care must be exercised when interpreting the fluorescence response of optically excited QDs at surfaces. This is largely because the so-called 'blinking' of QDs is still a poorly understood topic [74]. When the fluorescence response of a single quantum dot is measured under continuous illumination it can be seen that the QD can exist in both 'on states' and 'off states'. In an 'on state' the QD fluoresces normally under illumination, in an 'off state' the QD is dark under illumination. The reasons for 'on' and 'off' states are currently being hotly debated. It has been suggested that in an 'off' state, electrons trapped either in surface states or band-

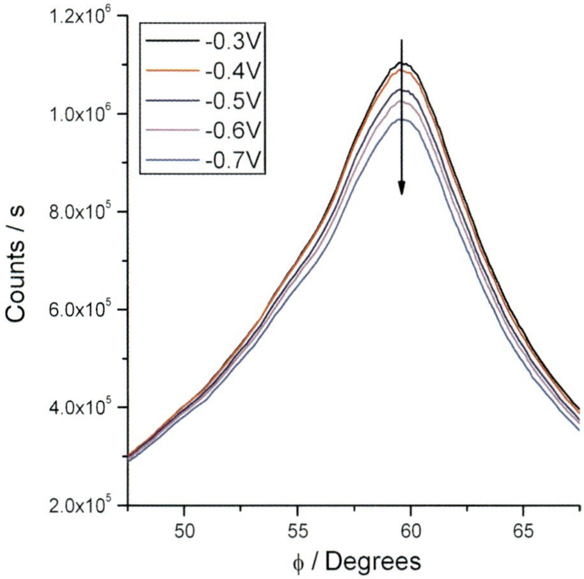

Figure 11 Quenching of surface-enhanced fluorescence from a film of CdZnSe QDs on a gold electrode as a negative bias is applied to the underlying gold substrate.

gap states block the further photoexcitation of electrons from the HOMO. It appears that the longer QDs are under illumination, the more probable it is that they will exist in an 'off state'. This leads to a steady reduction in fluorescence intensity with time that must be taken into account when QDs are being used as fluorescent markers in biosensing. Interpreting the kinetics of QDs binding to biological substrates must therefore be done with care [75, 76].

There are many biological applications where the unique photoelectric properties of the QDs can be exploited. Wilner *et al.* [77] attached a layer of gold nanoparticles to the surface of gold films; QDs were then coupled to the gold nanoparticles. Under illumination, electrons were excited in the QDs and transferred to the adjacent gold nanoparticles. Charging the gold nanoparticles led to a shift in the surface plasmon resonance response monitored in the gold substrate. The shift was thought to be due to changes in the localized plasmon waves in the gold nanoparticles upon charging, which in turn led to a change in coupling between the localized plasmon field and the continuous plasmon field in the gold substrate. The system was used as a sensor to monitor acetylcholine esterase inhibitors.

4.2.7 Outlook

Quantum dots have fascinating optical and electrochemical properties that are only now starting to be properly understood. QDs are regularly used as fluorescent labels for proteomics and genomics, allowing multicolor detection of cell components with a single excitation source [78]. Quantum-dot solids are being investigated for next-generation solid-state lasers that can potentially generate high-power low-noise optical pulses on the THz timescale [79]. QD solids are also being suggested as materials for LED-like semiconductor devices for quantum light generation [80]. QDs and QD films may have uses as circuit elements in photonic chips – next-generation computer chips that use light rather than electrons to store and transport information.

The effective use of QDs in any of these applications requires knowledge of their fundamental properties. Of particular importance is an understanding of how the QD changes when it is placed in a different dielectric environment; when it is placed in an electric field; when it is in close proximity to a metallic or semiconducting surface and when it is oxidized or reduced with electrons from the conducting surface. This chapter has focused on the changes induced in QDs when they are charged by a metallic electrode and highlights some of the current research into charged quantum dots, quantum-dot films and quantum dot solids.

References

1 Shipway, A.N., Katz, E. and Willner, I. (2000) *ChemPhysChem*, **1**, 18–52.
2 Katz, E., Willner, I. and Wang, J. (2004) *Electroanalysis*, **16**, 19–44.
3 Merkoçi, A., Aldavert, M., Marín, S. and Alegret, S. (2005) *Trends in Analytical Chemistry*, **24**, 341–349.
4 Girad, C. and Dujardin, E. (2006) *Journal of Optics A: Pure and Applied Optics*, **8**, S73–S86.
5 Vanmaelkelberg, D. and Liljeroth, P. (2005) *Chemical Society Reviews*, **34**, 299–312.
6 Nanda, K.K., Kruis, F.E., Fissan, H. and Behera, S.N. (2004) *Journal of Applied Physics*, **95**, 5035–5040.
7 Takagahara, T. (1993) *Physical Review B – Condensed Matter*, **47**, 4569–4583.
8 Baskoutas, S. and Terzis, A.F. (2006) *Journal of Applied Physics*, **99**, 013708/1–013708/4.
9 Diaz, J.G., Planelles, J., Bryant, G.W. and Aizpurua, J. (2004) *The Journal of Physical Chemistry B*, **108**, 17800–17804.
10 Harrison, W.A. (2002) *Solid State Communications*, **124**, 442–447.
11 Ramaniah, L.M. and Nair, S.V. (1993) *Physical Review B – Condensed Matter*, **47**, 7132–7139.
12 Delerue, C., Lannoo, M. and Allan, G. (2001) *Physica Status Solidi B – Basic Research*, **227**, 115–149.
13 Norris, D.J. and Bawendi, M.G. (1996) *Physical Review B – Condensed Matter*, **53**, 16338–16346.
14 Yoffe, A.D. (2001) *Advances in Physics*, **50**, 1–208.
15 Murray, C.B., Kagan, C.R. and Bawendi, M.G. (2000) *Annual Review of Materials Science*, **30**, 545–610.
16 Trindade, T., O'Brien, P. and Pickett, N.L. (2001) *Chemistry of Materials*, **13**, 3843–3859.

17 Yin, Y. and Alivisatos, A.P. (2005) *Nature*, **437**, 664–670.

18 Turner, E.A., Huang, Y. and Corrigan, J.F. (2005) *European Journal of Inorganic Chemistry*, 4465–4478.

19 Nobs, L., Buchegger, F., Gurny, R. and Allemann, E. (2004) *Journal of Pharmaceutical Sciences*, **93**, 1980–1992.

20 You, C., Verma, A. and Rotello, V.M. (2006) *Soft Matter*, **2**, 190–204.

21 Wilner, I., Baron, R. and Willner, B. (2007) *Biosensors & Bioelectronics*, **22**, 1841–1852.

22 Medintz, I.L., Tetsuo Uyeda, H., Goldman, E.R. and Mattoussi, H. (2005) *Nature Materials*, **4**, 435–446.

23 Merkoci, A., Aldavert, M., Marin, S. and Algret, S. (2005) *Trends in Analytical Chemistry*, **24**, 341–349.

24 Zin, M.T., Munro, A.M., Gungormus, M., Wong, N., Ma, H., Tamerler, C., Ginger, D.S., Sarikaya, M. and Jen, A.K.Y. (2005) *Journal of Materials Chemistry*, **17**, 866–872.

25 Shavel, A., Gaponik, N. and Eychmueller, A. (2005) *European Journal of Inorganic Chemistry*, 3613–3623.

26 Stellacci, F. (2005) *Nature Materials*, **4**, 113–114.

27 Wang, Y., Suna, A., Mahler, W. and Kasowski, R. (1987) *Journal of Chemical Physics*, **87**, 7315.

28 Albe, V., Jouanin, C. and Bertho, D. (1998) *Physical Review B – Condensed Matter*, **58**, 4713.

29 Wang, L.-W. and Zunger, A. (1994) *Physical Review Letters*, **73**, 1039.

30 Lannoo, M., Delerue, C. and Allan, G. (1995) *Physical Review Letters*, **74**, 3415.

31 Delerue, C. and Allan, G. (2006) *Applied Physics Letters*, **88**, 173117.

32 Cameron, P.J., Zhong, X. and Knoll, W. (2007) *The Journal of Physical Chemistry C*, **111**, 10313.

33 Koole, R., Luigjes, B., Tachiya, M., Pool, R., Vlugt, T.J.H., de Mello Donega, C., Meijerink, A. and Vanmaelkelberg, D. (2007) *The Journal of Physical Chemistry C*, **111**, 11208.

34 Zhou, J., Longtu, L., Gui, Z., Buddhudu, S. and Zhou, Y. (2000) *Applied Physics Letters*, **76**, 1540–1542.

35 Ogawa, S., Hu, K., Fan, F.F. and Bard, A. (1997) *The Journal of Physical Chemistry B*, **101**, 5707–5711.

36 Poznyak, S.K., Osipovich, N.P., Shavel, A., Talapin, D.V., Gao, M., Eychmueller, A. and Gaponik, N. (2005) *The Journal of Physical Chemistry B*, **109**, 1094–1100.

37 Chem, S., Truax, L.A. and Sommers, J.M. (2000) *Chemistry of Materials*, **12**, 3864–3870.

38 Haram, S.K., Quinn, B.M. and Bard, A.J. (2001) *Journal of the American Chemical Society*, **123**, 8860–8861.

39 Riley, D.J. (2002) *Current Opinion in Colloid & Interface Science*, **7**, 186–192.

40 Kuçur, E., Bueckling, W., Giernoth, R. and Nann, T. (2005) *The Journal of Physical Chemistry B*, **109**, 20355–20360.

41 Wang, D.Y.C., Wehrenberg, B.L. and Guyot-Sionnest, P. (2004) *Physical Review Letters*, **92**, 216802/1–216802/4.

42 Guyot-Sionnest, P. and Wang, C. (2003) *The Journal of Physical Chemistry B*, **107**, 7355–7359.

43 Greene, I.A., Wu, F., Zhang, J.Z. and Chen, S. (2003) *The Journal of Physical Chemistry B*, **107**, 5733–5739.

44 Gur, I., Fromer, N.A., Geier, M.L. and Alivisatos, A.P. (2005) *Science*, **310**, 462–465.

45 Nozik, A.J. (2002) *Physica E*, **14**, 115–120.

46 Kruger, J., Plass, R., Graetzel, M., Cameron, P.J. and Peter, L.M. (2003) *The Journal of Physical Chemistry B*, **107**, 7536–7539.

47 Bakkers, E.P.A.M., Hens, Z., Kouwenhoven, L.P., Gurevich, L. and Vanmaelkelberg, D. (2002) *Nanotechnology*, **13**, 258–262.

48 Bakkers, E.P.A.M., Kelly, J.J. and Vanmaelkelberg, D. (2000) *Journal of Electroanalytical Chemistry*, **482**, 48–55.

49 Bakkers, E.P.A.M., Marsman, A.W., Jenneskens, L.W. and Vanmaelkelberg, D. (2000) *Angewandte Chemie – International Edition*, **39**, 2297–2299.

50 Bakkers, E.P.A.M., Roest, A.L., Marsman, A.W., Jenneskens, L.W., de Jong-van Steensel, L.I., Kelly, J.J. and Vanmaelkelberg, D. (2000) *The Journal of Physical Chemistry B*, **104**, 7266–7272.

51 Houpten, A.J. and Vanmaelkelberg, D. (2005) *The Journal of Physical Chemistry B*, **109**, 19634–19642.
52 Shim, M. and Guyot-Sionnest, P. (2000) *Nature*, **407**, 981–983.
53 Wang, C., Shim, M. and Guyot-Sionnest, P. (2001) *Science*, **291**, 2390–2392.
54 Myung, N., Ding, Z. and Bard, A.J. (2002) *Nano Letters*, **2**, 1315–1319.
55 Wang, C., Wehrenberg, B.L., Woo, C.Y. and Guyot-Sionnest, P. (2004) *The Journal of Physical Chemistry B*, **108**, 9027–9031.
56 Shim, M., Wang, C. and Guyot-Sionnest, P. (2001) *The Journal of Physical Chemistry B*, **105**, 2369–2373.
57 Wehrenberg, B.L. and Guyot-Sionnest, P. (2003) *Journal of the American Chemical Society*, **125**, 7806–7807.
58 Jha, P.P. and Guyot-Sionnest, P. (2007) *The Journal of Physical Chemistry C*, **111**, 15440–15445.
59 Franceschetti, A. and Zunger, A. (2000) *Physical Review B – Condensed Matter*, **62**, 287–290.
60 Norris, D.J. and Bawendi, M.G. (1996) *Physical Review B – Condensed Matter*, **53**, 16338–16346.
61 Empedocles, S.A. and Bawendi, M.G. (1997) *Science*, **278**, 2114–2117.
62 Rothenberg, E., Kazes, M., Shaviv, E. and Banin, U. (2005) *Nano Letters*, **5**, 1581–1586.
63 Knoll, W. (1998) *Annual Review of Physical Chemistry*, **49**, 569–638.
64 Hutter, E., Fendler, J.H. and Roy, D. (2001) *Journal of Applied Physics*, **90**, 1977–1985.
65 Zayats, M., Kharitonov, A.B., Pogorelova, S.P., Lioubashevski, O., Katz, E. and Willner, I. (2003) *Journal of the American Chemical Society*, **125**, 16006–16014.
66 Brolo, A.G., Kwok, S.C., Cooper, M.D., Moffitt, M.G., Wang, C.W., Gordon, R., Riordon, J. and Kavanagh, K.L. (2006) *The Journal of Physical Chemistry B*, **110**, 8307–8313.
67 Ito, Y., Matsuda, K. and Kanemitsu, Y. (2007) *Physical Review B – Condensed Matter*, **75**, 033309/1–033309/4.
68 Neumann, T., Johansson, M., Kambhampati, D. and Knoll, W. (2002) *Advanced Functional Materials*, **12** (9), 575–586.
69 Okamoto, K., Vyawahare, S. and Scherer, A. (2006) *Journal of the Optical Society of America*, 1674–1678.
70 Komarala, V.K., Rakovich, Y.P., Bradley, A.L., Byrne, S.J., Gun'ko, Y.K., Gaponik, N. and Eychmueller, A. (2006) *Applied Physics Letters*, **89**, 253118/1–25118/2.
71 Gryczynski, I., Malicka, J., Jiang, W., Fischer, H., Chan, W.C.W., Gryczynski, Z., Grudzinski, W. and Lakowicz, J.R. (2005) *The Journal of Physical Chemistry B*, **109**, 1088–1089.
72 Song, J.-H., Atay, T., Shi, S., Urabe, H. and Nurmikko, A.V. (2005) *Nano Letters*, **5**, 1557–1561.
73 Knoll, W., Cameron, P., Caminade, A.M., Feng, C.L., Kim, D.H., Kreiter, M., Majoral, J.P., Müllen, K., Rocholz, H., Shumaker-Parry, J., Steinhart, M. and Zhong, X. (2007) *Proceedings of SPIE*, **6768**, 7680T.
74 Robelek, R., Stefani, F.D. and Knoll, W. (2006) *Physica Status Solidi A – Applications & Materials Science*, **203**, 3468–3475.
75 Stefani, F.D., Knoll, W., Kreiter, M., Zhong, X. and Han, M.Y. (2005) *Physical Review B – Condensed Matter*, **72**, 125304.
76 Stefani, F.D., Zhong, X., Knoll, W., Han, M.Y. and Kreiter, M. (2005) *New Journal of Physics*, **7**, 197.
77 Pardo-Yissar, V., Katz, E., Wasserman, J. and Willner, I. (2003) *Journal of the American Chemical Society*, **125** (3), 622–623.
78 Portney, N.G. and Ozkan, M. (2006) *Analytical and Bioanalytical Chemistry*, **384**, 620–630.
79 Rafailov, E.U., Cataluna, M.A. and Sibbett, W. (2007) *Nature Photonics*, **1**, 395–401.
80 Shields, A.J. (2007) *Nature Photonics*, **1**, 215–223.

4.3
Surface Engineering of Quantum Dots with Designer Ligands

Nikodem Tomczak, Dominik Jańczewski, Oya Tagit, Ming-Yong Han, and G. Julius Vancso

4.3.1
Introduction

Luminescence-based detection has traditionally played a dominant role in biosensing schemes and in studies of biologically relevant processes. This active field of research has a goal to visualize [1, 2], probe [3], and eventually steer *in-vivo* biological processes on the nanoscale, down to the single light-emitter level. Quantum dots (QDs) are nanocrystals made of semiconductor materials, which display size-dependent properties. As the size of the nanocrystals decreases to below a critical value, their energy bandgap becomes broader, and the energy levels near the band edges become discrete. Broad excitation spectra, narrow emission lines, and superior photostability make QDs an attractive choice as luminescence probes in biosensing, as compared to organic chromophores [4–7]. Additionally, the high chemical and photochemical stability, resulting from their high photobleaching thresholds, makes QDs a first choice as long-term markers and probes in biological applications [8–10].

Before considering applications of QDs as sensing elements, optical transducers in sensors or as nanoscale biological probes or tags [11], several requirements must be fulfilled regarding the QD dispersability in the surrounding medium. In addition, luminescence quantum yields (QY), colloidal, chemical and photochemical stability, and cytotoxicity of the complex QD–ligand assembly must be considered. The above listed requirements can be met by tailoring the ligands at the semiconductor nanoparticle surface. Design and chemical control of the QD surface provide quantum dots with functionality for dispersability in water, for molecular recognition or for environmental sensing. Introduction of surface reactivity by employing functionalized ligands is important for the coupling of QDs to surfaces and biomolecules. Molecular engineering of the periphery of the luminescent particles also enables dynamic control of the QD photophysical properties. As the QD surface influences its luminescence emission, proper surface passivation leads to high luminescence quantum yields by reducing the nonradiative recombination at the surface. This is usually accomplished by coating the nanocrystals with inorganic shells or by blocking the 'dangling' bonds of surface atoms.

Surface Design: Applications in Bioscience and Nanotechnology
Edited by Renate Förch, Holger Schönherr, and A. Tobias A. Jenkins

ISBN: 978-3-527-40789-7

The choice of the ligand will also determine the QD toxicity and shelf life. Therefore, the prospects for commercialization of QD-based materials will also be governed by the choice of the ligand.

In this contribution we will briefly review various aspects, including design principles and molecular engineering, of QD surfaces. We will show how the physicochemical and optoelectronic properties of QDs are modulated by the choice, chemistry, and electronic properties of the surface ligand. We will also address the issues related to QD solubility, dispersability in different solvents, and stability of the ligand shell. Strategies for surface functionalization with inorganic shells, small organic molecules, polymers, and biomacromolecules will be highlighted. As the literature for this subject is too large to be covered in a comprehensive fashion, we will limit ourselves mostly to CdSe nanocrystals synthesized with the method of Murray and coworkers [12]. These materials can be treated as prototypical QDs, and most of the findings can be applied to other types of QDs.

4.3.2 Historical Perspective

The investigation of the effects of the surface chemistry on the photochemistry of semiconductor particles has been necessitated by the early applications of such materials in photocatalysis [13–16]. The interest in this subject was extended by findings that the photoredox chemistry [16–18] and photophysical properties [19–21] of semiconductor nanocrystals varied as a function of the nanoparticle size. It was also found that the photophysical properties of the QDs might be influenced by the chemical composition of the QD surface [22]. For instance, it was shown that addition of amines [23] or creation of a layer of $Cd(OH)_2$ [24] on the surface of QDs influences the emission from the nanocrystals as a result of the passivation of surface sites, which may act as traps for photoexcited electrons or holes. In other cases, ferrocene derivatives [25], lanthanide complexes [26], aminocalixarenes [27] or polynucleotides [28] introduced onto the surface of QDs were shown to efficiently quench (or enhance) the QD luminescence depending on, e.g., the chemical substitution on the cyclopentadienyl rings of the ferrocenes [25]. This was thought to be due to a charge-transfer process involving the ligand and the photoexcited QD. Proper surface ligands were also necessary to stabilize the nanocrystals in solution and to control the nanoparticle size during the QD synthesis. It was thus very appealing to devise methods to simultaneously handle the stability of the nanoparticles in solution, their photophysical properties, and reactivity through proper surface derivatization.

The surface ligands were found to be relatively easily exchangeable, opening the way for the introduction of new ligands, which would also allow one to introduce new functionality [29–31] This allowed making of nanoparticles soluble in a broader range of solvents [12, 29, 31], or to be transferred from one solvent to another with markedly different polarity [32]. Soon, inorganic 'shells' were also grown on the initial QD 'core' [31, 33]. The presence of an additional inorganic shell made of semiconductor materials with a larger bandgap increased considerably the quantum yields of QDs by ~ 80%. Finally, with the introduction of QD synthesis from metallorganic precursors in the presence of high boiling point coordinating solvents, high-quality core-shell QDs coated with a surface ligand (usually trioctylphosphine oxide, TOPO) have become widely accessible [12]. The narrow and controllable size distribution of these nanocrystals and their well-defined luminescence properties allowed for progress in the understanding of the physical processes responsible for light emission. This was accompanied by systematic studies of the influence of the surface ligands on the electronic and optical properties of QDs.

Functionalization of the surface of semiconductor nanoparticles has opened the door to fascinating applications in the biomedical field. Two seminal papers, published in 1998 by Bruchez *et al.* and by Chan *et al.*, have demonstrated how upon proper surface chemistry the QDs can be used as luminescent probes in biological staining [2, 34, 35]. Accounts of the current state-of-the-art in QD applications in biomedical research can be found in a number of recent reviews [36–39]. Finally, surface functionalization, or lack thereof, was found to be very important in photovoltaic and electroluminescence applications, where charge transfer across QD/matrix interfaces influences the device performance [40–42]. These aspects, we believe, justify a review on molecular engineering of QD surface ligands.

4.3.3
Semiconductor Nanocrystals – Quantum Dots

Quantum dots are nanocrystals made of semiconductor materials with characteristic size in the range between 1 and 20 nm. As the size of the nanocrystals becomes smaller, QDs exhibit unique optical and electronic properties, which markedly differ from those of bulk materials.

Synthesis of high-quality samples of semiconductor nanocrystals has played a critical role in the field of QDs, firstly as precisely synthesized nanocrystals were needed to test predictions of many related theories, secondly because most of the promising applications of QDs were based on the high quality of these nanomaterials. In particular, monodispersity, surface functionality, and high QY have seen tremendous improvement during the last 20 years. Numerous methods have been developed for the preparation of semiconductor nanocrystals in the quantum-confinement regime, and related progress was summarized in a number of reviews [5, 43–50] and books [51, 52]. The work of Bawendi and coworkers [12] should be highlighted as it has opened the route to high-quality monodisperse QDs exhibiting relatively high luminescence QYs. The synthesis was based on high-temperature decomposition of organometallic precursors in the presence of a coordinating solvent. With some modifications of this method nanoparticles with polydispersity <5% and with quantum yields above 50% can now be routinely obtained. One crucial aspect of any nanoparticle synthesis is the stability of the nanocrystals during growth. The simplest way to stabilize the nanocrystals in the reaction

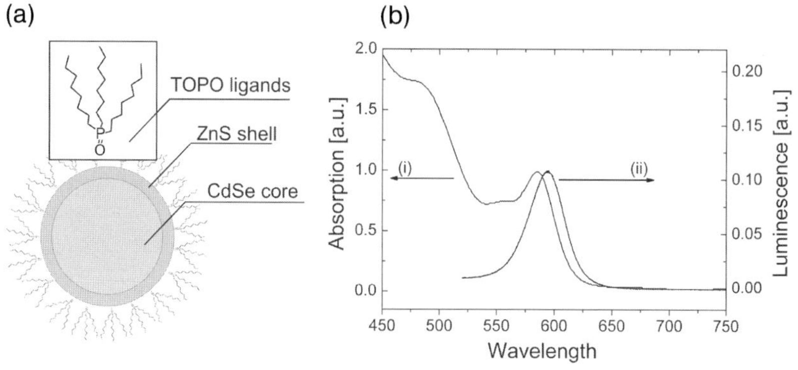

Figure 1 (a) Schematic representation of a CdSe-ZnS core shell QD with surface-bound TOPO ligands. (b) The unique feature of QDs is a combination of a broad absorption spectrum (i) with narrow emission lines (ii). The spectral position of the absorption and emission peaks can be tuned by controlling the size of the QDs.

mixture is to coat them with a shell of organic molecules. For this purpose, usually TOPO is used (Figure 1a). The surface coating also influences the ultimate size and shape of the nanocrystals [53, 54]. QDs can also be directly prepared in aqueous medium using functionalized thiols as stabilizers (e.g. 2-mercaptoethanol, mercaptopropionic acid, 1-thioglycerol or L-cysteine) [55–59], however, the nanocrystals obtained with these methods usually have lower QY than the QDs prepared by the 'TOPO method'.

Due to the unique effects confinement has on the electronic and optical properties, we briefly summarize the corresponding physics in the coming section. In bulk semiconductors the valence and conduction bands are separated by an energy gap (E_g) [60]. The magnitude of this bandgap is related to the fundamental materials properties. Depending on the material the values of E_g may vary from 0.3 (as in PbSe) to 3.8 eV (as in ZnS). The electrons in the valence band (VB) can be promoted to the conduction band (CB) by absorbing a quantum of energy ($h\nu$). The electron and the hole left behind in the VB may form a bound state, a so-called Mott–Wannier exciton. The Bohr radius (a_B), expressing the physical size of the formed exciton is given by:

$$a_B = \frac{\hbar^2 \varepsilon}{e^2}\left[\frac{1}{m_e} + \frac{1}{m_h}\right] \tag{1}$$

where ε is the dielectric constant, m_e and m_h are the effective masses of the electron and hole, respectively, and e is the elementary charge. When the size of the semiconductor nanoparticle approaches that of the exciton Bohr radius, the charge carriers in the QDs will experience a 'confinement' effect. This confinement results in a broadening of the energy gap between the VB and CB and in discretization of the energy levels near the band edges. In the early 1980s Brus developed a simple 'effective-mass approximation' model that relates the shift of the bandgap energy to the particle size [20, 43]:

$$\Delta E = \frac{\hbar^2 \pi^2}{2R^2}\left[\frac{1}{m_e} + \frac{1}{m_h}\right] - \frac{1.786 e^2}{\varepsilon R} + \text{smaller terms} \quad (2)$$

where R is the nanocrystal radius. The first term in (2), the quantum localization energy, shifts the bandgap energy state to higher energies. The second term in (2) is the 'Coulombic term', which shifts the first excited state to lower energies. In total, for small nanocrystals the quantum localization term dominates (as it is proportional to $1/R^2$) and the bandgap increases with decreasing size of the nanocrystals. This rather simple model gives only qualitative predictions for the bandgap shifts, but it nicely demonstrates the relation between the size of the QDs and the change in the bandgap energy. Equation (2) predicts well the bandgap broadening for larger crystallites. For smaller clusters with $R/a_B < 0.1$, some assumptions made for the parabolicity of the energy bands near the band edges, and for the values of the effective masses are no longer valid. Extensive presentation of different theories describing the size-related properties of QDs may be found in related books [52] and review articles.

To be able to understand the important factors affecting electronic and optical properties of QDs due to confinement, one must address the absorption, and more importantly the emission properties of QDs [51, 52, 61] (Figure 1b). Absorption of light by QDs is relatively well understood [62]. In direct-bandgap semiconductors (with conservation of the wave vectors for optical transitions) the oscillator strength of the optical transition increases with decreasing size of the QDs. In contrast to absorption, the luminescence mechanisms were much disputed [4, 63]. The luminescence emission originates from recombination of the hole and electron from the valence and conduction bands, respectively. For a long time the synthetic hurdles in obtaining highly monodisperse, high-quality samples of nanocrystals made interpretation of the luminescence behavior very difficult. In early studies of semiconductor nanocrystals the luminescence lifetime was usually long, in the range of microseconds, and the emission was strongly redshifted from the band edge. This behavior was widely attributed to surface effects and to the recombination of surface localized carriers. Although many features of the luminescence are now fairly well understood [64] a detailed picture of how surface ligands influence the luminescence is still lacking.

4.3.4 Surface Functionalization of Quantum Dots

Before describing the strategies for surface functionalization of QDs, we present in Scheme 1 some of the most important issues one would have to address and solve having a given application in mind. The requirements to be fulfilled include high optical performance, colloidal, chemical and photochemical stabilities in given solvents, long-term materials storage, functionality, biocompatibility, toxicity,

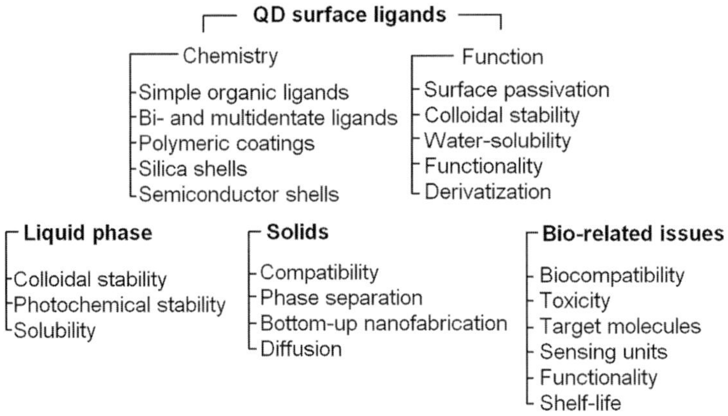

Scheme 1 Issues on the chemistry and function of QD surface ligands, and related topics of ligand-coated QDs in solution, in the solid state, and in bio-related applications.

and cost. However, although very tempting, it is virtually impossible to meet all requirements for a given QD system. For instance, it is rather straightforward that colloidal stability is interrelated with functionality. Although some general rules may be devised to achieve several requirements simultaneously, we must note that the details of such relationships are usually found by experience.

Various methods for surface functionalization exist. Inorganic shells can be grown directly onto the QD core to improve the luminescence and photostability of QDs. These shells can then be coated with specific ligands to stabilize the QDs in solution, or to supply functionality to the QDs. The stability of the ligand shell depends on the chemistry of the ligand and on the quality of its attachment to the QD surface. Introducing simple functional ligands onto the surface of QDs requires replacing the already existing ones, like TOPO. Monodentate, multidentate and finally polymeric ligands were used to improve the stability of the QD/ligand interface. Polymeric shells also improve biocompatibility, colloidal stability, reduce cytotoxicity, and offer the possibility to introduce multiple functionalities on the QD surface. Alternative methods based on hydrophobic–hydrophobic interactions were developed to avoid ligand-exchange reactions. In these methods the original TOPO shell is coated with an additional layer of molecules bearing usually hydrophobic alkyl chains. Such a method was most successfully applied to obtain polymer-coated QDs.

4.3.4.1 Passivation with Inorganic Shells

Most of the commonly used organic ligands are not able to entirely passivate the surface of the nanocrystals. This is due to their size, mobility, and packing ability. Their stability at a given interface is also important, as it will define the stability of the colloidal dispersions of the nanocrystals in a given solvent. 'Bare', as they are often called, QDs display relatively low QY, are highly unstable in solution, and also show besides the band-edge luminescence a pronounced redshifted emission originating presumably from the recombination from shallow and deep surface trap states. However, it was found that coating the nanocrystals with an additional layer (or 'shell' – such structures are therefore often called 'core-shell' structures) of an inorganic semiconducting material, which is able to passivate the surface of the 'core', is an excellent method to improve the properties of 'bare' QDs. A reduction of the exciton quenching at the QD surface by eliminating the energy levels inside the gap is observed for these core-shell structures [31, 33]. For example, covering CdSe QDs with ZnS [31, 65–67] or CdS [68] shells results in an increase of the QY by 50–70%. Numerous combinations of core-shell structures have been synthesized to date, and many reports describing the influence of the shell on the properties of QDs can be found in the literature [69–72].

The thickness of the inorganic shell can influence the QY of the QDs, and the highest QYs were achieved for shell thicknesses between 1 and 2 monolayers [67, 68]. Lower values of QY for thicker shells are believed to be related to the growing number of defects in the shell. The shell material is chosen such that the coating process is nearly epitaxial, and lattice mismatch between the core and the shell materials is small. However, it was also shown that if the coating is too perfect, the electron might delocalize over the additional shell thickness, suppressing the confinement effects. The bandgap of the second semiconductor, which makes the shell, can also be rationally chosen so as to achieve certain band alignment between the core and the shell [70, 73, 74]. This alignment results in structures that display different electronic and optical properties due to the different delocalization of electronic wave functions [75].

It should be remembered that the passivation of the surface with a second type of material invariantly changes the surface physicochemical properties. The affinity of the surface ligands to the surface atoms is therefore modified. For example, comparing to CdSe, the ZnS overcoat offers considerably 'harder' Lewis-basic surface sites for coordination of capping groups. Therefore the affinity of, e.g., TOPO is lower, and very often, harsh purification procedures lead to precipitation of the QDs. Thus, surface ligands, which are stronger Lewis bases [23] will result in better solubility of the QDs in a given organic solvent.

4.3.4.2
Encapsulation of Quantum Dots with Silica

The hydrophobic TOPO-coated QDs are not soluble in water. This prevents their use in biological applications. Coating the QDs with a silica shell is a simple method to render the QDs water soluble [34, 76–78]. Silica-coated QDs display higher colloidal stability in water, the thickness of the silica shell can be controlled, and surface functionalization (e.g. with amines, thiols, carboxylic acids, or phosphonic acid) for coupling to biomacromolecules can be easily introduced (Figure 2). Functionalized silica shells can be used to couple QDs to biomolecules directly or using bifunctional crosslinkers [34, 79, 80].

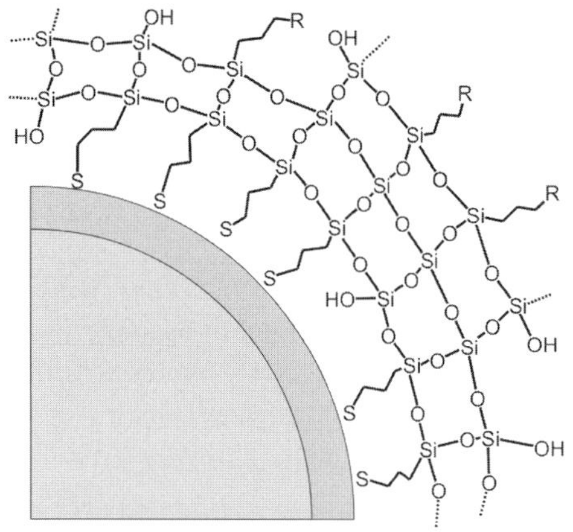

Figure 2 Silica shell formed on the surface of a QD. The shell renders the QDs soluble in water, protects against photo-oxidation, and reduces the toxicity of QDs.

Using the silica shell can solve some essential problems related to QD solubility, stability, and toxicity, while the procedures are relatively simple and cost effective. Functional silica coating of QDs give protection against photo-oxidation and acts as a barrier for diffusion of possibly toxic ions out of the nanocrystals into the environment, reducing therefore the QD toxicity. However, diffusion of chemical species through the silica layer to the QD surface may sometimes happen [78]. Functionalized silica shells also give stability in a wider range of buffer solutions and pH conditions. There are also some disadvantages of such methods of functionalization of QDs. During the synthesis QD crosslinking can occur, and one may also obtain large heterogeneity in size of the assemblies and in their surface charges.

To coat the QDs with silica, the QD surface is first functionalized with a primer silane-coupling agent (for example, 3-(mercaptopropyl) trimethoxysilane (MPS)) [34]. The role of the primer is to facilitate silicate deposition. The thiol functionality in the primer binds to the QD surface, and subsequently the shell is crosslinked, providing a siloxane coating. After the first layer of the silica shell is completed, the particle can be transferred to water or water/ethanol solution for further growth of the silica shell without particle coagulation. Thicker silica shells can then be grown using the classical Stöber process (base-catalyzed hydrolysis of tetraethoxysilane and subsequent condensation of the monomers onto the existing nuclei) [81]. Thicker and thinner [82] shells can be grown on the QDs depending on the QD type, the surface ligand, and the synthesis procedure. The thickness of the shell is usually in the range of 2–5 nm [82] but thicker shells can also be obtained [83]. Recently, a method to obtain single quantum dots in a silica sphere by microemulsion synthesis was reported. In comparison to the sol-gel approach described above, this method is simpler. However, the mechanism for the coating process is rather unclear at the moment [84].

4.3.4.3
Functionalization of Quantum Dots with Organic Ligands

The most commonly used surface ligands are low molar mass organic molecules possessing at least one functionality, which can bind efficiently to the nanocrystals' surface. The ligands can have a multiple number and type of functionalities. In such ligands, one (monodentate ligands) or more (multidentate ligands or polymeric ligands) functional groups bind to the QDs, while the remaining groups define the chemistry of the outermost part of the QDs, and therefore determine the QDs interfacial properties. Hydrophobic TOPO ligands prevent the QDs from being used in biological applications. Thus, ligand-exchange reactions from TOPO to a ligand with a functionality, which renders the QDs water soluble, have been actively pursued (Figure 3). These ligands of course should not have adverse effects on the photophysical properties of QDs.

For any type of ligand, the ligand binding strength, photochemical and chemical stability, surface coverage [64], and presence of charges, should be addressed. For example, for biological applications and coupling to biomacromolecules, the ligands should be bound to the surface of QDs such that they can withstand the common purification procedures and biochemical protocols. This also means solubility and stability in often complex biological buffers [85]. The presence of surface charges may render the QDs water soluble, and allow for functionalization via electrostatic interactions [86–88], but can also have adverse effects on the QD luminescence properties [89]. In general, it is known that adsorption of Lewis acids on the surface of CdSe quenches the band-edge luminescence, and adsorption of Lewis bases enhances the luminescence. The radiative recombination is not influenced by the ligand; the ligand modulates, however, the optical transitions from surface states. For instance, thiols influence the luminescence by hole

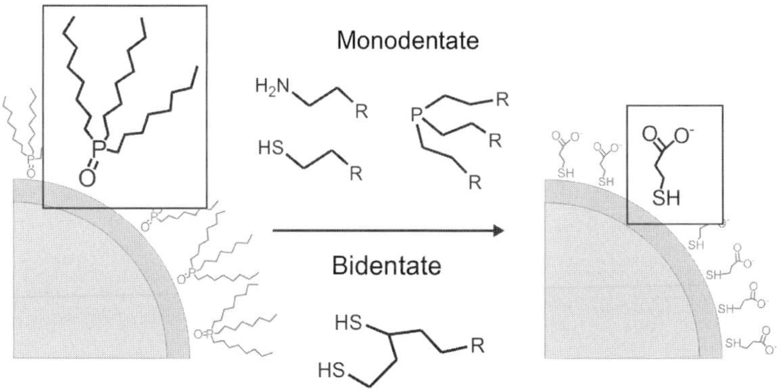

Figure 3 The TOPO ligands on the surface of the QD can be exchanged for a ligand-bearing functionality. Ligand-exchange reaction is a standard method to make the QD soluble in water and to provide functionality for coupling to, e.g., biomolecules. The most commonly used ligands are monodentate thiols, amines and phosphines. Bidentate or multidentate ligands improve the stability of the organic shell in solution.

trapping on the thiol molecule. This would explain different effects of thiols capping QDs with different energies of their valence bands with respect to the redox potential of the thiols [90].

Thiol-based ligands prevail in the QD literature to date, as ligand-exchange reactions from TOPO to thiols are easy to perform. Nevertheless, thiols bound to the surface of most semiconductors are relatively unstable and undergo catalytic photo-oxidation. Aldana *et al.* have conducted two systematic studies of the stability of hydrophilic thiols on the surface of CdSe as a function of their chemical structure and pH [91, 92]. When the QDs are illuminated with high-energy light, catalytic photo-oxidation of thiols to disulfides on the surface of QDs can occur (a process that is negligible in the dark). In this case the QDs act as photocatalysts. When the oxidized species become soluble, photo-oxidation would result in the loss of surface function. This would cause nanoparticle aggregation and precipitation. Simple monothiols were apparently more stable against photo-oxidation than dithiols or a bidentate ligand. Also, aromatic thiols showed lower stability compared to aliphatic thiols. Thiolates (deprotonated products of thiols) on the QD surface can be protonated by hydrogen ions competing with the surface cations if the pH decreases below a critical value. The thiols then detach from the surface and precipitation occurs. The dissociation pH was also reported to be QD size and composition dependent.

To render the QDs water soluble, mercaptoalcohols or mercaptoacids, like mercaptoacetic acid (MAA) [35] may be used (Figure 3). Hydroxylated QDs are obtained by exchanging the TOPO with e.g., dithiothreitol (DTT) [93]. The advantage of hydroxylated QDs is that hydroxyl groups have a reduced nonspecific binding to proteins or oligonucleotides. DTT-derivatized QDs have shown extended shelf life compared to QDs coated with MAA. Positively charged water-soluble QDs can be prepared with, e.g., 2-(dimethylamino)ethanethiol or thiocholine [94]. On some occasions it may be important to transfer water-soluble QDs to nonpolar organic solvents [95]. This is achieved by exchanging the original polar thiols by, e.g., dodecanethiol. The exchange is facilitated by addition of acetone to reduce the surface tension between water and dodecanethiol. Phase transfer can also be achieved by using specially tailored polymeric ligands [96, 97]. Thiol end-functionalized macromolecules [98–100] were also used to sterically stabilize the QDs and to fabricate complex multicomponent nanostructured materials. The size of the resulting QD/polymer assemblies is significantly increased, but the shell size is proportional to the molar mass of the polymer ligand [101].

As mentioned, thiol-based methods have been shown to be versatile and robust, however the stability of the ligands towards oxidation, and therefore the stability of the nanocrystals in solutions is still an issue. Some extensions of the thiol-based ligands have been reported; in particular, ligands based on carbodithioates [102, 103], and on *in-situ* formation of dithiocarbamates [104] have shown some promising results in this respect. In a series of papers Peng and coworkers have demonstrated that coating QDs with dendrons with functional groups in the focal point give simultaneous colloidal stability, functionality, and protection against oxidation [105, 106].

Another two popular groups of ligands are phosphines and amines. Alkyl phosphines were commonly used to passivate the surface of QDs, resembling their use in QD synthesis protocols. Oligomeric phosphine as a polydentate ligand was shown to better passivate the surface of QDs [107]. A number of specially tailored phosphine ligands were used for specific purposes. For instance, p-vinylbenzyl-di-n-octylphosphine oxide or p-bromobenzyl-di-n-octylphosphine oxides were used for polymerization from the QD surface [108–111], or tris(hydroxypropyl)phosphine was used to stabilize QDs in ethanol and to incorporate QDs into titania sol-gel matrices [112].

In the case of amines, very often exchange for pyridine is performed as an intermediate step in ligand-exchange reactions. Pyridine is a weakly coordinating molecule, which can be subsequently exchanged by other ligands, or stripped from the surface of QDs after, e.g. incorporation into polymeric matrices. A detailed comparison between the influences of several amine ligands on the QD properties was performed by Bullen and Mulvaney [113]. They compared various substituted amines with aliphatic chains. Primary amines displayed the best results, followed by secondary and tertiary amines. Amides and nitriles, as poor electron donors (poor Lewis bases), had no visible effect on the QD luminescence. Similarly, addition of $Y(phenyl)_3$ (Y = N,P,As,Sb) ligands had no visible effects. In the case of n-butylamine the solvent played some role, but this was not related to dissolution of oxygen in the solvent [114]. Polymeric compounds with amine functionality in the main or side chains were also used as ligands [115]. Pyridine end-functionalized polymers [116] or poly(amidoamine) dendrimers [117] were shown to effectively encapsulate the QDs.

Ligand-exchange reactions, although conceptually simple, are difficult to control, and often result in QD materials with lower QY. To avoid unwanted ligand exchange, and to retain the original QD properties, the TOPO aliphatic parts can be covered by hydrophobic parts of other molecules, forming therefore a secondary shell. Functionalization of TOPO-coated QDs via hydrophobic/hydrophobic interactions can be realized with small organic molecules as well as with specifically tailored amphiphilic polymers [118–122]. For example, cyclodextrines with hydrophobic pockets and hydroxyl functionalities were used to decorate the TOPO-coated QDs and transfer the QDs to water [123]. Also, phospholipids [124] and calixarene derivatives [125] were used to coat the QDs. Coating QDs with polymers via hydrophobic–hydrophobic interactions is a relatively easy and robust way of rendering the QDs water soluble, which also allows one to introduce multiple chemical functionalities at the QD surface [126].

Recently, we have developed a novel polymeric coating platform for functionalization and derivatization of hydrophobically coated nanoparticles (Figure 4) [127]. The polymer consisted of hydrophobic octyl side chains for interaction with TOPO, and hydrophilic carboxylic groups for water solubility. The polymer is obtained by opening an anhydride with suitable amines. TOPO-coated QDs could be subsequently dispersed in water by coating the QDs with the polymers. The unique feature of the anhydride approach lies in the ability to control the ratio between the octyl side chains, the number of carboxylic groups, and the number

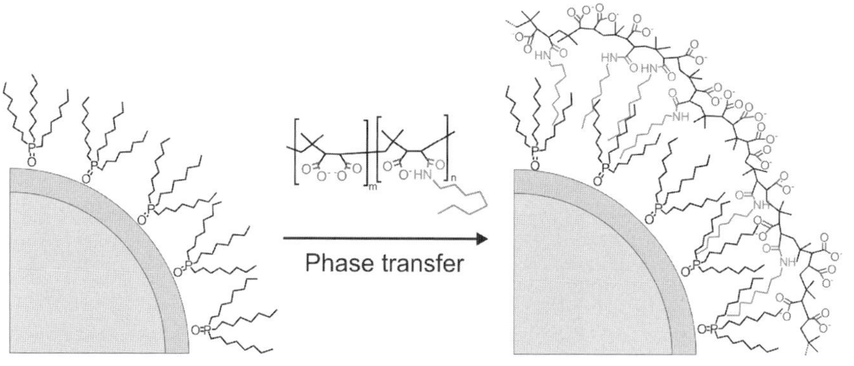

Figure 4 Hydrophobic TOPO-coated QDs can be dispersed in water by coating with specially tailored polymers bearing hydrophobic and hydrophilic units. The hydrophobic part (like, e.g., the octyl groups shown) interacts with TOPO via hydrophobic–hydrophobic interaction, while the hydrophilic part (e.g. carboxyl groups) render the QDs water soluble.

of functional units on the surface of QDs. By tuning this ratio one can find optimal conditions for solubilization of the QDs into water. The final dispersions of polymer-coated QDs are clear, transparent, and no aggregates are formed in solution [127]. The anhydride approach allows also for easy introduction of various chemical functionalities. In particular, we have functionalized QD with vinyl groups, PEG chains, acrylic esters [128], amines, and lanthanide complexes with kryptand units (Figure 5).

In summary, polymeric ligands have been shown to result in versatile and stable coatings of QDs, providing additionally the possibility to introduce various functionalities without the need for coupling agents [126–128].

4.3.5
Analysis and Characterization of QD Ligand Shells

It is inherently difficult to analyze 3D interfaces on the nanoscale. Despite the great importance of the QD surfaces, and of the structure and composition of the ligand zone in many applications, from optoelectronics through sensing to biology, there are relatively few studies regarding the exact structure, coverage and composition of the QD ligand shell. Even fewer studies explore the subject in detail. Regarding characterization, first, it is important to gather evidence that surface functionalization indeed occurred. There are also important questions regarding the surface coverage, accessibility, conformation, presence and amount of charges, and size of the assemblies.

An important tool to characterize surfaces and interfaces is the monitoring of the excited states, since the luminescence parameters report on the environment

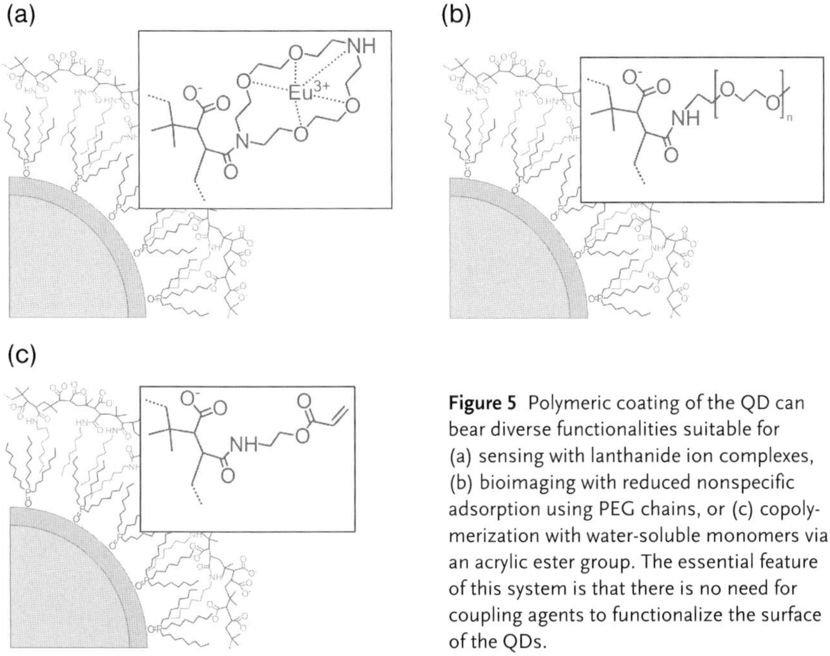

Figure 5 Polymeric coating of the QD can bear diverse functionalities suitable for (a) sensing with lanthanide ion complexes, (b) bioimaging with reduced nonspecific adsorption using PEG chains, or (c) copolymerization with water-soluble monomers via an acrylic ester group. The essential feature of this system is that there is no need for coupling agents to functionalize the surface of the QDs.

of the excited species, such as polarity, dielectric constant, or the presence of quenchers [129]. Steady-state and time-resolved luminescence spectra, or time-resolved fluorescence decay, can therefore give a wealth of information on the processes at the QD surface. For instance, the nature of the excited state can be probed and related to the presence of surface traps for charge carriers. Simple quenching experiments with a number of quenchers with different electron acceptor potentials can be used to detect these traps.

X-ray photoelectron spectroscopy (XPS) gave evidence for the formation of core-shell structures, e.g. in CdSe/ZnSe or CdSe/ZnS [33]. A detailed XPS study of CdSe nanocrystals was presented by Bowen-Katari [130]. XPS can also be applied to monitor ligand-exchange reactions by, e.g. following the P signals from TOPO. One can also use XPS to identify the bonds between the ligands and the surface, and to study the oxidation of, e.g., Se on the surface of QDs. When exposed to air Se oxidizes to SeO_2. These oxides are physisorbed on the surface of QDs but may evaporate in time.

For applications of QDs as cell markers, the knowledge of the physical size of the QDs is important. Transmission electron microscopy was used to estimate the size, and to assign a crystal structure of individual bare [131], or core-shell QD structures [132]. By proper staining of the organic shells, which are otherwise not visible on TEM images, these can also be visualized [118]. The hydrodynamic radius of polymer- or protein-coated QDs can be estimated by dynamic light scattering [133]. Determination of the hydrodynamic size of the polymer/QD assemblies was also reported by size exclusion chromatography (SEC) [101].

Nuclear magnetic resonance (NMR) was probably the most frequently used technique in the characterization of QD surfaces and ligand shells. This powerful technique enables chemical identification of the surface ligand shell [134–136], reports on local chemical environments [137] and allows for identification of the bonds between the ligands and the QD surface (for example the Se–Ph bond identified by ^{77}Se NMR) [29, 138]. The surface coverage and molecular conformation can also be deduced [139]. NMR as a characterization tool was important in ligand-exchange reactions [91], and in the monitoring of changes in the ligand shell, e.g. the photo-oxidation reactions of surface-bound thiolates and consequent dissociation [92].

NMR was often used in studies of surface structure and dynamics [134]. For example, one and two-dimensional ^{1}H NMR was used to characterize thiophenol ligands bound to the surface of CdS nanocrystals. Two-dimensional COSY (correlation spectroscopy) spectra were used to assign the chemical shifts and the identification of the *ortho, meta* and *para* protons of the thiophenol was performed. Comparing the chemical shift to the NMR spectra of model compounds it was deduced that the thiophenol binds to the Cd atoms on the surface of the QD. From the NMR experiments the surface coverage of the ligands was found to be rather low (7–20%) and it increased when the particle size decreased. In contrast, based on ^{1}H NMR, butanethiol was found to have much higher coverage and steric requirements were put forward as one of the reasons for such behavior [135].

^{31}P NMR revealed that all surface Cd atoms in CdSe QDs are passivated by TOPO [140]. Removal of TOPO and exchanging the ligand usually results in lower QY. Spin-echo experiments on ^{31}P showed that the average P–P distance matches the Cd–Cd distance on the surface of CdSe. This indicates that the ligands form a close-packed hydrophobic shell around the CdSe QDs. ^{1}H NMR spectra, however, gave evidence that the ligand-exchange reactions might not be complete, and some (10–15%) of the initial surface ligands can remain at the surface [64].

Many other techniques were used to characterize the QD ligand shell. For example, IR spectroscopy was used to probe the TOPO attachment to the CdSe surface [130, 141]. TOPO absorbs strongly at 1466 and 1146 cm^{-1}. The P=O stretching can be used as a useful probe for phosphine oxide molecules. Characteristic bands are shifted by 20 cm^{-1} to lower wave numbers upon attachment of TOPO to the surface of CdSe. Gel electrophoresis experiments can be used to investigate surface-charge properties [82]. X-ray absorption near-edge spectroscopy revealed structural disorder on the surface of the nanocrystallites [142]. Transformations on the surface of CdSe/ZnS nanocrystals at higher temperatures were observed and led to a decrease of photoluminescence intensity. Interestingly, these transformations could be reversed by cooling the sample back to room temperature followed by intense illumination of the nanocrystals [143]. On some occasions standard chemical characterization techniques can be used. For example, the number of thiols on silanized QDs can be estimated using Ellman's reagent method [82, 144].

4.3.6
Conclusion and Outlook

Surface chemical engineering can solve many common problems related to the optical properties and colloidal stability of quantum dots. Additionally, it offers the possibility to functionalize the QDs with suitable chemical groups. The functionalization strategies play an important role in applications. They may allow for the QD solubilization in water, incorporation of QDs into various matrices, coupling QDs with biomacromolecules or appropriate sensing units, and anchoring them to surfaces. It is often important to remember that surface ligands introduce two new, equally important interfaces, the ligand/QD interface and the ligand/environment interface. Both are from the practical point of view equally important, and in practical design of functionalized QD systems both should be carefully considered. Handling the ligands at the nanoparticles surface has to be performed with care. As the ligand influences the photophysical properties of the QDs, determines the QD surface functionality, and stabilizes the QDs in given solvents, loss of the ligand has dramatic consequences.

There are no established off-the-shelf protocols for ligand-exchange reactions. The stability of the surface ligands at high temperatures, at low and high pH values remains problematic. Oxidation and photo-oxidation of the ligands and of the QD semiconductor shells was not studied in detail for many ligands and QD combinations. Polymeric shells, although a promising alternative to low molecular weight monodentate ligands are still under investigation.

It is known that surface chemistry largely determines the toxicity of QDs [145, 146], however the size, concentration, and type of QDs are also important [147]. Coating QDs with polymeric shells holds great promise in reducing the nanoparticles cytotoxicity effects [148–150].

There are new emerging applications of QDs in self-assembly [151–153], also with other nanoscale materials [154–156]. The ligands may act as spacers and the spacer length may define the final properties of the assemblies. We believe that the chemical engineering of the QDs surface will also move to the design of responsive ligands, and towards controlled modulation of the optical properties of the QDs through the ligand shell.

Finally, investigations of QDs at the single nanocrystal level reveal phenomena otherwise impossible to observe when investigating an ensemble [157–164]. This holds great promise to make further progress in nanoscience and nanotechnology of single light emitters. For example, one of the obstacles for widespread application of single-QD imaging techniques in biological research is the 'blinking' effect, i.e. the interruption of emission [165, 166]. Chemical engineering of the ligand shell around the QDs offers the opportunity to reduce the blinking process, simultaneously endowing the QD with functionality.

References

1 Ballou, B., Lagerholm, B.C., Ernst, L.A., Bruchez, M.P. and Waggoner, A.S. (2004) *Bioconjugate Chemistry*, **15**, 79–86.

2 Michalet, X., Pinaud, F.F., Bentolila, L.A., Tsay, J.M., Doose, S., Li, J.J., Sundaresan, G., Wu, A.M., Gambhir, S.S. and Weiss, S. (2005) *Science*, **307**, 538–544.

3 Han, M.Y., Gao, X.H., Su, J.Z. and Nie, S. (2001) *Nature Biotechnology*, **19**, 631–635.

4 Alivisatos, A.P. (1996) *Science*, **271**, 933–937.

5 Grieve, K., Mulvaney, P. and Grieser, F. (2000) *Current Opinion in Colloid & Interface Science*, **5**, 168–172.

6 Eychmüller, A. (2000) *The Journal of Physical Chemistry. B*, **104**, 6514–6528.

7 Chan, W.C.W., Maxwell, D.J., Gao, X., Bailey, R.E., Han, M. and Nie, S. (2002) *Current Opinion in Biotechnology*, **13**, 40–46.

8 Mattoussi, H., Kuno, M.K., Goldmann, E.R., Anderson, G.P. and Mauro, J.M. (2002) Colloidal semiconductor quantum dot conjugates in biosensing, in *Optical Biosensors: Present and Future* (eds F.S. Ligler and C. Rowe Taitt), Elsevier Science, 10: 0444509747.

9 Åkerman, M.E., Chan, W.C.W., Laakkonen, P., Bhatia, S.N. and Ruoslahti, E. (2002) *Proceedings of the National Academy of Sciences of the United States of America*, **99**, 12617–12621.

10 Pinaud, F., Michalet, X., Bentolila, L.A., Tsay, J.M., Doose, S., Li, J.J., Iyer, G. and Weiss, S. (2006) *Biomaterials*, **27**, 1679–1687.

11 Riegler, J. and Nann, T. (2004) *Analytical and Bioanalytical Chemistry*, **379**, 913–919.

12 Murray, C.B., Norris, D.J. and Bawendi, M.G. (1993) *Journal of the American Chemical Society*, **115**, 8706–8715.

13 Grätzel, M. (1981) *Accounts of Chemical Research*, **14**, 376–384.

14 Kuczynski, J. and Thomas, J.K. (1982) *Chemical Physics Letters*, **88**, 445–447.

15 Kuczynski, J. and Thomas, J.K. (1983) *The Journal of Physical Chemistry*, **87**, 5498–5503.

16 Hagfeldt, A. and Grätzel, M. (1995) *Chemical Reviews*, **95**, 49–68.

17 Rossetti, R., Nakahara, S. and Brus, L.E. (1983) *Journal of Chemical Physics*, **79**, 1086–1088.

18 NedeljkoviJ.M., NenadoviM.T. and MiO.I. (1986) *The Journal of Physical Chemistry*, **90**, 12–13.

19 Brus, L.E. (1983) *Journal of Chemical Physics*, **79**, 5566–5571.

20 Brus, L.E. (1984) *Journal of Chemical Physics*, **80**, 4403–4409.

21 Brus, L.E. (1986) *IEEE Journal of Quantum Electronics*, **22**, 1909–1914.

22 Majetich, S.A. and Carter, A.C. (1993) *The Journal of Physical Chemistry*, **97**, 8727–8731.

23 Dannhauser, T., O'Neil, M., Johansson, K., Whitten, D. and McLendon, G. (1986) *The Journal of Physical Chemistry*, **90**, 6074–6076.

24 Spanhel, L., Haase, M., Weller, H. and Henglein, A. (1987) *Journal of the American Chemical Society*, **109**, 5649–5655.

25 Chandler, R.R., Coffer, J.L., Atherton, S.J. and Snowden, P.T. (1992) *The Journal of Physical Chemistry*, **96**, 2713–2717.

26 Chandler, R.R. and Coffer, J.L. (1991) *The Journal of Physical Chemistry*, **95**, 4–6.

27 Coffer, J.L., Chandler, R.R., Gutsche, C.D., Alam, I., Pinizzotto, R.F. and Yang, H. (1993) *The Journal of Physical Chemistry*, **97**, 696–702.

28 Bigham, S.R. and Coffer, J.L. (1992) *The Journal of Physical Chemistry*, **96**, 10581–10584.

29 Steigerwald, M.L., Alivisatos, A.P., Gibson, J.M., Harris, T.D., Kortan, R., Muller, A.J., Thayer, A.M., Duncan, T.M., Douglass, D.C. and Brus, L.E. (1988) *Journal of the American Chemical Society*, **110**, 3046–3050.

30 Herron, N., Wang, Y. and Eckert, H. (1990) *Journal of the American Chemical Society*, **112**, 1322–1326.

31 Kortan, A.R., Hull, R., Opila, R.L., Bawendi, M.G., Steigerwald, M.L., Caroll, P.J. and Brus, L.E. (1990) *Journal of the American Chemical Society*, **112**, 1327–1332.

32 Resch, U., Eychmüller, A., Haase, M. and Weller, H. (1992) *Langmuir*, **8**, 2215–2218.

33 Hoener, C.F., Allan, K.A., Bard, A.J., Campion, A., Fox, M.A., Mallouk, T.E., Webber, S.E. and White, J.M. (1992) *The Journal of Physical Chemistry*, **96**, 3812–3817.

34 Bruchez, M. Jr., Moronne, M., Gin, P., Weiss, S. and Alivisatos, A.P. (1998) *Science*, **281**, 2013–2016.

35 Chan, W.C.W. and Nie, S. (1998) *Science*, **281**, 2016.

36 Bailey, R.E., Smith, A.M. and Nie, S. (2004) *Physica E*, **1**, 1–12.

37 Alivisatos, P. (2004) *Nature Biotechnology*, **22**, 47–52.

38 Gao, X., Yang, L., Petros, J.A., Marshall, F.F., Simons, J.W. and Nie, S. (2005) *Current Opinion in Biotechnology*, **16**, 63–72.

39 Parak, W.J., Pellegrino, T. and Plank, C. (2005) *Nanotechnology*, **16**, R9–R25.

40 Greenham, N.C., Peng, X. and Alivisatos, A.P. (1996) *Physical Review B – Condensed Matter*, **54**, 17628–17637.

41 Milliron, D.J., Alivisatos, A.P., Pitois, C., Edder, C. and Frechet, J.M.J. (2003) *Advanced Materials*, **15**, 58–61.

42 Liu, J., Tanaka, T., Sivula, K., Alivisatos, A.P. and Frechet, J.M.J. (2004) *Journal of the American Chemical Society*, **126**, 6550–6551.

43 Brus, L.E. (1986) *The Journal of Physical Chemistry*, **90**, 2555–2560.

44 Steigerwald, M.L. and Brus, L.E. (1990) *Accounts of Chemical Research*, **23**, 183–188.

45 Wang, Y. (1991) *Accounts of Chemical Research*, **24**, 133–139.

46 Wang, Y. and Herron, N. (1991) *The Journal of Physical Chemistry*, **95**, 525–532.

47 Weller, H. (1993) *Angewandte Chemie – International Edition*, **32**, 41–53.

48 Green, M. and O'Brien, P. (1999) *Chemical Communications*, 2235–2241.

49 Eychmüller, A. (2000) *The Journal of Physical Chemistry. B*, **104**, 6514–6528.

50 Trindade, T., O'Brien, P. and Pickett, N.L. (2001) *Chemistry of Materials*, **13**, 3843–3858.

51 Woggon, U. (1996) *Optical Properties of Semiconductor Quantum Dot*, Springer Tracts in Modern Physics, 136.

52 Gaponenko, S.V. (1998) *Optical Properties of Semiconductor Nanocrystals*, Cambridge University Press, 10: 0521582415.

53 Manna, L., Scher, E.C. and Alivisatos, A.P. (2002) *Journal of Cluster Science*, **13**, 521–532.

54 Yin, Y. and Alivisatos, A.P. (2005) *Nature*, **437**, 664–670.

55 Gao, M., Richter, B., Kirstein, S. and Möhwald, H. (1998) *The Journal of Physical Chemistry. B*, **102**, 4096–4103.

56 Rogach, A., Kershaw, S., Burt, M., Harrison, M., Kornowski, A., Eychmüller, A. and Weller, H. (1999) *Advanced Materials*, **11**, 552–555.

57 Harrison, M.T., Kershaw, S.V., Burt, M.G., Rogach, A., Eychmüller, A. and Weller, H. (1999) *Journal of Materials Chemistry*, **9**, 2721–2723.

58 Gaponik, N., Talapin, D.V., Rogach, A.L., Hoppe, K., Shevchenko, E.V., Kornowski, A., Eychmüller, A. and Weller, H. (2002) *The Journal of Physical Chemistry. B*, **106**, 7177–7185.

59 Rogach, A.L., Franzl, T., Klar, T.A., Feldmann, J., Gaponik, N., Lesnyak, V., Shavel, A., Eychmüller, A., Rakovich, Y.P. and Donegan, J.F. (2007) *The Journal of Physical Chemistry.C*, **111**, 14628–14637.

60 Kittel, C. (2004) *Introduction to Solid State Physics*, 8th edn, John Wiley & Sons, 10: 047141526X.

61 Efros, Al.L. and Rosen, M. (2000) *Annual Review of Materials Science*, **30**, 475–521.

62 Norris, D.J. and Bawendi, M.G. (1996) *Physical Review B-Condensed Matter*, **53**, 16338–16346.

63 Nirmal, M. and Brus, L. (1999) *Accounts of Chemical Research*, **32**, 407–414.

64 Kuno, M., Lee, J.K., Dabbousi, B.O., Mikulec, F.V. and Bawendi, M.G. (1997) *Journal of Chemical Physics*, **106**, 9869–9882.

65 Gong, H.M., Zhou, Z.K., Song, H., Hao, Z.H., Han, J.B., Zhai, Y.Y., Xiao, S. and Wang, Q.Q. (2007) *The Journal of Fluorescence*, **17**, 715–720.

66 Hines, M.A. and Guyot-Sionnest, P. (1996) *The Journal of Physical Chemistry*, **100**, 468–471.

67 Dabbousi, B.O., Rodriguez-Viejo, J., Mikulec, F.V., Heine, J.R., Mattoussi, H., Ober, R., Jensen, K.F. and Bawendi, M.G. (1997) *The Journal of Physical Chemistry. B*, **101**, 9463–9475.

68 Peng, X., Schlamp, M.C., Kadavanich, A.V. and Alivisatos, A.P. (1997) *Journal of the American Chemical Society*, **119**, 7019–7029.

69 Wilson, W.L., Szajowski, P.F. and Brus, L.E. (1993) *Science*, **262**, 1242–1244.

70 Mews, A., Eychmüller, A., Giersig, M., Schooss, D. and Weller, H. (1994) *The Journal of Physical Chemistry*, **98**, 934–941.

71 Tian, Y., Newton, T., Kotov, N.A., Guldi, D.M. and Fendler, J.H. (1996) *The Journal of Physical Chemistry*, **100**, 8927–8939.

72 Danek, M., Jensen, K.F., Murray, C.B. and Bawendi, M.G. (1996) *Chemistry of Materials*, **8**, 173–180.

73 Eychmüller, A., Mews, A. and Weller, H. (1993) *Chemical Physics Letters*, **208**, 59–62.

74 Schooss, D., Mews, A., Eychmüller, A. and Weller, H. (1994) *Physical Review B – Condensed Matter*, **49**, 17072–17078.

75 Piryatinski, A., Ivanov, S.A., Tertiak, S. and Klimov, V.I. (2007) *Nano Letters*, **7**, 108–115.

76 Chang, S., Liu, L. and Asher, S.A. (1994) *Journal of the American Chemical Society*, **116**, 6739–6744.

77 Correa-Duarte, M.A., Giersig, M. and Liz-Marzán, L.M. (1998) *Chemical Physics Letters*, **286**, 497–501.

78 Mulvaney, P., Liz-Marzán, L.M., Giersig, M. and Ung, T. (2000) *Journal of Materials Chemistry*, **10**, 1259–1270.

79 Schroedter, A., Weller, H., Eritja, R., Ford, W.E. and Wessels, J.M. (2002) *Nano Letters*, **2**, 1363–1367.

80 Parak, W.J., Gerion, D., Zanchet, D., Woerz, A.S., Pellegrino, T., Micheel, C., Williams, S.C., Seitz, M., Bruehl, R.E., Bryant, Z., Bustamante, C., Bertozzi, C.R. and Alivisatos, A.P. (2002) *Chemistry of Materials*, **14**, 2113–2119.

81 Stöber, W., Fink, A. and Bohn, E. (1968) *Journal of Colloid and Interface Science*, **26**, 62–66.

82 Gerion, D., Pinaud, F., Williams, S.C., Parak, W.J., Zanchet, D., Weiss, S. and Alivisatos, A.P. (2001) *The Journal of Physical Chemistry. B*, **105**, 8861–8871.

83 Nann, T. and Mulvaney, P. (2004) *Angewandte Chemie – International Edition*, **43**, 5393–5396.

84 Darbandi, M., Thomann, R. and Nann, T. (2005) *Chemistry of Materials*, **17**, 5720–5725.

85 Boldt, K., Bruns, O.T., Gaponik, N. and Eychmüller, A. (2007) *The Journal of Physical Chemistry. B*, **110**, 1959–1963.

86 Mattoussi, H., Mauro, J.M., Goldman, E.R., Anderson, G.P., Sundar, V.C., Mikulec, F.V. and Bawendi, M.G. (2000) *Journal of the American Chemical Society*, **122**, 12142–12150.

87 Zhang, H., Wang, C., Li, M., Ji, X., Zhang, J. and Yang, B. (2005) *Chemistry of Materials*, **17**, 4783–4788.

88 Jaffar, S., Nam, K.T., Khademhosseini, A., Xing, J., Langer, R.S. and Belcher, A.M. (2004) *Nano Letters*, **4**, 1421–1425.

89 Shim, M., Wang, C. and Guyot-Sionnest, P. (2001) *The Journal of Physical Chemistry*, **105**, 2369–2373.

90 Wuister, S.F., de Mello Donegá, C. and Meijerink, A. (2004) *The Journal of Physical Chemistry. B*, **108**, 17393–17397.

91 Aldana, J., Wang, Y.A. and Peng, X.G. (2001) *Journal of the American Chemical Society*, **123**, 8844–8850.

92 Aldana, J., Lavelle, N., Wang, Y.J. and Peng, X.G. (2005) *Journal of the American Chemical Society*, **127**, 2496–2504.

93 Pathak, S., Choi, S.-K., Arnheim, N. and Thompson, M.E. (2001) *Journal of the American Chemical Society*, **123**, 4103–4104.

94 Torimoto, T., Yamashita, M., Kuwabata, S., Sakata, T., Mori, H. and Yoneyama, H. (1999) *The Journal of Physical Chemistry. B*, **42**, 8799–8803.

95 Gaponik, N., Talapin, D.V., Rogach, A.L., Eychmüller, A. and Weller, H. (2002) *Nano Letters*, **2**, 803–806.

96 Potapova, I., Mruk, R., Prehl, S., Zentel, R., Basché, T. and Mews, A. (2003) *Journal of the American Chemical Society*, **125**, 320–321.

97 Uyeda, H.T., Medintz, I.L., Jaiswal, J.K., Simon, S.M. and Mattoussi, H. (2005) *Journal of the American Chemical Society*, **127**, 3870–3878.

98 Mitchell, G.P., Mirkin, C.A. and Letsinger, R.L. (1999) *Journal of the American Chemical Society*, **121**, 8122–8123.

99 Willner, I., Patolsky, F. and Wasserman, J. (2001) *Angewandte Chemie – International Edition*, **40**, 1861–1864.

100 Patolsky, F., Gill, R., Weizmann, Y., Mokari, T., Banin, U. and Willner, I. (2003) *Journal of the American Chemical Society*, **125**, 13918–13919.

101 Krueger, K.M., Al-Somali, A.M., Mejia, M. and Colvin, V.L. (2007) *Nanotechnology*, **18**, Art. no. 475709.

102 Querner, C., Reiss, P., Bleuse, J. and Pron, A. (2004) *Journal of the American Chemical Society*, **126**, 11574–11582.

103 Querner, C., Benedetto, A., Demadrille, R., Rannou, P. and Reiss, P. (2006) *Chemistry of Materials*, **18**, 4817–4826.

104 Dubois, F., Mahler, B., Dubertret, B., Doris, E. and Mioskowski, C. (2007) *Journal of the American Chemical Society*, **129**, 482–483.

105 Wang, Y.A., Li, J.J., Chen, H. and Peng, X. (2002) *Journal of the American Chemical Society*, **124**, 2293–2298.

106 Guo, W., Li, J.J., Wang, Y.A. and Peng, X. (2003) *Journal of the American Chemical Society*, **125**, 3901–3909.

107 Kim, S. and Bawendi, M.G. (2003) *Journal of the American Chemical Society*, **125**, 14652–14653.

108 Skaff, H., Ilker, M.F., Coughlin, E.B. and Emrick, T. (2002) *Journal of the American Chemical Society*, **124**, 5729–5733.

109 O'Brien, P., Cummins, S.S., Darcy, D., Dearden, A., Masala, O., Pickett, N.L., Ryley, S. and Sutherland, A.J. (2003) *Chemical Communications*, 2532–2533.

110 Skaff, H., Sill, K. and Emrick, T. (2004) *Journal of the American Chemical Society*, **126**, 11322–11325.

111 Sill, K. and Emrick, T. (2004) *Chemistry of Materials*, **16**, 1240–1243.

112 Sundar, V.C., Eisler, H.-J. and Bawendi, M.G. (2002) *Advanced Materials*, **14**, 739–743.

113 Bullen, C. and Mulvaney, P. (2006) *Langmuir*, **22**, 3007–3013.

114 Landes, C., Burda, C., Braun, M. and El-Sayed, M.A. (2001) *The Journal of Physical Chemistry. B*, **105**, 2981–2986.

115 Wang, X.-S., Dykstra, T.E., Salvador, M.R., Manners, I., Scholes, G.D. and Winnik, M.A. (2004) *Journal of the American Chemical Society*, **126**, 7784–7785.

116 Skaff, H. and Emrick, T. (2003) *Chemical Communications*, 52–53.

117 Nann, T. (2005) *Chemical Communications*, 1735–1736.

118 Dubertret, B., Skourides, P., Norris, D.J., Noireaux, V., Brivanlou, A.H. and Libchaber, A. (2002) *Science*, **298**, 1759–1762.

119 Pellegrino, T., Manna, L., Kudera, S., Liedl, T., Koktysh, D., Rogach, A.L., Keller, S., Rädler, J., Natile, G. and Parak, W.J. (2004) *Nano Letters*, **4**, 703–707.

120 Luccardini, C., Tribet, C., Vial, F., Marchi-Artzner, V. and Dahan, M. (2006) *Langmuir*, **22**, 2304–2310.

121 Mulder, W.J.M., Koole, R., Brandwijk, R.J., Storm, G., Chin, P.T.K., Strijkers, G.J., de Mello Donegá, C., Nicolay, K. and Griffioen, A.W. (2006) *Nano Letters*, **6**, 1–6.

122 Yu, W.W., Chang, E., Falkner, J.C., Zhang, J.Y., Al-Somali, A.M., Sayes, C.M., Johns, J., Drezek, R. and Colvin, V.L. (2007) *Journal of the American Chemical Society*, **129**, 2871–2879.

123 Feng, J., Ding, S.Y., Tucker, M.P., Himmel, M.E., Kim, Y.-H., Zhang, S.B., Keyes, B.M. and Rumbles, G. (2005) *Applied Physics Letters*, **86**, Art no. 033108.

124 Giessbuehler, I., Hovius, R., Martinez, K.L., Adrian, M., Thampi, K.R. and Vogel, H. (2005) *Angewandte Chemie – International Edition*, **44**, 1388–1392.

125 Osaki, F., Kanamori, T., Sando, S., Sera, T. and Aoyama, Y. (2004) *Journal of the American Chemical Society*, **126**, 6520–6521.

126 Tomczak, N., Jańczewski, D., Han, M.Y. and Vancso, G.J. (2009) *Progress in Polymer Science*, **34**, 393–430.

127 Jańczewski, D., Tomczak, N., Khin, Y.W., Han, M.Y. and Vancso, G.J. (2009) *European Polymer Journal*, **45**, 3–9.

128 Jańczewski, D., Tomczak, N., Han, M.Y. and Vancso, G.J. (2009) *Macromolecules*, **42**, 1801–1804.

129 Thomas, J.K. (1987) *The Journal of Physical Chemistry*, **91**, 267–276.

130 Bowen Katari, J.E., Colvin, V.L. and Alivisatos, A.P. (1994) *The Journal of Physical Chemistry*, **98**, 4109–4117.

131 Shiang, J.J., Kadavanich, A.V., Grubbs, R.K. and Alivisatos, A.P. (1995) *The Journal of Physical Chemistry*, **99**, 17417–17422.

132 Mews, A., Kadavanich, A.V., Banin, U. and Alivisatos, A.P. (1996) *Physical Review B – Condensed Matter*, **53**, 13242–13245.

133 Pons, T., Uyeda, H.T., Medintz, I.L. and Mattoussi, H. (2006) *The Journal of Physical Chemistry. B*, **110**, 20308–20316.

134 Sachleben, J.R., Wooten, E.W., Emsley, L., Pines, A., Colvin, V.L. and Alivisatos, A.P. (1992) *Chemical Physics Letters*, **198**, 431–436.

135 Majetich, S.A., Carter, A.C., Belot, J. and McCullough, R.D. (1994) *The Journal of Physical Chemistry*, **98**, 13705–13710.

136 Schmelz, O., Mews, A., Basché, T., Herrmann, A. and Müllen, K. (2001) *Langmuir*, **17**, 2861–2865.

137 Berrettini, M.G., Braun, G., Hu, J.G. and Strouse, G.F. (2004) *Journal of the American Chemical Society*, **126**, 7063–7070.

138 Thayer, A.M., Steigerwald, M.L., Duncan, T.M. and Douglass, D.C. (1988) *Physical Review Letters*, **60**, 2673–2676.

139 Diaz, D., Rivera, M., Ni, T., Rodriguez, J.-C., Castillo-Blum, S.-E., Nagesha, D., Robles, J., Alvarez-Fregoso, O.-J. and Kotov, N.A. (1999) *The Journal of Physical Chemistry. B*, **103**, 9854–9858.

140 Becerra, L.R., Murray, C.B., Griffin, R.G. and Bawendi, M.G. (1994) *Journal of Chemical Physics*, **100**, 3297–3300.

141 Trindade, T., O'Brien, P. and Zhang, X. (1997) *Chemistry of Materials*, **9**, 523–530.

142 Hamad, K.S., Roth, R., Rockenberger, J., van Buuren, T. and Alivisatos, A.P. (1999) *Physical Review Letters*, **83**, 3474–3477.

143 Hess, B.C., Okhrimenko, I.G., Davis, R.C., Stevens, B.C., Schulzke, Q.A., Wright, K.C., Bass, C.D., Evans, C.D. and Summers, S.L. (2001) *Physical Review Letters*, **86**, 3132–3135.

144 Hermanson, G.T. (2008) *Bioconjugation Techniques*, 2nd edn, Academic Press, 10: 0123705010.

145 Hoshino, A., Fujioka, K., Oku, T., Suga, M., Sasaki, Y.F., Ohta, T., Yasuhara, M., Suzuki, K. and Yamamoto, K. (2004) *Nano Letters*, **4**, 2163–2169.

146 Guo, G.N., Liu, W., Liang, J.G., He, Z.K., Xu, H.B. and Yang, X.L. (2007) *Materials Letters*, **61**, 1641–1644.

147 Shiohara, A., Hoshino, A., Hanaki, K., Suzuki, K. and Yamamoto, K. (2004) *Microbiology and Immunology*, **48**, 669–675.

148 Derfus, A.M., Chan, W.C.W. and Bhatia, S.N. (2004) *Nano Letters*, **4**, 11–18.

149 Kirchner, C., Liedl, T., Kudera, S., Pellegrino, T., Javier, A.M., Gaub, H.E., Stölzle, S., Fertig, N. and Parak, W.J. (2005) *Nano Letters*, **5**, 331–338.

150 Zhang, T., Stilwell, J.L., Gerion, D., Ding, L., Elboudwarej, O., Cooke, P.A., Gray, J.W., Alivisatos, A.P. and Chen, F.F. (2006) *Nano Letters*, **6**, 800–808.

151 Coffer, J.L., Bigham, S.R., Li, X., Pinizzotto, R.F., Rho, Y.G., Pirtle, R.M. and Pirtle, I.L. (1996) *Applied Physics Letters*, **69**, 3851–3853.

152 Collier, C.P., Vossmeyer, T. and Heath, J.R. (1998) *Annual Review of Physical Chemistry*, **49**, 371–404.

153 Murray, C.B., Kagan, C.R. and Bawendi, M.G. (2000) *Annual Review of Materials Science*, **30**, 545–610.

154 Shenhar, R., Norsten, T.B. and Rotello, V.M. (2005) *Advanced Materials*, **17**, 657–669.

155 Shavel, A., Gaponik, N. and Eychmüller, A. (2005) *European Journal of Inorganic Chemistry*, 3613–3623.

156 Kolny, J., Kornowski, A. and Weller, H. (2002) *Nano Letters*, **2**, 361–364.

157 Empedocles, S.A., Norris, D.J. and Bawendi, M.G. (1996) *Physical Review Letters*, **77**, 3873–3876.

158 Blanton, S.A., Hines, M.A. and Guyot-Sionnest, P. (1996) *Applied Physics Letters*, **69**, 3905–3907.

159 Nirmal, M., Dabbousi, B.O., Bawendi, M.G., Macklin, J.J., Trautman, J.K.,

Harris, T.D. and Brus, L.E. (1996) *Nature*, **383**, 802–804.
160 Empedocles, S.A. and Bawendi, M.G. (1997) *Science*, **278**, 2114–2117.
161 Tittel, J., Göhde, W., Koberling, F., Basché, T., Kornowski, A., Weller, H. and Eychmüller, A. (1997) *The Journal of Physical Chemistry. B*, **101**, 3013–3016.
162 Efros, Al.L. and Rosen, M. (1997) *Physical Review Letters*, **78**, 1110–1113.
163 Empedocles, S.A., Neuhauser, R., Shimizu, K. and Bawendi, M.G. (1999) *Advanced Materials*, **11**, 1243–1256.
164 Neuhauser, R.G., Shimizu, K.T., Woo, W.K., Empedocles, S.A. and Bawendi, M.G. (2000) *Physical Review Letters*, **85**, 3301–3304.
165 Kuno, M., Fromm, D.P., Hamann, H.F., Gallagher, A. and Nesbitt, D.J. (2000) *Journal of Chemical Physics*, **112**, 3117–3120.
166 Kuno, M., Fromm, D.P., Hamann, H.F., Gallagher, A. and Nesbitt, D.J. (2001) *Journal of Chemical Physics*, **115**, 1028–1040.

4.4
Stimuli-Responsive Capsules

Yujie Ma, Mark A. Hempenius, E. Stefan Kooij, Wen-Fei Dong, Helmuth Möhwald, and G. Julius Vancso

4.4.1
Introduction

With the rapid development and multidisciplinary broadening of materials science, many structural platforms that feature unprecedented and characteristic functions have been introduced. New functional structures could be fabricated by the incorporation of classical materials using contemporary techniques. However, real breakthroughs may lie in the design and advance of novel functional materials [1]. One of the most important classes of these materials are stimuli-responsive polymers. These 'smart' polymeric materials can respond to specific external stimuli with drastic changes in their size and conformation [2–4]. In recent years, exploration of molecular structures based on stimuli-responsive polymers and their controlled properties in response to external *physical* stimuli, such as temperature, pH, ionic strength, solvent polarity variations, electric or magnetic fields and light has flourished. Nevertheless, structure–property manipulation using *chemical* stimuli still largely remains unexplored [5].

Among the various structural platforms, polyelectrolyte multilayers (PEMs) may offer new properties when constructed from stimuli-responsive polymeric materials [6]. Polyelectrolyte multilayers are fabricated using the electrostatic layer-by-layer (LBL) technique, based on the sequential adsorption of charged polyelectrolyte species. In this way, multilayered polymeric thin films are built up with controlled thickness and composition [7]. One of the most attractive features of this method is that thin films can be tailor-made to display specific chemical and physical properties by the choice of the polyelectrolyte material [7, 8]. Aqueous processing enables synthetic as well as natural polyelectrolytes [9, 10] to be assembled to thin films for a variety of applications, ranging from electroluminescent devices to biosensor arrays [7]. In order to broaden their application potentials, a number of responsive PEMs that employ different stimuli have been proposed during the last few years [8, 11].

The most widely explored are those based on traditional stimuli (e.g. pH [6e, 12, 13], ionic strength [13, 14], and solvent polarity [15]) to regulate intermolec-

Surface Design: Applications in Bioscience and Nanotechnology
Edited by Renate Förch, Holger Schönherr, and A. Tobias A. Jenkins

ISBN: 978-3-527-40789-7

ular interactions in the originally charge compensated multilayer structures. For example, pH responsiveness depends on the charge-density alteration of weak polyelectrolytes following external pH variations. This makes it possible to design applications regarding loading and releasing of drugs under an environmental pH different from pH values during the multilayer assembly [16]. Other types of stimuli, including temperature [17], light [18], magnetic field [19], ultrasound [20] and specific interactions [11] also attract growing interest. However, conventional, organic polyelectrolyte-based microcapsules have their limitations in some significant applications due to their slow response to trace amounts of trigger or restricted freedom of choice of stimuli. Chemical or electrochemical stimuli seem promising since e.g. certain small molecules may act as a signal. The redox potential near a desired organ where a pharmaceutical drug should be released may differ from other locations in the body, or, a corrosion pit may self-anneal due to the release of an inhibitor by local potential change [21]. Recently, Shchukin *et al.* have demonstrated an application of polyelectrolyte microcapsules as microcontainers with electrochemically reversible flux of redox-active materials [21, 22]. However, the construction of capsules featuring redox-responsive components in the wall has still rarely been discussed.

Poly(ferrocenylsilanes) (PFS), consisting of alternating ferrocene and alkylsilane units in the main chain, belong to the class of redox-responsive organometallic polymers. Following the discovery of thermal ring-opening polymerization by the group of Manners, transition-metal catalyzed [23] and anionic [24] ring-opening polymerization (ROP) of silicon-bridged ferrocenophanes has facilitated the synthesis of poly(ferrocenylsilane) homopolymers and block copolymers featuring corresponding organometallic blocks with well-defined molar mass and composition. Due to the presence of redox-active ferrocene units in the polymer backbone, PFS can be reversibly oxidized and reduced by chemical [25] as well as electrochemical means [26]. Previous studies in our group on self-assembled alkanethiol end-functionalized PFS monolayers on gold revealed electrochemically induced morphology and volume/thickness changes [27]. Atomic force microscopy (AFM)-based single-molecule force spectroscopy (SMFS) measurements on poly(ferrocenylsilane) single chains [28, 29] showed significantly increased Kuhn length and segment elasticity after oxidation, which is a direct proof of redox-induced changes of the torsional potential-energy landscape of the organometallic main chain.

Water-soluble poly(ferrocenylsilane) polycations and polyanions, belonging to the rare class of main-chain organometallic polyelectrolytes, have recently been reported by us and others [30, 31]. These compounds all bear certain charges on the polymer side chains. They are of great interest since they can be incorporated in the electrostatic layer-by-layer self-assembly process to form multilayer films and hollow capsules [8, 10, 30b, 31c, 32–34] with defined structures and directed functions due to the redox activity of the organometallic main chain. Here, we report on the use of a water-soluble poly(ferrocenylsilane) polyanion and polycation pair in the electrostatic layer-by-layer self-assembly process to form organometallic polyelectrolyte multilayer microcapsules. The organometallic capsules allowed us to study the effects of a new stimulus, i.e. changing the redox state, on the permeability of these multilayer structures [8]. Microcapsules made from layer-by-layer assembled PFS polyelectrolytes may serve as a unique polymeric microcontainer system that could respond specifically to redox stimuli. Composite-wall multilayer capsules featuring PFS components in the inner wall and organic redox-inert polyelectrolyte species poly(styrene sulfonate) (PSS^-) and poly (allylamine hydrochloride) (PAH^+) on the outer wall are discussed as a means to manipulate the responsive permeability of polyelectrolyte microcapsules on a molecular level.

4.4.2 Experimental Section

4.4.2.1 Materials

All chemicals were obtained from Aldrich and used as-received. The synthesis of the poly(ferrocenylsilane) polyions **1** and **2** (Chart 1) is described elsewhere [30c]. Side-group modifications of poly[ferrocenyl(3-iodopropyl)methylsilanes] with an average degree of polymerization DP ~70 and a polydispersity $M_w/M_n = 2.1$ produced the polyions. On the basis of repeating unit molar mass, polyanion and polycation $M_w = 53\ 000$ g/mol was estimated. Manganese carbonate ($MnCO_3$) particles were prepared according to the reported method [35].

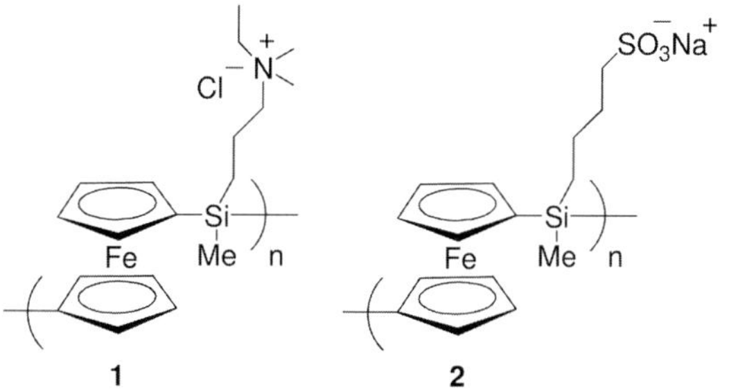

Chart 1

4.4.2.2 Multilayer Fabrication on Flat Substrates

Multilayers were deposited on quartz slides and silicon wafers. Quartz slides and silicon wafers were cleaned by immersion into a mixture of H_2O, H_2O_2 and NH_4OH with a volume ratio of 5:1:1 for 20 min followed by extensive rinsing with MilliQ and drying under a nitrogen stream. Before putting on poly(ferrocenylsilane) (PFS) polycation/polyanion multilayers, the substrates were first dipped into a PEI solution (~ 10 mM, based on repeating unit molar mass) for 30 min to impart positive charges onto the substrates. Multilayers were subsequently prepared by alternately dipping the (modified) substrates in the polycation and polyanion aqueous solutions (2 mg/mL, containing 0.5 M NaCl) for 10 min, with rinsing, dipping into pure MilliQ (2 min), secondary rinsing and drying with a stream of nitrogen between each deposition step.

4.4.2.3 Polyelectrolyte Multilayer Capsule Preparation

Alternate adsorption of polyelectrolytes (PFS, 1 mg/mL; PSS/PAH, 2 mg/mL) onto the $MnCO_3$ microparticles (~ 10% w/w in aqueous suspension) was carried out in 0.5 M NaCl solution for 10 min followed by centrifugation (1500 rpm, 2 min) and three MilliQ washing/centrifugation steps. Since the $MnCO_3$ particles possess positively charged surfaces, negatively charged polyelectrolyte species (PSS^- or PFS^-) were chosen as the first layer. After the deposition of a desired number of polyelectrolyte bilayers, the coated particles were subjected to core dissolution in 0.2 M EDTA (pH = 7) solution. After 60 min of agitation, the suspension was centrifuged (1500 rpm, 10 min), the supernatant was removed, and the capsules were resuspended in fresh EDTA. This washing procedure was repeated three times. The resulting capsules were washed thoroughly with MilliQ for three times and finally redispersed and stored in MilliQ.

4.4.3 Results and Discussion

4.4.3.1 Redox Characteristics of PFS Multilayers on Flat Substrates

PFS can be partially or fully oxidized by chemical methods, accompanied by a color change from orange to dark green and then to blue [25]. PFS oxidants include $FeCl_3$ [25a], I_2 [25d], tris(4-bromophenyl) ammonium hexafluorophosphate [25], and tetracyanoethylene (TCNE) [25c]. Since the fabrication and potential applications of PFS multilayers (planar films and free-standing microcapsules) involve aqueous environments, water-soluble oxidation/reducing agents are required to tune the redox state of PFS polyions. $FeCl_3$, a rather strong oxidant for PFS and used throughout this study [25e], may form insoluble precipitates in neutral water. The $FeCl_3$ solutions were therefore acidified slightly by adding dilute hydrochloric acid.

We first studied the redox behavior of PFS polyelectrolyte multilayers on flat interfaces upon chemical oxidation. Fully organometallic multilayers containing PFS^-/PFS^+ polyelectrolytes were assembled onto silicon wafers followed by subsequent oxidation using $FeCl_3$. The PFS polyelectrolytes used here (Chart 1) are both strong polyelectrolytes, i.e. their degree of ionization is independent on the pH value of the solution. UV/Vis absorption and spectroscopic ellipsometry spectra were recorded after each deposited bilayer. A linear increase of the characteristic PFS absorbance at 216 nm and in film thickness confirmed the formation of well-defined multilayer structures [32]. When deposited from PFS polyelectrolyte solutions at 2 mg/mL containing 0.5 M NaCl, a typical bilayer thickness of 4.5 nm was obtained by fitting of the ellipsometry spectra. The thickness of the as-assembled multilayer thin films was shown to depend not only on the deposited number of bilayers, but also on the salt concentration of the polyelectrolyte solution from which the films were deposited [32]. Figure 1 shows the ellipsometric spectra for a bare silicon substrate, after deposition of a ten-bilayer film (2 mg/mL PFS polyions, 0.25 M NaCl), and after subsequent oxidation by $FeCl_3$. The spectra were recorded at three different incident angles of 65°, 70° and 75°. The substantial change in the shape of the spectra as well as Δ values upon completion of layer-by-layer assembly (Figure 1b) indicated the existence of a thin film. When fitted, the spectra gave a thickness value of 29 nm, in agreement with thicknesses obtained earlier [32].

Following film formation, multilayers were oxidized using ferric chloride. The sample was immersed into $FeCl_3$ aqueous solution (10 mM, pH = 4) for five minutes followed by thorough rinsing and drying. The spectroscopic ellipsometry spectrum of the oxidized PFS multilayer sample (Figure 1c) is almost identical to that of the bare silicon (Figure 1a). From fittings on Figure 1(c), a film thickness of only 1 nm was obtained, indicating close to complete material loss by chemical oxidation. Further experiments on silicon wafers bearing PFS multilayers with different thicknesses show that the same fast disassembly of films can be achieved with even lower $FeCl_3$ concentrations (as low as 3 mM).

4.4.3.2 Microcapsule Formation and Permeability Threshold

PFS capsules were fabricated by the electrostatic LBL assembly of polyelectrolytes **1** and **2** onto colloidal templates, followed by template removal [6d, 8, 32, 34]. Melamine formaldehyde (MF) microparticles were initially used as templates for the deposition of polyanion/polycation pairs. These templating cores were removed by immersing the coated particles in HCl (pH = 1.0). In later experiments, manganese carbonate ($MnCO_3$) microparticles became templates of choice as they are readily synthesized [26, 27] and can be removed by ethylenediaminetetraacetic acid (EDTA) solutions under mild conditions [8, 32]. Capsules obtained using these cores were intact after core removal as opposed to capsules formed using MF cores. Use of the latter cores always led to a number of capsules having ruptured walls.

For permeability control purposes, a prerequisite is to establish a stable growth regime where the permeability of capsules upon formation can be predicted. In this study, we employed a fluorescence probe (TRITC-dextran, 4400 g/mol) to monitor the capsule-permeability dependence on capsule wall thickness by the use of confocal laser scanning microscopy (CLSM). Following a systematic study,

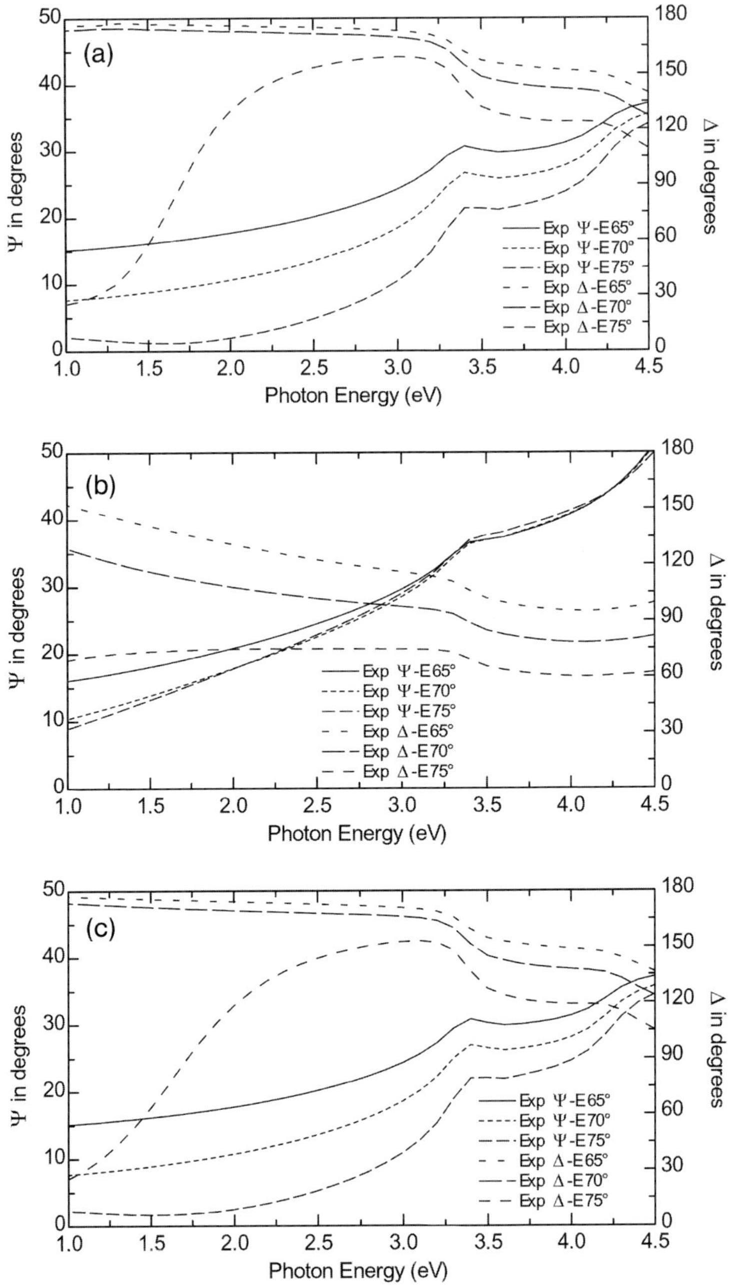

Figure 1 Ellipsometry spectra recorded for bare silicon (a), the same silicon with a $(PFS^-/PFS^+)_{10}$ film (b) and the same multilayer sample after oxidation (5 min in 10 mM $FeCl_3$, pH = 4) (c). The thickness deduced from (b) was 29 nm, while after oxidation it had decreased to 1 nm (c).

we found that microcapsules made from $MnCO_3$ microparticles and containing more than four bilayers of PFS polyion pairs are essentially impermeable to the molecular probe chosen [32]. The average bilayer thickness for a four bilayer PFS^-/PFS^+ capsule sample was measured by tapping-mode AFM to be around 6 nm, while for five bilayers an average bilayer thickness of 8 nm was found [8, 32]. The larger film thickness compared to multilayers deposited on silicon wafers (3 nm) obtained under the same deposition conditions may be related to influences of the substrate (e.g. surface roughness and charge density [36]).

4.4.3.3 Chemical Oxidation of $(PFS^-/PFS^+)_n$ Microcapsules

Ferric chloride was used for the study of redox-responsive permeability of PFS microcapsules that originally are impermeable to the probe dextran molecules. In the presence of the same reference molecular probe, CLSM was used to visualize the relative fluorescent intensity change in the capsule interior before and after mixing the capsule suspension with the aqueous oxidant solutions. $FeCl_3$ aqueous solution (1 mM, pH = 4, tuned by HCl) was on-site added to the capsule–dye mixture during CLSM imaging. With the diffusion of the Fe^{3+} ions originating from the $FeCl_3$ added, PFS capsules displayed a fast and continuous expansion. The shell expansion was accompanied by an accordingly increased permeability of the capsules. Figure 2 shows a series of CLSM micrographs of $(PFS^-/PFS^+)_5$ capsules following the different stages of $FeCl_3$ oxidation with time. The originally impermeable capsules started to grow in size and became permeable after mixing with $FeCl_3$ for seven minutes. The swelling continued for the majority of the capsules (over 90%) for approximately another five minutes, until their final disappearance. Before their disintegration, the oxidized capsules also became completely permeable to the 4400 g/mol TRITC-dextran molecules, as the fluorescence intensity from outside and inside the capsules could no longer be differentiated. In the final oxidized states of the capsules, there was an accompanying two- to three-fold increase in the capsule diameter. The speed of capsule expansion and permeability increase was found to depend on the concentration of the $FeCl_3$ solutions used. At a higher oxidant concentration (2 mM), the duration from the onset of adding the oxidants until the final capsule disintegration was shortened to only three minutes (*vide infra*). The remarkable swelling and disintegration of these capsules turned out to be independent of capsule wall thickness, since a similar redox response was also observed when the number of absorbed PFS bilayers was increased to six. Control experiments showed that the permeability and integrity of the capsules remain unchanged when solely mixed with dilute HCl (pH = 4) for several hours.

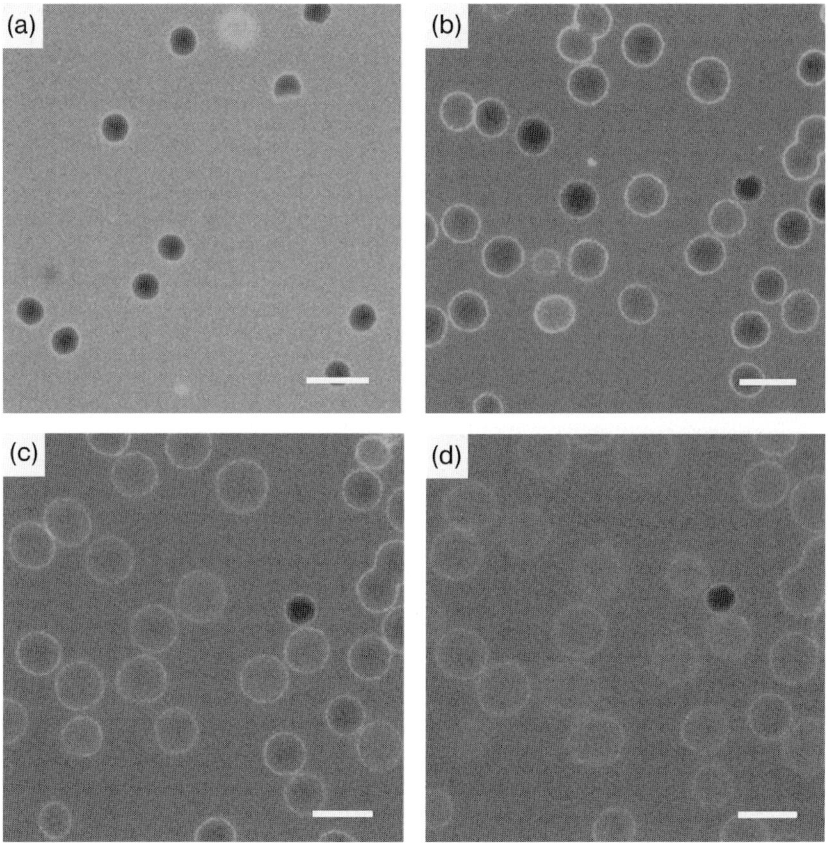

Figure 2 Representative confocal laser scanning micrographs of $(PFS^-PFS^+)_5$ microcapsules oxidized with time when exposed to $FeCl_3$ (1 mM, pH = 4) solutions. Capsules show increasing permeability to 4.4 kDa dextran molecules and increased size as the oxidation proceeds. Scale bar = 10 μm.

4.4.3.4 Chemical Reduction of $(PFS^-/PFS^+)_n$ Microcapsules

Recently, it was found that water-soluble ascorbic acid (vitamin C) and dithiothreitol (DTT) are two effective reducing agents for oxidized PFS polycations [25e]. The bulk reduction reaction is fast, as the solution color of oxidized PFS polycations instantly changes from green-blue to yellow-orange. The stoichiometry of oxidant ($FeCl_3$) and reducing agents is assumed to be 2:1 since the oxidation of vitamin C and DTT both involve a two-electron transfer process. In order to control the oxidation state of the PFS microcapsules, DTT was mixed at a certain point with the aqueous PFS microcapsule suspension during the $FeCl_3$ oxidation. Figure 3 shows

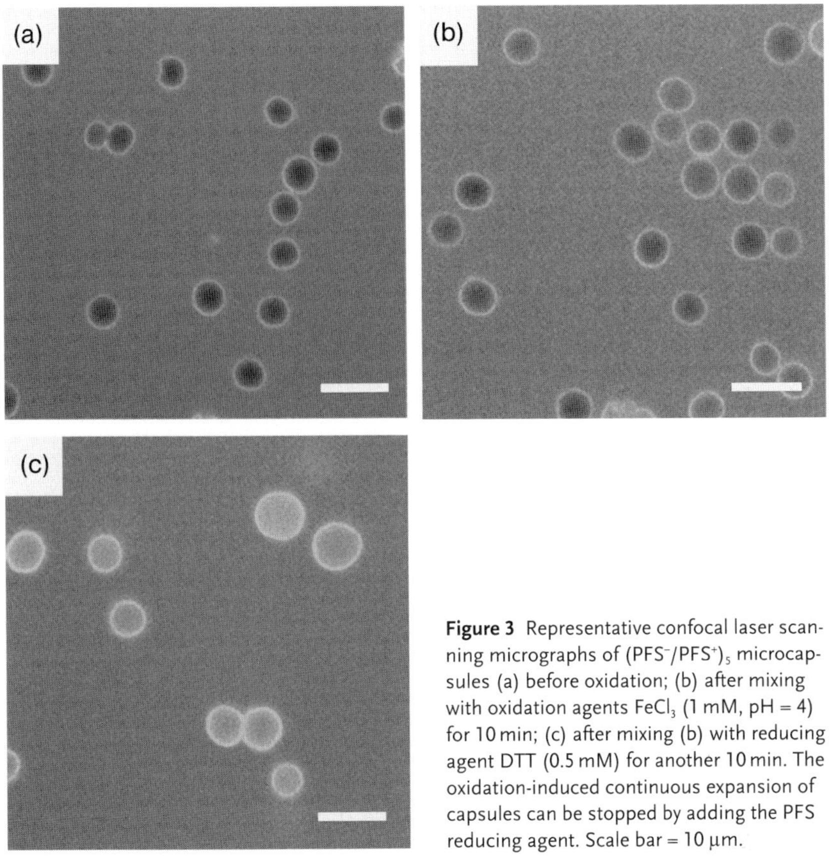

Figure 3 Representative confocal laser scanning micrographs of $(PFS^-/PFS^+)_5$ microcapsules (a) before oxidation; (b) after mixing with oxidation agents $FeCl_3$ (1 mM, pH = 4) for 10 min; (c) after mixing (b) with reducing agent DTT (0.5 mM) for another 10 min. The oxidation-induced continuous expansion of capsules can be stopped by adding the PFS reducing agent. Scale bar = 10 μm.

a series of representative CLSM micrographs recorded during the sequential oxidation-reduction process of $(PFS^-/PFS^+)_5$ microcapsules. As discussed, the whole process of oxidation–induced microcapsule permeability change and expansion until final disintegration will last for approximately 15 min under the afore-mentioned oxidation conditions ($FeCl_3$, 1 mM). In the experiment associated with Figure 3, the onset of capsule permeability change and expansion occurred after mixing with $FeCl_3$ solutions for 8 min. Figure 3(b) shows that when the reducing agent was added 10 min after the oxidation agent, many of the capsules had already become completely permeable to the probe dextran molecules (Figure 3c) and the mean diameter of the microcapsules increased to ~1.5 times its original value (averaged on ca. 20 capsules). As observed in Figure 3(c), instead of an accelerated expansion and final disintegration if only oxidants were added, these reducing-agent-treated oxidized capsules retained their spherical shape. The DTT treatment was continuously monitored for 30 min. Only a very small portion of capsules (< 5%, based on ca. 200 capsules) were observed to have disintegrated. The capsules reached a final diameter of close to twice the original value (averaged on ca. 30 capsules). The capsules that had undergone the oxidation–reduction treatment were stable and show excellent permeability to the probe molecules. Thus, chemical oxidation and reduction allows one to manipulate the permeability of the organometallic capsules.

4.4.3.5
Chemical Redox-Responsive Behavior of $(PSS^-/PFS^+)_5$ Microcapsules

In order to investigate the influence of multilayer composition on the redox-controlled permeability of layer-by-layer constructed microcapsules, microcapsules made from redox-sensitive PFS polycations and redox-insensitive polyanionic species poly(styrene sulfonate) (PSS^-) were fabricated. The PSS used here has a molar mass of 70 000 g/mol. $(PSS^-/PFS^+)_5$ microcapsules were fabricated and characterized following the same procedure as that of the $(PFS^-/PFS^+)_5$ system. Like the above-studied fully organometallic PFS microcapsules, these organic-organometallic mirocapsules also showed a redox-responsive permeability increase under the same chemical redox-trigger ($FeCl_3$, pH ~ 4). The originally impermeable capsules (towards 4400 g/mol TRITC-dextran) enlarged, became permeable to the probe, and finally disintegrated when exposed to the oxidant. However, the typical accompanying capsule expansion and disintegration proceeded at a lower speed compared to the $(PFS^-/PFS^+)_5$ microcapsules at the same oxidant concentration. In addition, the maximum duration from the onset of oxidation-induced capsule expansion until the complete disintegration of the capsules increased with decreasing the oxidant concentration. A similar trend of capsule expansion-rate dependence on oxidant concentration was also observed in the above $(PFS^-/PFS^+)_5$ capsule system. The recorded expansion and disintegration times of these two capsule systems are summarized in Table 1.

The redox-responsive properties of these organic-organometallic microcapsules indicate that PFS polycations alone are sufficient to provide characteristic redox activity to layer-by-layer constructed multilayer microcapsules. Moreover, replacing one of the redox-responsive PFS polyion components with another redox-inert polyelectrolyte species could serve as an alternative method for tuning the speed of the permeability response under a redox trigger of the corresponding microcapsules.

Table 1 Duration (in min) of oxidation induced capsule expansion until final disintegration for $(PSS^-/PFS^+)_5$ and $(PFS^-/PFS^+)_5$ systems, at varying oxidant concentrations.

$FeCl_3$ concentration (mM)	1	2	3	6
$(PSS^-/PFS^+)_5$	35		15	6
$(PFS^-/PFS^+)_5$	15	3	2	

4.4.3.6
Chemical Redox-Responsive Behavior of $(PFS^-/PAH^+)_5$ Microcapsules

Cationic and anionic PFS polyelectrolytes undergo different solution-solubility changes upon chemical oxidation. After oxidation, PFS picks up additional positive charges on the polymer main chain. With increasing positive charge density, the PFS polycations will become more soluble. On the contrary, the negative charges in the side groups of the PFS polyanions are compensated by the positive charges induced in their main chain. This may decrease the solubility of the anionic PFS polyelectrolyte species. Moreover, it may be expected that the negatively charged sulfonate groups (SO_3^-) on PFS polyanions contribute to their precipitation in aqueous media in the presence of Fe^{3+} due to ionic crosslinking. The solubility changes in exactly opposite directions of the polyelectrolytes may have consequences for the capsule behavior upon chemical oxidation by $FeCl_3$.

To probe the influence of multilayer composition on the oxidation response, planar multilayers and microcapsules based on PFS^-/PAH^+ were fabricated followed by studies on their chemical redox-responsive behavior. Multilayers of PAH^+/PFS^- were first absorbed onto quartz substrates featuring a predeposited PEI layer from 2 mg/mL aqueous solutions containing 0.5 M NaCl. The pH of the PAH^+ solutions was tuned to 6.3 using 0.1 M NaOH solutions. UV/Vis spectra recorded after each bilayer deposition confirmed a well-defined linear growth profile of the PAH^+/PFS^- multilayer system under current experimental conditions. Upon completion of the layer-by-layer deposition, a ten-bilayer sample of (PAH^+/PFS^-) on quartz was immersed into $FeCl_3$ solutions (1 mM) for different time intervals followed by UV/Vis characterization. Unlike the fast and complete desorption of PFS^+/PFS^- multilayers from the substrates, the change in the absorbance at 216 nm was minimal after oxidation for more than one hour. This shows that it is difficult to remove PAH^+/PFS^- multilayers by a similar chemical oxidation reaction.

Then, $(PAH^+/PFS^-)_5$ microcapsules were fabricated from polyelectrolyte solutions with the same concentration (2 mg/mL containing 0.5 M NaCl) onto colloidal templates ($MnCO_3$) followed by core removal. CLSM studies showed that the obtained capsules were also impermeable to the same molecular probe (4400 g/mol TRITC-dextran, Figure 4a). Similarly, $(PAH^+/PFS^-)_5$ microcapsules were mixed with $FeCl_3$ solutions (6 mM) to evaluate their permeability response. All the originally impermeable capsules became completely permeable to the same probe molecules after oxidation for over 15 to 20 min. During oxidation, the capsules underwent a continuous shrinkage, which is clearly demonstrated in the CLSM images shown in Figure 4. Subsequent addition of PFS reducing agent DTT into the reaction mixture did not reverse the capsule shrinkage.

The oxidized $(PAH^+/PFS^-)_5$ microcapsules were evaluated by atomic force microscopy (AFM) in tapping mode in the dry state. As a comparison, TM-AFM images of typical examples of $(PAH^+/PFS^-)_5$ microcapsules before and after chemical oxidation by $FeCl_3$ are shown in Figure 5. After oxidation, the original hollow structure of the microcapsules as prepared was lost. Instead, a collapsed and filled 'pancake'-like shape was adopted after oxidation, accompanied by a decrease in

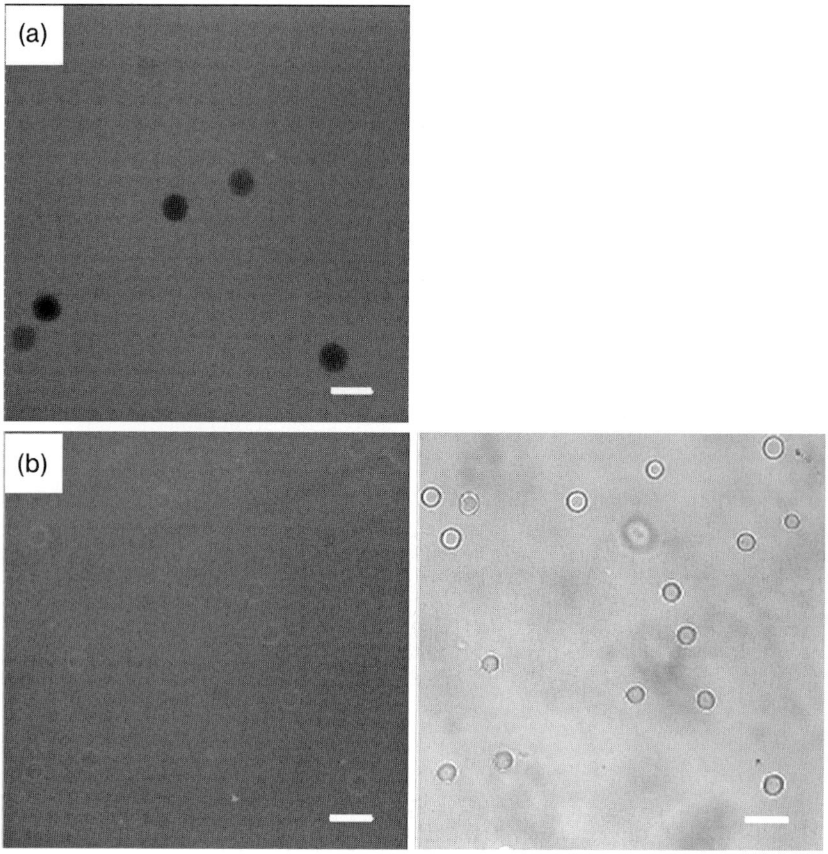

Figure 4 CLSM images of $(PAH^+/PFS^-)_5$ microcapsules oxidized by $FeCl_3$ (6 mM, pH 4.3) for (a) 4 minutes and (b) 20 min (right figure is the transmittance image). Fluorescence probe is TRITC-dextran (4400 g/mol). Scale bar = 10 μm for all the images.

the capsule diameter. The $(PAH^+/PFS^-)_5$ capsule-shrinking phenomenon is just contrary to previously observed swelling behavior of $(PFS^+/PFS^-)_5$ and $(PFS^+/PSS^-)_5$ microcapsules.

4.4.3.7
Redox-Responsive Permeability of Composite-Wall Microcapsules

In order to maintain the spherical shape and integrity of the capsules, and in the mean time effectively change the capsule permeability, capsules with PFS polyion pairs in the inner layers and redox-insensitive organic polyelectrolyte pairs in the outer layers were fabricated. After depositing five bilayers of PFS^-/PFS^+, different numbers of (PSS^-/PAH^+) 'capping' bilayers were coated onto the templates before core dissolution. The as-prepared 'composite-wall' capsules were visualized by

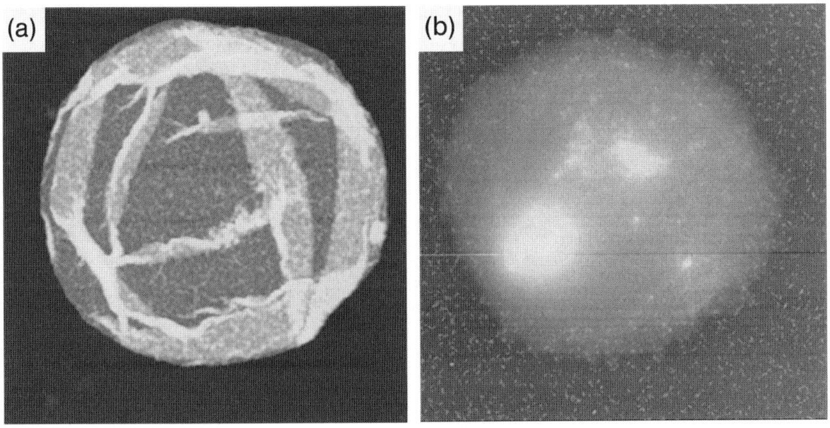

Figure 5 TM-AFM images of $(PAH^+/PFS^-)_5$ microcapsules in the dry state before (a) and after (b) chemical oxidation by $FeCl_3$ (6 mM, pH 4.3). Scan size is $9 \times 9\ \mu m^2$ and z range is 600 nm for (a) and 400 nm for (b).

CLSM. For example, $(PFS^-/PFS^+)_5(PSS^-/PAH^+)_1$ capsules demonstrated in Figure 6(a), are robust and essentially impermeable to 4400 g/mol TRITC-dextran molecules (with a population of ≥ 90%). Upon treatment with $FeCl_3$ solution (2 mM, pH = 4), the capsules started to become permeable, as was evident from the rise in fluorescent intensity in the capsule interior. Figure 6(b) shows representative CLSM images of $(PFS^-/PFS^+)_5(PSS^-/PAH^+)_1$ microcapsules after mixing with $FeCl_3$ solutions for 20 min. At this stage, the majority (more than 70%) of the capsules were completely permeable to the same probe molecules. Compared to the above-studied $(PFS^-/PFS^+)_5$ microcapsules, $(PFS^-/PFS^+)_5(PSS^-/PAH^+)_1$ composite-wall microcapsules have a slower response to the same concentration of chemical redox stimuli. By applying oxidants with the same concentration ($FeCl_3$, 1 mM), it took more than half an hour for the impermeable $(PFS^-/PFS^+)_5(PSS^-/PAH^+)_1$ capsules to show the same degree of permeability change as the full organometallic $(PFS^-/PFS^+)_5$ microcapsules. In addition to varying the oxidant concentration, the rate of permeability increase can be tuned by changing the number of capping (PSS^-/PAH^+) bilayers on the capsule wall. Oxidation experiments were performed on composite microcapsules containing different numbers of (PSS^-/PAH^+) bilayers. Increasing the number of 'capping' layers slows down the permeability change of the capsules. Obviously, these redox-insensitive polyelectrolyte bilayers were acting like 'blocking' layers on the capsule surface. On increasing the number of capping PSS^-/PAH^+ bilayers n from 1 to 3 on the same $(PFS^-/PFS^+)_5$ inner-layer structure and using oxidant solutions with the same concentration ($FeCl_3$, 2 mM), the recorded time scale for more than 80% of impermeable capsules to become permeable increased from 1 to 6 h.

'Composite-wall' PFS microcapsules were further characterized by AFM to check the integrity and wall thickness in the dry state. Examples of TM-AFM

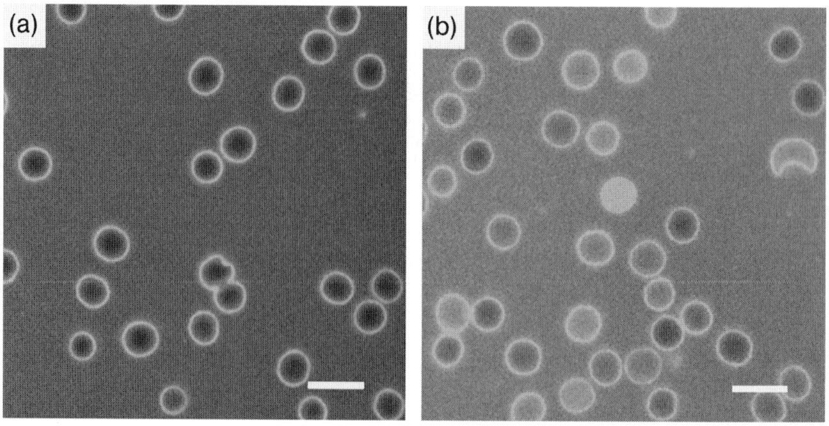

Figure 6 Local oxidation of $(PFS^-/PFS^+)_5(PSS^-/PAH^+)_1$ microcapsules by $FeCl_3$ (2 mM, pH = 4) monitored by CLSM. The majority (> 70%) of capsules that are originally impermeable (a) to 4.4 kDa dextran molecules show permeability (b) after oxidation for 20 min. Scale bar = 10 μm.

images of $(PFS^-/PFS^+)_5(PSS^-/PAH^+)_2$ capsules before and after oxidation are shown in Figure 7. The capsules show their integrity and characteristic hollow spherical structure. It is worth noting that even with only one capping bilayer of PSS^-/PAH^+, many of the capsules still preserved their integrity after oxidation for over two hours. In a sample imaging of nine $(PFS^-/PFS^+)_5(PSS^-/PAH^+)_1$ capsules with oxidation periods varying from two hours to overnight, none of them were found to be broken.

The double-wall thickness values of the capsules with $(PFS^-/PFS^+)_5(PSS^-/PAH^+)_n$ wall structures measured in the dry state by AFM before and after chemical oxidation are presented in Table 2. Previously, an average PFS^-/PFS^+ bilayer thickness of 6 nm for 4 bilayers was found. The PSS^-/PAH^+ bilayer thickness under the experimental conditions is 6 ±1 nm [37]. With these values at hand, one can estimate expected capsule wall thicknesses as a function of the number of bilayers. For $(PFS^-/PFS^+)_5(PSS^-/PAH^+)_1$ capsules a wall thickness of 36 nm is

Table 2 $(PFS^-/PFS^+)_5(PSS^-/PAH^+)_n$ capsule double wall thicknesses as measured by TM-AFM before and after oxidation for overnight (8 h) by $FeCl_3$ (1 mM, pH = 4).

No. of capping bilayers	Before oxidation (nm)	After oxidation (nm)
1	69 ± 3	22 ± 3
2	83 ± 3	38 ± 3
3	98 ± 2	70 ± 10

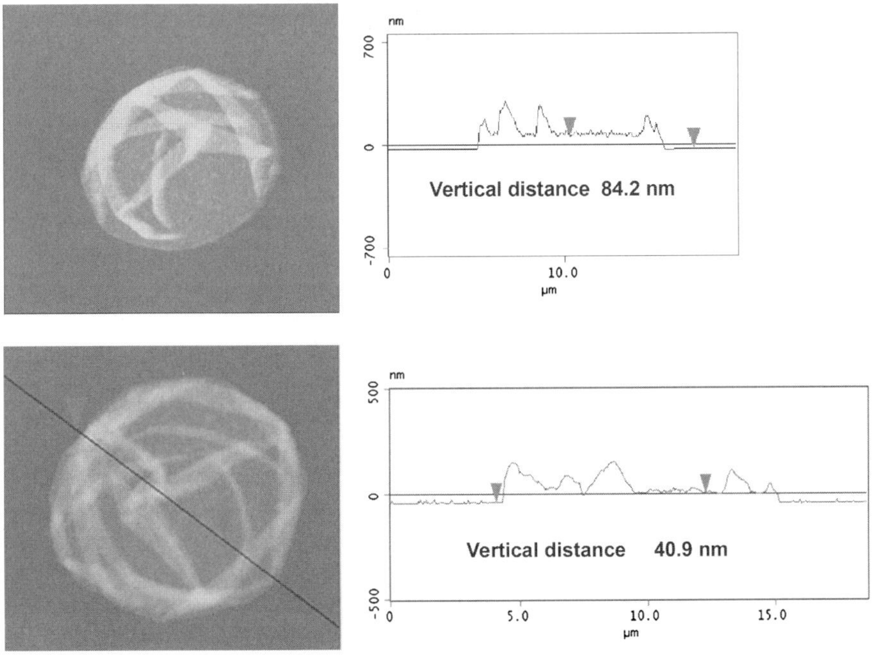

Figure 7 Representative TM-AFM height images $(PFS^-/PFS^+)_5$ $(PSS^-/PAH^+)_2$ capsules before and after oxidation for overnight (8 h) by $FeCl_3$ (1 mM, pH = 4). The integrity of the capsules has been preserved.

expected, which corresponds to a double wall thickness of 72 nm. Expected values for capsules with two and three (PSS^-/PAH^+) bilayers are 84 nm and 96 nm, respectively. These values correspond well with the double-wall thicknesses as measured by AFM.

Upon oxidation, the capsule walls decrease in thickness indicating that some material loss occurred. The oxidized capsule wall thickness values as summarized in Table 2 are higher than the expected capsule wall thickness values if only PSS^-/PAH^+ were present in the capsule wall. This suggests that although material loss occurred during the oxidation process, not all of the PFS polyelectrolytes were released from the capsule shell. Moreover, increasing the number of 'capping' layers seemed to decrease the degree of material loss. When the number of PSS^-/PAH^+ bilayers increased from 1 to 3, the decrease in capsule wall thickness upon oxidation dropped from ~70% to only ~30%. In addition, there was no substantial change in the root-mean-square (rms) roughness (5–6 nm) of the capsule wall in the fold-free regions after chemical oxidation. Both CLSM and AFM measurements show no measurable capsule diameter difference before and after oxidation.

4.4.3.8 Electrochemically Redox-Responsive Microcapsules

Depending on the electrostatic characteristics of the last-deposited polyelectrolyte layer, multilayer capsules bear positive or negative charges on their outer wall. This makes it possible to immobilize multilayer microcapsules onto charged surfaces through electrostatic interactions. In order to study the electrochemical responsiveness of PFS microcapsules, microcapsules with $(PFS^-/PFS^+)_5(PSS^-/PAH^+)_1$ wall structures were deposited onto gold substrates featuring a sodium 3-mercapto-1-propanesulfonate monolayer. The negatively charged surface is ideal for the adsorption of positively charged microcapsules [38]. Cyclic voltammograms (CV) were recorded for the organometallic-organic composite-wall polyelectrolyte microcapsules at different scan rates, as demonstrated in Figure 8.

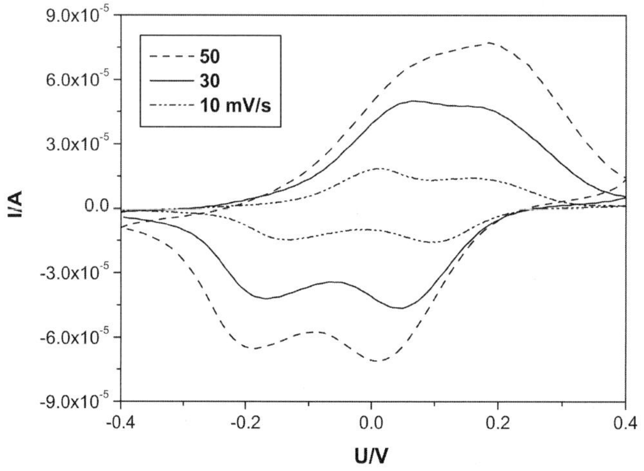

Figure 8 Cyclic voltammograms recorded on one layer of $(PFS^-/PFS^+)_5(PSS^-/PAH^+)_1$ capsules deposited on a gold electrode featuring a monolayer of sodium 3-mercapto-1-propanesulfonate. Scan rate 10–50 mV/s, 0.1 M $NaClO_4$ aqueous electrolyte solution, $Hg/HgSO_4$ reference and Pt counter electrodes.

The CVs of adsorbed PFS microcapsules show the two oxidation and reduction waves characteristic for PFS, resulting from intermetallic coupling between neighboring ferrocene centers [26]. Interestingly, the two oxidation waves could not be resolved when the scan rate is higher than 10 mV/s. This contrasts with former results on PFS multilayer films adsorbed on flat substrates, where the two oxidation and reduction waves always show up in the same range of scan rates [30b]. The capping PSS^-/PAH^+ layers may slow down electron transfer [39]. The CV scans can be repeated many times, showing the full reversibility of the electrochemical redox process.

4.4.4 Conclusions

In this work, we reported on the unique redox-responsive behavior of organometallic polyelectrolyte multilayers and microcapsules based on water-soluble poly(ferrocenylsilane) (PFS) polycations and polyanions. To our knowledge, this is the first study of a redox-responsive organometallic multilayer system. Optical absorbance based on UV/Vis spectroscopy and thickness values obtained from ellipsometric measurements show that PFS multilayers assembled on flat substrates could be removed by exposure to chemical oxidants, such as ferric chloride ($FeCl_3$).

PFS multilayer capsules were prepared using manganese carbonate ($MnCO_3$) as templates. In this way, stable and impermeable PFS capsules with an average size of 10 μm were obtained when the bilayer number was higher than four. The permeability of these stable PFS capsules could be tuned via chemical oxidation. Iodine (I_2) and iron chloride ($FeCl_3$) were both capable of oxidizing water-soluble PFS-based multilayer capsules. $FeCl_3$ proved to be more effective since a very dilute solution (in the sub-mM range) was sufficient to cause a fast increase in the capsule permeability.

The rising positive charges induced in the polymer main chain upon oxidation can bring coulomb repulsion within the chains, actually stretching the chain and making it stiffer [27, 28]. Moreover, since all the polyelectrolyte species on the capsule wall bear the same main-chain structure, electrostatic repulsion will occur both intrachain and interchain. With increasing repulsive forces along the chain as well as in the multilayer growth direction, a more loosened multilayer structure will be favored. Thus, one would accordingly expect increased pore sizes and enhanced permeability of the multilayer capsules. Enhanced permeability was demonstrated by I_2 oxidation and in the early stages of $FeCl_3$ oxidation.

As oxidation by $FeCl_3$ progressed, peculiar capsule expansion and final disintegration phenomena for capsules made solely from PFS were observed. When fully oxidized, cationic PFS will double its positive charges but anionic PFS will become charge neutral. The charge compensation originally present in the multilayer is lost upon increasing the degree of oxidation, leading to disassembly of the layer-by-layer structure [7, 41, 42]. It is worth noting that a very small trigger ($FeCl_3$ concentrations in the sub-mM range) is already sufficient to cause a fast and drastic permeability change and capsule disintegration. In addition to the electrostatic arguments presented above, PFS chains might undergo chain scission by nucleophilic attack at the ferrocenium sites in the oxidized PFS chains, which could cause the multilayer capsules to decompose. To establish if oxidation-induced macromolecular degradation plays a role, viscosity measurements were carried out on PFS polycations oxidized in solution. The $FeCl_3$-oxidized PFS showed a slow decrease in solution viscosity with time, which may indicate a slow chain scission. But the rate of chain scission (on the time scale of hours) is much slower when compared to the time scale of the capsule oxidation experiments. In other words, within an experimental time scale of 15 min up to one hour, there is no substantial viscosity change of the oxidized PFS polycation solution. This makes it very unlikely that polymer-chain scission caused the capsule disintegration. PFS polyelectrolyte multilayer removal on flat substrates by oxidation in less than five minutes further supports the view that variation in the electrostatic interactions within the multilayers most likely causes the capsule expansion–disintegration phenomenon.

Composite-wall microcapsules, fabricated by the additional adsorption of redox-insensitive polyelectrolyte species PSS^-/PAH^+ as the capsule-capping layers, were introduced to ensure the shape and integrity of redox-responsive PFS capsules. As demonstrated by CLSM, the permeability changing rate can be manipulated by varying the number of PSS^-/PAH^+ bilayers. Organic-organometallic microcapsules composed of PSS^-/PFS^+ and PFS^-/PAH^+ provided more insight into the mechanism of redox-induced permeability response. During the oxidation process, polyelectrolyte solubility variations occur that are different for the cationic and anionic PFS polyelectrolytes. Cationic PFS retained its water solubility as its positive charge density increased. On the contrary, the decrease of overall charge density and possible crosslinking effects induced by the Fe^{3+} ions will decrease the solubility of the anionic PFS polyelectrolyte species. Indeed, PSS^-/PFS^+ and PFS^-/PAH^+ behaved differently upon chemical oxidation by $FeCl_3$. Experiments showed that $(PSS^-/PFS^+)_5$ capsules underwent expansion and permeability changes similar to those of the $(PFS^-/PFS^+)_5$ capsules, while $(PAH^+/PFS^-)_5$ capsules experienced a shrinkage in size. TM-AFM showed collapsed $(PAH^+/PFS^-)_5$ capsules whose original 'hollow' structures appeared to be lost after oxidation, giving a further indication of the formation of more compact water-insoluble complexes. $(PAH^+/PFS^-)_5$ capsule shrinkage was driven by minimization of the surface area in contact with the aqueous medium. Similar capsule-shrinking phenomena have been observed before in other polyelectrolyte systems, when calcium (Ca^{2+}) ion was shown to effectively coordinate with the carboxylic acid groups of poly(methacrylic acid) (PMA) in the basic pH region [40].

Finally, the electrochemical oxidation of these composite-wall microcapsules featuring PFS polyelectrolytes was studied by fixing the capsules onto gold substrates. Cyclic voltammograms exhibited oxidation and reduction waves typical of PFS, verifying the reversible electrochemical addressability of these redox-responsive microcapsules. This novel organometallic multilayer system demonstrates that redox-stimulus responsive materials can be employed in controlled release applications. These we envisage in diverse areas, such as pharmacy and nanomaterials science.

Acknowledgments

We thank Anneliese Heilig for contributions to the AFM measurements in the Max Planck Institute of Colloids and Interfaces, Golm. The University of Twente, the $MESA^+$ Institute for Nanotechnology of the University of Twente, the Dutch Science Foundation for Chemical Research NWO-CW and NanoImpuls, a Nanotechnology Program of the Ministry of Economic Affairs of The Netherlands are acknowledged for financial support.

References

1 Barlow, S. and O'Hare, D. (1997) *Chemical Reviews*, **97**, 637.

2 Kwon, I.C., Bae, Y.H. and Kim, S.W. (1991) *Nature*, **354**, 291.

3 Ryan, A.J., Crook, C.J., Howse, J.R., Topham, P., Jones, R.A.L., Geoghegan, M., Parnell, A.J., Ruiz-Pérez, L., Martin, S.J., Cadby, A., Menelle, A., Webster, J.R.P., Gleeson, A.J. and Bras, W. (2005) *Faraday Discussions*, **128**, 55.

4 Hugel, T., Holland, N.B., Cattani, A., Moroder, L., Seitz, M. and Gaub, H.E. (2002) *Science*, **296**, 1103.

5 Liu, J. and Lu, Y. (2006) *Advanced Materials*, **18**, 1667.

6 (a) Decher, G. and Hong, J.-D. (1991) *Macromolecular Chemistry, Macromolecular Symposium*, **46**, 321; (b) Decher, G. and Hong, J.-D. (1992) *Thin Solid Films*, **210–211**, 831; (c) Decher, G. (1997) *Science*, **277**, 1232; (d) Donath, E., Sukhorukov, G.B., Caruso, F., Davis, S.A. and Möhwald, H. (1998) *Angewandte Chemie – International Edition*, **37**, 2202; (e) Hiller, J., Mendelsohn, J.D. and Rubner, M.F. (2002) *Nature Materials*, **1**, 59; (f) Tang, Z.Y., Kotov, N.A., Magonov, S. and Ozturk, B. (2003) *Nature Materials*, **2**, 413.

7 For recent reviews see (a) Decher, G. and Schlenoff F J.B. (eds) (2003) *Multilayer Thin Films, Sequential Assembly of Nanocomposite Materials*, Wiley-VCH, Weinheim; (b) Shi, X., Shen, M. and Möhwald, H. (2004) *Progress in Polymer Science*, **29**, 987; (c) Hammond, P.T. (2004) *Advanced Materials*, **16**, 1271; (d) Jiang, C. and Tsukruk, V.V. (2006) *Advanced Materials*, **18**, 829; (e) Tang, Z., Wang, Y., Podsiadlo, P. and Kotov, N.A. (2006) *Advanced Materials*, **18**, 3203.

8 Ma, Y., Dong, W.-F., Hempenius, M.A., Möhwald, H. and Vancso, G.J. (2006) *Nature Materials*, **5**, 724.

9 Yoo, P.J., Nam, K.T., Qi, J., Lee, S.-K., Park, J., Belcher, A.M. and Hammond, P.T. (2006) *Nature Materials*, **5**, 234.

10 Ma, Y., Dong, W.-F., Hempenius, M.A., Möhwald, H. and Vancso, G.J. (2007) *Angewandte Chemie – International Edition*, **46**, 1702.

11 Sukhishvili, S.A. (2005) *Current Opinion in Colloid and Interface Science*, **10**, 37.

12 Sukhorukov, G.B., Antipov, A.A., Voigt, A., Donath, E. and Möhwald, H. (2001) *Macromolecular Rapid Communications*, **22**, 44.

13 Sukhishvili, S.A. and Granick, S. (2000) *Journal of the American Chemical Society*, **122**, 9550.

14 Antipov, A.A., Sukhorukov, G.B. and Möhwald, H. (2003) *Langmuir*, **19**, 2444.

15 Dong, W.-F., Liu, S., Wan, L., Mao, G., Kurth, D.G. and Möhwald, H. (2005) *Chemistry of Materials*, **17**, 4992.

16 (a) Chung, A.J. and Rubner, M.F. (2002) *Langmuir*, **18**, 1176; (b) Burke, S.E. and Barrett, Ch.J. (2004) *Macromolecues*, **37**, 5375.

17 (a) Leporatti, S., Gao, C., Voigt, A., Donath, E. and Möhwald, H. (2001) *European Physical Journal E*, **5**, 13; (b) Glinel, K., Sukhorukov, G.B., Möhwald, H., Khrenov, V. and Tauer, K. (2003) *Macromolecular Chemistry and Physics*, **204**, 1784.

18 (a) Radt, B., Smith, T.A. and Caruso, F. (2004) *Advanced Materials*, **16**, 2184; (b) Skirtach, A.G., Dejugnat, C., Braun, D., Susha, A.S., Rogach, A.L., Parak, W.J., Möhwald, H. and Sukhorukov, G.B. (2005) *Nano Letters*, **5**, 1371.

19 Lu, Z., Prouty, M.D., Guo, Z., Golub, V.O., Kumar, C.S.S.R. and Lvov, Y.M. (2005) *Langmuir*, **21**, 2042.

20 (a) De Geest, B.G., Skirtach, A.G., Mamedov, A.A., Antipov, A.A., Kotov, N.A., De Smedt, S.C. and Sukhorukov, G.B. (2007) *Small*, **3**, 804; (b) Skirtach, A.G., De Geest, B.G., Mamedov, A., Antipov, A.A., Kotov, N.A. and Sukhorukov, G.B. (2007) *Journal of Materials Chemistry*, **17**, 1050.

21 Shchukin, D.G. and Möhwald, H. (2007) *Small*, **3**, 926.

22 Shchukin, D.G., Köhler, K. and Möhwald, H. (2006) *Journal of the American Chemical Society*, **128**, 4560.

23 Gomez-Elipe, P., Resendes, R., Macdonald, P.M. and Manners, I.

(1998) *Journal of the American Chemical Society*, **120**, 8348.
24 Ni, Y., Rulkens, R. and Manners, I. (1996) *Journal of the American Chemical Society*, **118**, 4102.
25 (a) Nguyen, M.T., Diaz, A.F., Dement'ev, V.V. and Pannell, K.H. (1993) *Chemistry of Materials*, **5**, 1389; (b) Pudelski, J.K., Foucher, D.A., Honeyman, C.H., Macdonald, P.M., Manners, I., Barlow, S. and O'Harc, D. (1996) *Macromolecules*, **29**, 1894; (c) Rulkens, R., Resendes, R., Verma, A., Manners, I., Murti, K., Fossum, E., Miller, P. and Matyjaszewski, K. (1997) *Macromolecules*, **30**, 8165; (d) Arsenault, A.C., Míguez, H., Kitaev, V., Ozin, G.A. and Manners, I. (2003) *Advanced Materials*, **15**, 503; (e) Giannotti, M.I., Lv, H., Ma, Y., Steenvoorden, M.P., Overweg, A.R., Roerdink, M., Hempenius, M.A. and Vancso, G.J. (2005) *Journal of Inorganic and Organometallic Polymers and Materials*, **15**, 527.
26 (a) Foucher, D.A., Tang, B.Z. and Manners, I. (1992) *Journal of the American Chemical Society*, **114**, 6246; (b) Foucher, D.A., Ziembinski, R., Petersen, R., Pudelski, J., Edwards, M., Ni, Y., Massey, J., Jaeger, C.R., Vancso, G.J. and Manners, I. (1994) *Macromolecules*, **27**, 3992; (c) Rulkens, R., Lough, A.J., Manners, I., Lovelace, S.R., Grant, C. and Geiger, W.E. (1996) *Journal of the American Chemical Society*, **118**, 12683.
27 Péter, M., Hempenius, M.A., Kooij, E.S., Jenkins, T.A., Roser, S.J., Knoll, W. and Vancso, G.J. (2004) *Langmuir*, **20**, 891.
28 (a) Zou, S., Ma, Y., Hempenius, M.A., Schönherr, H. and Vancso, G.J. (2004) *Langmuir*, **20**, 6278; (b) Zou, S., Hempenius, M.A., Schönherr, H. and Vancso, G.J. (2006) *Macromolecular Rapid Communications*, **27**, 103; (c) Zou, S., Korczagin, I., Hempenius, M.A., Schönherr, H. and Vancso, G.J. (2006) *Polymer*, **47**, 2483.
29 Shi, W., Cui, S., Wang, C., Wang, L., Zhang, X., Wang, X. and Wang, L. (2004) *Macromolecules*, **37**, 1839.
30 (a) Hempenius, M.A., Robins, N.S., Lammertink, R.G.H. and Vancso, G.J. (2001) *Macromolecular Rapid Communications*, **22**, 30; (b) Hempenius, M.A., Robins, N.S., Péter, M., Kooij, E.S. and Vancso, G.J. (2002) *Langmuir*, **18**, 7629; (c) Hempenius, M.A., Brito, F.F. and Vancso, G.J. (2003) *Macromolecules*, **36**, 6683.
31 (a) Power-Billard, K.N. and Manners, I. (2000) *Macromolecules*, **33**, 26; (b) Wang, Z., Lough, A. and Manners, I. (2002) *Macromolecules*, **35**, 7669; (c) Ginzburg, M., Galloro, J., Jäkle, F., Power-Billard, K.N., Yang, S., Sokolov, I., Lam, C.N.C., Neumann, A.W., Manners, I. and Ozin, G.A. (2000) *Langmuir*, **16**, 9609.
32 Ma, Y., Dong, W.-F., Kooij, E.S., Hempenius, M.A., Möhwald, H. and Vancso, G.J. (2007) *Soft Matter*, **3**, 889.
33 Manners, I. (2004) *Synthetic Metal-Containing Polymers*, Wiley-VCH, Weinheim.
34 Ma, Y., Hempenius, M.A. and Vancso, G.J. (2007) *Journal of Inorganic and Organometallic Polymers and Materials*, **17**, 3.
35 Antipov, A.A., Shchukin, D.G., Fedutik, Y., Petrov, A.I., Sukhorukov, G.B. and Möhwald, H. (2003) *Colloids and Surfaces A – Physicochemical and Engineering Aspects*, **224**, 175.
36 Zhang, H. and Rühe, J. (2003) *Macromolecular Rapid Communications*, **24**, 576.
37 Ma, Y. and Dong, W.-F., unpublished results.
38 Hodak, J., Etchenique, R., Calvo, E.J., Singhal, K. and Bartlett, P.N. (1997) *Langmuir*, **13**, 2708.
39 Gosser, D.K. (1993) *Cyclic Voltammetry: Simulation and Analysis of Reaction Mechanisms*, Wiley-VCH, p. 75.
40 Mauser, T., Déjugnat, C., Möhwald, H. and Sukhorukov, G.B. (2006) *Langmuir*, **22**, 5888.
41 Hoogeveen, N.G., Cohen Stuart, M.A., Fleer, G.J. and Böhmer, M.R. (1996) *Langmuir*, **12**, 3675.
42 Bertrand, P., Jonas, A., Laschewsky, A. and Legras, R. (2000) *Macromolecular Rapid Communications*, **21**, 319.

4.5
Nanoporous Thin Films as Highly Versatile and Sensitive Waveguide Biosensors

K.H. Aaron Lau, Petra Cameron, Hatice Duran, Ahmed I. Abou-Kandil, and Wolfgang Knoll

4.5.1
Introduction

Nanoporous oxide thin films prepared by bottom-up approaches have garnered considerable attention for a range of nanotechnology applications. For example, nanoporous anodic aluminum oxide (AAO) films and membranes have been employed as nanopatterning masks [1, 2] and as templates for nanotubes and nanorods for a wide range of metallic, semiconductor and organic materials [3–7]. As another example, oxide particle thin films deposited from colloidal suspensions have been exploited for their photonic properties and also as templates for fabricating other nanoarchitectures [8–10]. In this chapter, we show that thin films of these nanoporous oxide structures are suitable for optical waveguiding at visible wavelengths, and that the refractive index of the nanoporous structure is described very well by effective medium theory (EMT). Moreover, high-sensitivity biosensing can be accomplished by waveguide mode analysis and the EMT characterization of refractive-index changes of the nanostructure during analyte binding to the (suitably functionalized) oxide surface. Since our original publication in 2004 [11], a number of research groups have applied the concept to a variety of sensing applications [12–19].

Advantages to sensing with nanoporous oxide thin-film waveguides are the high sensitivities achieved, the ability to differentiate processes that occur within the pore structure from those occurring on the top surface, and the ability to functionalize the oxide surface by established protocols. High sensitivity is achieved due to the very sharp resonances of waveguide modes intrinsic to the physical phenomenon of optical waveguiding [20, 21], and because the nanoporous structure amplifies the optical response by providing a vast internal surface area for processes to occur [11]. The different electric-field distributions within the waveguiding film for different waveguide modes confer the ability to differentiate processes occurring in different regions of the film [22]. Surface functionalization of the nanoporous oxide film may be achieved by a variety of techniques, including silane chemistry [9, 23, 24], and physical adsorption or layer-by-layer deposition of functional macromolecules [3, 25–27].

Surface Design: Applications in Bioscience and Nanotechnology
Edited by Renate Förch, Holger Schönherr, and A. Tobias A. Jenkins

ISBN: 978-3-527-40789-7

Nanoporous oxide thin films may be conveniently prepared by bottom-up approaches that can be accomplished on a laboratory bench. We prepared nanoporous AAO by electrochemical etching (anodization), in which the pore size and pore separation are controlled by the choice of acid electrolyte and anodizing voltage [28–30]. We also employed nanoporous TiO_2 particle thin films prepared by layer-by-layer deposition, for which the pore structure is determined by the particle size [12, 31]. Waveguide sensing with nanoporous films prepared by lithographic approaches has also been reported [14, 15, 17]. Since waveguiding is conducted at wavelengths much longer than the pore dimensions, sensing abilities are not significantly affected by the perfection of order in the nanoporous structure and a wide range of fabrication routes may be employed.

This contribution is organized as follows: (1) Experimental techniques necessary for applying nanoporous thin films as waveguide sensors are first reviewed. Summaries on nanoporous AAO and TiO_2-particle thin-film fabrication, optical waveguide spectroscopy (OWS), and EMT are presented. (2) We demonstrate the principles of waveguide sensing for both static and kinetic measurements, and the high sensitivity of the system, through a simple example: avidin adsorption on nanoporous AAO and biotin binding. (3) In the last section, we discuss some advanced applications of the nanoporous waveguide sensing platform. We present the recently reported technique of simultaneous waveguide and electrochemical measurements [12], and recent results on the characterization of nanoporous AAO functionalized with peptide brushes [13] and AAO-templated layer-by-layer deposited dendrimers nanotubes.

4.5.2
Experimental Techniques

4.5.2.1
Fabrication of Nanoporous Oxide Thin-Film Waveguides by Bottom-Up Approaches

We prepared nanoporous oxide films by layer-by-layer deposition of TiO_2 particles to form foam-type particle thin films (Figure 1A) [12], and by anodization of aluminum metal films, to form nanoporous AAO with uniform, cylindrical pores that run straight through the thickness of the film (Figure 1B) [13]. As will be elaborated below, a simple waveguide configuration is to prepare the nanoporous film on top of a thin, semitransparent metal layer on glass substrates. Although Au is typically used [20], Ag and Al are also suitable metal layers [32]. We generally used films $\sim 1\ \mu m$ thick for waveguide sensing.

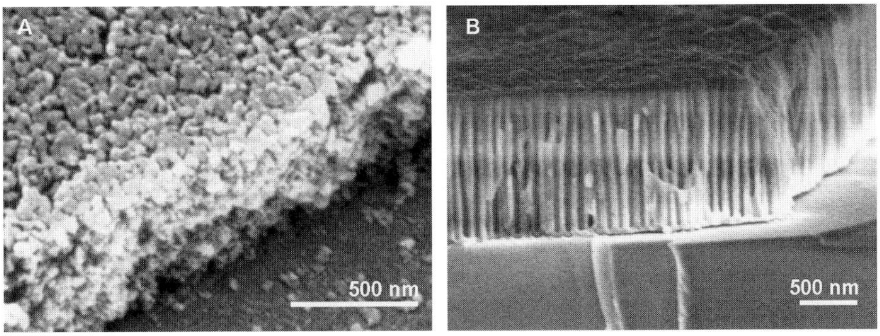

Figure 1 Scanning electron micrographs of (A) a nanoporous TiO_2 thin film deposited on a Au-coated glass slide, and (B) a nanoporous AAO thin film with a thin layer of Al in between the AAO and the glass substrate.

4.5.2.2
TiO_2 Particle Thin Films

A 50-nm Au film (with 2 nm Cr as an adhesion layer) was vacuum deposited on a high refractive index glass slide (LaSFN9 composition). Then, a self-assembled monolayer of mercapto propionic acid (MPA) was formed on the Au-coated slide by immersing it overnight in a 1 mM solution of MPA in ethanol. The slide was then rinsed in ethanol, dried in a stream of nitrogen and a layer of TiO_2 particles was deposited by dipping the slide into a colloidal TiO_2 solution (10 times diluted with respect to the 31.5 wt% stock solution in deionized water). The coated slide was rinsed in deionized water and dipped into phytic acid solution (40 mM acidified to pH 3 with perchloric acid) and again rinsed in deionized water to prepare for dip-coating of the next layer of TiO_2 particles. The number of TiO_2–phytic acid cycles dictated the film thickness. 6-nm TiO_2 particles were used for the film shown in Figure 1(A), and the films, as measured by waveguide mode analysis, had 53% porosity, but the porosity can be changed by using differently sized TiO_2 nanoparticles.

4.5.2.3
Nanoporous Anodic Aluminum Oxide (Nanoporous AAO)

1 μm thick Al films were deposited by sputtering on LaSFN9 glass substrates. The Al films were placed in a beaker of 0.3 M oxalic acid in deionized water at 2 °C, connected as the anode opposite a Pt mesh counter electrode, and anodized at a constant potential of 40 V. These conditions gave an average center-to-center pore spacing of 90–100 nm. When anodization was nearly complete (after ~ 30 min), the current was switched off to leave behind a thin Al metal layer ~ 40 nm thick. The pores were then widened after anodization to ~ 60 nm in diameter by immersing the samples in 5 wt% H_3PO_4 in deionized water (etch rate ~ 0.3 nm/min). Dif-

Table 1 AAO dimensions under different anodization conditions.

Anodizing acid, V	Pore repeating period	Accessible pore diameters
Sulfuric acid, 25 V	~ 60 nm	10–30 nm
Oxalic acid, 40 V	~ 100 nm	20–75 nm
Phosphoric acid, 160–195 V	300–450 nm	150–350 nm

ferent pore diameters and center-to-center spacings can be achieved with different anodizing acids and voltages (Table 1) [28, 29].

4.5.2.4 Optical Waveguiding and Optical Waveguide Spectroscopy (OWS)

The basic optical waveguide is a core layer of material, with a refractive index (n) higher than its surroundings and with good optical transmission properties at the guided wavelength. Light at this specific wavelength can then propagate along the core layer at certain combinations of core thickness and refractive indices of the system. For example, a one-dimensional slab waveguide transmitting light at 633 nm may be prepared by placing a layer of Al_2O_3 on top of a 40-nm Al film on a glass substrate. The Al_2O_3 film ($n_{633\,nm} = 1.64$) then acts as the waveguiding core and the underlying Al ($n_{633\,nm} = 1.37 + i7.6$)[1] and the surrounding air ($n = 1.0$) or aqueous buffer ($n_{633\,nm} = 1.33$) act as the cladding. For an Al_2O_3 thickness of ~ 1 µm, several waveguide modes can be excited at visible wavelengths – it is a multimode waveguide.

The three most important methods used to couple light into a waveguide are prism, grating and end-fire coupling [20]. We employed the Kretschmann configuration, shown in Figure 2, which uses a prism to couple light both in and out of the waveguide. Waveguide modes are indicated by minima appearing in reflectivity vs. incidence angle (R vs. θ) measurements as a laser (633 nm) is reflected off the metal layer on the substrate side of the sample. Further information on the Kretschmann configuration and OWS, as well as the grating coupling, are given in the technical glossary.

In an OWS measurement, different waveguide modes are excited in turn as the incidence angle (θ) is scanned in a θ–2θ configuration. Particular attention should be paid to the facts that different sets of waveguide modes can be excited for light polarized in different orientations. Light propagating through the waveguiding thin film (waveguide modes) with its magnetic field in the plane of the film are transverse magnetic (TM) modes, while waveguide modes with its electric field polarized in the plane of the film are transverse electric (TE) modes.

1) Refractive-index values are highly dependent on vacuum-deposition conditions.

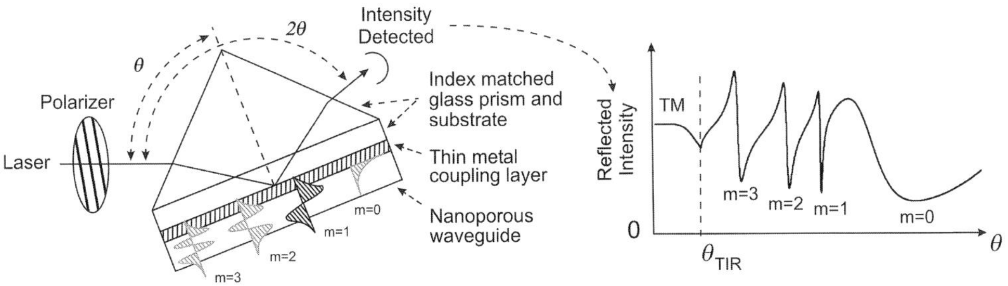

Figure 2 Schematic of optical coupling in the Kretschmann configuration and the waveguide spectroscopy detection scheme. θ_{TIR} is the critical angle for total internal reflection. The reflected intensity (R) vs. θ trace shown indicates a measurement of TM modes, for which the schematics of the electric-field distributions are overlaid on the waveguiding thin film.

The refractive index and the thickness of the waveguide film layer sensitively control the incidence angles at which waveguide mode resonances can be excited. Thus, analyte binding, which causes refractive-index and thickness changes in a nanoporous thin film, are sensitively detected as shifts in the waveguide mode resonance angles.

The electric-field distributions and the reflectivity vs. incidence angle (R vs. θ) trace shown in Figure 2 were calculated by solving the Maxwell equations for light incident on a 1-dimensional layer system described by the thickness and refractive indices of the different layers. The solutions are the Fresnel equations [20, 21], which describe the slab waveguide sample prepared on glass substrates exactly; in experiments, the thickness and refractive index of the waveguide are measured by finding those values that best fit R vs. θ measurements.

4.5.2.5
Effective Medium Theory (EMT)

In the case of a porous thin film, light can be guided with minimum scattering if the pores are much smaller (< 1/10) than the wavelength of the propagating radiation [11, 33]. As described in Sections 4.5.2.2 and 4.5.2.3, nanoporous AAO and TiO_2 thin films can be easily prepared such that the pores are much smaller than the 633-nm laser wavelength employed. Although the guided light cannot directly resolve the nanostructure, it responds to the nanoscale heterogeneity through the effective refractive index (n_{eff}) of the composite pore-oxide film. For dielectric materials, EMT based on the Maxwell–Garnett approach can predict n_{eff} with high accuracy based on the pore structure, the volume fractions (f) occupied by the pores and the solid matrix, and the refractive indices of the component pore and matrix materials. In other words, measurement of n_{eff} of a nanostructure reveals information about its nanostructure, component volume fractions, and the refractive indices of the nanostructure components [33]. n_{eff} of the nanoporous oxide film is given by [33, 34]:

$$n_{\mathrm{eff(film)}} = \sqrt{\varepsilon_{\mathrm{eff(film)}}}, \ \varepsilon_{\mathrm{eff(film)}} = \varepsilon_{\mathrm{oxide}} \frac{\varepsilon_{\mathrm{oxide}} + (f_{\mathrm{pore}} + f_{\mathrm{oxide}} q)(\varepsilon_{\mathrm{pore}} - \varepsilon_{\mathrm{oxide}})}{\varepsilon_{\mathrm{oxide}} + f_{\mathrm{oxide}} q(\varepsilon_{\mathrm{pore}} - \varepsilon_{\mathrm{oxide}})} \tag{1a}$$

$$\text{for } q = 0, \ \varepsilon_{\mathrm{eff(film)}} = \varepsilon_{\mathrm{oxide}} + f_{\mathrm{pore}}(\varepsilon_{\mathrm{pore}} - \varepsilon_{\mathrm{oxide}}) \tag{1b}$$

$$\text{for } q = 1, \ \varepsilon_{\mathrm{eff(film)}} = \frac{\varepsilon_{\mathrm{oxide}}\varepsilon_{\mathrm{pore}}}{f_{\mathrm{pore}}\varepsilon_{\mathrm{oxide}} + f_{\mathrm{oxide}}\varepsilon_{\mathrm{pore}}} \tag{1c}$$

where ε is the dielectric constant at infinite frequency. q is a parameter describing the electromagnetic field response due to the shape and orientation of the nanopores ($0 \geq q \geq 1$). $\varepsilon_{\mathrm{oxide}}$ is generally obtained by other bulk measurements, and $\varepsilon_{\mathrm{pore}}$ may correspond to $\varepsilon_{\mathrm{air}}$ or $\varepsilon_{\mathrm{buffer}}$.

The random arrangement of particles in the nanoporous TiO_2 films confers the film an isotropic optical response. In such a case, $q = 1/3$. In the case of nanoporous AAO films with cylindrical pores aligned to the surface normal (Figure 1B), there is a clear structural anisotropy and there is a corresponding anisotropy in the refractive index description. For an electric field oriented parallel to the surface normal and to the pore/oxide interfaces, the field probes the different materials in turn and does not cut across the interfaces. Thus, $n_{\mathrm{eff(film)}}$ in this direction is simply the volume fraction weighted average of the refractive indices of the pore and oxide materials ($q = 0$, (1b)). On the other hand, the cylindrical pores are arranged at random in the plane of the surface, and there is radial symmetry in the pore distribution in the plane of the film, thus $q = 1/2$ for $n_{\mathrm{eff(film)}}$ along the surface normal. In summary, taking the surface normal as the z-direction, the refractive indices describing the nanoporous films are:

$$TiO_2 \text{ particle assembly: } n_x = n_y = n_z = n_{\mathrm{eff(film)}} \ (q = 1/3) \tag{2}$$

$$\text{Nanoporous anodic alumina: } n_x = n_y = n_{\mathrm{eff(film)}} \ (q = 1/2); \ n_z = n_{\mathrm{eff(film)}} \ (q = 0) \tag{3}$$

The optical response to biomolecule binding on the pore walls of nanoporous thin films is described by applying (1) in an iterative manner. We first adapt (1) to describe $n_{\mathrm{eff(pore)}}$ within the pores, which now consist of both biomolecules on the pore walls and the surrounding media (buffer):

$$n_{\mathrm{eff(pore)}} = \sqrt{\varepsilon_{\mathrm{eff(pore)}}}, \ \varepsilon_{\mathrm{eff(pore)}} = \varepsilon_{\mathrm{buffer}} \frac{\varepsilon_{\mathrm{buffer}} + (f_{\mathrm{biomolecule}} + f_{\mathrm{buffer}} q)(\varepsilon_{\mathrm{biomolecule}} - \varepsilon_{\mathrm{buffer}})}{\varepsilon_{\mathrm{buffer}} + f_{\mathrm{buffer}} q(\varepsilon_{\mathrm{biomolecule}} - \varepsilon_{\mathrm{buffer}})} \tag{4}$$

$\varepsilon_{\mathrm{eff(pore)}}$ is then substituted for $\varepsilon_{\mathrm{pore}}$ in (1) to calculate $n_{\mathrm{eff(film)}}$ of the entire film–biomolecule nanostructure. Since the amount of biomolecules bound to the pore surface in the context of high-sensitivity sensing is expected to be small (e.g., a monolayer), and the biomolecule and the buffer are both dielectrics with relatively small differences in refractive index, the choice of q in (4) does not significantly impact on the numerical accuracy of the model. For convenience, $q = 0$ may be assumed.

For isotropic TiO_2 particle films with thick adsorbed layers, the Bruggeman equations [33] may be more appropriate. A fully anisotropic model [35] is also available for AAO: $q = 1/2$ for $n_{\text{eff(pore)}}$ in the x and y directions, and $q = 0$ for $n_{\text{eff(pore)}}$ in the z direction, which are substituted in:

$$n_{x,y} = \sqrt{\varepsilon_{x,y}}, \varepsilon_{x,y} = \varepsilon_{\text{Al}_2\text{O}_3} \frac{\varepsilon^{q=1/2}_{\text{pore-eff}} + \varepsilon_{\text{Al}_2\text{O}_3}\Delta) + f_{\text{cycl.}}(\varepsilon^{q=1/2}_{\text{pore-eff}} - \varepsilon_{\text{Al}_2\text{O}_3}\Delta)}{\varepsilon^{q=1/2}_{\text{pore-eff}} + \varepsilon_{\text{Al}_2\text{O}_3}\Delta) + f_{\text{cycl.}}(\varepsilon^{q=1/2}_{\text{pore-eff}} - \varepsilon_{\text{Al}_2\text{O}_3}\Delta)}, \Delta = \sqrt{\frac{\varepsilon^{q=1/2}_{\text{eff-pore}}}{\varepsilon^{q=0}_{\text{eff-pore}}}} \quad (5a)$$

$$n_z = \sqrt{\varepsilon_z}, \varepsilon_z = \varepsilon_{\text{oxide}} + f_{\text{pore}}(\varepsilon^{q=0}_{\text{pore(eff)}} - \varepsilon_{\text{oxide}}) \quad (5b)$$

to calculate the effective index for the entire AAO–buffer–adsorbed layer composite film. Finally, applications involving components of much higher optical density than either the alumina or buffer (e.g. quantum dots and metal nanoparticles), may be better described by models that include results from numerical simulations [34].

4.5.3 Nanoporous Waveguide Sensing

We use the example of avidin adsorption and subsequent biotin binding to illustrate the basic technique. Avidin [36] is a glycoprotein with strong ($K_a = 10^{15}\ M^{-1}$) noncovalent interaction with biotin, which is an important cofactor in metabolic reactions related to cell growth. Avidin has a molecular mass of 67–68 kDa and biotin is a small molecule with a molecular mass of 244 g/mol.

Figure 3(a) shows the R vs. θ measurements of a blank nanoporous AAO thin film, and the same film after 1.5 h adsorption from 0.5 mg/ml avidin in pH 7.4 phosphate buffered saline (PBS). The AAO was immersed in PBS throughout this set of measurements and 3 TE and 2 TM waveguide mode minima could be observed. (In fact, a third TM waveguide mode minimum near 47° was convoluted with changes in reflectivity relating to the phenomenon of total internal reflection at $\theta_{TIR} = 47.33°$, and was not used for analysis.) The first point to note is the excellent fit between the waveguide measurements and the Fresnel calculations using an effective refractive index description of the nanoporous film. For example, the pore fraction and pore diameter of the blank nanoporous AAO calculated according to the measured n_{eff} are 33% and 54 nm, respectively, which are in excellent agreement with scanning electron micrographs (Figure 1B).

Avidin adsorption concentrated a layer of protein on the pore surfaces of the waveguide, thereby displacing an amount of PBS in the pores corresponding to the volume of the avidin layer. Avidin has a refractive index, $n_{\text{avidin}} = 1.52$ [37], which is higher than the refractive index of PBS, $n_{\text{PBS}} = 1.33$. Therefore, n_{eff} of the nanoporous film also increased with adsorption and waveguide mode resonances were shifted to higher incidence angles [11]. n_{eff} before and after avidin adsorption

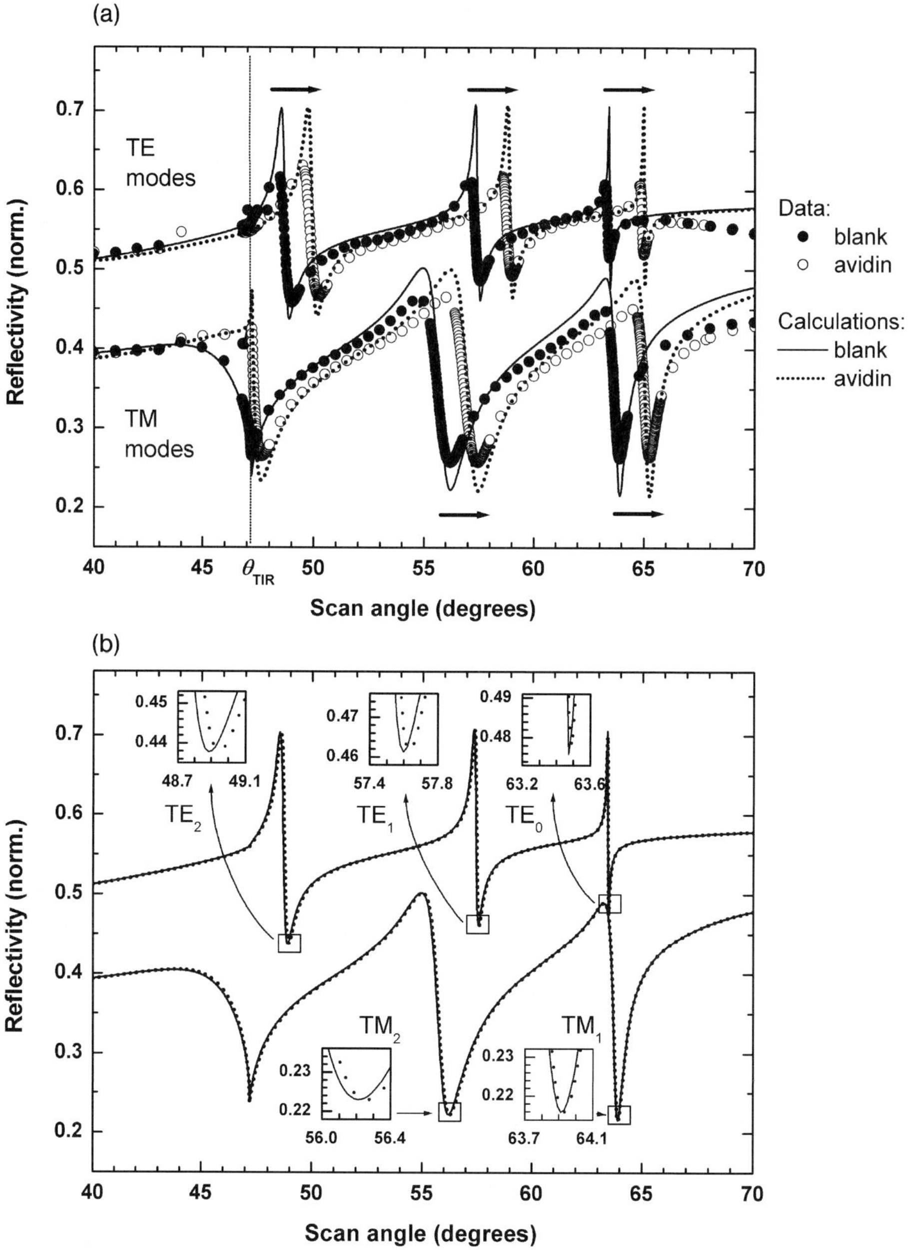

Figure 3 R vs. θ response for (a) adsorption of a 2.7-nm layer of avidin throughout a nanoporous AAO waveguide, and (b) an identical layer on only the top surface of the waveguide. In (a), solid symbols refer to waveguide measurements of the blank nanoporous AAO and open symbols refer to the measurement after avidin adsorption. In both (a) and (b), solid lines refer to Fresnel calculations fitting the blank waveguide, while dotted lines refer to calculations for the response after avidin adsorption.

are given in Table 2 and the avidin layer thickness, calculated by assuming a uniform layer of protein covering the pores, is 2.7 nm. This is in excellent agreement with the dimensions of avidin (2 × 2.8 × 2.8 nm) [37] and also with the results of Clerc and Lukosz, who measured avidin adsorption on the surface of a solid oxide waveguide [37].

Table 2 Effective refractive indices of the nanoporous AAO before and after avidin adsorption.

$n_{PBS} = 1.332$ $n_{avidin} = 1.52$ $n_{Al2O3} = 1.637$		Measured refractive index	Modeled refractive index	Index of the pores only (PBS + avidin)
Blank film	$n_{x,y} =$	1.530	1.530	$= n_{PBS}$
immersed in PBS	$n_z =$	1.541	1.543	(no avidin)
Avidin adsorbed	$n_{x,y} =$	1.544	1.544	1.370
	$n_z =$	1.553	1.554	

An important point to note in Figure 3(a) is that all the waveguide modes shifted by approximately the same amount. As depicted in Figure 2, the different waveguide modes have different spatial distributions in field amplitudes within the film. Individual modes are most sensitive to changes occurring in the region of the film where the field intensities are highest. Therefore, the quasiparallel shifts in all the waveguide modes in Figure 3(a) indicated that adsorption was uniform along the cross-section of the film and avidin had diffused efficiently down the pores of the nanoporous AAO [11].

Another point to note is that the ~ 1.3° shift in waveguide mode resonances corresponding to the 2.7-nm monolayer of avidin adsorbed was more than 3 orders of magnitude larger than the 0.001° angular resolution allowed for by the sharpness of the reflectivity minima, and translates to an excellent signal-to-noise ratio. For comparison, Figure 3(a) shows the calculated optical response for the adsorption of a hypothetical monolayer of avidin on only the top surface of a solid waveguide, which produces much smaller angle shifts < 0.05°. Figure 3(b) also shows that, unlike the quasiparallel shifts resulting from uniform avidin adsorption throughout the nanoporous structure, the presence of a layer of film on only the top surface of the waveguide produces highly asymmetric waveguide mode shifts. These shifts are larger for higher-order modes at lower incidence angles. In other words, surface adsorption within the pores and on the top surface of the waveguide can be distinguished by analysis of the different degrees of shifts for the different waveguide modes. This kind of mode analysis was employed for the characterization of layer-by-layer deposited dendrimer polyelectrolytes to be described in Section 4.5.4.2.

Other than measuring the waveguide responses before and after a surface reaction had occurred, real-time kinetic measurements of the surface reaction may also be performed by *in-situ* tracking of the waveguide mode angle change induced by the reaction. This may be accomplished either by directly tracking the resonance angle change or by monitoring the concomitant reflectivity change at a fixed angle where the R vs. θ trace is quasilinear (Figure 4a). In the latter case, the angle shift can be calculated from the slope of the R vs. θ trace. The layer thickness change can then be estimated by fitting the Fresnel and EMT equations to effective refractive-index changes that reproduce the measured angle change. The

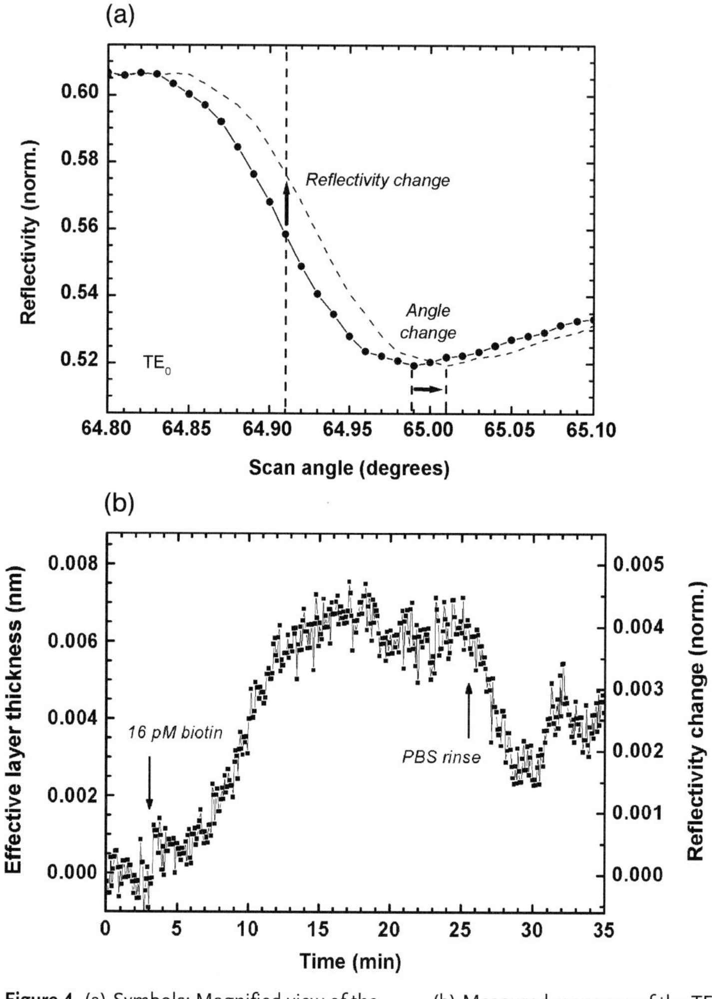

Figure 4 (a) Symbols: Magnified view of the TE_0 mode after avidin adsorption shown in Figure 3(a). Dashed line: hypothetic angle shift of the mode in response to surface binding of molecules. The arrows indicate how the shift can be monitored in real time. (b) Measured response of the TE_0 mode shown in (a) to biotin binding from 16 pM solution. The reflectivity change was monitored at $\theta = 64.91°$ where the R vs. θ trace near TE_0 is quasilinear and has the highest slope.

refractive index of the molecule layer is needed for this calculation, but if it is not known, the Fresnel and EMT equations may be solved simultaneously for measurements in multiple environments (e.g. in air and buffer, or different solvent mixtures) for consistent values of both molecular layer thickness and refractive index.

Figure 4(b) shows the kinetics of biotin binding to the avidin-functionalized nanoporous AAO from 16 pM biotin in pH 7.4 PBS. At such dilute concentrations, diffusion and fluid handling become significant factors in the binding kinetics. The slow rise in reflectivity during the first 5 min after the biotin solution was introduced, probably reflected the mixing of the biotin solution with the pure PBS that had initially filled the sample chamber and the porous thin film. Hohlbein *et al.* applied single-molecule fluorescence correlation spectroscopy to characterize the diffusion within nanoporous AAO and found 1-dimensional diffusion behavior within the straight pores [38]. Applying their results to the present experiment, small molecules like biotin would have a dwell time of the order of 1 ms within individual cylindrical pores of the AAO waveguiding thin films, which is an order of magnitude longer in duration than unrestricted 3-dimensional diffusion.

In Figure 4(b) saturation in binding was reached after ~10 min. Rinsing with PBS resulted in a final reflectivity change of ~0.002. This translates to a ~0.003 nm effective layer thickness change and ~10^2 biotin molecules bound within each pore. Sensing in such dilute concentrations is possibly close to the limit of *label-free* detection of such a small molecule.

4.5.4 Advanced Applications

4.5.4.1 Functionalization of Nanoporous AAO with Polypeptide Brushes

The high sensitivity possible in label-free detection with nanoporous waveguide sensing shown in the previous section was possible in part because of the strong affinity between biotin and avidin. For a number of biosensing applications, it is desirable to maximize the number of binding sites in order to enhance sensing capabilities. The use of a nanoporous matrix such as nanoporous AAO or TiO_2 to provide an extra internal surface for binding is one step towards that goal. Functionalization of the surface with polymer brushes to create more binding sites per unit area represents another strategy.

We modified nanoporous AAO with poly(γ-benzyl-L-glutamate) (PBLG) polypeptide brushes by surface-initiated polymerization. PBLG is an interesting polypeptide because its ester side chains may be easily modified, and because it possesses nonlinear optical properties [13]. To prepare PBLG brushes, the surface of AAO was first functionalized with a layer of 3-aminopropyl-triethoxysilane (APTES) by immersing the nanoporous AAO in 1 wt% APTES solution in deionized water.

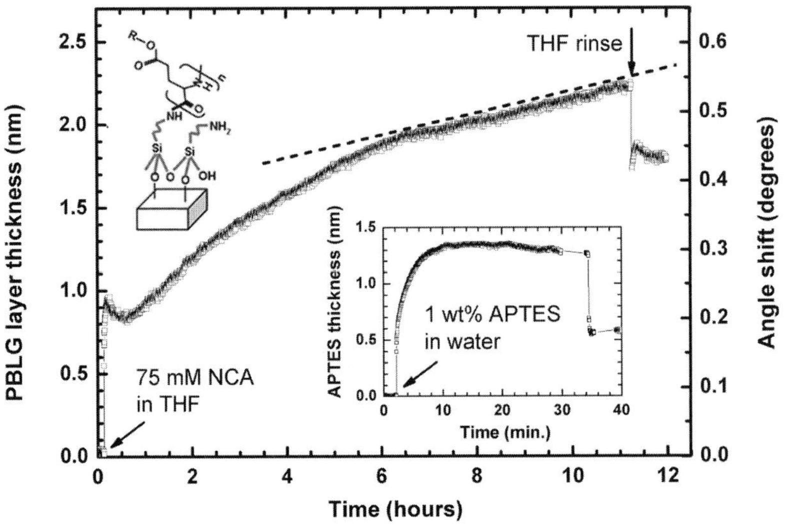

Figure 5 Kinetics of NCA polymerization to form PBLG brushes on nanoporous AAO, as measured by tracking the waveguide mode resonance angle of the TM_1. The inset shows the preceding functionalization of the oxide surface with APTES. The large drops in layer thickness near the end of the measurements were caused by rinsing with pure solvent. The schematic shows the basic molecular structure of the resulting polymer system.

Preparations in organic solvents and in the gas phase are also possible [9]. The resulting primary amines on the oxide surface act as initiators for N-carboxy anhydride (NCA) polymerization to form PBLG.

Figure 5 shows the polymerization kinetics from 75 mM NCA in anhydrous tetrahydrofuran and the inset shows the kinetics of APTES functionalization. Both processes were monitored *in situ* by tracking the resonance-angle change of the TM_1 waveguide mode. APTES deposition kinetics was best fit by assuming a 2-step model in which the first ~1 min of deposition was diffusion limited, but in which subsequent deposition was described by simple Langmuir kinetics. This initial transient might indicate the time necessary for APTES to permeate the nanoporous thin film. For PBLG growth, after an initial spike, the average PBLG chain length reached ~2 nm after 10 h. There appeared to be two regimes to the polymerization process. In particular, the second stage appeared to follow a linear growth regime (dashed line in Figure 5). Detailed analysis of the polymerization process is ongoing. Nonetheless, the ability to characterize, *in situ*, the surface-initiated polymerization in the nanoporous AAO thin film is expected to yield information on the 3-dimensional functional architecture, and to reveal the kinetics of brush growth within a nanoconfined geometry.

4.5.4.2
In-Situ Characterization of Layer-by-Layer (LbL) Deposition of Dendrimer Polyelectrolyte

Another strategy for increasing the number of binding sites per surface area is the layer-by-layer (LbL) deposition of polyelectrolyte multilayers. The LbL process is simple and flexible [39], and the multilayer assembly can be functionalized to obtain a permeable, 3-dimensional mesh with a high density of ligands [25]. Moreover, such functionalized multilayers prepared on a nanoporous AAO thin film also result in a nanotube geometry. Such a geometry has been applied as nanotest tubes for drug delivery [40] and has been suggested for separation processes [3]. If the nanotubes remained immobilized in the nanoporous AAO matrix, sensing configurations with high lateral resolution may also be envisioned [3].

We prepared LbL assemblies from alternating depositions of *N,N*-disubstituted hydrazine phosphorus-containing dendrimers having 96 terminal functional groups with either cationic ($G_4(NH^+Et_2Cl^-)_{96}$) or anionic ($G_4(CHCOO^-Na^+)_{96}$) character [41–43]. The nanoporous AAO thin films used were 1.3 μm thick with pores ~60 nm in diameter. The dendrimers have an unperturbed diameter of ~4 nm and they were deposited from 1 mg/ml deionized water containing a small amount of NaCl. LbL deposition [3] was preceded by functionalization of the AAO surface with 3-aminopropyl-dimethylethoxysilane (APDMES) from the gas phase to give a positively charged surface in water. The increase in total dendrimer thickness was monitored *in situ* over a fixed sample area by measuring the R vs. θ waveguide response. The thickness values were calculated from the waveguide resonance angle shifts by the Fresnel and EMT equations (see Sections 4.5.2.4 and 4.5.2.5) and are shown in Figure 6.

As discussed earlier, waveguide mode analysis can distinguish between processes occurring on the internal surface within pores (Figure 6a) from those on the top surface of the nanoporous waveguide (Figure 6b). The much lower uncertainties in the measurement within pores compared to that on the top surface are a reflection of the order of magnitude higher internal surface area of the porous structure over which the optical response was integrated. Figure 6 shows that the LbL process within the pores stopped after 2.5 bilayers, but assembly continued on the top surface in a linear fashion. Moreover, the per layer thickness increase within the pores was ~2 nm, which, when compared with the ~4 nm diameter of the dendrimer, indicates ~50% coverage per layer. This would be consistent with the 55% theoretical 'jamming' limit of random sequential adsorption [44]. The decrease in pore size after deposition measured by scanning electron microscopy also agrees with the internal layer thickness increase measured by waveguide mode analysis. On the other hand, the per layer thickness increase on the top surface was ~4 nm, and might indicate a different deposition process.

The halt in deposition within the pores may be indicative of charge repulsion effects as the pores become smaller with layer buildup. Further LbL deposition experiments on nanoporous AAO are being performed to probe the size of the dendrimer charge barrier in a variety of conditions. In any case, the present

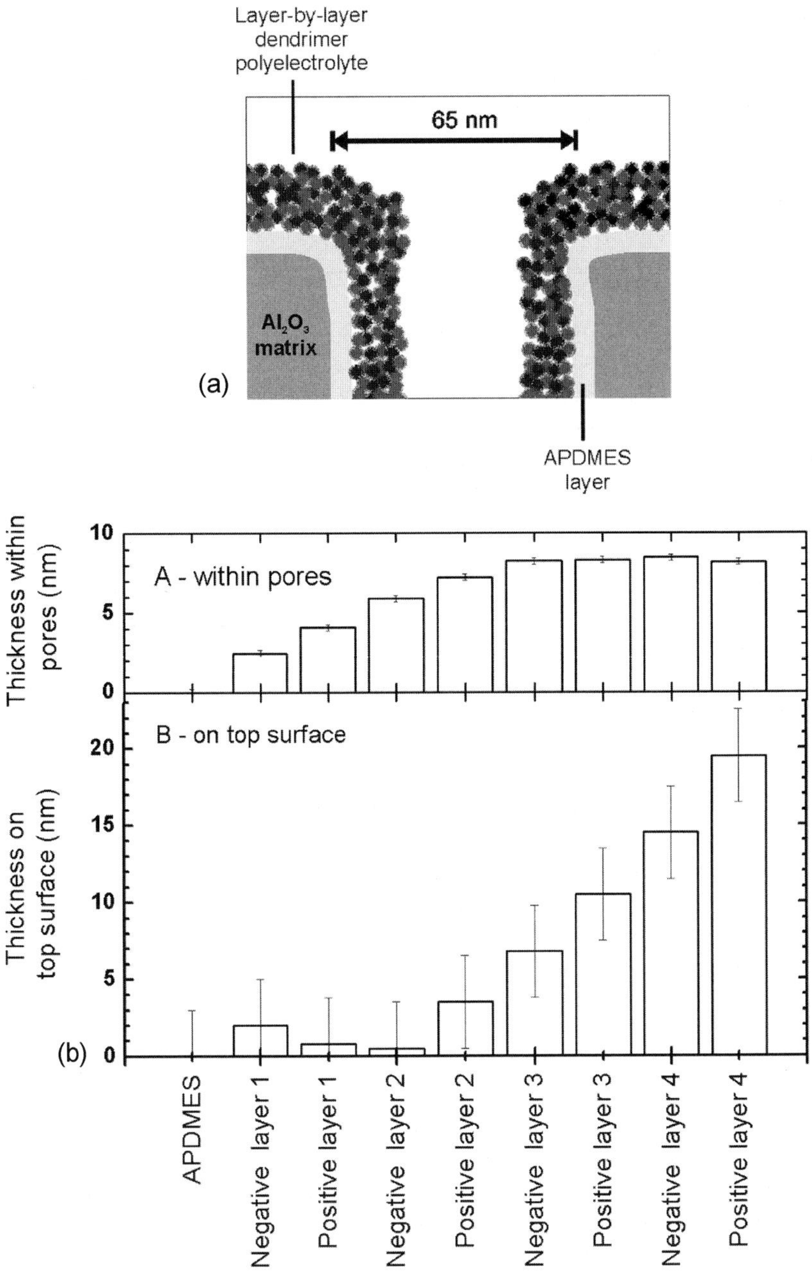

Figure 6 Dendrimer thickness measured by waveguide mode analysis as successive layers were deposited onto nanoporous AAO surface; (a) shows the thickness on the pore wall and (b) shows the thickness on the top surface.

experiment has clearly demonstrated the selectivity and sensitivity of nanoporous waveguide measurement to distinguish, *in situ*, processes occurring within and on top of a nanoporous thin film, which provides an approach to guide the preparation of nanoarchitectures using nanoporous templates.

4.5.4.3
Simultaneous Waveguide and Electrochemistry Measurements

The previous examples have shown that waveguide sensing with nanoporous AAO thin films is a highly sensitive technique. It is also able to characterize the nanoporous thin-film structure on which the technique is based by measurement of the effective refractive index of the film, and is able to distinguish processes occurring within the pore structure from those occurring on its top surface via waveguide mode analysis. Nonetheless, it is desirable to have an independent technique to verify the transport of analyte through the porous structure and to independently measure the processes occurring in the nanoporous thin film. To this end, we performed simultaneous waveguide and electrochemical measurements with a 1.6 μm thick nanoporous TiO_2 thin film deposited on a bottom Au electrode that is in direct contact with the buffer-filled porous network. As a biosensing example, we measured the adsorption of the redox active cytochrome c protein [12].

Cytochrome c has been shown to readily permeate a TiO_2 porous network [45], which provides adsorption pockets for the 2.5 × 2.5 × 3.7 nm protein [12]. Analogous to the previous examples on avidin, PBLG and dendrimer assemblies, cytochrome c adsorption within the nanoporous TiO_2 thin film can be measured by waveguide angle tracking. However, an electrochemical current would only be measured when the protein has diffused through the porous network to the bottom of the film, where it can exchange redox electrons with the bottom Au electrode. In addition, the kinetics of adsorption and the time lag between registering the waveguide response and the electrochemical signal would give valuable information about the diffusion characteristics within the nanoporous film. Results for adsorption from 0.1 mM cytochrome c in PBS are shown in Figure 7.

The presence of the electrochemical signal (Figure 7b) showed conclusively that cytochrome c was able to permeate the nanoporous network composed of TiO_2 particles only 6 nm in diameter. Moreover, this electrochemical signal lagged behind the waveguide signal (Figure 7a) by ~30 min. Analysis of the adsorption kinetics reveal that the diffusion coefficient of cytochrome c within the nanoporous network was $\sim 1 \times 10^{-16}$ m^2/s, which is 5 orders of magnitude lower than in solution when there is no geometric confinement [12]. Moreover, comparison of Figure 7(a) and (b) also showed that waveguide sensing recorded 2 orders of magnitude higher amounts of cytochrome c than the electrochemical measurement. This indicated that electron transfer occurred only via the bottom Au electrode and not via the TiO_2 matrix. It also demonstrated the sensitivity of the nanoporous waveguide sensing technique imparted by the vast internal pore surface area over which surface processes can occur.

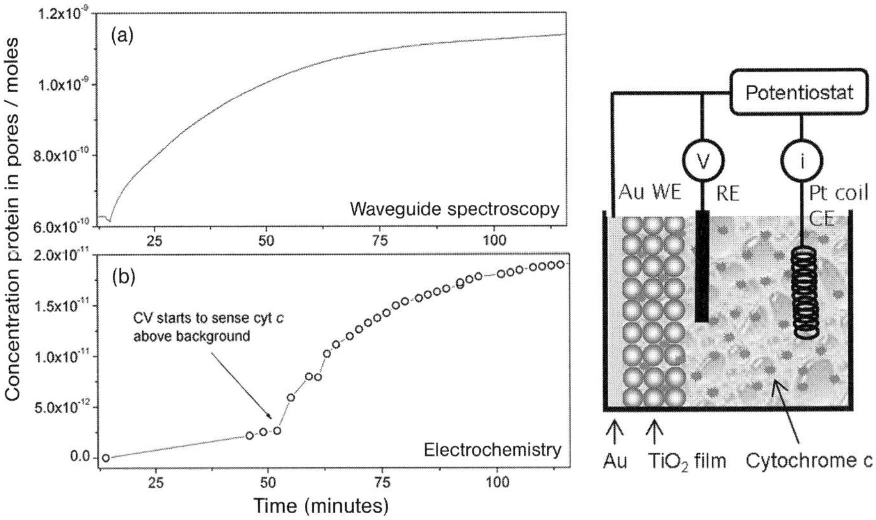

Figure 7 Adsorption from 100 μM cytochrome c in PBS in a nanoporous TiO_2 thin film, simultaneously measured by (a) waveguide mode tracking, and (b) cyclic voltammetry (CV). The schematic on the right shows the electrochemical setup and the diffusion of cytochrome c to the Au electrode at the bottom of the nanoporous TiO_2 film. The concentration of protein adsorbed was calculated from the angle shift of a TM_3 waveguide mode in (a) and from integrating the area under repeated cyclic voltammetry scans at 10 mV/s in (b).

4.5.5 Conclusions

We have demonstrated biosensing with thin-film waveguides of nanoporous anodic aluminum oxide (nanoporous AAO) as well as with nanoporous TiO_2 particle films. We employed optical waveguide spectroscopy as the detection scheme, but other methods may be used according to the needs of individual laboratories. The optical properties of the nanoporous thin films as well as the quantitative description of analyte binding were analyzed by Fresnel equations and effective medium theory, and the model calculations were in excellent agreement with measurements.

High sensitivities were achieved for label-free detection. For example, binding from 16 pM solution of a small molecule, biotin, to a nanoporous AAO waveguiding film functionalized with avidin was detected. This was achieved due to the very sharp resonances of waveguide modes, and because the nanoporous structure amplifies the optical response by providing a vast internal surface area over which surface processes can occur and the signal integrated.

Waveguide mode analysis also enables the differentiation of processes occurring inside the nanoporous thin film from those occurring on the top surface of the film. Using nanoporous AAO as a template, we have characterized, *in situ*, the transition from layer-by-layer deposition of dendrimer polyelectrolytes within the cylindrical pores, to deposition of extraneous material on top of the template. We also demonstrated the functionalization of nanoporous AAO with poly(γ-benzyl-L-glutamate) peptide brushes. Such 3-dimensional molecular architectures can provide high densities of ligands per area for further enhancement of sensing capabilities.

In designing sensing architectures with nanoporous waveguide platforms, attention should be paid to the size ratio between the pores and the analyte. Transport of redox active cytochrome c in a nanoporous TiO_2 thin film was studied by simultaneous waveguide and electrochemical measurements. Cytochrome c concentration at an electrode placed at the bottom of the film was directly quantified by cyclic voltammetry.

In summary, we have demonstrated sensing and characterization of a variety of surface processes relevant to biosensing using nanoporous waveguiding thin films. This approach has opened up a broad avenue for label-free, high-sensitivity biosensing. At the same time, other interesting systems and sensitivity enhancements may be explored in combination with fluorescent- and nanoparticle-labeling.

Acknowledgments

We wish to thank Dr. Frank Marken at the Department of Chemistry, University of Bath, and Gunnar Glasser at the Max-Planck-Institute for Polymer Research, Mainz, for their scanning electron microscopy work. We are also grateful to Stefan Schmitt at the Institut für Mikrotechnik Mainz GmbH, for the deposition of Al films, and the Molecular, Macromolecular and Supramolecular Heterochemistry group of Jean-Pierre Majoral and Anne-Marie Caminade at the Laboratoire de Chimie de Coordination CNRS, Toulouse, for the provision of dendrimer polyelectrolytes. H.D. gratefully acknowledges financial support from the European Union for the Marie Curie Intra-European Fellowship (MEIF-CT-2005-024731).

References

1 Shin, S.W., Lee, S.G., Lee, J., Whang, C.N., Lee, J.H., Choi, I.H., Kim, T.G. and Song, J.H. (2005) *Nanotechnology*, 1392.

2 Sander, M.S. and Tan, L.S. (2003) *Advanced Functional Materials*, **13**, 393.

3 Kim, D.H., Karan, P., Göring, P., Leclaire, J., Caminade, A.-M., Majoral, J.-P., Gösele, U., Steinhart, M. and Knoll, W. (2005) *Small*, **1**, 99.

4 Law, M., Goldberger, J. and Yang, P. (2004) *Annual Review of Materials Research*, **34**, 83.

5 Gong, D.G., Varghese, C.A., Hu, O.K., Singh, W., Chen, R.S., Dickey, Z. and E.C. (2001) *Journal of Materials Research*, **16**, 3331.

6 Huczko, A. (2000) *Applied Physics A-Materials Science & Processing*, **70**, 365.

7 Hornyak, G., Kröll, M., Pugin, R., Sawitowski, T., Schmid, G., Bovin,

J.-O., Karsson, G., Hofmeister, H. and Hopfe, S. (1997) *Chemistry – A European Journal*, **3**, 1951.
8 Brauna, P.V. and Wiltzius, P. (2002) *Current Opinion in Colloid & Interface Science*, **7**, 116.
9 Jonas, U. and Krueger, C. (2002) *Journal of Supramolecular Chemistry*, **2**, 255–270.
10 Velev, O.D. and Lenhoff, A.M. (2000) *Current Opinion in Colloid & Interface Science*, **5**, 56.
11 Lau, K.H.A., Tan, L.S., Tamada, K., Sander, M.S. and Knoll, W. (2004) *The Journal of Physical Chemistry. B*, **108**, 10812.
12 Cameron, P.J.J., Jenkins, A.T.A., Knoll, W., Marken, F., Milsom, E.V. and Williams, T.L. (2008) *Chemistry of Materials*, **18**, 4304.
13 Duran, H., Lau, K.H.A., Lübbert, A., Jonas, U., Steinhart, M. and Knoll, W. (2008) in *Polymers for Biomedical Applications*, Vol. 977 (ed. A. Mahapatro), Oxford University Press, p. 371.
14 Awazu, K., Rockstuhl, C., Fujimaki, M., Fukuda, N., Tominaga, J., Komatsubara, T., Ikeda, T. and Ohki, Y. (2007) *Optics Express*, **15**, 2592.
15 Fujimakia, M., Rockstuhla, C., Wanga, X., Awazua, K., Tominagaa, J., Ikedab, T., Ohkib, Y. and Komatsubarac, T. (2007) *Microelectronic Engineering*, **84**, 1685.
16 Lau, K., Knoll, W. and Kim, D. (2007) *Macromolecular Research*, **15**, 211.
17 Reimhult, E., Kumar, K. and Knoll, W. (2007) *Nanotechnology*, 275303.
18 Kim, D.H., Lau, K.H.A., Joo, W., Peng, J., Jeong, U., Hawker, C.J., Kim, J.K., Russell, T.P. and Knoll, W. (2006) *The Journal of Physical Chemistry B*, **110**, 15381.
19 Kim, D.H., Lau, K.H.A., Robertson, J.W.F., Lee, O.J., Jeong, U., Lee, J.I., Hawker, C.J., Russell, T.P., Kim, J.K. and Knoll, W. (2005) *Advanced Materials*, **17**, 2442.
20 Knoll, W. (1998) *Annual Review of Physical Chemistry*, **49**, 569.
21 Fowles, G.R. (1989) *Introduction to Modern Optics*, Dover Publications, Inc., New York.
22 Beines, P.W., Klosterkamp, I., Menges, B., Jonas, U. and Knoll, W. (2007) *Langmuir*, **23**, 2231.
23 Crampton, N., Bonass, W.A., Kirkham, J. and Thomson, N.H. (2005) *Langmuir*, **21**, 7884.
24 Simon, A., Cohen-Bouhacinaa, T., Portéb, M.C., Aiméa, J.P. and Baquey, C. (2003) *Journal of Colloid and Interface Science*, **251**, 278.
25 Feng, C.-L., Zhong, X.H., Steinhart, M., Caminade, A.M., Majoral, J.-P. and Knoll, W. (2007) *Advanced Materials*, **19**, 1933.
26 Yu, F., Ahl, S., Caminade, A.M., Majoral, J.P., Knoll, W. and Erlebacher, J. (2006) *Analytical Chemistry*, **78**, 7346.
27 Kim, B.S., Lebedeva, O.V., Koynov, K., Gong, H., Caminade, A.M., Majoral, J.P. and Vinogradova, O.I. (2006) *Macromolecules*, **39**, 5479.
28 Jessensky, O., Muller, F. and Gosele, U. (1998) *Applied Physics Letters*, **72**, 1173.
29 Li, A.P., Muller, F., Birner, A., Nielsch, K. and Gosele, U. (1998) *Journal of Applied Physics*, **84**, 6023.
30 O'Sullivan, J.P. and Wood, G.C. (1970) *Proceedings of the Royal Society of London. Series A: Mathematical and Physical Sciences*, **317**, 511.
31 McKenzie, K.J. and Marken, F. (2003) *Langmuir*, **19**, 4327.
32 Raether, H. (1988) *Surface-Plasmons on Smooth and Rough Surfaces and on Gratings*, Springer Verlag, Berlin, Heidelberg.
33 Aspnes, D.E. (1982) *Thin Solid Films*, **89**, 249.
34 Maldovan, M., Bockstaller, M.R., Thomas, E.L. and Carter, W.C. (2003) *Applied Physics B – Lasers and Optics*, **76**, 877.
35 García-Vidal, F.J., Pitarke, J.M. and Pendry, J.B. (1997) *Physical Review Letters*, **78**, 4289.
36 Green, N.M. (1975) in *Advances in Protein Chemistry*, Vol. 29 (ed. C.B. Anfinsen), Academic Press Inc., New York, p. 85.
37 Clerc, D. and Lukosz, W. (1997) *Biosensors & Bioelectronics*, **12**, 185.
38 Hohlbein, J., Steinhart, M., Schiene-Fischer, C., Benda, A., Hof, M. and Hübner, Christian G. (2007) *Small*, **3**, 380.
39 Decher, G. (1997) *Science*, **277**, 1232.

40 Hillebrenner, H., Buyukserin, F., Kang, M., Mota, M.O., Stewart, J.D. and Martin, C.R. (2006) *Journal of the American Chemical Society*, **128**, 4236.

41 Sebastián, R.-M., Magro, G., Caminade, A.-M. and Majoral, J.-P. (2000) *Tetrahedron*, **56**, 6269.

42 Loup, C., Zanta, M.-A., Caminade, A.-M., Majoral, J.-P. and Meunier, B. (1999) *Chemistry – A European Journal*, **5**, 3644.

43 Launay, N., Caminade, A.-M., Lahana, R. and Majoral, J.-P. (1994) *Angewandte Chemie – International Edition*, **33**, 1589.

44 Talbot, J., Tarjus, G., Van Tassel, P.R. and Viot, P. (2000) *Colloids and Surfaces A-Physicochemical and Engineering Aspects*, **165**, 287.

45 Topoglidis, E., Lutz, T., Willis, R.L., Barnett, C.J., Cass, A.E.G. and Durrant, J.R. (2000) *Faraday Discussions*, **116**, 35.

5
Surface and Interface Analysis

Surface Design: Applications in Bioscience and Nanotechnology
Edited by Renate Förch, Holger Schönherr, and A. Tobias A. Jenkins

ISBN: 978-3-527-40789-7

5.1
Stretching and Rupturing Single Covalent and Associating Macromolecules by AFM-Based Single-Molecule Force Spectroscopy

Marina I. Giannotti, Weiqing Shi, Shan Zou, Holger Schönherr, and G. Julius Vancso

5.1.1
Single-Molecule Force Spectroscopy Using AFM

Single-molecule manipulation methods are particularly important in areas where temporal and spatial averaging is to be avoided, mechanical forces must be measured and individual molecular species are to be tracked. The development of experimental tools has allowed the precise application and measurement of minute forces, thus opening up new perspectives in life sciences as well as in materials science [1, 2]. Single-molecule devices contribute to the discovery, identification and description of temporally and spatially distinct states of molecular species. They render it possible to follow, in real time and at an individual molecular level, the movements, forces and strains developed during the course of a reaction, and even conformational changes induced by external forces can be detected. The knowledge obtained through single-molecule experiments is primarily fundamental, and provides essential evidence for existing molecular principles, as well as information regarding intra- and intermolecular interactions directly at the single-chain level. Among others, these tools have made possible observations of biological processes that could not otherwise be directly detected, for example, protein folding [3, 4], elasticity of macromolecules [5–7], DNA mechanics [8, 9], mechanical work generated by motor proteins [10], identification of individual molecules [11], and binding potential of host–guest pairs [6, 12, 13].

The methods for single-molecule handling and testing require a probe capable of generating and detecting forces and displacements, together with a capability for spatial location of the molecules. Various techniques are used for single-molecule manipulation and mechanical characterization according to the force range, minimum displacements, and applications, considering also practical advantages and disadvantages of the processes used (for a review see refs. [2] and [14]). Atomic force microscopy (AFM)-based single-molecule force spectroscopy (SMFS) applies, or senses, forces through the displacement of a bendable beam (cantilever). With the advantages of high spatial range sensitivity and versatility, as well as the possibility of locating and probing single molecules under environmentally controlled conditions (water or organic solvent, temperature, salt, electrochemical

Surface Design: Applications in Bioscience and Nanotechnology
Edited by Renate Förch, Holger Schönherr, and A. Tobias A. Jenkins

ISBN: 978-3-527-40789-7

potential, etc.), this technique has extensively been used for the detection and mechanical characterization of biomacromolecules like DNA [8, 9], proteins [3], polysaccharides [15–17], as well as synthetic polymers [17, 18]. Intermolecular and intramolecular interactions can be directly probed by AFM-SMFS. The technique allows the determination of molecular conformations through the uncoiling and rupturing of molecular and supramolecular structures in biomacromolecules as well as in synthetic polymer systems. Moreover, the capability of AFM to resolve nanometer-sized details, together with its force-detection sensitivity, have led to the development of molecular recognition imaging [19]. Through a combination of topographical imaging and force measurements, receptor sites can be localized with nanometer accuracy, thus making it possible to identify specific components in a complex biological sample while retaining high-resolution imaging.

Colton *et al.* are considered the first to describe an experimental approach using force spectroscopy to determine the interaction between complementary strands of DNA and intrachain forces associated with the elasticity of single DNA strands [8]. In another early pioneering work, Gaub *et al.* described the use of force spectroscopy to determine intermolecular forces and energies between ligands and receptors, specifically between avidin (or streptavidin) and biotin analogs [20]. This new experimental platform has been extended since those early days of applications to the understanding of molecular mechanisms in biological processes and in life sciences, and of material properties in soft-matter physics.

The principles of AFM-SMFS have been thoroughly described in numerous reviews [5, 6, 12, 17, 21]. In brief, in a typical AFM-SMFS configuration the force is measured through the detection of the AFM-cantilever deflection by using the optical beam principle (Figure 1a). Laser light is focused at the back of the cantilever, which is terminated by a sharp tip (with typical radius values ranging from a few to tens of nanometers) that is used to pick up and stretch single macromolecules. The movement of the piezoelectric positioning tube during an approach–retract cycle is schematized in Figure 1(b). During an experiment, the cantilever initially stays free, as long as there are no long-range interactions (Figure 1b, position 1). Subsequently, the piezo, together with the substrate containing the physically or chemically adsorbed molecules, moves towards the cantilever (approach; Figure 1b, position 2). While tip and sample are in contact, a force is applied to the sample, the cantilever is deflected upwards (positive deflection), and the macromolecules on the substrate can adsorb onto the tip (Figure 1b, position 3). Another strategy encompasses the chemical tethering of the macromolecules to the AFM tip. In this case, when the tip and surface are in contact, the polymer molecules make a bridge while getting adsorbed to the substrate. Upon separation (retraction), the linking macromolecule is first uncoiled (Figure 1b, position 4) and stretched, which results in the deflection of the cantilever towards the substrate (Figure 1b, position 5). A negative deflection is registered due to an attraction, and as the chain is fully stretched, the weakest point of the structure breaks. In other words, the macromolecule desorbs either from the surface or from the tip, and the cantilever rapidly returns to its relaxed state (Figure 1b, position 6).

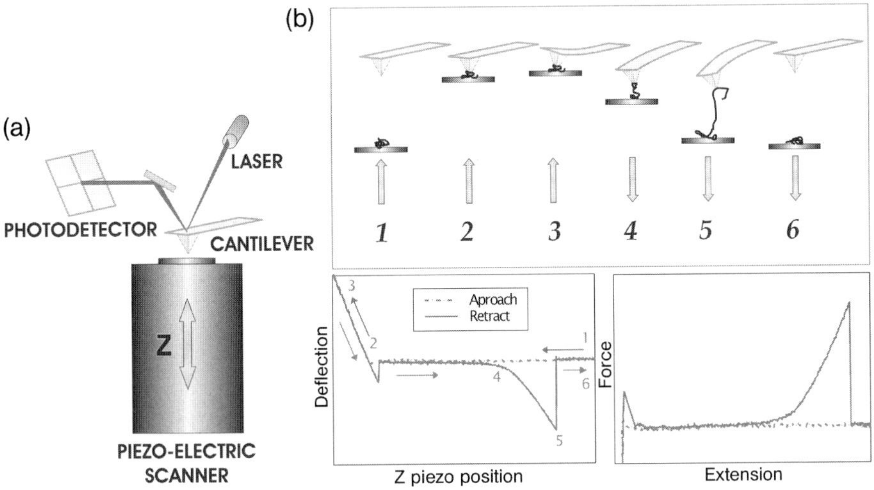

Figure 1 (a) The principle of AFM-based SMFS. (b) Schematic illustration of a single-molecule deflection–displacement (piezo position) experiment, the registered deflection-Z piezo position data, which is subsequently transformed to force–extension curves (on the right). (Reproduced with permission from [17]).

The deflection–displacement (Z piezo position) (D–z_0) registered profile (Figure 1b) describes the elasticity of the macromolecule, and can be easily converted into force vs. the actual distance between the AFM tip and the substrate (extension) (F–z) when the spring constant of the cantilever (k_s) is known [22]. This procedure is described in ref. [17].

It is important to take into consideration possible nonequilibrium effects that may be caused by fast piezo retraction (i.e. bond loading) rates. In a force spectroscopy experiment chemical bonds (covalent, supramolecular, or adhesive bonds) are loaded and stressed upon increasing the tip–substrate distance bridged by the chain. In SMFS, usually the rupture force value of the 'weakest link' is measured. The force that is needed to rupture a chain is governed by thermomechanically activated bond dissociation kinetics and mechanochemistry [23]. For supramolecular bonds, e.g. in associating polymers [24], mechanical loading increases bond dissociation rates compared to the mechanical stress-free case, as was first emphasized by Bell [25], in far from equilibrium situations. This complicates the use of AFM-SMFS for complex molecular architectures, however, it also presents opportunities to study well-defined systems for new perspectives, as from loading-rate dependences important parameters of the potential-energy landscape along the unbinding reaction coordinate can be determined [26]. For stretching of covalent chains, prior to rupture, one needs to consider the rate of chain-segment fluctuation (between the tip and the substrate) and the bond-loading rate of the AFM experiment. If the bond-loading rate (chain stretching rate) is much slower than the chain fluctuations, the macromolecule is in quasiequilibrium during stretch;

i.e. no loading-rate dependence is anticipated. However, kinetic aspects of the internal conformational transitions during chain stretch may give rise to a loading-rate dependence, in particular when the experimental time scale becomes comparable to the molecular kinetics (single-chain viscoelasticity, i.e. energy dissipation). In equilibrium, force curves (loading and unloading) should be fully reversible with no 'snapping' of the cantilever upon bond separation. However, when single chains are stretched (loaded) that exhibit conformational transformations (e.g. for polymers with glycan rings 'unclicked' – shorter – and 'clicked' – longer – states, see Section 5.1.4) a loading-rate dependence can occur [27].

5.1.2
Stretching of Individual Macromolecules

The restoring force profiles (force–extension curves) recorded during stretching experiments encompass valuable information about the elasticity of the macromolecule. Two statistical-mechanical models are commonly used to describe the entropic elasticity of a polymer chain during extension, i.e. the freely jointed chain (FJC) [28, 29] and the worm-like chain (WLC) [30–32] models.

In the FJC model, a polymer chain is treated as N number of rigid Kuhn segments with lengths of l_k (Kuhn's segment length), connected through flexible joints without any long-range interactions. The FJC model assumes that each segment has a fixed length (l_k), and considers a fully elongated chain when the extension z equals the value $N.l_k$. As a consequence, if the chain were infinitely strong, the tension F would become infinite. This approach is thus a good approximation for small elongations (small applied forces) of a molecule [29]. To cover the whole force range, a generic approach was developed to take into account the enthalpic contribution of the deformation of bonds and bond angles. This model is known as the modified, or extended, FJC (m-FJC or FJC$^+$) [33], and describes the molecule as n identical elastic springs in series, characterized by the segment elasticity parameter K_s. Table 1 summarizes the interpolation formulas used to describe these models.

Alternatively, the WLC model describes a polymer chain as a homogeneous string with a constant bending elasticity, and treats the molecule as a continuous entity with a persistence length l_p [30]. The persistence length reflects the sum of the average projections of all chain segments in a specific direction described by a given segment. Below this length the polymer is considered to be linear. Although this approach takes into account both entropic and enthalpic contributions, the extension is limited by the contour length of the polymer (L_c), and the model fails for macromolecules under conditions of high stress. To broaden the range of applications of the WLC model, the extended WLC models by Odijk [34] (Odijk WLC) and Wang [35] (modified Marko–Siggia), among other extensions of WLC (see Table 1), include a stretching term in the F–z equation that is appropriate for intrinsic enthalpic contributions, and includes a parameter that denotes the specific stiffness (K) of the polymer chain (Table 1).

Table 1 Elasticity models describing the force-extension relationship for single polymer chains.

Model	Force-extension expression	Fitting parameters
Freely jointed chain FJC [28]	$z(F) = L_c\left[\coth\left(\frac{Fl_k}{k_B T}\right) - \frac{k_B T}{Fl_k}\right]$	l_k, L_c
Worm-like chain WLC (Marko–Siggia) [32]	$F(z) = \frac{k_B T}{l_p}\left[\frac{1}{4}\left(1 - \frac{z}{L_c}\right)^{-2} + \frac{z}{L_c} - \frac{1}{4}\right]$	l_p, L_c
Extended FJC (FJC$^+$ or m-FJC) [33]	$z(F) = L_c\left[\coth\left(\frac{Fl_k}{k_B T}\right) - \frac{k_B T}{Fl_k}\right]\left(1 + \frac{F}{K_s l_k}\right)$	l_k, L_c, K_s
Odijk WLC [34]	$z(F) = L_c\left[1 - \frac{1}{2}\left(\frac{k_B T}{Fl_p}\right)^{1/2} + \frac{F}{K}\right]$	l_p, L_c, K
Modified Marko–Siggia WLC (WLC$^+$) [35]	$F(z) = \frac{k_B T}{l_p}\left[\frac{1}{4}\left(1 - \frac{z}{L_c} + \frac{F}{K_o}\right)^{-2} + \frac{z}{L_c} - \frac{F}{K_o} - \frac{1}{4}\right]$	l_p, L_c, K_o
Hooke spring modified WLC [50]	$f(z) = \frac{k_B T}{l_p}\left[\frac{1}{4}\left(1 - \frac{z}{L_c}\right)^{-2} + \frac{z}{L_c} - \frac{1}{4}\right] + K'z$	l_p, L_c, K'

F: force; z: extension; k_B: Boltzmann's constant; T: temperature;
L_c: contour length; l_k: Kuhn segment length; l_p: persistence length;
K_s, K, K_0, K': elasticity parameters.

In the following section we describe the use of AFM-SMFS to prove the principle of a single-macromolecular nanomotor, by means of the variations in the elasticity of a polymer chain induced by the application of external stimuli. As is discussed later in the study of H-bond-mediated structures in polysaccharide single chains, deviations from the force–extension behavior from the previously described models can reflect structural transitions that require further analysis and modeling. These deviations are due to events of an enthalpic nature experienced by the macromolecule during stretching and may reveal thorough knowledge regarding intermolecular [8, 9, 36–38] and macromolecule–solvent interactions [36, 39–43] as well as intramolecular rearrangements [37, 43–49]. Furthermore, in the final section of this chapter we illustrate the AFM-SMFS as a powerful tool to address interaction forces and directly probe individual H-bonded supramolecular polymer chains.

5.1.3 Realization of a Single-Macromolecular Motor

In nature, several classes of molecular motors fulfil different functions in the living cells, e.g. proton pump in membranes [51], motor proteins like myosin [52], DNA and RNA polymerases [53], and flagellar motors in bacteria [54], which convert chemical energy into mechanical work. A molecular motor is a molecule that is operated in a controlled cyclic fashion to perform mechanical movement (work output) as a consequence of appropriate external stimulation (energy input). During a cycle, the molecule runs through a number of different structural conformations/configurations and/or chemical states. Eventually, the molecular motor must be reset to its initial state. The external stimulus, and thus the source of energy, can be of different nature, such as scission of chemical bonds (chemical energy), light (electromagnetic waves), or electrical potential.

Pioneering work led to insight into the fundamental molecular-scale processes and to the exploitation of natural motors in fascinating biomimetic applications; yet the proposed energy-conversion mechanisms and molecular aspects remain in many cases controversial [55]. Research on the synthetic counterpart of natural molecular motors [56], including molecular switches [57] and machines [58], has also recently witnessed significant advances [59]. Stimuli-responsive polymers are defined as polymers that undergo relatively large and abrupt chemical and/or physical changes in response to small changes in the environmental conditions [60]. These polymers recognize a stimulus as a signal and subsequently alter their chain conformation as a direct response and thus are prime candidates as model systems to study synthetic chain motors.

Surface-immobilized molecular motors can be addressed by various techniques, such as AFM-SMFS, as they can unveil mechanical characteristics of isolated macromolecules. The combination of this tool and the great potential of stimuli-responsive polymers has opened up a new area of research on artificial molecular motors, aimed at obtaining a fundamental understanding of the relevant molecular-scale processes and at realizing and exploiting the smallest man-made machine [61–64].

The first demonstration of photochemical energy conversion in an individual macromolecule was reported by the group of Gaub [61]. Single polyazopeptides were reversibly shortened against an external force in SMFS experiments by photochemical switching between *trans*- and *cis*-azobenzene isomeric configurations. Under the experimental circumstances, light was used as an external stimulus to tune the elasticity of the macromolecule. The maximum efficiency of the cycle at the molecular level was estimated to be about 10%.

On the other hand, Vancso and coworkers have systematically investigated the stimulus-responsive poly(ferrocenylsilane) (PFS) polymer [65] as a model system for the realization of (macro)molecular motors powered by a redox process, which is an alternative way to switch the flexibility of the chain, where the stimulus can be easily confined to a small number of macromolecules [62, 63, 66]. The PFS polymers consist of alternating ferrocene and substituted silane units in the main chain (Figure 2a), and can be reversibly oxidized and reduced by an external potential. By means of SMFS, different segment lengths and elasticities (l_k and K_s, from the m-FJC model) for the neutral and oxidized forms of individual surface-confined PFS chains were observed, induced by electrochemical or wet-chemical redox chemistry [62, 63]. Differences in the elasticity of PFS polymers were also reported by the group of Zhang [67]. The studies revealed that the elasticity of PFS could be reversibly controlled *in situ* by adjusting the applied potential in electrochemical SMFS experiments, setting the basis for the demonstration of a single-macromolecular motor, operated in a cyclic manner.

A proposed possible experimental cycle to realize the molecular motor was first defined by keeping the deflection of the cantilever constant in SMFS measurements, i.e. applying a constant force, during the transition from the oxidized to the neutral state (*vice versa*) (Figure 2b). The two branches (neutral and oxidized states) were determined by the corresponding elasticities of the polymer using AFM-SFMS, and were quantitatively described by the m-FJC model. Starting from a low force (20 pN) (point 1 in Figure 2b) under an applied constant external potential of +0.5 V, an individual, oxidized PFS polymer chain with 50 nm contour length is pulled to a force of 140 pN (point 2). At a constant force of 140 pN, the PFS chain is reduced to its neutral state by controlling the external potential back to 0 V (point 3), thus giving rise to a change in the elasticity of the polymer chain. Subsequently, the force on the polymer is reduced back to 20 pN (point 4) and finally, the cycle is completed by applying an external potential of +0.5 V to completely oxidize the whole PFS chain. By periodically controlling the external potential, the corresponding oxidized and/or neutral PFS chains can be created to realize the operating cycle, with a mechanical work of about 3.4×10^{-19} J as calculated from the enclosed area.

Such electrochemically driven, single-chain-based motors are potentially interesting for the realization of single-molecule devices, as they can in principle be addressed on the single-molecule level by using miniaturized electrodes and can be repeatedly run in cycles in a reversible manner. In this context, it is important

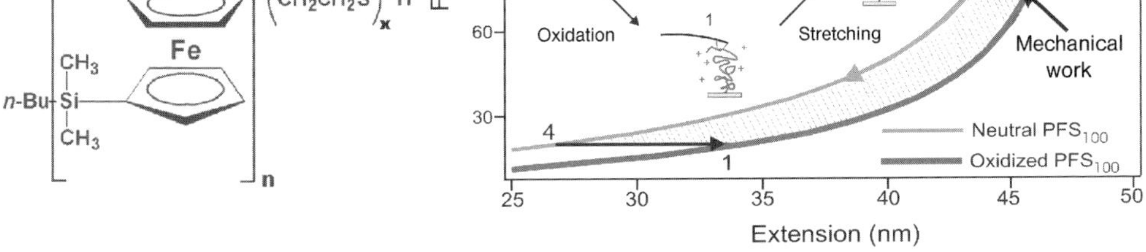

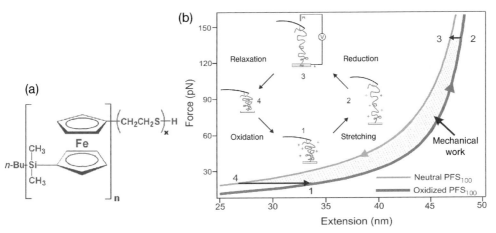

Figure 2 (a) Chemical structure of ethylene sulfide end-capped poly(ferrocenylsilane) (PFS_{100}). (b) Force–extension curves of a single-molecule motor based on one PFS macromolecule driven by an electrochemical potential. The two force–extension curves are plotted based on an m-FJC fit with an l_K of 0.38 nm (neutral PFS) and 0.65 nm (oxidized PFS), and K_s of 30 nN nm^{-1} (neutral PFS), and 45 nN nm^{-1} (oxidized PFS), respectively, as observed experimentally. Inset: a schematic illustration of a single-molecule operating cycle with redox-active macromolecules. (Reproduced with permission from [62]).

to determine how the efficiency of single motors depends on various experimental and (macro)molecular design parameters. Hence, the analysis of closed mechano-electrochemical cycles of individual macromolecules is required, including the localization and addressing of a single macromolecule by the AFM tip, and the stretching and relaxing of the molecule *in situ* under different applied electrochemical potentials.

The first experimental realization of a closed-loop mechano-electrochemical cycle of individual PFS chains in electrochemical AFM-based SMFS was recently reported by Vancso and coworkers [66]. Individual PFS chains kept in an extended state between the AFM tip and the electrode surface were electrochemically oxidized at constant z position by applying a potential of +0.5 V (cycle 1, 1-2-3-1′, as shown in Figure 3a). Cyclic voltammetry was performed to ensure the complete oxidation of the PFS chains on the gold working electrode. Correspondingly, single chains were stretched in the oxidized state (by keeping the potential at +0.5 V), followed by electrochemical reduction at constant z position and continued force spectroscopy (cycle 2, 1-3-2-1′, as shown in Figure 3b). The two data sets display entire mechano-electrochemical experimental cycles for individual PFS molecules in both possible directions.

In the experiments, the force at fixed maximum extension was observed to decrease upon oxidation and increase upon reduction to the neutral state due to the lengthening of the oxidized chain with respect to the neutral one. This is attributed to the electrostatic repulsion among the oxidized ferrocene centers along the chain [68]. Therefore, the change in the redox state is directly coupled to a mechanical output signal of the force sensor (for cycle 1: ≈ 200 pN).

For the molecular motor described here, the work performed per cycle can be obtained by measuring the area enclosed between the force–extension curves for the oxidized and neutral PFS. While in cycle 1, this corresponds to work input (W_{in}), in cycle 2 it is the work output (W_{out}) of the electrochemical cycle. The elec-

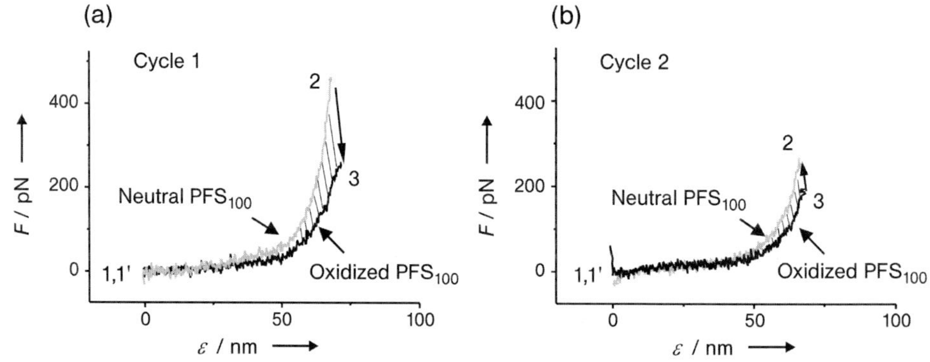

Figure 3 Force–extension curves of (a) cycle 1 and (b) cycle 2. The enclosed areas of the cycles correspond to the mechanical work input (or output) of the single-polymer-chain mechano-electrochemical cycle. (Reproduced with permission from [66]).

trochemical work required to switch the redox state of the molecule (electrochemical potential) is:

$$\text{cycle 1: } \Delta\mu = \bar{\mu}_i - \mu_i = z_i F \Delta\phi; \quad \text{cycle 2: } \Delta\mu = \mu_i - \bar{\mu}_i = z_i F \Delta\phi \tag{1}$$

where $\bar{\mu}_i$ is the chemical potential of the electrochemically oxidized PFS chain, μ_i is the chemical potential of the neutral PFS chain, z_i is the number of charges per redox center ($z_i = 1$), F is the Faraday constant (96 486 C mol^{-1}), and $\Delta\phi$ is the electric potential required to complete the oxidation/reduction of the entire PFS chain. Consequently, the efficiency η of the single-molecule mechano-electrochemical cycle for ΔN number of transfer charges, is obtained as [$W = W_{out} = W_{in}$]:

$$\eta = \frac{W}{|\Delta\mu|\Delta N} \tag{2}$$

For cycle 1, W was estimated to be 2.2×10^{-18} J for an extension from 40 to 68 nm, with a corresponding efficiency of 26%, while for cycle 2, W was estimated to be 0.7×10^{-18} J for an extension from 40 to 66 nm, with a corresponding efficiency of 8%.

In this work, the influence of the attainable single-molecule forces on the efficiency of the motor was unraveled, encouraging the future development of optimized, electrochemically driven single-chain polymer devices. For the above-mentioned experiments, the difference in the numerical values of efficiency for cycles 1 and 2 originates from the different maximum forces (and corresponding degree of extension) applied during stretching. Assuming a constant potential, the relation between stretching ratio (defined as the ratio between extension ε for the neutral chain and maximum extension realized experimentally for the neutral chain in cycle 1, ε^{1}_{max}) was evaluated from Figure 3(a). The resulting plot showed a monotonically increasing relationship between efficiency and stretching ratio (Figure 4). The mean attainable efficiency under the experimental conditions used was estimated to be of the order of 20% (based on the determined mean rupture force of 0.35 nN). The efficiency of cycle 2 was also plotted in the corresponding graph (by assuming identical contour lengths). A good agreement of the efficiencies measured for these individual molecules within experimental error proved the same efficiency for both directions of the closed electromechanical cycle.

The maximum attainable efficiency for a certain extension is also influenced by the molecular structure and external parameters, which should be taken into consideration when choosing the appropriate macromolecular system to build a molecular motor. The inset in Figure 4 shows a logarithmic plot calculated based on the modified freely jointed chain (m-FJC) model of efficiency versus stretching ratio for cycles of a macromolecule that changes from state 1 to different final states 2, for which different elasticity parameters of the m-FJC model were assumed (20, 50 and 80% increase in the Kuhn length l_K and a corresponding decrease in the segment elasticity K_s). The maximum efficiency is critically influenced by the variation in elasticity that can be achieved through the external sti-

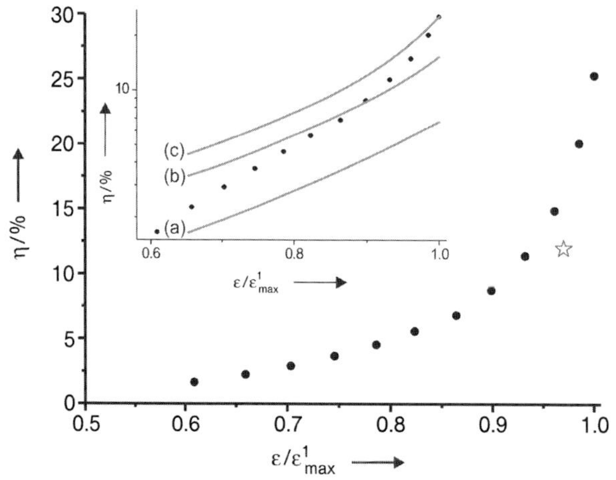

Figure 4 Efficiency of the closed-loop mechano-electrochemical cycle of an individual PFS_{100} macromolecule versus stretching ratio ($\varepsilon/\varepsilon^1_{maz}$) calculated from the data of cycle 1 by integration of the area enclosed between the curves for different ε values. The star shows the efficiency of a different macromolecule (data converted from cycle 2). Inset: logarithmic plot of the predictions of the efficiency versus stretching ratio for cycles of a macromolecule that changes from state 1 to different final states 2, where the Kuhn length l_K and segment elasticity K_s parameters from the m-FJC model for state 2 were varied by: (a) 20, (b) 50, and (c) 80% (see text for details). (Reproduced with permission from [66]).

mulus, that is, 7, 15 and 26% for a systematic change in l_K (increase) and K_s (decrease) of 20, 50 and 80%, respectively. The graph also shows the experimental results from cycle 1 for PFS_{100}.

5.1.4 Stretching of Individual Polysaccharide Filaments

In addition to the application of external stimuli to induce elasticity changes in an AFM-SMFS experiment, this powerful technique has proven to be very valuable in studying macromolecular conformations and their relation to variation in the external conditions (such as temperature, solvent, etc) [8, 9, 36–49]. In an SMFS experiment, the macromolecule under stress may be driven into higher energetic levels and overcome conformational barriers, or break intramolecular interactions i.e. of an H-bond nature. These transitions are normally revealed in the force–extension data as deviations from the random-coil behavior of macromolecules (described by the models mentioned in Table 1), due to the enthalpic contributions of conformational rearrangement events taking place during stretching.

Rief *et al.* [44] were the first to identify a transition into a stiffer conformation when stretching dextran molecules, revealed as a plateau in the force–extension signal at 250–350 pN for carboxymethylated dextran, and at 700–850 pN for native dextran. This transition was predicted by molecular dynamics simulations and was ascribed to a purely elastic and reversible conformational change of the polysaccharide, due to rotation of an exocyclic bond. Marszalek *et al.* [45] observed the same plateau region in the force–extension response for dextran and amylose. In accordance with theoretical calculations, they identified it as a chair-to-boat conformational transition of the pyranose rings induced by the stretching force; into a thermodynamically accessible conformation that is normally less populated due to its high conformational energy. Independently, the groups of Gaub and Zhang [46] simultaneously reported on the chair-to-twisted-boat transition in α-(1,4)-linked polysaccharides. Contrarily, β-(1,4)-linked polysaccharides demonstrate a stiff and extended conformation, and consequently do not present this force-induced conformational transition, which is thus called a nanomechanical 'fingerprint' of the α-(1,4)-linked glycans. Subsequently, these observations were expanded to include chair inversions for pectin molecules [47], which consist of a resolved two-step conversion of the pyranose ring conformation from a chair to a boat and then to an inverted chair. A brief yet complete overview of fingerprinting polysaccharides has been published by Marszalek and Fernandez [16] in an article that reports on the use of AFM–SMFS for identifying individual polysaccharide molecules in solution.

SMFS has been used in many cases to expose secondary structures of individual macromolecules. In 1999, the group of Gaub [40] reported on marked deviations in the transition region from entropic to enthalpic elasticity for SMFS experiments of PEG in water. These deviations indicated the deformation of a secondary structure (referred to as 'superstructure') within the polymer, stabilized by water binding, via hydrogen-bond interactions.

Hydrogen bonds play an important role in the local conformation of stereoregular polysaccharides, as many of them adopt a helical conformation in the solid state, held together by corresponding specific supramolecular forces. In aqueous solution, under specific thermodynamic conditions, they adopt an 'ordered conformation', assumed to be helical in many cases, as unveiled by circular dichroism, optical rotation, or NMR spectroscopy [69], traditionally used for conformation analysis. For many of these polysaccharides, a helix-coil conformational transition is induced when increasing the temperature, as a result of the diminution of the attraction by H bonds when the temperature increases. Another option is to decrease the ionic concentration, giving rise to electrostatic repulsion between ionic groups of the chain [70]. Conclusions concerning the structural conformation of individual aqueous polysaccharide chains have generally been drawn via extrapolation of ensemble-average results to infinite dilution, implicitly removing intermolecular contacts and specific interactions. The molecular structure of polysaccharides is at the basis of their properties and the development of AFM-SMFS made possible the characterization and comprehension of the conformational behavior of the polysaccharide chains in solution directly at the single-molecule level. For example, the ordered, helical, secondary structure observed for xanthan [38, 71], or the inter-residue H bonds that amylose forms under various solvent conditions [36, 49, 72].

Recently, the group of Vancso reported on one of the first AFM-SMFS experiments performed at variable temperature, investigating the influence of temperature on the hyaluronan (HA) single-chain conformation in aqueous NaCl solution [43]. Hyaluronan is a glycosaminoglycan that is present in the extracellular matrix of most vertebrate connective tissues as well as in some bacterial capsules [73], and most of its functions are based on its physical properties.

It is well known that the HA polysaccharide adopts a helical conformation in the solid state [74]. Meticulously performed infrared (IR) spectroscopy studies on HA films in the dried and hydrated states have demonstrated the formation of a structure going from an intramolecular H-bonded organization (dried state, Figure 5a) to a H-bonded intermolecular structure where 'water wires' bridge the chains (hydrated state) [75].

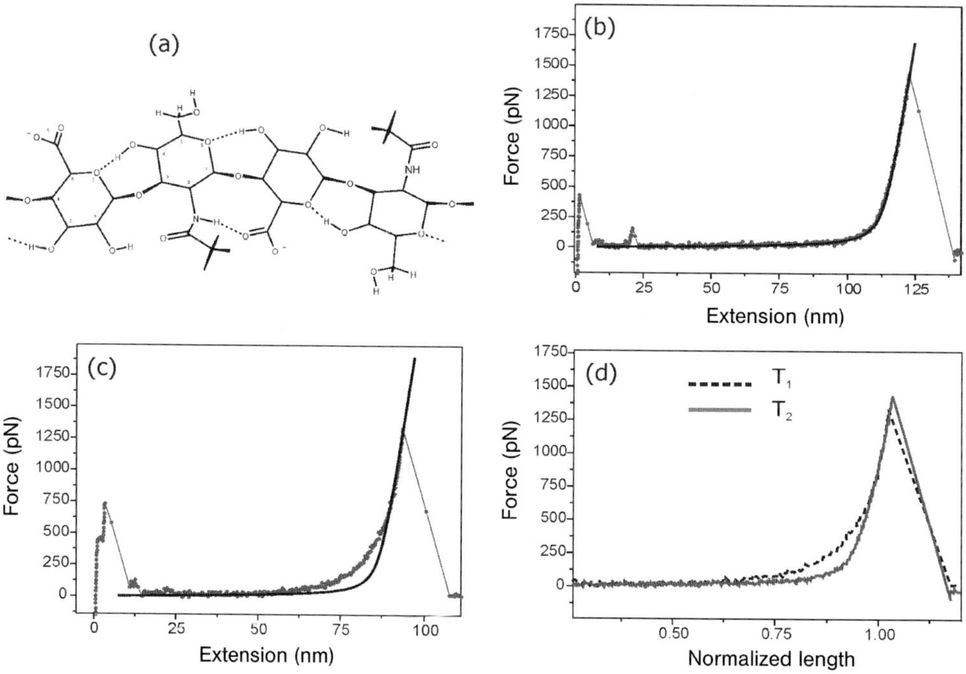

Figure 5 (a) The local conformation of a fragment of HA with a representation of intrachain H-bonds existing in the dried state. (b) Representative force–extension trace of an individual HA molecule measured at 46 ± 1 °C (T_2) in 0.1 M NaCl. The continuous dark line corresponds to the fitting with the m-FJC model ($l_k = 0.53$ nm, $K_s = 30.85$ nN/nm, $L_c = 113.40$ nm). (c) Representative force–extension traces of individual HA molecules measured at 29 ± 1 °C (T_1) in 0.1 M NaCl. The continuous dark line corresponds to the simulation with m-FJC model, using the segment parameters obtained in the fitting of the results at 46 ± 1 °C, and the corresponding contour length ($l_k = 0.53$ nm, $K_s = 30.85$ nN/nm, $L_c = 87.10$ nm). (d) Superposition of the normalized force curves of individual HA molecules measured at 29 ± 1 °C (T_1) (dotted line) and 46 ± 1 °C (T_2) (continuous line) in 0.1 M NaCl (normalized at 900 pN). (Reproduced in part with permission from [43], Copyright 2007 by American Chemical Society).

The first depiction of HA in aqueous solution was suggested after performing studies by classical methods of physical chemistry of polymer solution, where HA was represented as a comparatively stiff random coil [76]. Later, certain indications, gathered by Scott [77, 78], suggested that the formerly proposed random coil HA (aqueous) had a secondary structure stabilized by extensive intraresidue hydrogen bonding. Moreover, the formation of a tertiary structure of aqueous HA was shortly thereafter discussed by Scott and Heatley on the basis of ^{13}C NMR data [79]. Additionally, Haxaire *et al.* [80] reported on ^{1}H NMR and viscosity measurements of HA aqueous solutions in the presence of NaCl, that revealed a semi-rigid structure for the HA chains related to the formation of an H-bonded network structure. Furthermore, a progressive decrease of stiffness on increasing the temperature beyond ~40 °C was detected, and ascribed to a destabilization of the H-bonded network.

The elastic response of HA single chains was thus studied by the group of Vancso at two different temperatures, below and above 40 °C, in order to analyze the influence of the temperature on the structure at the single-molecule level. When the elastic response was compared with the predictions from the m-FJC model (Figure 5b and c), a change from a nonrandom coil to a random coil behavior on increasing the temperature from 29 to 46 °C was easily evidenced. As is depicted in Figure 5(b), the m-FJC model describes very well the elastic behavior of the HA polymer chains in aqueous NaCl solution for the high-temperature range (around 46 °C). Nevertheless, no theoretical model for random-coiled polymer elasticity (neither the FJC nor the WLC) could describe the force–extension results of the polysaccharide in aqueous NaCl at room temperature (around 29 °C), suggesting that its behavior differs from an isolated single-chain random coil in these conditions. However, the m-FJC behavior was simulated at this temperature using the elasticity parameters obtained for the experiments at 46 °C (l_k and K_s), with the corresponding contour length values (L_c) for the results at 29 °C. When the simulation and experimental data (for 46 °C) were compared (see Figure 5c), it was clearly evidenced that, in the high force regime (after ~700 pN), the m-FJC model offered a very good description of the stretching behavior also at 29 °C.

Accordingly, the conformation of HA at room temperature differs from an isolated, single-macromolecule random coil. Instead, it adopts a superstructure in aqueous NaCl solution, i.e. a local structure involving an H-bonded network along the polymeric chain, with H bonds between the polar groups of HA and possibly water as well as water-mediated intramolecular bonds. This structure clearly becomes increasingly destabilized when the temperature is raised to 46 °C. Furthermore, the deviation in the middle-force regime suggests that, at low temperature (29 °C), the superstructure (tertiary structure) breaks progressively when increasing the force, leading to the same conformational structure that HA chains adopt at high temperature, as is evidenced by the superposition of the normalized force curves at high force values (Figure 5d).

The use of DMSO as a solvent that breaks the H bonds demonstrated that the superstructure of HA single chains in aqueous NaCl solution had an H-bonded water-bridged nature, i.e. an H-bonded network along the polymeric chain, with H bonds between the polar groups of HA and possibly water. In Figure 6, a representative normalized force curve obtained in DMSO is compared to the ones corresponding to 29 and 46 °C in aqueous NaCl solution. The behavior of the single HA chains in DMSO at 29 °C is in excellent accordance with the behavior in NaCl aqueous solution at 46 °C, when the tertiary structure was destabilized. In other words, the presence of DMSO suppressed the formation of water-bridged intramolecular tertiary H bonds, and led to a structure identical to the one the macromolecule adopts in NaCl (aq) at 46 °C. These results were in agreement with the structure of HA described by the group of Scott [78], where intramolecular H bonds are present in the structure of HA in DMSO solution (in the same way as in Figure 5a), and 'water bridges' (that lead to tertiary structures) are generated when DMSO is partly replaced by water. In water, the H bond between the NH and C=O groups of HA is modified because water molecules are insert between the two groups. The same water-mediated long-range H-bonded structure starts to break when temperature increases over 40 °C, while the intrachain H bonds between adjacent groups remain.

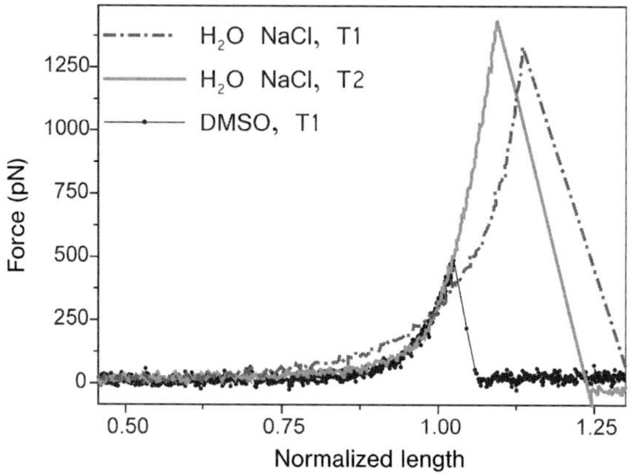

Figure 6 Superposition of normalized force curves of individual HA molecules measured in DMSO at 29 ± 1 °C (T1), as well as in 0.1 M NaCl at 29 ± 1 °C (T1) and 46 ± 1 °C (T2) (normalized at 300 pN). (Reproduced in part with permission from [43], Copyright 2007 by American Chemical Society).

Analogous types of superstructures were identified using AFM-SMFS. A SMFS study of poly(ethylene glycol) (PEG) in various solvents carried out by the group of Gaub [40] showed the deformation of the superstructure within the polymer, i.e. a nonplanar superstructure stabilized by water bridges. Marked deviations from the ideal behavior in the transition region from the entropic to enthalpic elasticity were found. A similar behavior was reported for poly(*N*-vinyl-2-pyrrolidine) in water [42], and was also ascribed to the formation of H-bonds between the solvent molecules and the polymer, thereby establishing a water bridge between the carbonyl groups of pyrrolidone rings in water. For poly(vinyl alcohol) (PVA) in water [41, 81], the deviation in the intermediate force regime of the force curves was attributed to the existence of a H-bonded superstructure (supramolecular assembly). In a comparable way, the group of Marszalek [49] used AFM-SMFS to study and determine an inter-residue H-bonded structure for amylose in a nonaqueous solution.

5.1.5 Rupture of Host–Guest Complexes and Supramolecular Polymers

Inspired by nature, supramolecular chemistry uses multiple, reversible, cooperative intermolecular interactions, such as hydrogen bonding, to fabricate self-organized structures with a very rich structural hierarchy [82]. The corresponding intermolecular interactions (forces) govern the interplay of order and mobility, as well as the properties and functions in such systems [83].

The strength of intermolecular interactions and the corresponding intermolecular forces have been traditionally assessed by ensemble thermodynamic approaches with thermodynamic potentials (enthalpy, free energy). However, as a result of the growing interest in bottom-up molecular nanotechnologies, there is a growing need to know bond strengths and molecular stability from the single-molecule perspective, as well. As atomic force microscopy (AFM) techniques can now be used to manipulate molecules with nanometer precision, adequate knowledge of supramolecular forces is also crucially important for molecular nanofabrication by AFM.

Supramolecular polymers are formed by bifunctional monomeric units that are reversibly aggregated through relatively strong noncovalent interactions [84, 85]. A prime example encompasses the supramolecular polymers formed through self-complementary recognition of the quadruple hydrogen-bonded 2-ureido-4[1H]-pyrimidinone (UPy) motif, introduced by Sijbesma, Meijer and coworkers [85–87]. AFM-SMFS has proved to be suitable to address interaction forces and to directly probe individual supramolecular polymer chains, as has been shown by Vancso *et al.* in systematic studies that targeted such artificial, supramolecular H-bonded species [88–90]. Formerly, variable-temperature SMFS in organic media was used to study the unbinding forces of dimeric $(UPy)_2$ complexes and its dependence on temperature and solvent [88], under dynamic loading conditions covering thermodynamic quasiequilibrium (loading-rate independence), the transition from quasiequilibrium to nonequilibrium, and the nonequilibrium (loading-rate dependent) domain [89].

The experimental setup consisted of a self-assembled monolayer (SAM) platform on Au comprising poly-(ethylene glycol) (PEG)-linked UPy disulfides (as shown in Figure 7a). The PEG spacer efficiently decouples the complexation process from the surface and hence suppresses contributions resulting from secondary interactions. In addition, the PEG spacer allows one to unequivocally identify single-molecule stretching and, thus, bond-rupture events (Figure 7b) [91]. The

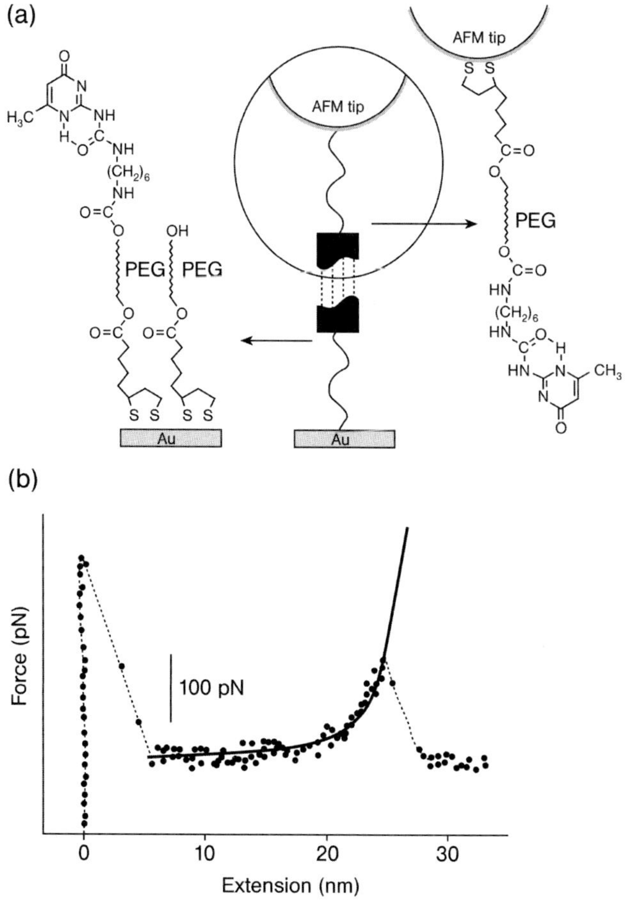

Figure 7 (a) Scheme of PEG-UPy disulfide self-assembled both on Au-coated AFM tip and Au substrate. (b) Representative force–extension curve measured between PEG-$(UPy)_2$ complexes by SMFS. The fit of the data to the m-FJC model is shown as a solid line. (Reproduced in part with permission from [89], Copyright 2005 by American Chemical Society).

unbinding forces of single, quadruple H-bonding $(UPy)_2$ complexes, as observed by AFM-SMFS, exhibited the loading-rate dependence anticipated in nonequilibrium conditions for loading rates in the range of 5 to 500 nN s^{-1} at 301 K in hexadecane. By contrast, these rupture forces were independent of the loading rate from 5 to 200 nN s^{-1} at 330 K. By plotting the two dynamic rupture force spectra in one graph, a single master curve for the rupture force at a reference temperature was built, as shown in Figure 8. The crossover force (from loading rate independent to dependent) was observed at ~ 145 pN. These results indicate that the unbinding behavior of individual supramolecular complexes can be directly probed under both thermodynamic nonequilibrium and quasiequilibrium conditions.

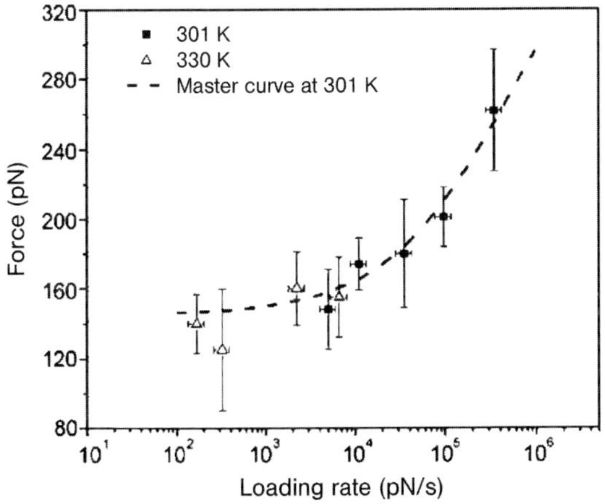

Figure 8 Master plot of crossover from loading-rate independence to loading-rate dependence of rupture forces of single $(UPy)_2$ complexes in hexadecane at $T_{ref} = 301$ K (the dashed line serves to guide the eye). (Reproduced in part with permission from [89], Copyright 2005 by American Chemical Society).

Using the same platform, individual supramolecular polymers were probed, composed of reversibly aggregated self-complementary building blocks. The systematic investigation of the single-chain mechanical properties of the supramolecular UPy polymers was based on the measurements of the individual molecular interactions present in these reversible polymers. The UPy groups were immobilized through reaction of UPy isocyanate with the alcohol group of surface-immobilized ω-hydroxy PEG chains onto Au-coated substrates (see Figure 9). The supramolecular polymers were formed by self-complementary association of the telechelic PEG derivative **4** (bis(UPy)) in hexadecane, and were probed by AFM utilizing a gold-coated AFM tip functionalized with the asymmetrically substituted UPy disulfide. Owing to the high complexation constant of $\geq 10^9$ M^{-1} and a complex lifetime greater that 1 s [87, 92], bis(UPy) forms supramolecular polymers with an effective degree of polymerization of approximately 40 in solution, and can be tested by AFM-SMFS with restoring forces observed for extensions higher than 30 nm, that were fitted by the m-FJC model. Several representative force–extension curves are displayed in Figure 10(a) exhibiting different ultimate extension lengths. Based on the fitting parameter results (l_k of 0.68 ± 0.08 nm and K_s of 7.6 ± 2.5 nN nm^{-1}), it was possible to confirm that single supramolecular polymer 'chains' were stretched, as the elasticity parameters were within the error margins for the data measured for a single PEG chain [89, 90], and the stretching length was significantly longer that the contour length of a single PEG segment (9.5 ± 1.0 nm).

The mean value of the rupture forces observed for the supramolecular polymers in hexadecane was of 172 ± 23 pN, within the error margins for the rupture force of the dimeric complex. However, the authors observed that rupture forces

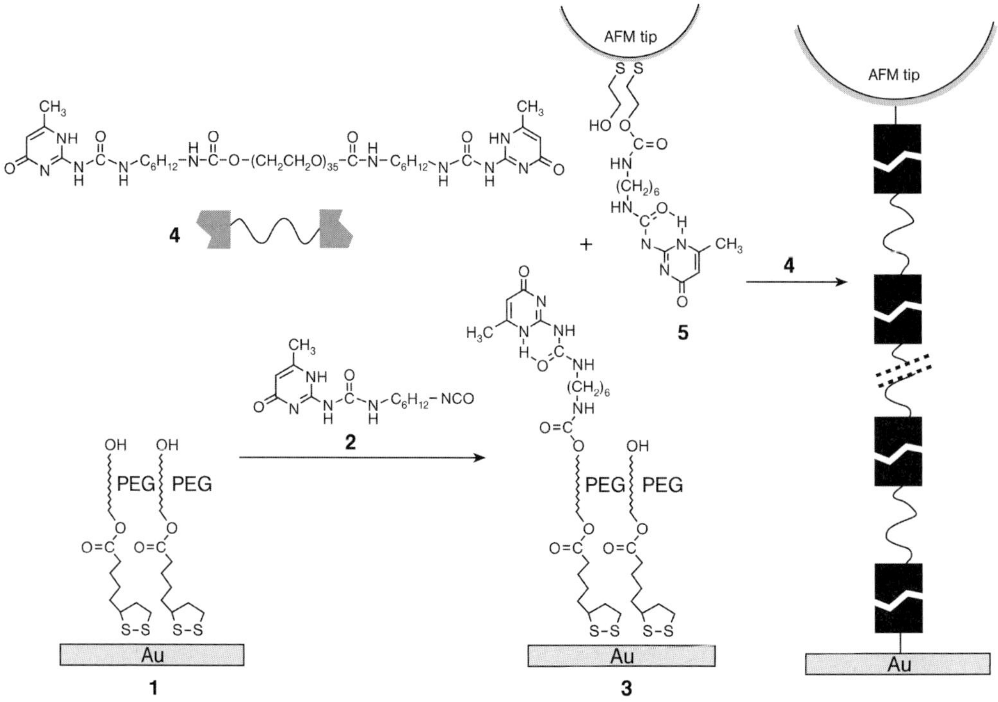

Figure 9 Molecular thin layers of ω-hydroxy-terminated PEG are obtained by self-assembly of disulfide **1** from solution in chloroform onto Au(111) surfaces, followed by reaction with isocyanate **2** to form a layer **3** that exposes UPy moieties. Supramolecular polymers, which are formed by self-complementary association of the telechelic PEG derivative **4** in hexadecane solution, can be probed by AFM (right) by utilizing gold-coated AFM tips that are functionalize with a self-assembled monolayer of the asymmetrically substituted UPy disulfide **5** in hexadecane. (Reproduced with permission from [90]).

decreased slightly with the increasing lengths of the stretched supramolecular polymer chains, in agreement with the theory developed by Evans *et al.* [93] who predicted a decrease in the magnitude of the single-complex rupture force with increasing spacer length and an increasing number of consecutive bonds along one molecule [94]. Figure 10(b) shows a plot of the observed rupture forces (at a loading rate of 35 nN/s) vs. the number of PEG-linkers N. The rupture-force values decrease with increasing number of stretched PEG linkers. Similarly, the AFM-SMFS experiments were performed in DMSO, where the hydrogen bonds are disrupted and consequently, the stretching and rupture of long supramolecular polymers was not observed, but only very short chains were stretched.

The AFM data proved that supramolecular polymers, and bond stability and rupture can be investigated on the single-molecule level. In the presence of building blocks that form reversible polymer chains, individual supramolecular polymer chains with up to 15 repeat units have been stretched successfully. In PEG-based telechelic bis(UPy) materials, the individual reversible linking sites along

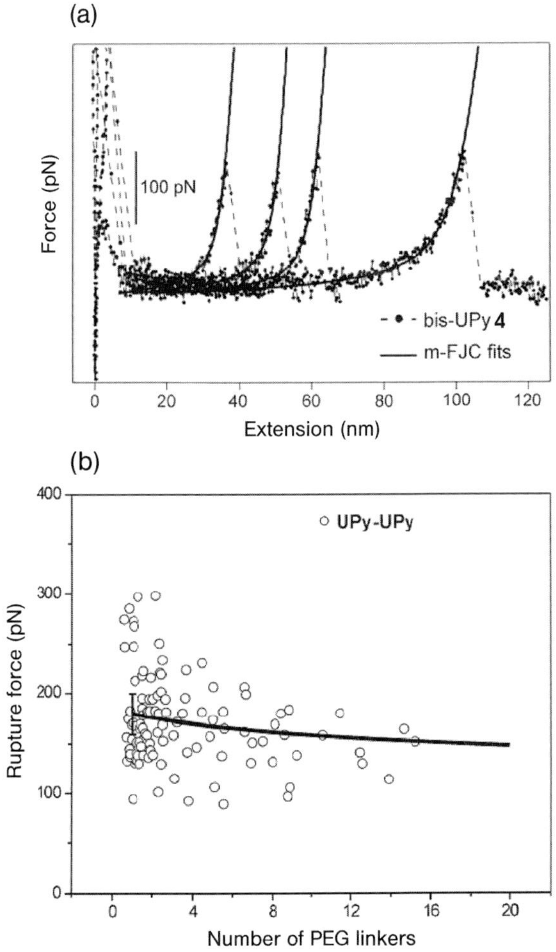

Figure 10 (a) Force–extension curves measured between gold-coated AFM tip and Au(111) sample functionalized both with layers of the short chain UPy-disulfide in the presence of the bifunctional bis-UPy PEG derivative **4** in hexadecane at 301 K, loading rate ~35 nN/s. The fits of the data to the m-FJC model are shown as solid lines. (b) Plot of rupture forces observed for UPy supramolecular polymers in hexadecane vs. the number of PEG linkers, *N*. The solid line is the predicted relation for rupture forces of N independent bonds in series. (Reproduced with permission from [90]).

the supramolecular polymer chain, that is, the complexes based on self-complementary recognition of 2-ureido-4[1H]-pyrimidinone, act as independent bonds in series. As the material properties, including viscosity, composition, or chain length, are functions of various external parameters and stimuli, which can be investigated *in situ*, highly useful information for the design and construction of nanometer-scale devices and stimuli-responsive systems will now become directly accessible from SMFS experiments.

References

1 Weisenhorn, A.L., Hansma, P.K., Albrecht, T.R. and Quate, C.F. (1989) *Applied Physics Letters*, **54**, 2651; Williams, P. (2006) in *Scanning Probe Microscopies Beyond Imaging: Manipulation of Molecules and Nanostructures* (ed. P. Samorì), Wiley-VCH, Weinheim, p. 250.

2 Bustamante, C., Macosko, J.C. and Wuite, G.J.L. (2000) *Nature Reviews. Molecular Cell Biology*, **1**, 130.

3 Mitsui, K., Harab, M. and Ikai, A. (1996) *FEBS Letters*, **385**, 29; Li, H., Linke, W.A., Oberhauser, A.F., Carrion-Vazquez, M., Kerkvliet, J.G., Lu, H., Marszalek, P.E. and Fernandez, J.M. (2002) *Nature*, **418**, 998.

4 Carrion-Vazquez, M., Oberhauser, A.F., Fowler, S.B., Marszalek, P.E., Broedel, S.E., Clarke, J. and Fernandez, J.M. (1999) *Proceedings of the National Academy of Sciences of the United States of America*, **96**, 3694.

5 Janshoff, A., Neitzert, M., Oberdörfer, Y. and Fuchs, H. (2000) *Angewandte Chemie – International Edition*, **39**, 3212.

6 Zlatanova, J., Lindsay, S.M. and Leuba, S.H. (2000) *Progress in Biophysics and Molecular Biology*, **74**, 37.

7 Strick, T.R., Dessinges, M.-N., Charvin, G., Dekker, N.H., Allemand, J.-F., Bensimon, D. and Croquette, V. (2003) *Reports on Progress in Physics*, **66**, 1.

8 Lee, G.U., Chrisey, L.A. and Colton, R.J. (1994) *Science*, **266**, 771.

9 Rief, M., Clausen-Schaumann, H. and Gaub, H.E. (1999) *Nature Structural Biology*, **6**, 346; Krautbauer, R., Clausen-Schaumann, H. and Gaub, Hermann E. (2000) *Angewandte Chemie – International Edition*, **39**, 3912; Krautbauer, R., Rief, M. and Gaub, H.E. (2003) *Nano Letters*, **3**, 493; Bustamante, C., Bryant, Z. and Smith, S.B. (2003) *Nature*, **421**, 423.

10 Kishino, A. and Yanagida, T. (1988) *Nature*, **334**, 74; Rayment, I., Holden, H.M., Whittaker, M., Yohn, C.B., Lorenz, M., Holmes, K.C. and Milligan, R.A. (1993) *Science*, **261**, 58; Vale, R.D. and Milligan, R.A. (2000) *Science*, **288**, 88; Keller, D. and Bustamante, C. (2000) *Biophysical Journal*, **78**, 541; Bustamante, C., Keller, D. and Oster, G. (2001) *Accounts of Chemical Research*, **34**, 412.

11 Samorì, P., Surn, M., Palermo, V., Lazzaroni, R. and Leclère, P. (2006) *Physical Chemistry Chemical Physics*, **8**, 3927; Kellermayer, M.S.Z. (2005) *Physiological Measurement*, **26**, R119.

12 Hugel, T. and Seitz, M. (2001) *Macromolecular Rapid Communications*, **22**, 989.

13 Zou, S., Schönherr, H. and Vancso, G.J. (2006) in *Scanning Probe Microscopies Beyond Imaging: Manipulation of Molecules and Nanostructures* (ed. P. Samorì), Wiley-VCH, Weinheim, p. 315.

14 Clausen-Schaumann, H., Seitz, M., Krautbauer, R. and Gaub, H.E. (2000) *Current Opinion in Chemical Biology*, **4**, 524.

15 Rief, M., Oesterhelt, F., Heymann, B. and Gaub, H.E. (1997) *Science*, **275**, 1295; Abu-Lail, N.I. and Camesano, T.A. (2003) *Journal of Microscopy*, **212**, 217.

16 Marszalek, P.E., Li, H.B. and Fernandez, J.M. (2001) *Nature Biotechnology*, **19**, 258.

17 Giannotti, M.I. and Vancso, G.J. (2007) *ChemPhysChem*, **8**, 2290.

18 Liu, C.J., Shi, W.Q., Cui, S.X., Wang, Z.Q. and Zhang, X. (2005) *Current Opinion in Solid State & Materials Science*, **9**, 140.

19 Kienberger, F., Ebner, A., Gruber, H.J. and Hinterdorfer, P. (2006) *Accounts of Chemical Research*, **39**, 29.

20 Oesterhelt, F., Oesterhelt, D., Pfeiffer, M., Engel, A., Gaub, H.E. and Müller, D.J. (2000) *Science*, **288**, 143.

21 Merkel, R. (2001) *Physics Reports*, **346**, 343; Zhang, W. and Zhang, X. (2003) *Progress in Polymer Science*, **28**, 1271; Carrion-Vazquez, M., Oberhauser, A.F., Fisher, T.E., Marszalek, P.E., Li, H. and Fernandez, J.M. (2000) *Progress in Biophysics and Molecular Biology*, **74**, 63; Wang, K., Forbes, J.G. and Jin, A.J. (2001) *Progress in Biophysics and Molecular Biology*, **77**, 1; Butt, H.-J., Cappella,

B. and Kappl, M. (2005) *Surface Science Reports*, **59**, 1.
22 Hodges, C.S. (2002) *Advances in Colloid and Interface Science*, **99**, 13.
23 Beyer, M.K. and Clausen-Schaumann, H. (2005) *Chemical Reviews*, **105**, 2921.
24 Vancso, G.J. (2007) *Angewandte Chemie – International Edition*, **46**, 3794.
25 Bell, G.I. (1978) *Science*, **200**, 618.
26 Evans, E. and Ritchie, K. (1997) *Biophysical Journal*, **72**, 1541.
27 Haverkamp, R.G., Marshall, A.T. and Williams, M.A.K. (2007) *Physical Review E*, **75**.
28 Smith, S.B., Finzi, L. and Bustamante, C. (1992) *Science*, **258**, 1122.
29 Bueche, F. (1962) *Physical Properties of Polymers*, Interscience, New York.
30 Porod, G. (1949) *Monatshefte fur Chemie*, **80**, 251; Kratky, O. and Porod, G. (1949) *Recueil des Travaux Chimiques des Pays*, **68**, 1106.
31 Flory, P.J. (1898) *Statistical Mechanics of Chain Molecules*, Hanser, Munich.
32 Bustamante, C., Marko, J.F., Siggia, E.D. and Smith, S. (1994) *Science*, **265**, 1599; Marko, J.F. and Siggia, E.D. (1995) *Macromolecules*, **28**, 8759.
33 Smith, S.B., Cui, Y. and Bustamante, C. (1996) *Science*, **271**, 795.
34 Odijk, T. (1995) *Macromolecules*, **28**, 7016.
35 Wang, M.D., Yin, H., Landick, R., Gelles, J. and Block, S.M. (1997) *Biophysical Journal*, **72**, 1335.
36 Zhang, Q.M., Lu, Z.Y., Hu, H., Yang, W.T. and Marszalek, P.E. (2006) *Journal of the American Chemical Society*, **128**, 9387.
37 Zhang, Q.M. and Marszalek, P.E. (2006) *Polymer*, **47**, 2526.
38 Li, H.B., Rief, M., Oesterhelt, F. and Gaub, H.E. (1998) *Advanced Materials*, **10**, 316.
39 Li, H.B., Zhang, W.K., Zhang, X., Shen, J.C., Liu, B.B., Gao, C.X. and Zou, G.T. (1998) *Macromolecular Rapid Communications*, **19**, 609; Zou, S., Zhang, W.K., Zhang, X. and Jiang, B.Z. (2001) *Langmuir*, **17**, 4799; Xu, Q.B., Zou, S., Zhang, W.K. and Zhang, X. (2001) *Macromolecular Rapid Communications*, **22**, 1163.
40 Oesterhelt, F., Rief, M. and Gaub, H.E. (1999) *New Journal of Physics*, **1**, 6.
41 Li, H.B., Zhang, W.K., Xu, W.Q. and Zhang, X. (2000) *Macromolecules*, **33**, 465.
42 Liu, C.J., Cui, S.X., Wang, Z.Q. and Zhang, X. (2005) *The Journal of Physical Chemistry. B*, **109**, 14807.
43 Giannotti, M.I., Rinaudo, M. and Vancso, G.J. (2007) *Biomacromolecules*, **8**, 2648.
44 Rief, M., Oesterhelt, F., Heymann, B. and Gaub, H.E. (1997) *Science*, **275**, 1295.
45 Marszalek, P.E., Oberhauser, A.F., Pang, Y.P. and Fernandez, J.M. (1998) *Nature*, **396**, 661.
46 Li, H.B., Rief, M., Oesterhelt, F., Gaub, H.E., Zhang, X. and Shen, J.C. (1999) *Chemical Physics Letters*, **305**, 197.
47 Marszalek, P.E., Pang, Y.P., Li, H.B., El Yazal, J., Oberhauser, A.F. and Fernandez, J.M. (1999) *Proceedings of the National Academy of Sciences of the United States of America*, **96**, 7894.
48 Xu, Q.B., Zhang, W. and Zhang, X. (2002) *Macromolecules*, **35**, 871; Marszalek, P.E., Li, H.B., Oberhauser, A.F. and Fernandez, J.M. (2002) *Proceedings of the National Academy of Sciences of the United States of America*, **99**, 4278; Marszalek, P.E., Oberhauser, A.F., Li, H.B. and Fernandez, J.M. (2003) *Biophysical Journal*, **85**, 2696; Lu, Z.Y., Nowak, W., Lee, G.R., Marszalek, P.E. and Yang, W.T. (2004) *Journal of the American Chemical Society*, **126**, 9033; Lee, G., Nowak, W., Jaroniec, J., Zhang, Q.M. and Marszalek, P.E. (2004) *Biophysical Journal*, **87**, 1456; Lee, G., Nowak, W., Jaroniec, J., Zhang, Q. and Marszalek, P.E. (2004) *Journal of the American Chemical Society*, **126**, 6218; Zhang, Q.M., Lee, G.R. and Marszalek, P.E. (2005) *Journal of Physics – Condensed Matter*, **17**, S1427; Haverkamp, R.G., Williams, M.A.K. and Scott, J.E. (2005) *Biomacromolecules*, **6**, 1816; Zhang, Q.M. and Marszalek, P.E. (2006) *Journal of the American Chemical Society*, **128**, 5596; Walther, K.A., Brujic, J., Li, H.B. and Fernandez, J.M. (2006) *Biophysical Journal*, **90**, 3806.

49 Zhang, Q.M., Jaroniec, J., Lee, G. and Marszalek, P.E. (2005) *Angewandte Chemie – International Edition*, **44**, 2723.

50 Zhang, B. and Evans, J.S. (2001) *Biophysical Journal*, **80**, 597.

51 Boyer, P.D. (1997) *Annual Review of Biochemistry*, **66**, 717; Wang, H.Y. and Oster, G. (1998) *Nature*, **396**, 279.

52 Howard, J. (2001) *Mechanics of Motor Proteins and the Cytoskeleton*, Sinauer Associates, Massachusetts.

53 Lipowski, R. and Klumpp, S. (2005) *Phys A*, **352**, 53112.

54 Atsumi, T., McCarter, L. and Imae, Y. (1992) *Nature*, **355**, 182.

55 Alonso, M.C., Drummond, D.R., Kain, S., Hoeng, J., Amos, L. and Cross, R.A. (2007) *Science*, **316**, 120; Hackney, D.D. (2007) *Science*, **316**, 58; Hess, H., Bachand, G.D. and Vogel, V. (2004) *Chemistry – A European Journal*, **10**, 2110.

56 Davis, A.P. (1999) *Nature*, **401**, 120; Saha, S. and Stoddart, J.F. (2007) *Chemical Society Reviews*, **36**, 77.

57 Pease, A.R., Jeppesen, J.O., Stoddart, J.F., Luo, Y., Collier, C.P. and Heath, J.R. (2001) *Accounts of Chemical Research*, **34**, 433.

58 Browne, W.R. and Feringa, B.L. (2006) *Nature Nanotechnology*, **1**, 25.

59 Fletcher, S.P., Dumur, F., Pollard, M.M. and Feringa, B.L. (2005) *Science*, **310**, 80.

60 Gil, E.S. and Hudson, S.M. (2004) *Progress in Polymer Science*, **29**, 1173.

61 Hugel, T., Holland, N.B., Cattani, A., Moroder, L., Seitz, M. and Gaub, H.E. (2002) *Science*, **296**, 1103; Holland, N.B., Hugel, T., Neuert, G., Cattani-Scholz, A., Renner, C., Oesterhelt, D., Moroder, L., Seitz, M. and Gaub, H.E. (2003) *Macromolecules*, **36**, 2015.

62 Zou, S., Hempenius, M.A., Schonherr, H. and Vancso, G.J. (2006) *Macromolecular Rapid Communications*, **27**, 103.

63 Zou, S., Korczagin, I., Hempenius, M.A., Schonherr, H. and Vancso, G.J. (2006) *Polymer*, **47**, 2483.

64 Ryan, A.J., Crook, C.J., Howse, J.R., Topham, P., Jones, R.A.L., Geoghegan, M., Parnell, A.J., Ruiz-Pérez, L., Martin, S.J., Cadby, A., Webster, J.R.P., Gleeson, A.J. and Bras, W. (2005) *Faraday Discussions*, **128**, 55; Butt, H.J. (2006) *Macromolecular Chemistry and Physics*, **207**, 573.

65 Foucher, D.A., Ziembinski, R., Tang, B.-Z., Macdonald, P.M., Massey, J., Jaeger, C.R., Vancso, G.J. and Manners, I. (1993) *Macromolecules*, **26**, 2878; Schubert, U.S., Newkome, G.R. and Manners, I. (eds) (2006) *Metal-Containing and Metallosupramolecular Polymers and Materials*, American Chemical Society, Washington, DC.

66 Shi, W., Giannotti, M.I., Zhang, X., Hempenius, M.A. and Schonherr, H. (2007) *Angewandte Chemie – International Edition*, **46**, 8400.

67 Shi, W.Q., Cui, S., Wang, C., Wang, L., Zhang, X., Wang, X.J. and Wang, L. (2004) *Macromolecules*, **37**, 1839.

68 Odijk, T. (1979) *Macromolecules*, **12**, 688; Skolnick, J. and Fixman, M. (1977) *Macromolecules*, **10**, 944.

69 Rinaudo, M. (2006) in *Kirk-Othmer Encyclopedia of Chemical Technology*, 5th edn, vol. **20**, John Wiley & Sons, p. 549.

70 Rinaudo, M. (2006) *Macromolecular Bioscience*, **6**, 590.

71 Li, H., Rief, M., Oesterhelt, F. and Gaub, H.E. (1999) *Applied Physics A: Materials Science & Processing*, **68**, 407.

72 Liu, C.J., Wang, Z.Q. and Zhang, X. (2006) *Macromolecules*, **39**, 3480.

73 Hascall, V.C. (1981) *Biology of Carbohydrates*, **1**, 1; Hardingham, T. (1981) *Biochemical Society Transactions*, **9**, 489.

74 Guss, J.M., Hukins, D.W.L., Smith, P.J.C., Winter, W.T., Arnott, S., Moorhouse, R. and Rees, D.A. (1975) *Journal of Molecular Biology*, **95**, 359; Winter, W.T., Smith, P.J.C. and Arnott, S. (1975) *Journal of Molecular Biology*, **99**, 219.

75 Haxaire, K., Marechal, Y., Milas, M. and Rinaudo, A. (2003) *Biopolymers*, **72**, 10; Haxaire, K., Marechal, Y., Milas, M. and Rinaudo, M. (2003) *Biopolymers*, **72**, 149; Marechal, Y., Milas, M. and Rinaudo, M. (2003) *Biopolymers*, **72**, 162.

76 Meyer, K. and Palmer, J.W. (1934) *The Journal of Biological Chemistry*, **107**, 629; Nichol, L.W., Ogston, A.G. and Preston, B.N. (1967) *The Biochemical Journal*, **102**, 407.

77 Scott, J.E. and Tigwell, M.J. (1978) *The Biochemical Journal*, **173**, 103.

78 Scott, J.E., Heatley, F. and Hull, W.E. (1984) *The Biochemical Journal*, **220**, 197; Heatley, F. and Scott, J.E. (1988) *The Biochemical Journal*, **254**, 489.

79 Scott, J.E. and Heatley, F. (1999) *Proceedings of the National Academy of Sciences of the United States of America*, **96**, 4850; Scott, J.E. and Heatley, F. (2002) *Biomacromolecules*, **3**, 547.

80 Haxaire, K., Buhler, E., Milas, M., Perez, S. and Rinaudo, M. (2002) in *Hyaluronan: Chemical, Biochemical and Biological Aspects*, Vol. 1 (eds J.F. Kennedy, G.O. Phillips, P.A. Williams and V.C. Hascall), Woodhead Publishing Ltd, p. 37.

81 Li, H.B., Zhang, W.K., Zhang, X., Shen, J.C., Liu, B.B., Gao, C.X. and Zou, G.T. (1998) *Macromolecular Rapid Communications*, **19**, 609.

82 Lehn, J.-M. (1995) *Supramolecular Chemistry – Concepts and Perspectives*, VCH, Weinheim,(1999) *Supramolecular Materials and Technologies*, John Wiley & Sons, New York.

83 Ringsdorf, H., Schlarb, B. and Venzmer, J. (1988) *Angewandte Chemie – International Edition*, **27**, 113; Buckingham, A.D., Legon, A.C. and Roberts, S.M. (1993) *Principles of Molecular Recognition*, Blackie, London; Buckingham, A.D., Legon, A.C. and Roberts, S.M. (1994) *The Lock and Key Principle: The State of the Art*, John Wiley & Sons, Chichester; Cooke, G. and Rotello, V.M. (2002) *Chemical Society Reviews*, **31**, 275; Lehn, J.-M. (2002) *Science*, **295**, 2400; Reinhoudt, D.N. and Crego-Calama, M. (2002) *Science*, **295**, 2403.

84 Lehn, J.-M. (1993) *Macromol Chem Macromol Symp*, **69**, 1.

85 Sijbesma, R.P., Beijer, F.H., Brunsveld, L., Folmer, B.J.B., Hirschberg, J.H.K.K., Lange, R.F.M., Lowe, J.K.L. and Meijer, E.W. (1997) *Science*, **278**, 1601.

86 Beijer, F.H., Kooijman, H., Spek, A.L., Sijbesma, R.P. and Meijer, E.W. (1998) *Angewandte Chemie – International Edition*, **37**, 75; Beijer, F.H., Sijbesma, R.P., Kooijman, H., Spek, A.L. and Meijer, E.W. (1998) *Journal of the American Chemical Society*, **120**, 6761; Brunsveld, L., Folmer, B.J.B., Meijer, E.W. and Sijbesma, R.P. (2001) *Chemical Reviews*, **101**, 4071.

87 ten Cate, A.T. and Sijbesma, R.P. (2002) *Macromolecular Rapid Communications*, **23**, 1094.

88 Zou, S., Zhang, Z., Förch, R., Knoll, W., Schönherr, H. and Vancso, G.J. (2003) *Langmuir*, **19**, 8618.

89 Zou, S., Schönherr, H. and Vancso, G.J. (2005) *Journal of the American Chemical Society*, **127**, 11230.

90 Zou, S., Schönherr, H. and Vancso, G.J. (2005) *Angewandte Chemie – International Edition*, **44**, 956.

91 Hinterdorfer, P., Kienberger, F., Raab, A., Gruber, H.J., Baumgartner, W., Kada, G., Riener, C., Wielert-Badt, S., Borken, C. and Schindler, H. (2000) *Single Mol*, **1**, 99.

92 Söntjens, S.H.M., Sijbesma, R.P., van Genderen, M.H.P. and Meijer, E.W. (2000) *Journal of the American Chemical Society*, **122**, 7487.

93 Evans, E. and Ritchie, K. (1997) *Biophysical Journal*, **72**, 1541; Evans, E., Ritchie, K. and Merkel, R. (1995) *Biophysical Journal*, **68**, 2580; Evans, E. and Williams, P. (2002) in *Physics of Biomolecules and Cells* (eds H. Flyvbjerg, F. Julicher, P. Ormos and F. David), EDP Sciences, Springer, Berlin, p. 145.

94 Evans, E. (2001) *Annual Review of Biophysics and Biomolecular Structure*, **30**, 105.

5.2
Quantitative Lateral Force Microscopy

Holger Schönherr, Ewa Tocha, Jing Song, and G. Julius Vancso

5.2.1
Introduction

Despite its central role for a large range of technological processes, as well as for everyday life, a fundamental understanding of friction has not yet emerged [1, 2]. In the most common situation normal friction (kinetic friction accompanied by wear and/or plastic deformation) takes place when two rough surfaces slide with respect to each other. The surface asperities may deform elastically or plastically. When a strong force is applied to the surfaces, damage (or wear) of the shearing substrates occurs. Under certain conditions (low load, completely elastic interactions, smooth tip shape, atomically flat substrate, unreactive surfaces, etc.) a single-asperity contact may be formed and wearless friction can be observed. This situation is often referred to as interfacial or boundary friction [3]. In this regime it has been observed that friction is proportional to the contact area [4].

Since the interplay of the many asperities found on both surfaces of contacting macroscopic bodies that are in relative movement with each other, together with other factors, determines macroscopic friction, the single-asperity behavior and its extrapolation to macroscopic behavior are of central interest. Lateral force microscopy (LFM) and friction force microscopy thus possess fundamental relevance in (nano)tribology to experimentally address friction forces on the level of single or a few asperities, since the sharp tip exploited in LFM may be considered a single asperity in some cases [4, 5].

LFM represents a particular mode of conventional contact-mode atomic force microscopy (AFM) [6, 7]. In this mode, forces in the surface-normal direction, as well as forces inplane (lateral forces) are measured simultaneously. The sharp AFM tip (typical radius of 10–100 nm) that is microfabricated on the end of a flexible cantilever (typical normal spring constant k_N of 0.01 to 1.0 N m^{-1}) experiences attractive and/or repulsive forces [8]. These forces are detected by monitoring the deflection of the cantilever in the normal direction. When the sample is scanned perpendicular to the main cantilever axis, lateral forces act on the tip and cause the cantilever to twist (Figure 1) [9]. The magnitude of the corresponding lateral deflection is proportional to the friction forces of the tip–sample contact. The typi-

Surface Design: Applications in Bioscience and Nanotechnology
Edited by Renate Förch, Holger Schönherr, and A. Tobias A. Jenkins

ISBN: 978-3-527-40789-7

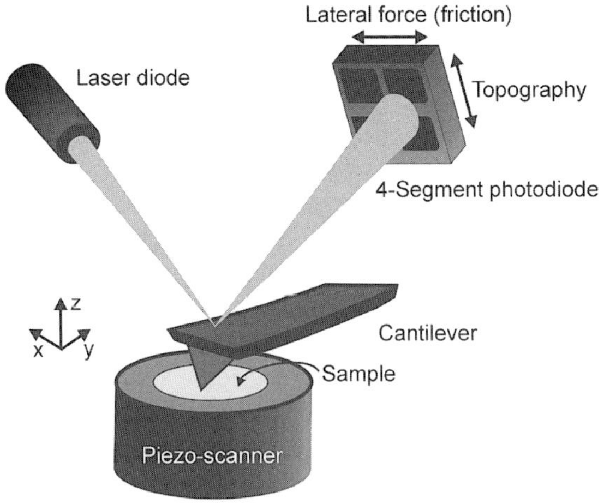

Figure 1 Schematic diagram of contact-mode AFM. The photodiode response to the bending, or the twist angle, of the cantilever (ΔU_N and ΔU_L), which corresponds to the movement of the laser spot in the photodiode in normal and lateral directions, respectively, are recorded.

cally applied optical detection method measures the deflection and bending angles of the cantilever, which are linearly proportional to the cantilever deflections (in the normal and lateral direction, respectively) for small angles [10, 11].

In the context of soft matter, organic or polymeric materials and micro- and nanostructured and patterned surfaces for biomaterial and life-science applications LFM also plays an important role. As one of the important imaging modes in AFM that provide materials contrast for surface characterization, LFM represents a key technique for the assessment of the dimensions, quality and homogeneity of lithographically fabricated patterns and structures.

In addition to self-assembled systems [12], including self-assembled monolayers (SAMs) [13] patterned by soft lithography (Figure 2a) [14, 15] or scanning probe lithography (SPL, Figure 3) [16, 17] (compare also Chapters 2.4 and 3.2), this also refers to patterned platforms for sensing or cell-surface interaction studies based on polymers (Chapter 3.4) [18, 19], micro- and nanostructured (hetero)materials (Chapters 4.3 and 4.5), gradient surfaces (Figure 4) or stimulus-responsive materials (Chapter 2.2).

The patterns obtained in microcontact printing [20] of suitable adsorbate molecules onto high-affinity substrates followed by backfilling with differently functionalized adsorbate molecules are conveniently imaged by LFM under ambient conditions (Figure 2). This early work of Wilbur [21] and coworkers demonstrated that compositional mapping is feasible in SAMs that expose three different terminal functional groups (i.e. $-CH_3$, $-COOH$ and $-OH$). The friction signal plotted in Figure 2(b) as 'relative' force clearly varies among the distinct regions. In addition, these authors analyzed mixed SAMs successfully.

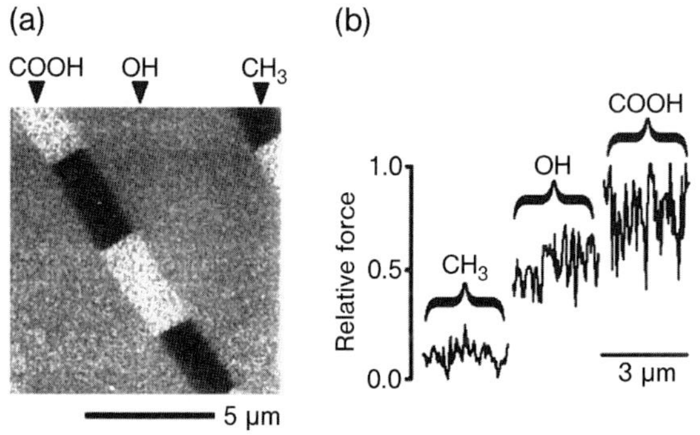

Figure 2 (a) Lateral force microscopy image of regions of SAMs terminated by CH_3 (dark), OH (gray), or COOH (light). (b) The relative lateral force between the tip and the surface, and the fluctuations in this force, increased with an increase in the surface free energy of the SAM. (Reproduced with permission from [21]; J. L. Wilbur, H. A. Biebuyck, J. C. MacDonald, G. M. Whitesides, Langmuir 1995, 11, 825, Copyright 1995 by American Chemical Society).

The interaction of the AFM tip (oxidized silicon or silicon nitride) with the respective substrate area differs due to the difference in surface energy. The contrast observed under ambient conditions may be dominated by the effect of capillary forces [10, 11], however, as discussed below, imaging with controlled chemical contrast is feasible in some cases by exploiting chemically functionalized probe tips in LFM measurements in liquid media.

Compositional analysis of patterned monolayers at the length scales shown in Figure 2 can obviously be afforded by complementary techniques, such as time-of-flight secondary ion mass spectrometry (ToF-SIMS) or high-resolution X-ray photoelectron spectroscopy (XPS) [31]. Alternatively, fluorescence microscopic analyses may be applicable, if the surface is functionalized with fluorophores and radiationless energy transfer from the fluorophore to the substrate is negligible. However, for feature sizes well below the optical diffraction limit, such as patterns fabricated with high-resolution scanning probe lithography approaches, LFM may be regarded advantageous (Figure 3).

In dip-pen nanolithography (DPN) (Figure 3a) [22, 23], AFM tip-mediated molecular transfer of low molar mass thiols via the water meniscus formed between tip and substrate is exploited to deposit molecules on gold with high precision. In the absence of pronounced diffusion, patterning on sub-100 nm length scales is feasible [24–26]. The LFM data displays the pattern with good contrast, however, the local coverage of the substrate with thiols in the areas surrounding the lines is difficult to discern.

Figure 3(b) shows the results of an oxidation reaction carried out with a near-field probe using a 244-nm laser on a silicon surface functionalized with chloromethylphenylsiloxane [27]. In the LFM images three parallel lines with a width of 45 nm can be observed. This feature width is well below the diffraction limit. The contrast in the LFM image was attributed to the higher surface free energy of the carboxylic-acid-functionalized regions.

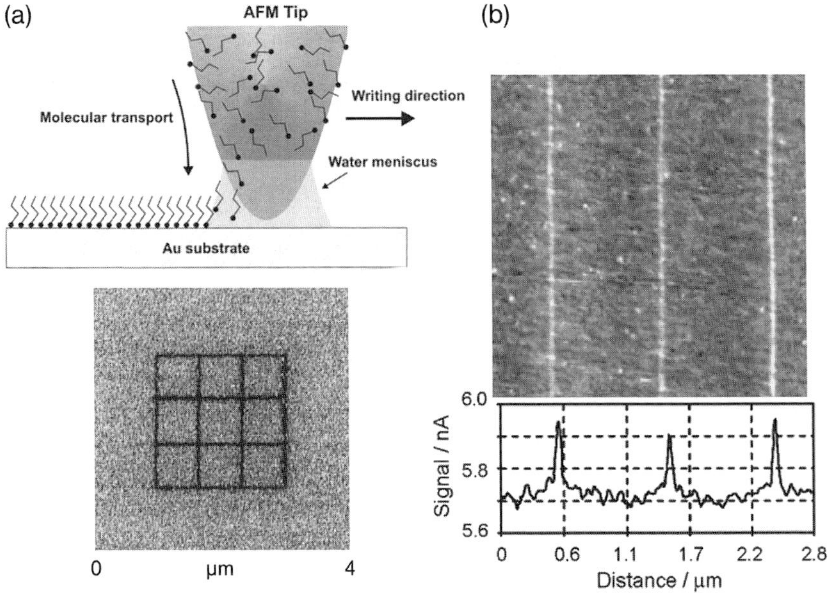

Figure 3 (a) Top: Schematic of AFM tip-mediated molecular transfer, dip-pen nanolithography (DPN); bottom: LFM image of a line pattern of octadecane thiol on gold, each line is 100 nm in width and 2 μm in length. (From R. D. Piner, J. Zhu, F. Xu, S. H. Hong, C. A. Mirkin, Science 1999, 283, 661. Reprinted with permission from AAAS). (b) LFM image of pattern of carboxylic acid groups created inside a chloromethylphenylsiloxane SAM on silicon by scanning near-field lithography using UV light with λ = 244 nm. (Reproduced with permission from [27]; S. Q. Sun, M. Montague, K. Critchley, M. S. Chen, W. J. Dressick, S. D. Evans, G. J. Leggett, NanoLett. 2006, 6, 29, Copyright 2006 by American Chemical Society).

A third instructive example of the application of LFM in the area relevant for this book are gradient surfaces, as reported by the group of Spencer [28]. The gradients are typically prepared by a two-step immersion method. In the first step, a gradual immersion of a clean gold surface into a solution of thiol A gives rise to islands of ~25 nm diameter at the end that had only been briefly immersed, whereas an increasingly continuous film is formed along the gradient. This type of structural evolution is well documented in the topographic and friction images shown in Figure 4. The gradients are further tailored by backfilling with a different thiol B that is terminated with a different end group. According to Spencer and coworkers the initially deposited structure of step 1 was found to persist after backfilling as well. In this study LFM was instrumental in unraveling the mechanism.

As alluded to above, the chemical specificity of the friction force contrast obtained was shown to be tailored by using chemically modified AFM tips [29–31]. In principle, chemical contrast can be obtained by using nonmodified tips, as shown among others in an early landmark paper of Frommer and cowork-

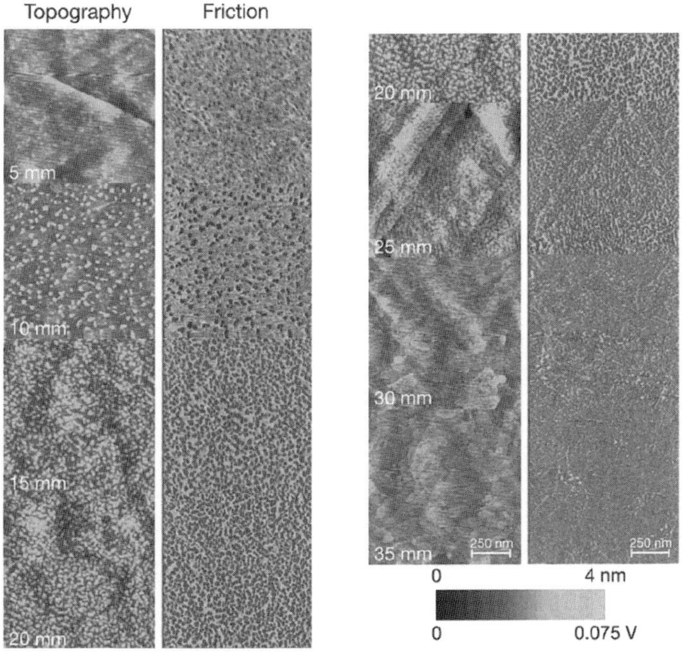

Figure 4 Topography (left) and friction (right) images (1 μm × 1 μm) from contact-mode AFM along a single-component (dodecane-thiol) gradient from the briefly to the lengthily immersed end. Island structures are visible both in topography and friction images. The island coverage increases toward the end that was immersed for a longer time (bottom). Note that, while the same dynamic range is employed in each topographical or friction image, the offset has been automatically adjusted by the SPM software to preserve the same mean value. Thus, absolute colors cannot be compared from one end of the gradient to the other. (Adapted/reproduced with permission from [28]; S. M. Morgenthaler, S. Lee, N. D. Spencer, Langmuir 2006, 22, 2706, Copyright 2006 by American Chemical Society).

ers [12]. In the approach termed originally chemical force microscopy (CFM),[1] the interaction between functional groups exposed at the surface of a SAM-functionalized AFM tip and the sample surface depends, in addition to the medium, on the nature and strength of the intermolecular interactions. Thereby, predictable friction force contrast between domains of different distinct functionalities can be obtained (Figure 5). In addition, it was demonstrated that friction forces are sensitive to pH [32] and may unveil local surface pKa and surface charge changes [33, 34].

1) Chemical force microscopy (CFM) is used as a synonym for 'AFM using defined surface chemistry, for instance self-assembled monolayer functionalization, on AFM probe tips in order to measure differences in surface chemical composition' (using friction or adhesion differences related to interactions between functional groups or atoms exposed on both tip and sample surface as contrast) throughout this chapter.

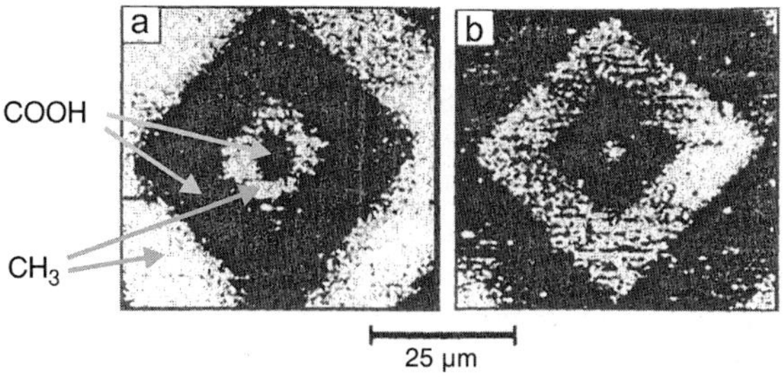

Figure 5 Friction-force image recorded on a patterned CH_3 / COOH monolayer sample with (a) a CH_3-terminated tip and (b) a COOH-terminated tip in ethanol. The friction forces, which increase from dark to bright contrast, show a contrast reversal. (From C. D. Frisbie, L. F. Rozsnyai, A. Noy, M. S. Wrighton, C. M. Lieber, Science 1994, 265, 2071. Reprinted with permission from AAAS).

A noteworthy shortcoming in many LFM experiments is the absent or inaccurate calibration of the AFM that prevented and still prevents the acquisition of friction forces in a truly quantitative manner [35–37]. The situation is complicated by the fact that friction forces scale with externally applied load, they may be rate dependent on the nanometer scale, and alignment on the molecular scale may lead to pronounced differences in friction forces. In fact, a dependence of friction forces on the relative orientation of this alignment with respect to the fixed scan direction in AFM was observed for aligned systems, including LB films [38, 39], molecular crystals [40], polymer chains in crystalline areas [41] or at the fold surface of solution-grown lamellar crystals [42, 43] (friction anisotropy). Furthermore, unlike for macroscopic contacts friction force for single-asperity contact is proportional to the contact area, thus, in some way to the size of the probe tip [4]. Thus, quantification of the forces is essential in order to be able to compare data from different labs and to allow one to assess desired information, such as coverage in patterned monolayers, more reliably.

In this chapter, we will hence summarize a recently introduced and refined improved wedge calibration method [35, 36] that relies on a universal calibration specimen. In addition to the extension of the calibration method to encompass the correct calibration of LFM for measurements in liquid media, representative data obtained in our laboratory will be discussed to illustrate the role of LFM as important *quantitative* imaging and surface analytical tool in the context of analysis required for designed nanostructured biomaterials in particular for life-science applications.

5.2.2
Results and Discussion

The problem of accurate calibration of friction forces is three-fold: (i) there is no generally applicable calibration method for the torsional or lateral spring constant of cantilevers with arbitrary shape; (ii) the calibration of the photodiode sensitivity (the factor that relates photodiode output signal to displacement) is very challenging; (iii) a number of factors, such as laser alignment, laser-spot asymmetry, additional sample tilt, the effect of the tip being located off the central axis of the cantilever, as well as the effect of different feedback-loop settings have significant effects on the calibration, depending on the method.

To discuss the new calibration approach/specimen, let us first have a look at a cantilever-tip assembly on which normal (Figure 6) and lateral forces (Figure 7) act [44]. The schematic drawings in these figures consider the effect of the medium, i.e. the refraction of light that passes from the cantilever through air or liquid via the liquid cell and air to the detector. The lateral shifts of the optical paths due to refraction at the medium-quartz and quartz-air surfaces for air and in liquid are negligible and will not be considered here. Upon deflection of the cantilever in the normal direction by an angle of θ in air, the exit angle α is deflected to $\alpha + 2\theta$ (Figure 6). When the same experiment is carried out in liquid, refraction results in altered exit angles β and $\beta + 2\theta^{L}$. The relations between α, β, as well as, $\alpha + 2\theta$ and $\beta + 2\theta^{L}$ are described by Snell's law using refractive indexes of air and the liquid, n_{air} and n, respectively. Similarly, for the torsion of the cantilever (Figure 7) of an angle ω results in beam reflected at 2ω (in air) and $2\omega^{L}$ (in liquid), which are also related by Snell's law. Consequently, the photodiode response to the bending and torsion angles of the cantilever in liquid medium and air are different due to beam refraction. The extension of these calibration procedures for measurement in liquids is discussed further in this chapter.

The photodiode response to the bending, or the torsion angle, of the cantilever (ΔU_N and ΔU_L), which corresponds to the movement of the laser spot in the photodiode in the normal and lateral directions, respectively, is expressed by the photodiode sensitivities in the normal and lateral directions (S_N and S_L). The normal and lateral forces (F_N and F_L) acting on the cantilever can, in general, be expressed as:

$$F_N = k_N S_N \Delta U_N \tag{1}$$

$$F_L = k_L S_L \Delta U_L = \vartheta \cdot \Delta U_L \tag{2}$$

where ϑ denotes the lateral calibration factor, which transforms the measured lateral difference signal [V] into friction force [nN].

The force constants of single-beam cantilevers (normal spring constant k_N, torsional spring constant k_ω, and lateral spring constant k_L) can be calculated according to continuum elasticity mechanics of isotropic solids [44, 45]:

$$k_N = \frac{Ewt^3}{4l^3} \qquad k_\omega = \frac{Gwt^3}{3l} \qquad k_L = \frac{k_\omega}{h^2} = \frac{Gwt^3}{3lh^2} \tag{3}$$

with cantilever length l, cantilever thickness t, cantilever width w, tip height h, Young's modulus E, Poisson's ratio ν, and shear modulus $G = E/2(1+\nu)$.

Since there are in practice significant variations of the cantilever thickness t the experimental calibration of k_N by, e.g., the thermal tune method [45] is advisable. S_N can be readily obtained from a force–displacement curve (Figure 6b) acquired in air or in a liquid on a hard substrate in the linear compliance region, where the tip and the piezo are in hard wall contact and move together.

The calculation of the lateral force is in principle carried out analogously to the calculation of the normal force (2). However, the separate determination of k_L and S_L is in practice problematic. The challenges alluded to arise from the fact that the available reference and two-step methods [46, 47] suffer from systematic errors introduced by contaminations on the reference samples and that a separate calibration of the lateral force constant k_L and the photodiode sensitivity for lateral deflection S_L is hampered by a number of problems, respectively. The accuracy of the determined value of k_L is limited due to large errors in the determination of

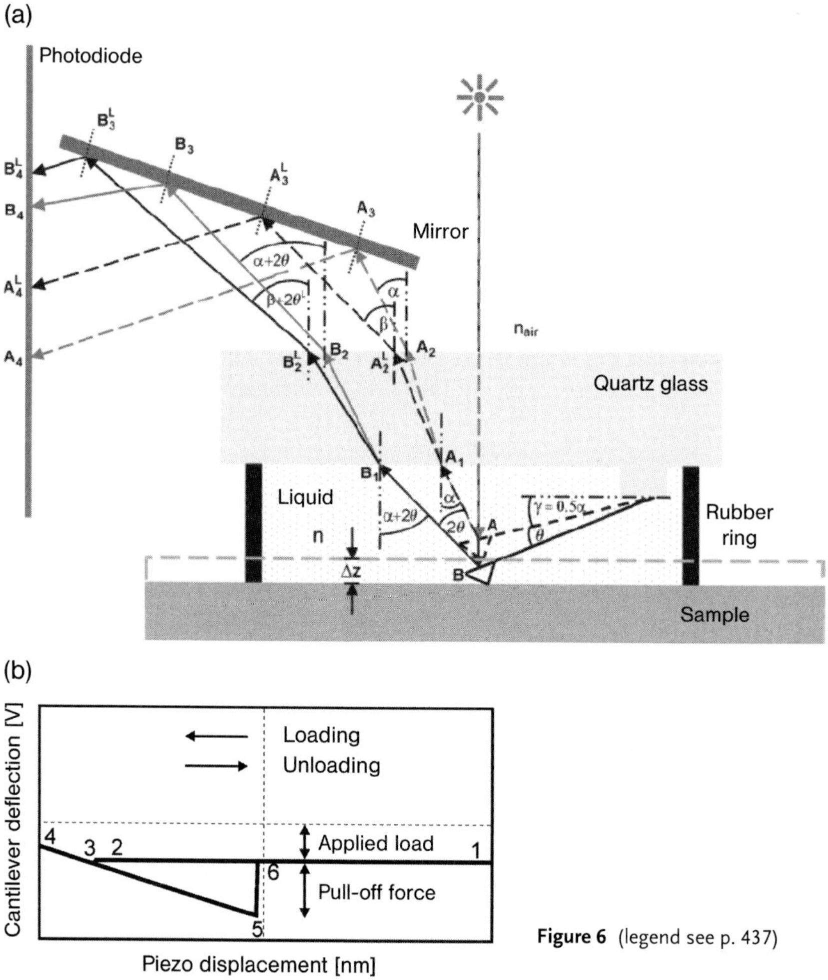

Figure 6 (legend see p. 437)

the cantilever dimensions and the uncertainty in the values for Young's moduli and Poisson's ratios for Si_3N_4 (if applicable). The unavailability of a reliable *in-situ* method to calibrate the photodiode sensitivity S_L and its dependence on factors, including laser-beam position on the lever, spot size and asymmetry, etc., represent additional complications [48].

These problems are circumvented in the so-called improved wedge calibration method. This calibration technique is typically applied in air, however, it can be extended to liquid medium using suitable calibration specimen. Here, a tip/cantilever is scanned across a special calibration sample with two well-defined slopes (Figure 8). These slopes possess an angle δ that is preferably in the range of 20–35 degree. From the captured friction and height data the calibration factor ϑ can be estimated with an error as low as ~ 5%.

It can be shown that

$$\sin\delta(L\cos\delta + A)\cdot\mu_s^2 - \frac{\Delta_s - \Delta_f}{W_s}(L + A\cos\delta)\cdot\mu_s + L\sin\delta\cos\delta = 0 \tag{6}$$

$$\mu_f = \frac{\vartheta\cdot W_f}{L + A} \tag{7}$$

where μ denotes the friction coefficient, W denotes the half-width of the friction loop $W = (M_u - M_d)/2$), Δ denotes the friction-loop offsets ($\Delta = (M_u + M_d)/2$) and the subscripts s and f denote sloped and flat surfaces, respectively (see also Figure 9 below). By solving (6) for μ_s two possible mathematical solutions are provided (for any given load and adhesion), corresponding to two values of the friction calibration factor ϑ. Since ϑ must be identical for sloped and flat surfaces, we obtain μ_f from (7). The physical solution stands for μ_s and $\mu_f < 1/\tan\delta$.

Figure 6 (a) Simplified side-view schematic (not to scale) of an AFM experiment carried out using a liquid cell: The light paths for two cantilever deflections are indicated for media with different refractive indices (normal bending signals: black line: light path in liquid; gray line: light path in air). We denote the cantilever bending angle θ, and the angle between cantilever and holder $\gamma = 1/2\alpha$. n, n_{air} and α, β stand for the refractive indices of the liquid and air, and the entry/exit angles for light propagating through the liquid/quartz and quartz/air interfaces, respectively. (Reprinted with permission from [44]; E. Tocha, J. Song, H. Schönherr, G. J. Vancso, Langmuir 2007, 23, 7078, Copyright 2006 by American Chemical Society). (b) Typical force–displacement curve for adhesive contact: During the approach (loading) part (position 1–2), no interactions occur between the tip and the sample surface. As the tip–surface distance becomes sufficiently small, the gradient of the attractive force overcomes the cantilever spring constant and brings the tip in contact with the sample surface (position 3). Further approaching causes a deflection of the cantilever (position 3–4). The unloading part of the force–displacement curve starts from position 4, the deflection of the cantilever is decreased as the sample surface retracts from the tip. When the sample surface is further withdrawn from the tip, the cantilever is deflected owing to adhesive forces. At position 5, the elastic force in the cantilever overcomes the force gradient and the tip snaps off from the surface (position 6). From position 6 to 1, the cantilever returns to its equilibrium position. The adhesion between tip and sample is characterized by the so-called pull-off or pull-out force (snap off).

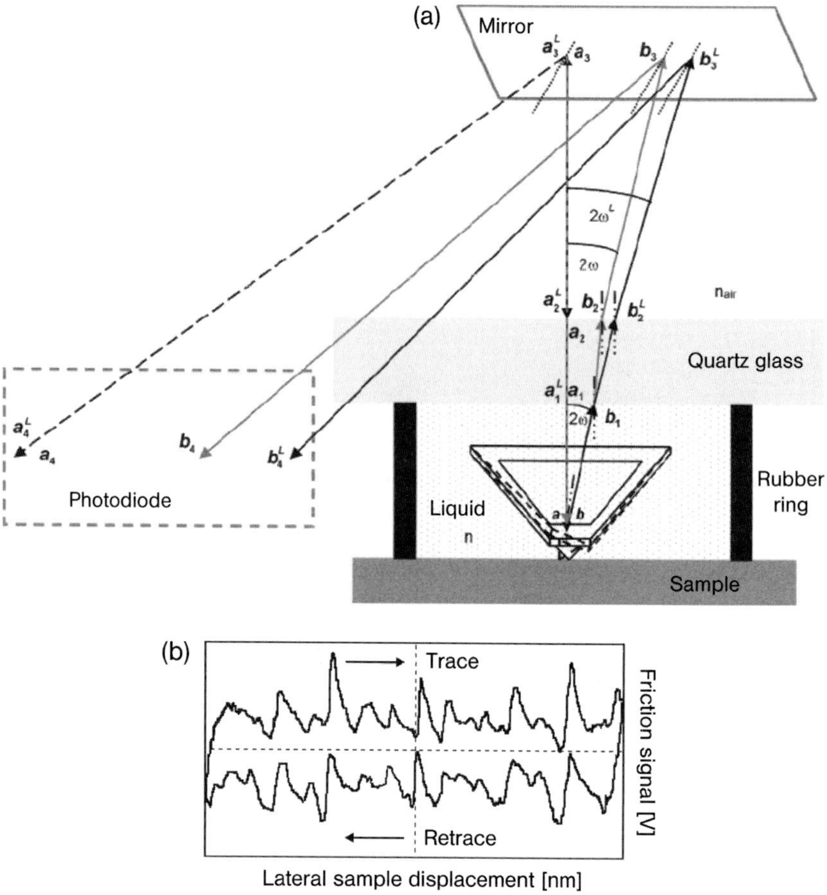

Figure 7 (a) Simplified schematic (not to scale) of an AFM experiment carried out using a liquid cell: The light paths for cantilever deflection are indicated for media with different refractive indices (lateral torsion signals: black line: light path in liquid; gray line: light path in air). The cantilever torsion angle is denoted as ω. (Reprinted with permission from [44]; E. Tocha, J. Song, H. Schönherr, G. J. Vancso, Langmuir 2007, 23, 7078, Copyright 2006 by American Chemical Society). (b) Typical friction loop: In the friction loop, the difference friction signal (also called friction signal) is plotted against the trace (forward scan) and retrace (backward scan) lateral sample displacement. At the beginning of each trace and retrace, the sample remains in static contact until the shear force increases and overcomes the static friction force. The signal changes sign for the retrace with respect to the trace scan. For a given load (normal force), the friction force can be determined as one half of the difference between the corresponding friction signals for trace and retrace scans.

In Figure 9, topographic and lateral force data obtained on a universally applicable standard specimen is shown that enables one to accurately calibrate all types of AFM cantilevers and tips for quantitative friction-force measurements. Using this standard and the improved wedge calibration method, calibration factors ϑ

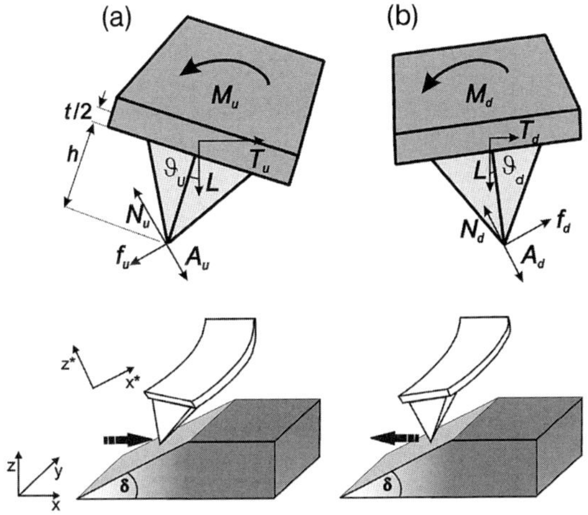

Figure 8 Schematic illustration of cantilever torsion while (a) sliding up and (b) sliding down on a sloped surface (in the x direction). While sliding across a sloped surface with angle δ, the acting forces (the applied load L, the horizontal tractive force T_i, the adhesion force A_i, the reaction force from the surface acting on the tip with a component N_i in the surface normal direction and a component f_i (friction force) parallel to the surface) and the torsion momentum M_i are in equilibrium and depend on the direction of motion – uphill and downhill, denoted here with subscripts u and d, respectively. ω represents the torsion angle of the cantilever, which is proportional to the friction force. h and t stand for tip height and cantilever thickness, respectively. (Adapted/reprinted with permission from [48]; E. Tocha, H. Schönherr, G. J. Vancso, Langmuir 2006, 22, 2340, Copyright 2006 by American Chemical Society).

can be calculated with an error of ~5%. The approach is not affected by an additional small sample tilt, different feedback settings, and a possible tip position off the central cantilever axis, as is frequently observed. Only laser light interference and nonspherical tip apex shapes must be taken into account [48].

To demonstrate the validity of the approach, two Si_3N_4 probes with different cantilever geometry were used for the nanotribological analysis of a micropatterned binary SAM. This monolayer was prepared according to the procedures published by the Whitesides group and exposed hydrophilic and hydrophobic headgroups (–COOH and –CH_3, respectively). The measurements shown in Figures 10 and 11 were carried out in a controlled environment (50% RH and 25 °C). The circular carboxylic-acid-terminated areas of the micropatterned SAM are clearly recognized as high friction areas in the LFM difference images in Figure 10. The effect of increasing load on the contrast in this two-component SAM is obvious. With increasing load, the difference friction images show higher contrast between –COOH and –CH_3 head groups. The corresponding friction-force distributions reveal that both the difference between the mean friction forces, as well as the peak widths, increase with higher applied force.

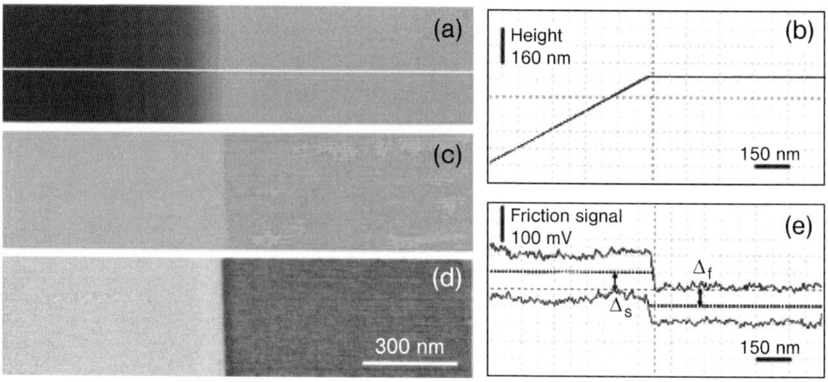

Figure 9 Example of experimental data measured with a Si_3N_4 tip on both sloped and flat surfaces: (a) topography image (vertical scale from black to white 800 nm); (b) cross-section of topography (vertical scale 800 nm); (c) difference friction image (trace – retrace, vertical scale 0.5 V); (d) offset of the friction loops (trace + retrace, vertical scale 0.5 V); (e) friction loop corresponding to cross-section shown in panel (b) (the off-sets for sloped and flat surface, Δ_s and Δ_f, respectively, have been marked). (Reprinted with permission from [48]; E. Tocha, H. Schönherr, G. J. Vancso, Langmuir 2006, 22, 2340, Copyright 2006 by American Chemical Society).

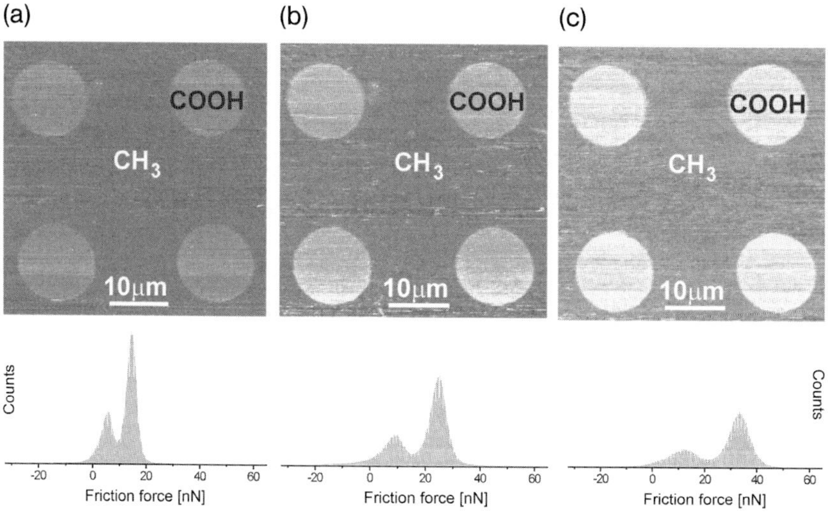

Figure 10 Difference friction images and corresponding friction force distribution of binary SAM measured with a Si_3N_4 tip (single beam cantilever) for different applied loads: (a) 5 nN, (b) 18 nN and (c) 30 nN. Bright and dark colors correspond to high and low friction forces and are attributed to –COOH and –CH_3 headgroups, respectively. The sample was scanned with a velocity of 200 μm/s at 50% RH and 25 °C. (Adapted/reprinted with permission from [48]; E. Tocha, H. Schönherr, G. J. Vancso, Langmuir 2006, 22, 2340, Copyright 2006 by American Chemical Society).

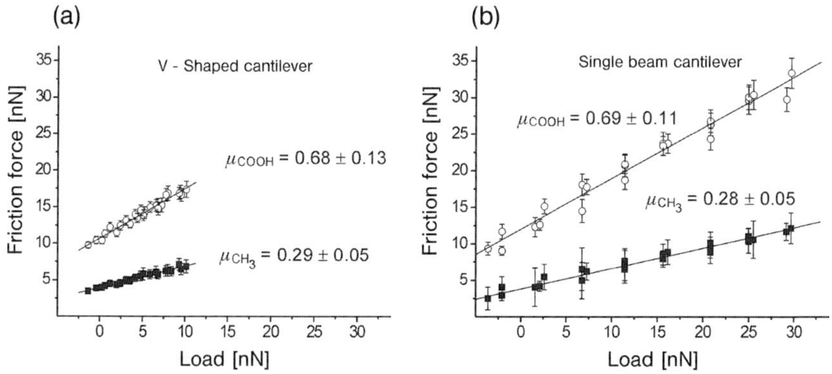

Figure 11 Friction force versus load for a micropatterned SAM sample (exposing $-CH_3$ and $-COOH$ headgroups, respectively) measured using two different Si_3N_4 cantilevers: (a) V-shaped and (b) single beam, with scanning velocity of 40 μm/s at 50% RH and 25 °C. The solid lines correspond to linear least squares fits. The error bars indicate the standard deviation (n = 128) of the data analyzed for a given load. (Adapted/reprinted with permission from [48]; E. Tocha, H. Schönherr, G. J. Vancso, Langmuir 2006, 22, 2340, Copyright 2006 by American Chemical Society).

This observation is due to a significantly higher friction coefficient of the hydrophilic areas compared to the hydrophobic areas, in line with data discussed in the introduction. Figure 11 shows the friction force as a function of load obtained with (a) a V-shaped and (b) a single-beam cantilever. The pull-off forces for $-COOH$ and $-CH_3$ groups were 21 nN and 14 nN measured with the V-shaped cantilever and 15 nN and 9 nN measured with the single-beam cantilever, respectively. The ratio between the pull-off forces for hydrophobic and hydrophilic groups was about 0.6 for both probes, also consistent with the different surface energies and the resulting capillary forces.

The friction coefficients determined with the two cantilever probes with different geometry were identical to within the error (μ_{Si3N4} (CH_3) = 0.29±0.05 and 0.28±0.05 and μ_{Si3N4} (COOH) = 0.68±0.13 and 0.69±0.11, using V-shaped and single-beam cantilevers, respectively). Green *et al.* reported similar values of friction coefficients (μ_{Si3N4} (CH_3) = 0.34 and μ_{Si3N4} (COOH) = 0.76) [49].

The results discussed above show that the wedge calibration method, in conjunction with the new universal calibration specimen, enables one to perform truly quantitative nanotribology. This calibration approach is a convenient, yet robust and precise, method and can be expected to open the pathway to improved fundamental LFM work in the area of tribology. In addition, it allows an improved quantitative compositional mapping and imaging in nanoscience and nanotechnology-related surface characterization (compare Chapters 2.4 and 3.2).

The extension of this calibration procedure for LFM measurement in liquids (Figures 5 to 7) necessitates a correction for the refraction of the laser light that is reflected off the cantilever. The direct application of the calibration factor ϑ (or, if k_L is known, S_L) determined in air for the friction-force calibration of the experi-

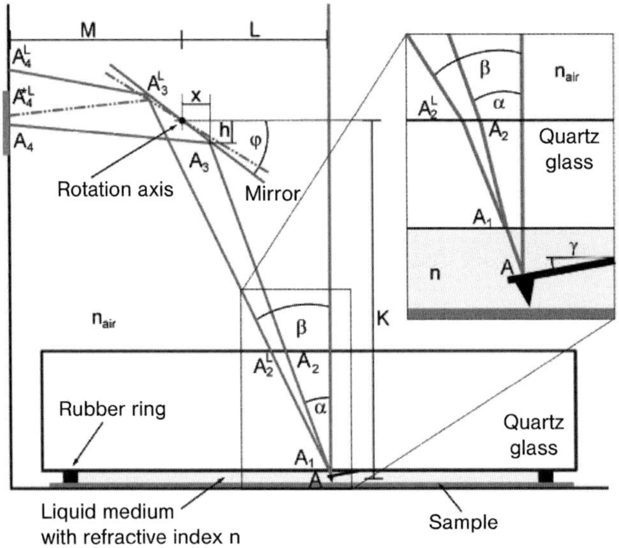

Figure 12 Schematic (approximately to scale) of an AFM experiment carried out using a liquid cell. The light paths for normal cantilever deflection are indicated for media with different refractive indices. n, n_{air} and α, β denote the refractive indices of the liquid and air, and the entry/exit angles for light propagating through the liquid/quartz and quartz/air interfaces, respectively. We denote the mirror rotation angle as φ, the vertical distance between the cantilever apex and the mirror rotation axis as K, the horizontal distance between the photodiode and the mirror rotation axis as M, the horizontal distance between the incident beam and the mirror rotation axis as L, and the angle between the cantilever and the holder as $\gamma = \frac{1}{2}\alpha$. x and h stand for the horizontal and vertical distances of the reflected laser beam on the mirror relative to the rotation axis, respectively. (Reprinted with permission from [44]; E. Tocha, J. Song, H. Schönherr, G. J. Vancso, Langmuir 2007, 23, 7078, Copyright 2006 by American Chemical Society).

mental data obtained in liquids results in a severe overestimate of the friction forces. As shown in Figure 12 for the normal deflection in more detail, the light is refracted depending on the refractive index of the imaging medium n. For measurements in air ($n_{air} = 1.00$) or water ($n_{water} = 1.36$) this would result in the paths of the light A–A_1–A_2 and A–A_1–A_2^L, respectively. Thus, the photodiode response to the bending and torsion angles ω of the cantilever will be altered significantly.

In conventional contact-mode AFM height imaging and normal (adhesion) force data acquisition, this effect can be corrected on not too soft samples *in situ* by determining the photodiode sensitivity S_N in *f–d* curves in the imaging liquid (see above). However, as mentioned above, the corresponding sensitivity S_L cannot be easily obtained *in situ* for lateral force microscopy (LFM).

It can be shown and experimentally confirmed [44] that

$$\frac{\omega^L}{\omega} = \frac{n}{n_{air}} \tag{8}$$

and thus

$$\frac{S_L}{S_L^L} = \frac{n}{n_{air}} \tag{9}$$

Since the calibration factors ϑ are proportional to S_L (2), values of S_L determined in air can be converted to the correct calibration factors for liquid by multiplication with n_{air}/n_{liquid}. This corresponds to a correction of as much as 25–74% for perfluorohexane and diiodomethane. For water as the relevant imaging liquid in view of biological applications the correction is 36%.

5.2.3 Outlook

The LFM calibration approaches discussed show that the once notoriously difficult calibration with errors approaching in some cases 50–100% (in air) and another 36% if refraction is neglected for measurements in water, is no longer a bottleneck. In addition, the data discussed demonstrate that a reliable calibration of friction forces in LFM is feasible using a simple one-step method in conjunction with a microfabricated reference sample.

For the widespread application in the area of designed nanostructured biomaterials, in particular for life-science applications, the question remains whether the acquisition of qualitative friction data is sufficient or whether truly quantitative data is required. Friction mode is included in most commercial AFM instruments. Unfortunately, due to various reasons discussed above and in some cases instrumental limitations, LFM has been often merely used as a qualitative tool in the target areas of life-science applications. More recently (bio)chemically modified tips have become commercially available, however, they still reside in a small niche of the probe market.

For the assessment of pattern definition, quality control, but more so for the elucidation of surface coverages, e.g. in nanometer-sized patterns generated by SPL, reproducibility and quantification is in our opinion a necessity. This holds also and in particular for the determination of the relation between surface coverages, surface energies and measured friction coefficients in CFM applications. Since the tools are now at hand, there are no further technical issues that may hamper this development and we expect that the application of LFM in the context of biomaterials, life-science applications and beyond will flourish in the future.

Acknowledgment

This work has been financially supported by the Dutch Technology Foundation STW (STW-project 5287, Nanoscale wear-resistant ceramic materials with low friction).

References

1 Dowson, D. (1998) *History of Tribology*, Professional Engineering Publishing, London.
2 Singer, I.L. and Pollock, H.M. (eds) (1992) *Fundamentals of Friction*, Kluwer, The Netherlands.
3 Homola, A.M., Israelachvili, J.N., McGuiggan, P.M. and Gee, M.L. (1990) *Wear*, **136**, 65.
4 Carpick, R.W. and Salmeron, M. (1997) *Chemical Reviews*, **97**, 1163.
5 Meyer, E., Overney, R.M., Dransfeld, K. and Gyalog, T. (1998) *Nanoscience: Friction and Rheology on the Nanometer Scale*, World Scientific.
6 Binnig, G., Quate, C.F. and Gerber, C. (1986) *Physical Review Letters*, **56**, 930.
7 Colton, R.J., Engel, A., Frommer, J.E., Gaub, H.E., Gewirth, A.A., Guckenberger, R., Rabe, J., Heckl, W.M. and Parkinson, B. (1998) *Procedures in Scanning Probe Microscopies*, John Wiley & Sons, New York.
8 Israelachvili, J.N. (1991) *Intermolecular Surface Forces*, 2nd edn, Academic Press, London.
9 Perry, S.S. (2004) *MRS Bulletin*, **29**, 478.
10 Cappella, B. and Dietler, G. (1999) *Surface Science Reports*, **34**, 1.
11 Butt, H.-J., Cappella, B. and Kappl, M. (2005) *Surface Science Reports*, **59**, 1.
12 Overney, R.M., Meyer, E., Frommer, J., Brodbeck, D., Lüthi, R., Howald, L., Güntherodt, H.-J., Fujihira, M., Takano, H. and Gotoh, Y. (1992) *Nature*, **359**, 133.
13 Ulman, A. (1991) *An Introduction to Ultrathin Organic Films: From Langmuir-Blodgett to Self-Assembly*, Academic Press, New York.
14 Xia, Y.N. and Whitesides, G.M. (1998) *Angewandte Chemie – International Edition*, **37**, 551.
15 Michel, B., Bernard, A., Bietsch, A., Delamarche, E., Geissler, M., Juncker, D., Kind, H., Renault, J.P., Rothuizen, H., Schmid, H., Schmidt-Winkel, P., Stutz, R. and Wolf, H. (2001) *IBM Journal of Research and Development*, **45**, 697.
16 Krämer, S., Fuierer, R.R. and Gorman, C.B. (2003) *Chemical Reviews*, **103**, 4367.
17 Wouters, D. and Schubert, U.S. (2004) *Angewandte Chemie – International Edition*, **43**, 2480.
18 Feng, C.L., Embrechts, A., Bredebusch, I., Schnekenburger, J., Domschke, W., Vancso, G.J. and Schönherr, H. (2007) *Advanced Materials*, **19**, 286.
19 Feng, C.L., Vancso, G.J. and Schönherr, H. (2006) *Advanced Functional Materials*, **16**, 1306.
20 Kumar, A., Biebuyck, H.A. and Whitesides, G.M. (1994) *Langmuir*, **10**, 1498.
21 Wilbur, J.L., Biebuyck, H.A., MacDonald, J.C. and Whitesides, G.M. (1995) *Langmuir*, **11**, 825.
22 Jaschke, M. and Butt, H.J. (1995) *Langmuir*, **11**, 1061.
23 Piner, R.D., Zhu, J., Xu, F. Hong, S.H., and Mirkin, C.A. (1999) *Science*, **283**, 661.
24 McKendry, R., Huck, W.T.S., Weeks, B., Florini, M., Abell, C. and Rayment, T. (2002) *Nano Letters*, **2**, 713.
25 Degenhart, G.H., Dordi, B., Schönherr, H. and Vancso, G.J. (2004) *Langmuir*, **20**, 6216.
26 Salazar, R.B., Shovsky, A., Schönherr, H. and Vancso, G.J. (2006) *Small*, **2**, 1274.
27 Sun, S.Q., Montague, M., Critchley, K., Chen, M.S., Dressick, W.J., Evans, S.D. and Leggett, G.J. (2006) *Nano Letters*, **6**, 29.
28 Morgenthaler, S.M., Lee, S. and Spencer, N.D. (2006) *Langmuir*, **22**, 2706.
29 Frisbie, C.D., Rozsnyai, L.F., Noy, A., Wrighton, M.S. and Lieber, C.M. (1994) *Science*, **265**, 2071.
30 Noy, A., Vezenov, D.V. and Lieber, C.M. (1997) *Annual Review of Materials Science*, **27**, 381.
31 Vancso, G.J., Hillborg, H. and Schönherr, H. (2005) *Advances in Polymer Science*, **182**, 55.
32 Marti, A., Hahner, G. and Spencer, N.D. (1995) *Langmuir*, **11**, 4632.
33 Vezenov, D.V., Noy, A., Rozsnyai, L.F. and Lieber, C.M. (1997) *Journal of the American Chemical Society*, **119**, 2006.

34 van der Vegte, E.W. and Hadziioannou, G. (1997) *Langmuir*, **13**, 4357.

35 Ogletree, D.F., Carpick, R.W. and Salmeron, M. (1996) *Review of Scientific Instruments*, **67**, 3298.

36 Varenberg, M., Etsion, I. and Halperin, G. (2003) *Review of Scientific Instruments*, **74**, 3362.

37 Cannara, R.J., Eglin, M. and Carpick, R.W. (2006) *Review of Scientific Instruments*, **77**, 053701.

38 Overney, R.M., Takano, H., Fujihira, M., Paulus, W. and Ringsdorf, H. (1994) *Physical Review Letters*, **72**, 3546.

39 Liley, M., Gourdon, D., Stamou, D., Meseth, U., Fischer, T.M., Lautz, C., Stahlberg, H., Vogel, H., Burnham, N.A. and Duschl, C. (1998) *Science*, **280**, 273.

40 Schönherr, H., Kenis, P.J.A., Engbersen, J.F.J., Harkema, S., Hulst, R., Reinhoudt, D.N. and Vancso, G.J. (1998) *Langmuir*, **14**, 2801.

41 Vancso, G.J., Förster, S. and Leist, H. (1996) *Macromolecules*, **29**, 2158.

42 Nisman, R., Smith, P. and Vancso, G.J. (1994) *Langmuir*, **10**, 1667.

43 Vancso, G.J. and Schönherr, H. (1999) in *Microstructure and Microtribology of Polymer Surfaces*, ACS Symposium Series, Vol. 741 (eds V.V. Tsukruk and K.J. Wahl), American Chemical Society, New York, p. 317.

44 Tocha, E., Song, J., Schönherr, H. and Vancso, G.J. (2007) *Langmuir*, **23**, 7078.

45 Butt, H.J. and Jaschke, M. (1995) *Nanotechnology*, **6**, 1.

46 Schwarz, U.D., Köster, P. and Wiesendanger, R. (1996) *Review of Scientific Instruments*, **67**, 2560.

47 Liu, E., Blanpain, B. and Celis, J.P. (1996) *Wear*, **192**, 141.

48 Tocha, E., Schönherr, H. and Vancso, G.J. (2006) *Langmuir*, **22**, 2340.

49 Green, J.B.D., McDermott, M.T., Porter, M.D. and Siperko, L.M. (1995) *The Journal of Physical Chemistry*, **99**, 10960.

5.3
Long-Range Surface Plasmon Enhanced Fluorescence Spectroscopy as a Platform for Biosensors

Amal Kasry, Jakub Dostálek, and Wolfgang Knoll

5.3.1
Introduction

Over the last two decades, great strides have been achieved in the development of biosensors for fast and sensitive detection of numerous compounds relevant to important areas such as environmental monitoring, food control and medical diagnostics [1–3]. Optical biosensors based on surface plasmon resonance (SPR) are devices that exploit surface plasmon (SP) waves excited at a metallic surface [4]. These surface waves probe the interactions between target molecules present in a liquid sample and biomolecular recognition elements anchored to the metallic surface. The capture of target molecules on the metallic surface causes an increase in the refractive index that can be observed using the spectroscopy of SPs. This label-free approach enables the direct detection of large and medium-sized molecules (typically >10 kDa) that can produce sufficiently high refractive index changes. In order to increase the sensitivity of the measurement of the binding of target molecules, surface plasmon enhanced fluorescence spectroscopy (SPFS) was introduced to SPR biosensors [5]. SPFS-based biosensors take advantage of the increased intensity of the electromagnetic field on the metallic surface occurring upon the excitation of SPs. The SP field is used to excite the chromophore-labeled molecules bound to the surface, thus providing large enhancement of the fluorescence signal. In comparison with SPR biosensors relying on the measurement of refractive index changes, SPFS-based biosensors typically enable detection of analytes with several orders of magnitude lower detection limits [6].

With the advent of SPR biosensors, various surface plasmon modes were employed for the refractive index and fluorescence spectroscopy-based observation of biomolecular binding events. These include long-range surface plasmons (LRSP) [7], SPs coupled to a dielectric waveguide [8], Bragg-scattered SPs [9] and localized SPs supported by metallic nanoparticles [10] and nanostructured metallic surfaces [11]. LRSP is a special surface plasmon mode that originates from the coupling of surface plasmons (SPs) propagating along opposite surfaces of a thin metal film. LRSPs can propagate with an order of magnitude lower damping compared to conventional SPs [12]. Therefore, LRSPs have attracted attention for the

Surface Design: Applications in Bioscience and Nanotechnology
Edited by Renate Förch, Holger Schönherr, and A. Tobias A. Jenkins

ISBN: 978-3-527-40789-7

design of high-resolution SPR sensors [13, 14] and currently an ultrahigh refractive-index resolution (smallest detectable refractive index change) of 2.5×10^{-8} was reported [15]. Only recently, LRSPs were applied in SPFS-based biosensors [16, 17]. Due to their low damping, the excitation of LRSPs is associated with a large enhancement of field intensity at the metal/dielectric interface that can be directly translated into an increase in the fluorescence signal.

Within this chapter, we describe recent advancements in SPFS-based biosensors achieved through the excitation of LRSPs. In the following, we describe the characteristics of LRSPs, the arrangements used for their excitation and their implementation in a biosensor. A comparison of the performance of SPFS-based biosensors relying on the excitation of conventional SPs and LRSPs is presented.

5.3.2
Surface Plasmon Modes Propagating on a Thin Metal Film

A surface plasmon (SP) is an optical wave trapped on a metallic surface that originates from coupled collective oscillations of the electron plasma and the associated electromagnetic field, see Figure 1(a). This optical wave propagates along an interface between a semi-infinite metal and a dielectric medium with the complex propagation constant β:

$$\beta = k_0 \sqrt{\frac{n_{\mathrm{m}}^2 n_{\mathrm{d}}^2}{n_{\mathrm{m}}^2 + n_{\mathrm{d}}^2}} \tag{1}$$

where $k_0 = 2\pi/\lambda$ is the wave vector of light in vacuum, λ is the wavelength, n_{d} is the refractive index of the dielectric and n_{m} is the (complex) refractive index of the metal. The optical field of SP is transverse magnetic (TM) and by using the Cartesian coordinates shown in Figure 1(a) it is described by the following nonzero components: magnetic intensity parallel to the interface H_y, electric intensity parallel to the interface E_z and electric intensity perpendicular to the interface E_x. The SP field exponentially decays from the metal/dielectric interface with the penetra-

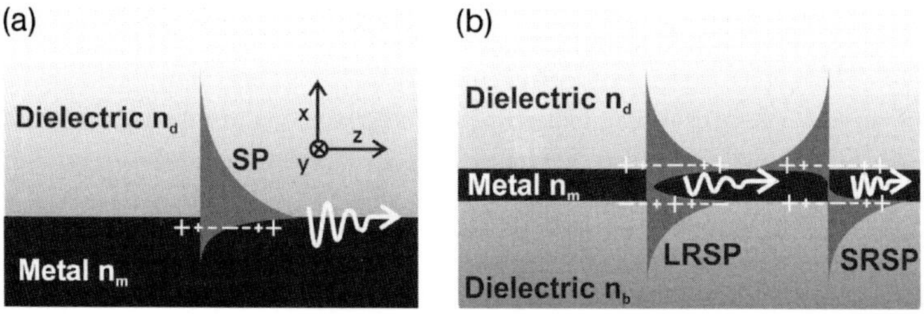

Figure 1 (a) Surface plasmon propagating on a metal/dielectric interface; (b) symmetrical (LRSP) and antisymmetrical (SRSP) surface plasmon modes coupled though a thin metal film embedded in a dielectric.

tion depth $L_{pen} = (\beta^2 - k_0^{\,2} n^2)^{-1/2}$ where n is equal to n_m for the metal and n_d for the dielectric. The penetration depth is defined as the distance perpendicular to the surface at which the field amplitude decreases by a factor 1/e. The energy of SP dissipates, while it propagates along the metal surface due to the losses within the metal film. This damping can be described by the propagation length $L_{pro} = (2\mathrm{Im}\{\beta\})^{-1}$ as the distance along the metallic surface at which the intensity of the SP mode drops to 1/e.

In order to illustrate typical characteristics of SPs, let us assume a wavelength in the visible part of the spectrum $\lambda = 0.633\,\mu\mathrm{m}$ and the interface between gold with the refractive index of $n_m = 0.1 + 3.5\mathrm{i}$ and a dielectric with $n_d = 1.33$. For these parameters, the field distribution of SP is depicted in Figure 2. The penetration depth of SP into the dielectric is equal to $L_{pen} = 0.183\,\mu\mathrm{m}$ and that into the gold is $L_{pen} = 0.027\,\mu\mathrm{m}$. The propagation length of the SP reaches $L_{pro} = 7.2\,\mu\mathrm{m}$.

Further, let us investigate SP modes propagating along a thin metal film embedded between two dielectrics with refractive indices of n_d and n_b, see Figure 1(b). In general, this geometry with two metallic interfaces supports two SP modes with propagation constants that obey the following equation:

$$\tan(\kappa d_m) = \frac{\gamma_d n_m^2/\kappa n_d^2 + \gamma_b n_m^2/\kappa n_b^2}{1 - (\gamma_d n_m^2/\kappa n_d^2)(\gamma_b n_m^2/\kappa n_b^2)} \qquad (2)$$

where d_m is the thickness of the metal film, $\kappa^2 = (k_0^{\,2} n_m^{\,2} - \beta^2)$ and $\gamma_{d,b} = \beta^2 - k_0^{\,2} n^2_{d,b}$.

If the refractive indices of the dielectrics are identical, $n_d = n_b$, SPs on the two metallic interfaces can couple, giving rise to two new SP modes. Owing to the symmetry of the configuration, these two modes exhibit symmetrical and anti-symmetrical distribution of the magnetic intensity H_y. By solving (2), it can be shown that Re$\{\beta\}$ of the symmetrical mode is smaller than that of a regular SP on

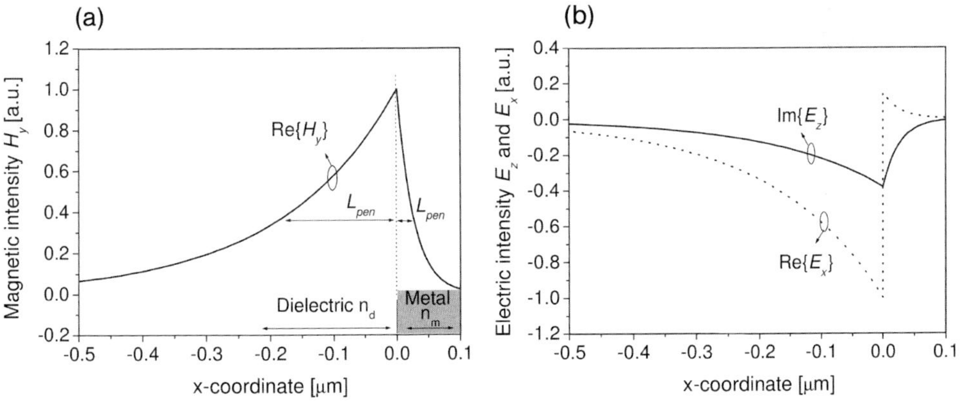

Figure 2 Distribution of (a) the magnetic intensity H_y and (b) the electric intensities E_x and E_z of SP propagating along the interface of gold ($n_m = 0.1 + 3.5\mathrm{i}$) and a dielectric ($n_d = 1.33$); wavelength of $\lambda = 0.633\,\mu\mathrm{m}$.

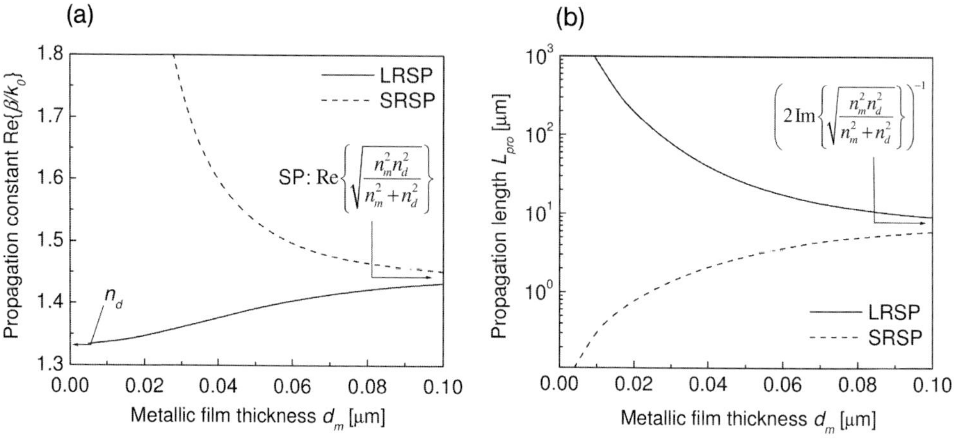

Figure 3 (a) The real part of the propagation constant and (b) the propagation length L_{pro} for LRSP and SRSP on a gold film with $d_m = 0–0.1\,\mu m$, $n_m = 0.1+3.5i$, $n_d = n_b = 1.33$ and $\lambda = 0.633\,\mu m$.

an individual interface and it decreases upon decreasing the metal slab thickness d_m. On the contrary, Re{β} of the antisymmetrical mode is larger than that of a regular SP and it increases when decreasing the thickness d_m, see Figure 3(a). The propagation length L_{pro} of the symmetrical mode is larger and that of antisymmetrical mode is smaller than the one of SP, see Figure 3(b). Therefore, the symmetrical and antisymmetrical modes are referred to as long-range surface plasmon (LRSP) and short-range surface plasmon (SRSP), respectively. For thicknesses of the metallic film that are much larger than the penetration depth of SP into the metal $d_m >> L_{pen}$, the coupling across the metal film disappears and thus the propagation constant β of LRSP and SRSP approaches that for the SP on a single metal/dielectric interface. For small thicknesses d_m, SPs on opposite interfaces are coupled strongly and the LRSP propagation constant approaches the one of light in the dielectric with refractive index $n_d = n_b$ and the propagation constant of SRSP diverges.

For instance, on a gold film with the thickness of $d_m = 20\,nm$ and a refractive index of $n_m = 0.1+3.5i$, dielectrics with $n_d = n_b = 1.33$ and a wavelength of $\lambda = 0.633\,\mu m$, the propagation length of LRSP and SRSP reaches $L_{pro} = 190\,\mu m$ and $0.78\,\mu m$, respectively. If the geometry is not exactly symmetrical $n_d \neq n_b$, there exist a thickness of the metal film d_m below which the LRSP modes cease to exist. This thickness is referred to as the cutoff thickness: for example, at a wavelength of $\lambda = 0.633\,\mu m$ and a gold film embedded in dielectrics with $n_b = 1.333$ and $n_d = 1.340$ it is equal to $d_m = 8\,nm$.

The electromagnetic fields of both LRSP and SRSP exponentially decay into the dielectrics and exhibit the maximum intensity at the metal/film interfaces. As seen in Figure 4, the LRSP mode exhibits a symmetrical distribution of H_y and E_x and an antisymmetrical profile for E_z. Complementary to that, the SRSP mode

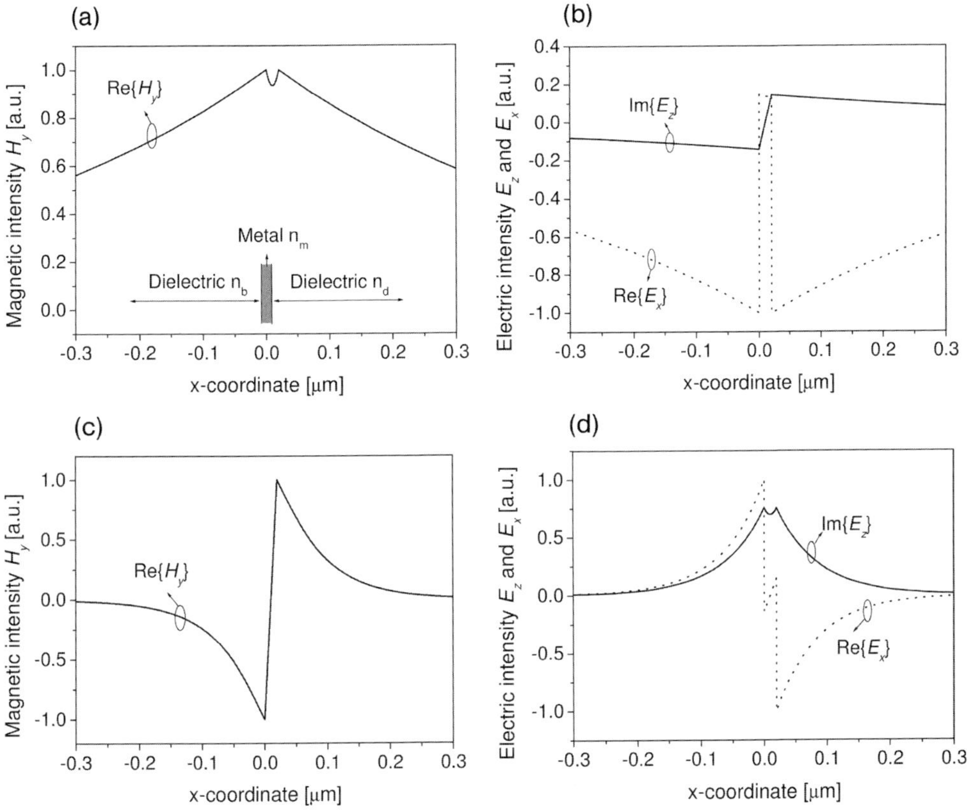

Figure 4 Components of the electromagnetic field components H_y, E_x and E_z calculated for (a) and (b) a LRSP and (c) and (d) a SRSP mode, $d_m = 20$ nm, $n_m = 0.1+3.5i$, $n_d = n_b = 1.33$ and $\lambda = 633$ nm.

has an antisymmetrical profile of H_y and E_x and a symmetrical profile of the E_z. In general, SP modes are strongest coupled to the oscillations of electron plasma density through the component E_z. Because this component for LRSP is antisymmetrical across the metallic film (Figure 4a), this mode is weakly coupled to the electron plasma oscillations and thus its damping is lowered and the penetration depth into the dielectrics L_{pen} is enlarged. Contrary to this, the E_z distribution of SRSP is symmetrical (Figure 4b) leading to a stronger coupling with the electron plasma oscillations and, hence, an increased damping and lowered penetration depth L_{pen}.

The electromagnetic-field distribution of LRSP and SRSP presented in Figure 4 was calculated for a gold film with a thickness of $d_m = 20$ nm, dielectrics with refractive indices of $n_b = n_d = 1.33$ and a wavelength of $\lambda = 0.633$ μm. For such a structure, the penetration depth of the LRSP into the dielectric is of $L_{pen} = 556$ nm and that of SRSP is of $L_{pen} = 65$ nm.

5.3.3 Optical Excitation of LRSPs

In order to efficiently couple light to a LRSP, the photon and LRSP waves need to be phase matched along the metallic surface. As the Re{β} of LRSP is always larger than the wave vector of a light wave in the dielectric $k_0 n_{b,d}$, a prism coupler [12, 13] can be used to enhance the momentum of the incident light and thus enable an efficient transfer of energy to LRSP.

In the prism coupler, a light beam is launched into a prism with a refractive index n_p that is higher than that of the buffer layer n_b and the top dielectric n_d (Figure 5a). Upon incidence of the light beam at the interface between the prism and buffer layer, the light beam is totally reflected and coupled through its evanescent tail to LRSP if the following phase-matching condition is fulfilled:

$$k_0 n_p \sin(\theta) = \mathrm{Re}\{\beta\} \tag{3}$$

As illustrated in the angular reflectivity spectrum in Figure 5(b), the excitation of LRSPs is manifested as a resonant dip centered at the angle of incidence for which (3) holds. The strength of the coupling between the optical wave and LRSP can be tuned by varying the thickness of the buffer layer d_b in order to achieve the full coupling of light to LRSPs. For example, for a gold film thickness of d_m = 22.5 nm, a buffer layer with n_b = 1.340, a dielectric with n_b = 1.333 and a wavelength of λ = 633 nm, this optimum buffer layer thickness is close to d_b = 900 nm, see Figure 5(b).

The comparison of the reflectivity spectra measured for the excitation of LRSPs and a regular SP in Figure 5(b) reveals that LRSPs are excited at lower angles of

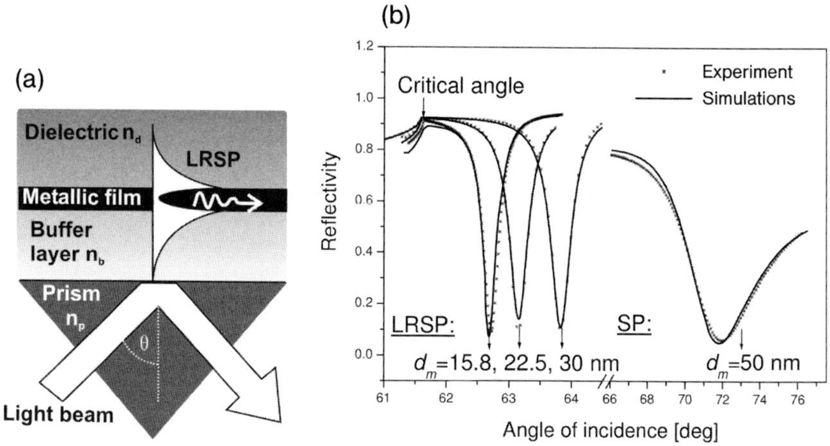

Figure 5 (a) Scheme of a prism coupler for the excitation of LRSPs. (b) Comparison of the angular reflectivity spectra measured for the excitation of SP and LRSPs using a BK7 prism (n_p = 1.515), a Cytop buffer layer (n_b = 1.340, d_b = 900 nm), a gold film (d_m indicated in the graph and n_m = 0.18+3.1i for d_m = 22.5 nm), water as a dielectric medium (n_d = 1.333) and a light beam with a wavelength of λ = 633 nm. (Reproduced from [17]).

incidence than SP due to their lower real part of the propagation constant Re{β} (see Figure 3a). In addition, the width of the resonant dip associated with the excitation of LRSP is more than an order of magnitude lower that that for SP due to the larger propagation length of LRSP (see Figure 3b). For instance, the full width in half-minima (FWHM) of the LRSP resonance on a gold film with $d_m = 22.5$ nm is $\Delta\theta = 0.4$ deg compared to $\Delta\theta = 5.1$ deg for the excitation of a regular SP.

Upon excitation of LRSP, the energy carried by the incident wave is accumulated within the LRSP leading to the enhancement of the field intensity in the vicinity of the metallic film. In general, the magnitude of the field-intensity enhancement is increasing with the propagation length L_{pro}. In Figure 6(a), the distribution of the electric-field intensity is presented for a LRSP excited on a 22.5 nm thick with a plane wave incident at the resonant angle of $\theta = 63.1$ deg (parameters of the layer structure: $n_p = 1.515$, $d_b = 900$ nm, $n_b = 1.340$, $n_m = 0.18+3.1i$ and $n_d = 1.333$, wavelength of $\lambda = 633$ nm). For such a geometry, the magnetic intensity $|H_y|^2$ on the top of the gold surface is 60 times larger than that of the incident wave. For comparison, the field-intensity enhancement achieved through the excitation of SPs on a gold film at the same wavelength is 16 [17]. In addition, the field-intensity enhancement is achieved within a larger region adjacent to the metallic film due to the higher penetration depth of LRSP, see Figure 6(b).

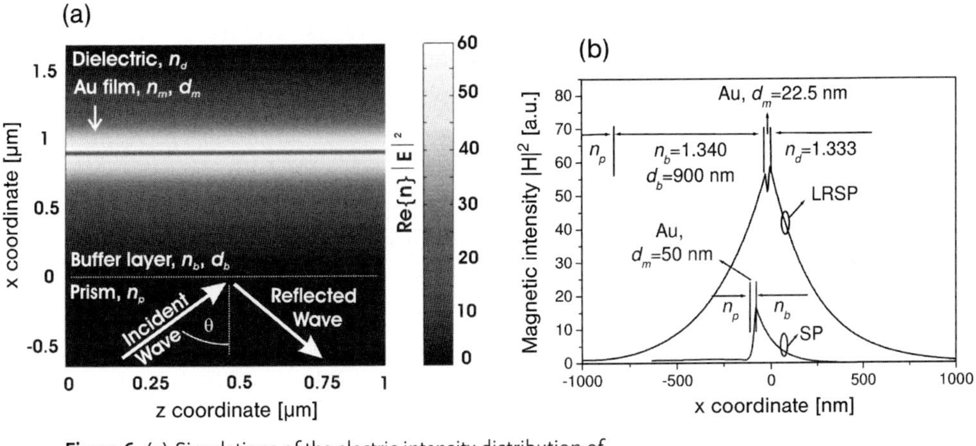

Figure 6 (a) Simulations of the electric intensity distribution of LRSPs excited on a gold film with $d_m = 22.5$ nm (electric intensity of the incident plane wave is equal to 1). (b) Comparison of the magnetic intensity profile for the excitation of LRSP and conventional SP. The same parameters as in Figure 5(b) were used and the light wave was incident on the prim base at an angle of incidence of $\theta = 63.1$ deg.

5.3.4
Implementation of LRSPs in a SPFS-Based Biosensor

In SPFS-based biosensors, LRSPs are used for the excitation of chromophore-labeled molecules captured by biorecognition elements anchored on a metallic sensor surface. These devices incorporate two key components: (i) an optical structure providing the enhancement of the electromagnetic field through the excitation of LRSPs, and (ii) detection assay and the binding matrix with immobilized ligands for the capture of target molecules.

In the LRSP-enhanced fluorescence spectroscopy, mostly prism couplers utilizing the attenuated total internal reflection method (ATR) were used [16, 17], see Figure 7. A monochromatic light beam with the wavelength matching the absorption band of a chromophore is coupled to a high refractive index prism. On the base of the prism, a sensor chip with a layer coating supporting LRSPs is optically matched. The sensor chip consists of a glass slide, a low refractive index buffer layer and a thin metal film. For biosensor applications, on the top of the metal film an additional layer with biorecognition elements capable to capture target molecules from aqueous samples is anchored. Against the sensor chip, a transparent flow cell with the analyte solution is attached. The fluorescence light is detected using a lens optics, a bandpass filter to suppress the background due to the scattered light at the excitation wavelength and a photomultiplier or CCD-based detector.

In order to achieve a refractive-index-symmetrical structure needed for the excitation of LRSPs, the refractive index of the buffer layer n_b needs to match that of aqueous samples, which is close to that of water $n_d = 1.333$ (at $\lambda = 633$ nm). Up to date, different commercially available Teflon-based materials [13, 14] (Teflon AF from Dupont Inc., USA, with $n_b = 1.31$ and Cytop from Asahi Inc., Japan, with $n_b = 1.34$), low refractive index dielectrics such as aluminum fluoride (AlF_3, $n_b = 1.34$) [18] or magnesium fluoride (MgF_2, $n_b = 1.38$) [14] and nanoporous silicates (the refractive index can be tuned by the size of pores) were used. These materials can be spin coated (Teflon) or deposited by vacuum thermal evaporation (magnesium and aluminum fluoride). For the excitation of surface plasmons in the visible and NIR part of the spectrum, noble metals are employed. Among these, gold is preferably used owing to its stability. Gold films are mostly prepared by sputtering or vacuum thermal evaporation. As shown before [17], gold films with thicknesses smaller than $d_m \sim 20$ nm exhibit an island morphology if deposited by these techniques on surfaces with low surface energy such as Teflon. This effect leads to the deteriorating of their optical properties

In LRSP-enhanced fluorescence spectroscopy, various surface chemistries developed for SPR or SPFS biosensors can be used [19] For LRSP-based sensors, particularly three-dimensional binding matrices are of interest by which the whole extended evanescent field of LRSP can be used for the sensing. Currently, research in novel hydrogel-based materials that can be used for the construction of three-dimensional binding matrices is being carried out [20, 21].

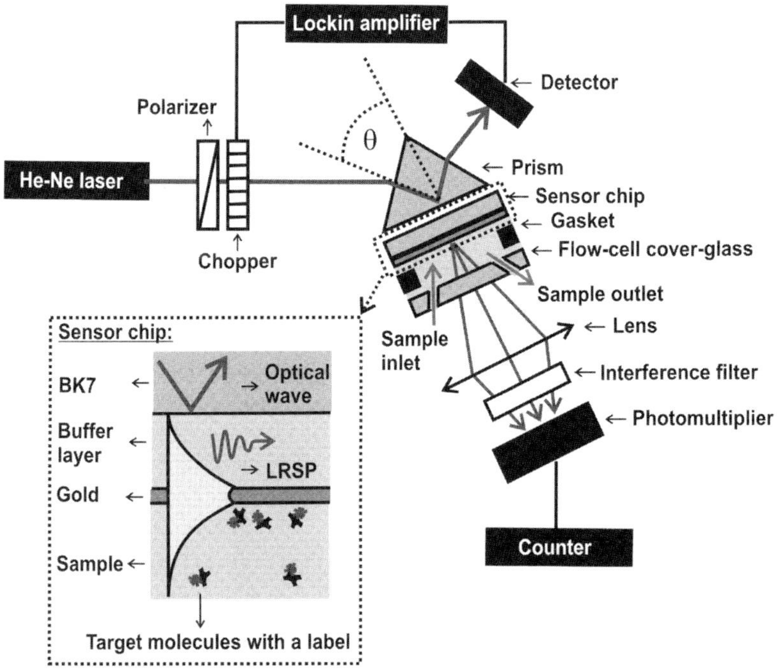

Figure 7 Optical setup of a sensor based on the spectroscopy of LRSPs for the detection of biomolecular binding events on a metallic surface using the label-free detection principle and LRSP-enhanced fluorescence spectroscopy. (Reproduced from [17]).

5.3.5 Comparison of LRSP and SP-Enhanced Fluorescence Spectroscopy

As illustrated in the previous sections, the excitation of LRSPs results in an enhancement of the intensity of the electromagnetic field at a metallic surface. If compared to conventional surface plasmons, the enhancement in close proximity to the gold surface by a factor of up to 4 was reported [17]. In addition, the evanescent field of LRSP can extend into the medium on top of the gold surface with up to an order of magnitude higher depth. These features enables increasing the fluorescence signal due to the capture of chromophore-labeled molecules on the sensor surface as a larger number of molecules can be excited with a higher excitation rate.

Experimental results confirmed these predictions [16, 17]. Using a low refractive index spacer layer, a monolayer of chromophore-labeled molecules was bound at different distances from the gold sensor surface. The dependence of the fluorescence intensity on the distance was measured upon the excitation of chomophores via SP (on a 50-nm thick gold film) and via LRSP (propagating along a 16-nm thick gold film between a Cytop buffer layer and the aqueous sample). As seen in the measured reflectivity and fluorescence intensity spectra in Figure 8(a), the maximum fluorescence signal is detected during the resonat excitation of the surface plasmon modes. For the chromophore excitation by LRSPs, the fluorescence intensity is detected within a narrower range of angles of incidence and exhibits a higher peak intensity than found upon the excitation by conventional SPs. The comparison of the fluorescence signal in Figure 8(b) reveals that the fluorescence intensity decays exponentially for the chromophore-labeled monolayer deposited at distances larger than 40 nm. For smaller distance, we observed a decrease in the peak fluorescence intensity due to the nonradiative decay of a dye induced by the presence of the metal [22]. Assuming the excitation of dyes distributed within the whole evanescent field of LRSP and SP, the overall enhancement in the collected fluorescence intensity can be assumed to be the product of the enhance-

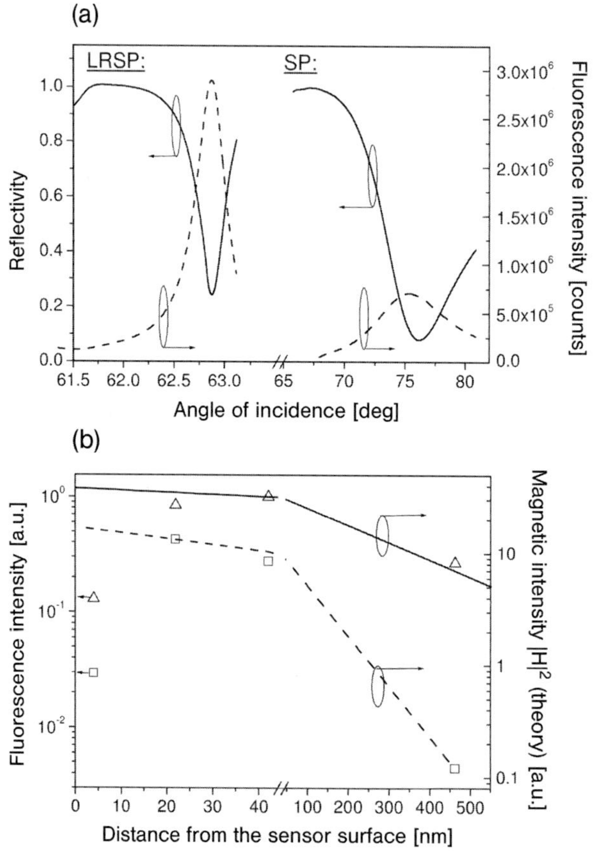

Figure 8 Comparison of the fluorescence signal recorded via the excitation of LRSP and SP. (a) Angular reflectivity and fluorescence intensity spectra measured for a monolayer of chromophore-labeled molecules deposited at a distance of 42 nm from the sensor surface. (b) The dependence of the maximum fluorescence intensity on the distance of the chromophore from the sensor surface. (Reproduced from [17]).

ment in the proximity to the sensor surface (4.4 measured at a distance of 4 nm) and the ratio of the penetration depths (for LRSP with d_{Au} = 15.8 nm and conventional SP this ratio is equal to 2.9). By this means, we can estimate that more than an order of magnitude increase in the fluorescence intensity can be achieved by taking advantage of the higher enhancement and the more extended field of LRSP when compared with conventional SPS.

5.3.6
LRSP-Enhanced Fluorescence Spectroscopy: Biomolecular Binding Studies

In SPFS-based biosensors, the binding of chromophore-labeled analyte molecules from a sample to their biomolecular partners on the sensor surface is measured through an induced fluorescence signal. The biomolecular partners (biorecognition elements) can be immobilized on the sensor surface using various architectures [23] including monolayer assemblies or three-dimensional binding matrices. In general, a three-dimensional binding matrix attached to the sensor surface allows for a higher sensor response as the whole evanescent field of a surface plasmon mode can be exploited. Within such a binding matrix, a higher amount of ligand molecules can be immobilized and thus stronger fluorescence signal can be observed. As was reported by Yu *et al.*, a three-dimensional dextrane brush binding matrix (CM5 sensor chip from Biacore Inc., USA) loaded with a mouse immunoglobulin G (IgG) enabled the detection of antimouse IgG antibodies at concentrations as low as 500 aM (80 fg/mL) [24] by using surface plasmon enhanced fluorescence spectroscopy. In this experiment, the dextrane brush with thickness of ~100 nm was used that matches the penetration depth of surface plasmon (see Figure 2).

In order to let SPFS benefit from LRSP that exhibits a more extended evanescent field (see Figures 2 and 4), a three-dimensional binding matrix with up to micrometer thickness needs to be developed. A first attempt to achieve this goal was carried out using a NIPAAM [N-(isopropylacrylamide)]-based hydrogel [21] deposited on a LRSP-supporting layer structure (Teflon AF with thickness of d_b = 500 and a gold film with a thickness d_{Au} = 40 nm), see Figure 9(a). Briefly, carboxyl groups within the polymer were activated by incubating in a mixture of trifluoroacetyl-N-succinimidyl ester (TFA-NHS) in CH_2C_{12} and $N(Et)_3$ for 24 h under Ar at room temperature. Afterwards, the polymer was precipitated in Et_2O, spin coated on a gold surface that was modified with an adhesion promoter containing a thiol group (synthesized in the lab), crosslinked with UV light and swelled in phosphate buffer saline (PBS). The thickness and the refractive index of the swollen hydrogel were determined as 402 nm and 1.345, respectively. A purified mouse IgG (from Invitrogen, USA) was *in-situ* covalently bound into the hydrogel from a solution with a concentration of mouse IgG of 20 μg/mL. After the immobilization of mouse IgG, the sensor surface was washed out with PBS. The refractive index of the hydrogel after the immobilization of mouse IgG was of 1.3463, which indicates an IgG surface coverage of 2.6 ng mm^{-2}.

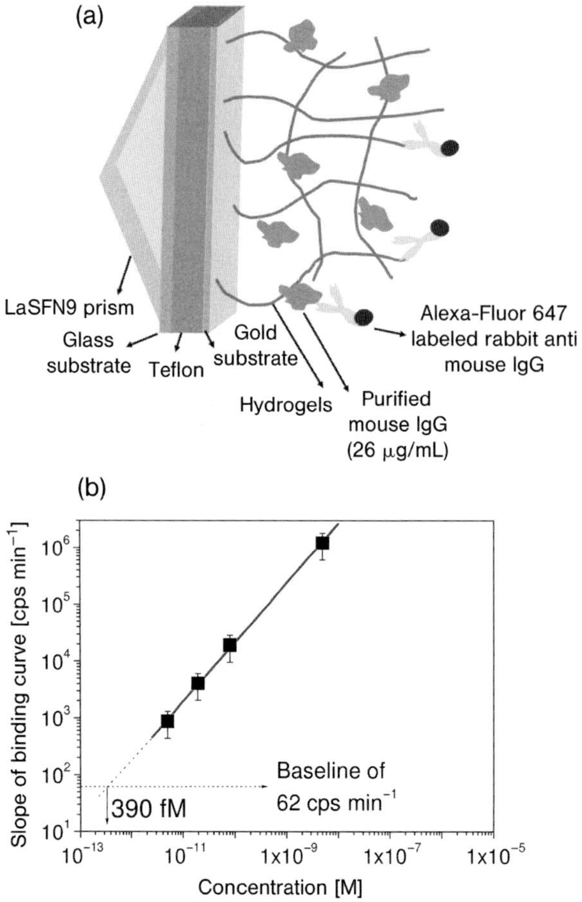

Figure 9 (a) A scheme of the LRSPR sample with the hydrogel matrix indicating the protein binding to the matrix. (b) The calibration curve of the sensor measured for the binding of antimouse IgG to mouse IgG anchored within the hydrogel-binding matrix.

Using such prepared hydrogel matrix, the binding of antimouse IgG (from Invitrogen Inc., USA) labeled with Alexa Fluor dye (Alexa Fluor 647 from Molecular Probes Inc., USA) was measured using LRSP-enhanced fluorescence spectroscopy. A light beam at the wavelength of 633 nm was coupled to a LRSP wave to probe the binding in the hydrogel matrix. The LRSP wave excited the chromophore-labeled antimouse IgG molecules that were affinity captured within the binding matrix. The fluorescence signal was measured in time, while a series of samples with antimouse IgG dissolved at concentrations of (40 aM, 5 fM, 5 pM, 20 pM, 80 pM, and 5 nM) were successively injected. Each sample was flowed along the sensor surface for 10 min and the increase in the fluorescence signal due to antimouse IgG capture was determined. For each concentration, the slope of time evolution of the fluorescence signal due to the binding of target molecule was determined. Preliminary results presented in Figure 9(b) revealed that the NIPAAM-based hydrogel is suitable for a construction of binding matrix and the estimated limit of detection (LOD) was of 400 fM (60 pg/mL).

5.3.7
Conclusions and Future Outlook

We described recent advances in surface plasmon enhanced fluorescent spectroscopy (SPFS) through the implementation of special surface plasmon modes – long-range surface plasmons (LRSPs). The characteristics of LRSPs, their excitation and implementation in SPFS biosensors were discussed. We showed that the excitation of LRSPs allows for a dramatic enhancement of the intensity of electromagnetic field at the sensor surface that can be directly translated into an increase in the fluorescence signal. In addition, LRSPs enables probing the medium adjacent to the sensor surface with its extended evanescent field up to a micrometer in depth. This feature allows for the detection of the fluorescence signal from a larger sensing volume in which a higher amount of chromophore-labeled molecules can be captured. The presented data indicates that this approach holds the potential to increase the fluorescence signal intensity by more than an order of magnitude compared to conventional SPFS. Future research will include improvements of LRSP-based optics to reach even larger enhancement of electromagnetic field and higher yield of detected fluorescence light. Furthermore, novel materials for three-dimensional binding matrices are under investigation to fully exploit the potential of LRSP-enhanced fluorescence spectroscopy for the design of new ultrasensitive biosensors. In addition, the extended evanescent field of LRSPs can provide advantages for the detection of large analytes such as bacterial pathogens.

Acknowledgments

We would like to thank and appreciate the help of Cathrin Corten and Dirk Kuckling (Technische Universität, Dresden) for supplying us with the NIPAAM polymer, Robert Roskamp (MPIP) for helping in the chemistry of the polymer activation, Maria Gianneli (MPIP) for helping in the polymer characterization by FCS, and Fang Yu for the useful discussions. Partial support by the Deutsche Forschungsgemainschaft (Priority Program SPP 1259: "Intelligente Hydrogele", KN 224/18-1) is greatly acknowledged. In addition, this publication was partially financially supported by the European Commission in the Communities 6th Framework Programme, Project TRACEBACK (FOOD-CT-036300), and Coordinated by Tecnoalimenti. It reflects the authors' views and the Community is not liable for any use that may be made of the information contained in this publication.

References

1 Gonzalez-Martinez, M.A., Puchades, R. and Maquieira, A. (2007) Optical immunosensors for environmental monitoring: How far have we come? *Analytical and Bioanalytical Chemistry*, **387**, 205–218.

2 Rasooly, A. and Herold, K.E. (2006) Biosensors for the analysis of food- and waterborne pathogens and their toxins. *Journal of Aoac International*, **89**, 873–883.

3 Vo-Dinh, T. and Cullum, B. (2000) Biosensors and biochips: advances in biological and medical diagnostics. *Fresenius Journal of Analytical Chemistry*, **366**, 540–551.

4 Rather, H. (1983) *Surface Plasmons on Smooth and Rough Surfaces and on Gratings*, Springer Verlag, Berlin.

5 Liebermann, T. and Knoll, W. (2000) Surface-plasmon field-enhanced fluorescence spectroscopy. *Colloids and Surfaces A – Physicochemical and Engineering Aspects*, **171**, 115–130.

6 Healy, D.A., Hayes, C.J., Leonard, P., McKenna, L. and O'Kennedy, R. (2007) Biosensor developments: application to prostate-specific antigen detection. *Trends in Biotechnology*, **25**, 125–131.

7 Matsubara, K., Kawata, S. and Minami, S. (1990) Multilayer System for a High-Precision Surface-Plasmon Resonance Sensor. *Optics Letters*, **15**, 75–77.

8 Chien, F.C. and Chen, S.J. (2006) Direct determination of the refractive index and thickness of a biolayer based on coupled waveguide-surface plasmon resonance mode. *Optics Letters*, **31**, 187–189.

9 Dostalek, J., Adam, P., Kvasnicka, P., Telezhnikova, O. and Homola, J. (2007) Spectroscopy of Bragg-scattered surface plasmons for characterization of thin biomolecular films. *Optics Letters*, **32**, 2903.

10 Lal, S., Link, S. and Halas, N.J. (2007) Nano-optics from sensing to waveguiding. *Nature Photonics*, **1**, 641–648.

11 Genet, C. and Ebbesen, T.W. (2007) Light in tiny holes. *Nature*, **445**, 39–46.

12 Sarid, D. (1981) Long-Range Surface-Plasma Waves on Very Thin Metal-Films. *Physical Review Letters*, **47**, 1927–1930.

13 Nenninger, G.G., Tobiska, P., Homola, J. and Yee, S.S. (2001) Long-range surface plasmons for high-resolution surface plasmon resonance sensors. *Sensors and Actuators B – Chemical*, **74**, 145–151.

14 Wark, A.W., Lee, H.J. and Corn, R.M. (2005) Long-range surface plasmon resonance imaging for bioaffinity sensors. *Analytical Chemistry*, **77**, 3904–3907.

15 Slavik, S. and Homola, J. (2007) Ultrahigh resolution long-range surface plasmon-based sensor. *Sensors and Actuators B – Chemical*, **123**, 10–12.

16 Kasry, A. and Knoll, W. (2006) Long range surface plasmon fluorescence spectroscopy. *Applied Physics Letters*, **89**, 101106.

17 Dostalek, J., Kasry, A. and Knoll, W. (2007) Long range surface plasmons for observation of biomolecular binding events at metallic surfaces. *Plasmonics*, **2**, 97–106.

18 Vala, M., Dostalek, J. and Homola, J. (2007) Diffraction Grating-Coupled Surface Plasmon Resonance Sensor Based on Spectroscopy of Long-Range and Short-Range Surface Plasmons. *SPIE*, **6585**, 658522–658521.

19 Löfås, S. and McWhirter, A. (2006) in *Surface Plasmon Resonance Based Sensors* (ed. J. Homola), Springer, pp. 117–151.

20 Beines, P.W., Klosterkamp, I., Menges, B., Jonas, U. and Knoll, W. (2007) Responsive thin hydrogel layers from photo-cross-linkable poly(N-isopropylacrylamide) terpolymers. *Langmuir*, **23**, 2231–2238.

21 Giannelli, A., Beines, P.W., Roskamp, R.F., Koynov, K., Fytas, G. and Knoll, W. (2007) Local and global dynamics of transient polymer networks and swollen gels anchored on solid surfaces. *Journal of Physical Chemistry C*, **111**, 13205–13211.

22 Vasilev, K., Knoll, W. and Kreiter, M. (2004) Fluorescence intensities of chromophores in front of a thin metal film. *Journal of Chemical Physics*, **120**, 3439–3445.

23 Knoll, W., Park, H., Sinner, E.K., Yao, D.F. and Yu, F. (2004) Supramolecular interfacial architectures for optical biosensing with surface plasmons. *Surface Science*, **570**, 30–42.

24 Yu, F., Persson, B., Lofas, S. and Knoll, W. (2004) Attomolar sensitivity in bioassays based on surface plasmon fluorescence spectroscopy. *Journal of the American Chemical Society*, **126**, 8902–8903.

Appendices: Surface Analytical Tools

Surface Design: Applications in Bioscience and Nanotechnology
Edited by Renate Förch, Holger Schönherr, and A. Tobias A. Jenkins

ISBN: 978-3-527-40789-7

Appendix A
Material Structure and Surface Analysis

The structure and composition of a sample and the reactions at its surface may be examined with different levels of resolution. A summary of some commonly applied analytical methods and the types of information and resolution obtained are given in the tables below.

Thickness

Name	Quantity measured	Analysis in solvent	Lateral resolution	Analysis depth
Profilometry	Step height at edges	no	> nm–μm	interface
Ellipsometry	Film thickness and optical constants	possible	μm	nm–μm
Surface plasmon resonance spectroscopy (SPR)	Optical thickness ($n\ d$)	yes	> μm	nm – ~ 150 nm
Optical waveguide mode spectroscopy	Resonance angle, % reflectivity	yes	> μm	nm–μm

Surface topography: surface roughness

Name	Lateral resolution
Optical microscopy	μm–mm
Normarski microscope	μm–mm
Laser beam scattering	> μm
Optical interferometry	μm
Scanning electron microscopy (SEM)	~ 5 nm – > μm
Atomic force microscopy (AFM)	~ 1 nm – > 100 μm

Surface Design: Applications in Bioscience and Nanotechnology
Edited by Renate Förch, Holger Schönherr, and A. Tobias A. Jenkins

ISBN: 978-3-527-40789-7

Chemical bonding and composition

Name	Analysis in solvent	Lateral resolution	Analysis depth
Infrared absorption spectroscopy	possible	> μm	μm
Raman spectroscopy	possible	< μm	μm
X-ray photoelectron spectroscopy (XPS)	no	1–5 μm	5–10 nm
Secondary ion mass spectroscopy (SIMS)	no	< 1 μm	< 5 nm
Auger spectroscopy	no	1–5 μm	5–10 nm

Mechanical properties

Name	Analysis in solvent	Lateral resolution	Depth resolution
Scanning probe microscopy (modulus)	possible	nm	nm
Nanoindentation (modulus)	possible	nm	nm
Microcantilever sensors	possible	–	–
Scratch-peel tests	no	mm	–
Shear tests	no	mm	–

The choice of analytical techniques to be used for the analysis of a particular sample is always governed by:

1. availability of a technique;
2. nature of the substrate – metallic, polymeric, hydrogel, a 2D thin film or a 3D object or device, transparent or oblique;
3. size of substrate;
4. compatibility – for example, vacuum techniques are not suitable for all samples and may lead to the denaturing of the material (e.g. biological materials immobilized on surface) or a change in structure (e.g. hydrogels tend to collapse). In such cases the results obtained do not reflect the true nature of the materials.

In most cases there is not one single analytical technique that will provide sufficient information on the properties of a material. The analysis of thin films and materials therefore always requires the application of several different complementary techniques and the results obtained by these techniques must be compared with each other.

The following pages briefly introduce a number of 'standard' surface analytical techniques that are suitable for thin-film analysis. Some of the techniques discussed are particularly suitable for the analysis in solvent, which allows them to be applied in the study of biological interactions at surfaces.

Appendix B
Atomic Force Microscopy

Atomic force microscopy (AFM) is a routine nonoptical microscopy technique that allows one to interrogate surfaces and nanometer- to micrometer-scale structures in different media and atmospheres. In AFM a sharp probe tip is scanned across the sample to collect highly localized information about the sample surface and its structure. Noteworthy is the broad range of lateral scan sizes covered by the technique, which ranges from the >100 μm down to the molecular scale. The most commonly utilized imaging modes, i.e. contact and intermittent (tapping) mode AFM, thereby provide insight into topography, as well as surface properties. In specialized AFM modes discussed in Chapters 5.1 and 5.2, the use of force measurements to unravel the mechanical properties of single molecules and to enable surface compositional analysis by friction force measurements are discussed.

B.1
Principles of AFM

B.1.1
Contact Mode AFM

In contact mode AFM (Figure B.1a and b) a sharp tip, which is mounted on a flexible cantilever beam with a spring constant of ~0.01 to ~1.00 N/m, is brought into contact with the sample specimen. Exploiting repulsive and/or attractive tip–surface forces the surface profile can be mapped quantitatively with up to molecular-scale resolution by raster scanning using an x,y,z transducer the sample and keeping the measured tip–sample force constant using a feedback loop (Figure B.1c and d). The corresponding z travel of the transducer used as a function of lateral displacement represents the surface topology as an isoforce image. When the sample is scanned perpendicular to the long axis of the cantilever, friction forces can be recorded by measuring the corresponding cantilever twist (Figure B.1b). The true resolution for imaging is limited by the shape and the radius of the tip, as well as the range of intermolecular interactions.

Surface Design: Applications in Bioscience and Nanotechnology
Edited by Renate Förch, Holger Schönherr, and A. Tobias A. Jenkins

ISBN: 978-3-527-40789-7

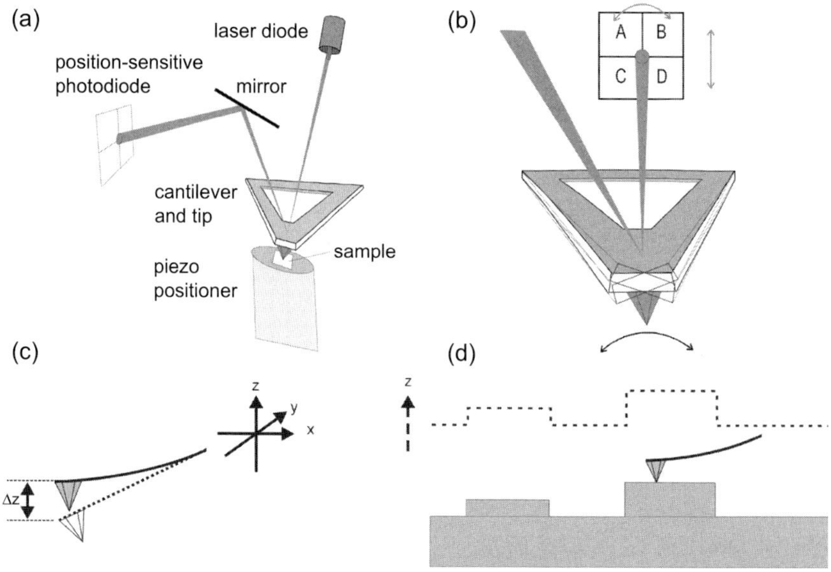

Figure B.1 (a) Schematic of AFM with all important components. In the AFM shown the sample is moved in the *x,y,z* directions by a piezopositioner; other AFM types utilize a scanned tip. (b) The deflection of the cantilever in the surface normal and in the lateral direction can be measured using a segmented photodetetcor. (c) During scanning, the cantilever deflection Δz (or force $F = k_c\,\Delta z$, with cantilever spring constant k_c) is maintained constant; thereby the surface is tracked as shown in panel (d).

AFM can be performed in ambient conditions, as well as in controlled environments such as liquids (including buffered or aqueous solutions), controlled gas atmospheres or vacuum. Since the tip probes the surface of interest via tip–surface forces, the sample of interest can often be analyzed directly. Thus, elaborate sample-preparation methods, such as those required to render samples amenable to electron microscopic investigations, are circumvented.

In addition to laterally resolved information on, e.g., topography (quantitative surface profile), the composition in heterogeneous systems, elasticity, and other properties of interest can be mapped in suitable modes. For instance, contact mode AFM can be used to measure both repulsive and attractive forces between sample and surface with a high accuracy by analyzing force–displacement curves ('force spectroscopy' and 'force mapping'). In *f–d* curve data acquisition, the sample is moved in and out of contact with the sample surface and the corresponding forces are measured by monitoring the cantilever beam deflection. These *f–d* curves can be then used to estimate forces as a function of tip–sample separation, mechanical properties in the contact regime or adhesive interaction upon tip retraction, see Chapters 5.1 and 5.2.

B.1.2
Intermittent Contact Mode AFM

The lateral (shear) forces between tip and sample in contact mode AFM may limit the applicability of contact AFM in investigations of soft matter as the surface structures may be deformed or even destroyed. The lateral forces are practically eliminated in intermittent contact modes, such as tapping mode AFM, in which an oscillating cantilever–tip assembly (~ 100–300 kHz in air, ~ 10–50 kHz in liquid) is employed to track the surface. Tapping mode AFM relies on a similar setup as contact mode AFM (Figure B.1), however, for measurements in air a stiffer cantilever is used (k = 20–80 N/m). This cantilever is excited to vibrate near its resonance frequency and during scanning the feedback loop adjusts the z-position of

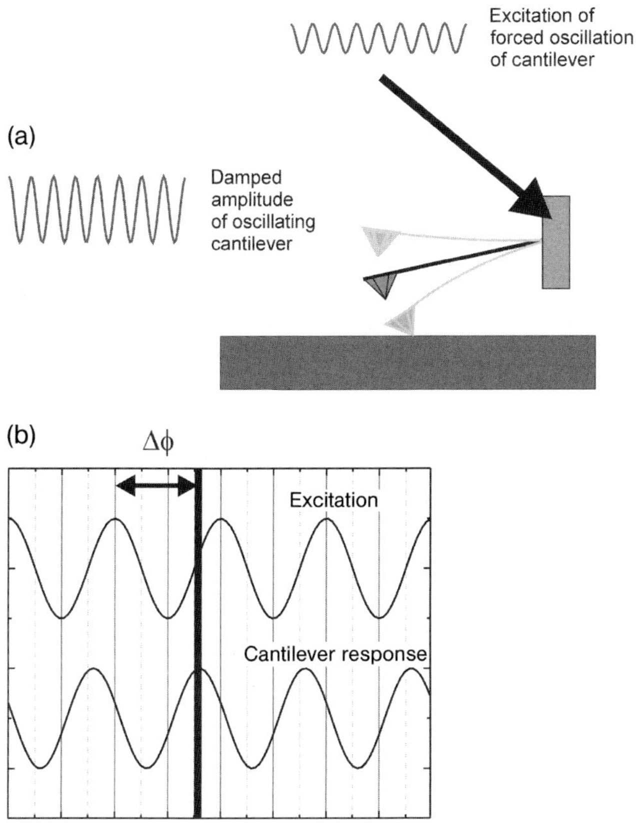

Figure B.2 (a) The reduced amplitude of a forced oscillating cantilever–tip assembly is exploited to the surface with nanometer accuracy. (b) The phase shift between the excitation and lever response provides a sensitive contrast mechanism for materials with different energy dissipation.

the transducer to keep the amplitude of the resonating lever constant (Figure B.2a). In addition to this topography data the change in the phase difference between excitation vibration and vibration of the cantilever can be used to obtain contrast due to different surface characteristics related to energy dissipation and thus materials properties (Figure B.2b).

References

1 R. J. Colton, A. Engel, J. E. Frommer, H. Gaub, A. Gewirth, R. Guckenberger, W. Heckl, B. Parkinson, J. Rabe, *Procedures in Scanning Probe Microscopies.* Wiley, New York, 1998.

2 L. C. Sawyer, D. T. Grubb, G. F. Meyers, *Polymer Microscopy: Characterization and Evaluation of Materials,* 3rd edn. Springer-Verlag, Berlin, 2008.

3 *SPM Beyond Imaging: Manipulation of Molecules and Nanostructures,* P. Samori (ed.), Wiley-VCH, Weinheim, 2006.

4 G. J. Vancso, H. Hillborg, H. Schönherr, *Adv. Polym. Sci.* 2005, *182,* 55.

Appendix C
Contact Angle Goniometry

Contact goniometry is a method to determine the wettability of a surface. It can also be used to calculate the surface energy. The contact angle is the angle at which a liquid/vapor interface meets the solid surface. It is specific for any given system and is determined by the interactions across the three interfaces – solid/liquid, solid/gas and liquid/gas.

The term 'wetting' describes the contact between a liquid and a solid surface and is a result of intermolecular interactions when the two are brought together. The amount of wetting depends on the energies (or surface tensions) of the interfaces involved such that the total energy is minimized. The degree of wetting is described by the contact angle. This is the angle at which the liquid/vapor interface meets the solid/liquid interface.

For flat surfaces the contact angle is measured from a drop of a suitable liquid resting on a surface (Figure C.1). If the liquid is very strongly attracted to the solid surface, as for example, water on a hydrophilic solid, the droplet will completely spread out on the solid surface and the contact angle will be close to zero degrees. Less strongly hydrophilic solids will have a contact angle up to 90°. If the solid surface is hydrophobic, the contact angle will be larger than 90°. Surfaces with a contact angle greater than 150° are called superhydrophobic surfaces. On these surfaces, water droplets simply rest on the surface, without actually wetting to any significant extent. The contact angle thus directly provides information on the interaction energy between the surface and the liquid.

The theoretical description of contact arises from the consideration of a thermodynamic equilibrium between the three phases: the liquid phase of the droplet (L), the solid phase of the substrate (S), and the gas/vapor phase of the ambient (V). The latter will be a mixture of ambient atmosphere and an equilibrium concentration of the

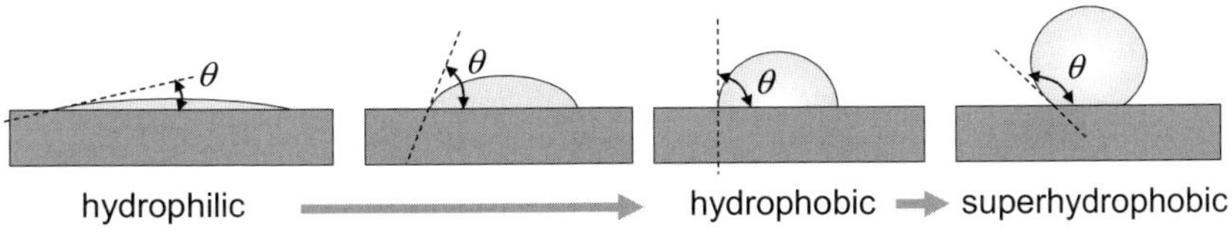

Figure C.1 Schematic of different levels of wettability of surfaces.

Surface Design: Applications in Bioscience and Nanotechnology
Edited by Renate Förch, Holger Schönherr, and A. Tobias A. Jenkins

ISBN: 978-3-527-40789-7

liquid vapor or could alternatively be another immiscible liquid phase. At thermodynamic equilibrium, the chemical potential in the three phases should be equal.

It is generally convenient to consider the interfacial energies involved. Using the simplified planar geometry Young's equation [1, 2] can be written as:

$$0 = \gamma_{SG} - \gamma_{SL} - \gamma \cos\theta$$

where γ_{SG}, γ_{SL} and γ are the interfacial energy at the solid/gas, solid/liquid and liquid/gas interface. Young's equation assumes a perfectly flat surface, and in many cases surface roughness and impurities cause a deviation in the equilibrium contact angle from the contact angle predicted by Young's equation. Even in a perfectly smooth surface a drop will assume a wide spectrum of contact angles between the highest (advancing) contact angle, θ_A, and the lowest (receding) contact angle, θ_R.

There are several methods available to measure the contact angle of surfaces [3]:

1. The most convenient is the static sessile drop method where the angle formed between the liquid/solid interface and the liquid/vapor interface is measured using a microscope optical system or, with high-resolution cameras and software to capture and analyze the contact angle.

2. The dynamic sessile drop is similar to the static sessile drop but requires the drop to be modified. The largest contact angle possible without increasing its solid/liquid interfacial area is measured by adding volume dynamically (Figure C.2). This maximum angle is the advancing angle, θ_A. Volume is then removed to produce the smallest possible angle, the receding angle, θ_R. The difference between the advancing and receding angle is the contact angle hysteresis.
   The dynamic contact angle can also be measured for a drop travelling down a sloped flat surface as shown in Figure C.2.

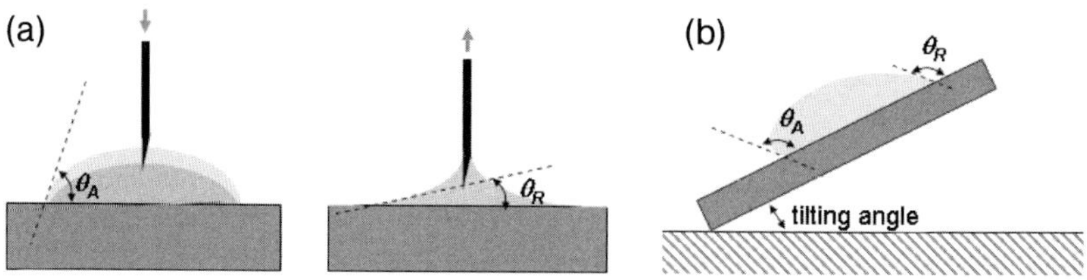

Figure C.2 Different methods to measure the advancing and receding contact angle at solid surfaces (a) dynamic sessile drop and (b) advancing drop method.

3. The dynamic Wilhelmy method can be used to calculate the average advancing and receding contact angles on solids of uniform geometry. Both sides of the solid must have the same properties. The wetting force on the solid is measured as the solid is immersed in or withdrawn from a liquid of known surface tension (Figure C.3).

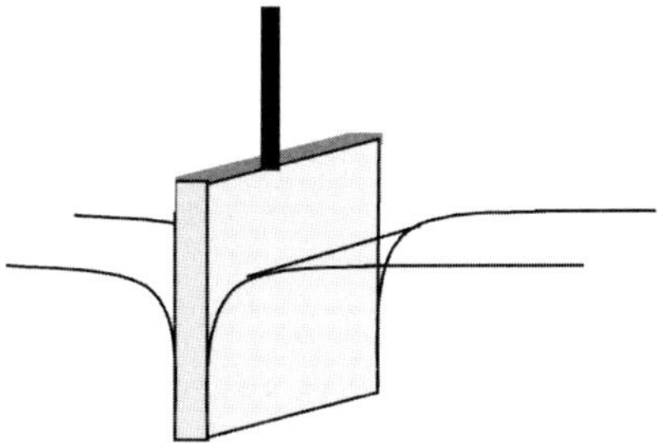

Figure C.3 Schematic of the dynamic Wilhelmy method.

4. The single-fiber Wilhelmy method can be applied to single fibers for measuring the advancing and receding contact angles.

References

1 T. Young (1805). An Essay on the Cohesion of Fluids. *Philos. Trans. R. Soc. Lond.* **95**, 65–87.

2 T. S. Chow (1998). Wetting of rough surfaces. *J. Phys.: Condens. Matter* **10** (27), 445.

3 P.G. de Gennes. Wetting: statics and dynamics. *Rev. Mod. Phys.* 57, 3 (part I), July 1985, pp. 827–863.

Appendix D
Ellipsometry

Ellipsometry is a nondestructive optical technique that allows one to determine the thickness of, e.g., organic thin films with subnanometer sensitivity in the surface normal direction and the optical constants (refractive indices of the material). In addition, by employing suitable models, surface roughness, composition and optical anisotropies can be unraveled.

In practice, a monochromatic light beam with known state of polarization (e.g., linear polarization, see Figure D.1) is reflected off the sample surface in specular reflectance (Figure D.1). Here one can define an angle of the incoming light as θ_i, the plane spanned by the incident and reflected light defines the p plane, while the s-plane is perpendicular to that plane. The state of polarization is changed for p-polarized light in a different manner from that of s-polarized light, resulting in an elliptical polarization of the reflected light.

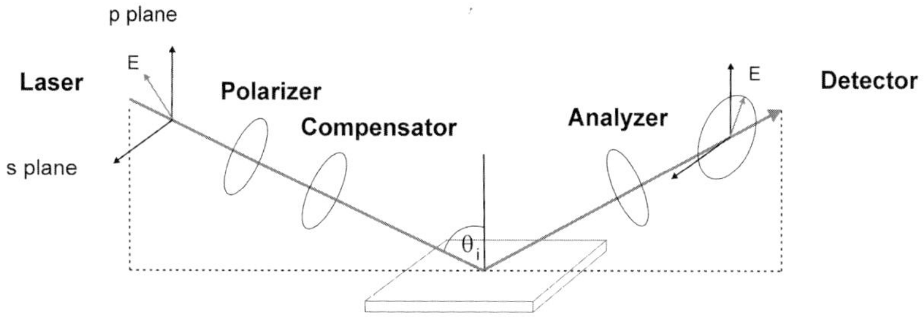

Figure D.1 Geometry of an ellipsometry experiment and important components of the ellipsometer.

Ellipsometry measures this change in polarization of the reflected light. The reflected light is characterized by two reflection coefficients, R_s and R_p, respectively. In the experiment one determines the ellipsometry angles Ψ and Δ that are related to the ratio of R_p and R_s

$$\tan(\Psi)\, e^{i\Delta} = R_p \setminus R_s$$

Surface Design: Applications in Bioscience and Nanotechnology
Edited by Renate Förch, Holger Schönherr, and A. Tobias A. Jenkins

ISBN: 978-3-527-40789-7

In practice, the measurement can be performed in different measurement schemes, namely null-ellipsometry, rotating analyzer or rotating polarizer, as well as phase-modulation mode.

Variations of the wavelength and the angle of incidence generate sufficiently large data sets that can be used to calculate the required data. These calculations are based on an optical model that accounts for all the layers that are present in the sample. Typically, the experimental data are fitted to the model to obtain, e.g., the best-fit value for the film thickness. For thin organic films on solid supports the refractive index of the film material must be known to calculate the desired thickness. The high accuracy of ellipsometry is based in part on the measurement of the ratio of the two values of the reflection coefficients.

References

1 R. W. Collins, D. L. Allara, Y.-T. Kim, Y. Lu, J. Shi in *Characterization of Thin Organic Films*, A. Ulman, ed., pp 35, Butterworth-Heineman, Boston, **1994**.

2 J. L. Keddie, *Curr. Opin. Colloid Interf. Sci.* **2001**, *6*, 102.

3 J. A. Woollam, B. Johs, C. Herzinger, J. Hilfiker, R. Synowicki, C. Bungay, *SPIE Proceedings* **1999**, *CR72*, 3.

Appendix E
Fourier Transform Infrared Spectroscopy

Infrared spectroscopy is an experimental tool for obtaining information about the chemical structure of materials. Molecules can absorb infrared light of frequencies that match the energy of changes in molecular vibrations. Infrared spectroscopy is therefore often also called *vibrational spectroscopy*. In the context of infrared spectroscopy, wavelength is measured in 'wavenumbers', which have the units cm^{-1}. The most useful I.R. region lies between 4000–670 cm^{-1}.

This absorption of light can be very useful because a particular functional group (such as O–H bonds, C=O bonds...) will absorb infrared light within a small range of frequencies regardless of other structural features. So, for example, all compounds containing carbonyls will show absorptions at the same place in an infrared spectrum.

Photon energies associated with this part of the infrared (from 1–15 kcal/mole) are not large enough to excite electrons, but may induce vibrational excitation of covalently bonded atoms and groups. The covalent bonds in molecules are not rigid sticks or rods, but are like stiff springs that can be stretched and bent. In addition to the rotation of groups about single bonds, molecules experience a wide variety of vibrational motions, characteristic of their component atoms. Consequently, virtually all organic compounds will absorb infrared radiation that corresponds in energy to these vibrations. Infrared spectrometers permit chemists to obtain absorption spectra of compounds that are a unique reflection of their molecular structure.

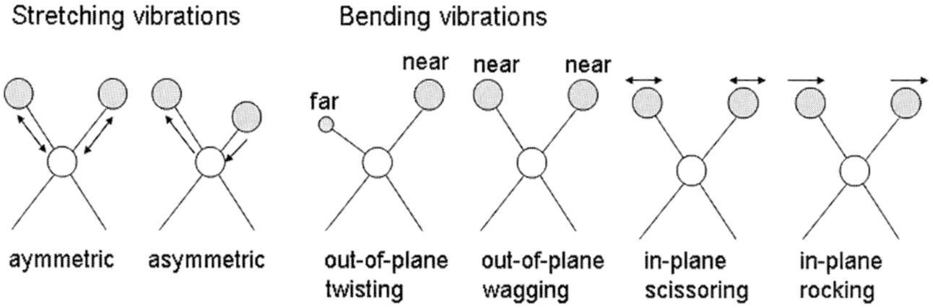

Figure E.1 Some typical stretching and bending vibrations in molecules.

Surface Design: Applications in Bioscience and Nanotechnology
Edited by Renate Förch, Holger Schönherr, and A. Tobias A. Jenkins

ISBN: 978-3-527-40789-7

Infrared spectra may be obtained from samples in all phases (liquid, solid and gaseous). Liquids are usually examined as a thin film sandwiched between two polished salt plates. If solvents are used to dissolve solids, care must be taken to avoid obscuring important spectral regions by solvent absorption. Perchlorinated solvents such as carbon tetrachloride, chloroform and tetrachloroethene are commonly used. Alternatively, solids may either be incorporated in a thin KBr disk, prepared under high pressure, or mixed with a little nonvolatile liquid and ground to a paste (or mull) that is smeared between salt plates. For the analysis of thin films in the transmission mode double-sided polished Si wafers (which are transparent to IR) are often used. Here, the limit in analyzable thickness is around 100 nm in modern spectrometers.

IR spectroscopy is also possible on a surface or an ultra thin surface layer. The IR beam can be reflected off an underlying metal substrate in the grazing incidence mode, also called infrared reflection absorption spectroscopy, IRRAS. This allows for the characterization of monolayer thin films with excellent sensitivity.

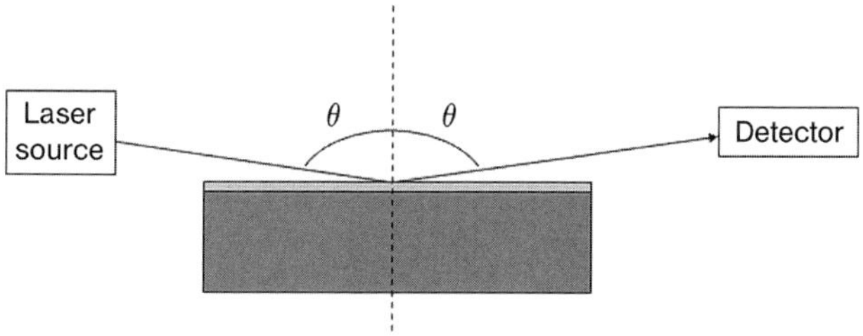

Figure E.2 Schematic of IRRAS.

Another advantage of the technique is that because of the grazing incident angle there is no bulk signal to worry about. The surface signal is readily distinguishable from gas-phase absorptions using polarization effects.

Alternatively, the IR beam may be introduced via a crystal to give multiple reflections at the interface between two dielectrics (attenuated total reflection, ATR). In this case the material to be analyzed may be either deposited directly onto the crystal or it may be pressed against it. Alternatively, the ATR may occur at a metal/dielectric interface as shown in Figure E.3.

The totally internally reflected laser light couples with electrons in the gold and creates a finite electric field on the metal surface that decays exponentially away from the surface. The advantages of this are:

- sensitivity – can measure monolayers;
- information on molecular functionality and surface orientation;
- all solvent problems.

The same principles of IR in bulk, gas or liquids apply on surfaces: i.e. the functional groups must change their net dipole moment on vibration to be infrared active. Thus, in most cases, the same absorption bands can be observed. Tables of typical wave numbers are readily available online and in numerous books.

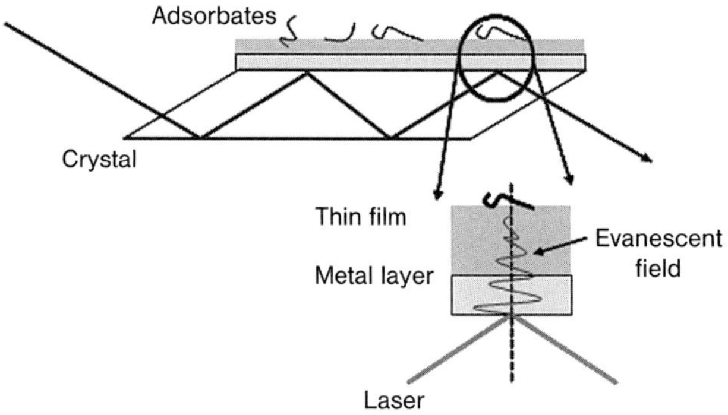

Figure E.3 Schematic of ATR-FTIR.

Further Reading

1 Peter R. Griffiths, James A. De Haseth. Fourier Transform Infrared Spectrometry, 2nd edn, Wiley-Interscience, 2007, ISBN 978-0471194040.

2 Brian C. Smith. Fundamentals of Fourier Transform Infrared Spectroscopy. CRC, 1st edn, 1995, ISBN 978-0849324611.

3 Brian C. Smith. Infrared Spectral Interpretation: A Systematic Approach. CRC, 1st edn, 1998, ISBN 978-0849324635.

Appendix F
Impedance Spectroscopy

Impedance spectroscopy on thin-film surface architectures can yield a great deal of information about the film, particularly the film resistance and coverage. In the context of this book, its description and utilization will be confined to considering how it can be used, and what can be measured, when studying thin organic or inorganic films on conducting metal surfaces such as gold.

Impedance measurements on thin films are usually made using two or three electrodes in a classical electrochemical cell. The underlying metal substrate is used as the working electrode, and counter and reference electrodes are also employed. A sinusoidal potential wave form, typically of around 10 mV amplitude is applied between the working and counter electrodes. The resultant current that flows is measured and the difference in potential amplitude ΔE and current amplitude Δi, together with any phase shift between the potential and current recorded. In general, many data points over a range of applied frequencies are measured, with the resultant spectrum being plotted and analyzed. Impedance, given the symbol Z, is a generalized form of resistance, so we can write:

$$Z = \frac{\Delta E}{\Delta i} \qquad \text{and a phase term, } \phi$$

Impedance is a vector quantity, containing both a magnitude, Z and a direction ϕ. Consequently, it is often convenient to plot impedance spectra on a complex plane, also called a Nyquist plot, or plotting impedance and phase separately but as functions of frequency in a Bode plot. The impedance response of a thin film, measured as a function of frequency can be fitted to an electrical circuit model of that film. In general, most thin films can be modeled (with varying degrees of simplification) as a combination of resistors and capacitors. To obtain a better understanding of how measured impedance relates to the properties of the thin film, it is often useful to fit to an electrical circuit model, which can allow one to obtain more precise information about a surface.

Surface Design: Applications in Bioscience and Nanotechnology
Edited by Renate Förch, Holger Schönherr, and A. Tobias A. Jenkins

ISBN: 978-3-527-40789-7

F.1
Electrical Equivalent Circuit Models of Interfaces

These models are all simplifications of reality, but allow useful information about interfaces and thin films to be obtained. The inherent simplification of this approach should be kept in mind when using impedance spectroscopy to measure systems.

F.1.1
Resistor and Capacitor in Series

The RC series model is the simplest model of an interface, and is often used on two distinct systems: to model a noble metal such as gold in a nonredox media such as sodium chloride solution, or as a model of a fully intact, pinhole-free film on a metal. In both cases, the resistance term relates to the resistance of electrolyte between the electrode and the reference electrode, while capacitance relates to the double-layer capacitance of the metal and the capacitance of the thin film, respectively, in the two cases. Although a simple model, measurement of capacitance change over time can be a useful way of following thin-film formation on electrodes, as will be shown later.

F.1.2
Resistor in Series with a Parallel Resistor and Capacitor (RC)

This model is commonly used to model metal surfaces that are in a redox active electrolyte, or surfaces that are dissolving/corroding. The series resistance again models the electrolyte resistance between the working and reference electrode, and the capacitor also models the double-layer capacitance. However, the parallel resistor is now inversely related to the rate of electron transfer taking place at the electrode interface, wither by reduction/oxidation of the redox mediator or the oxidation of the electrode itself.

More complex equivalent circuit models exist and are discussed in the literature. These include models of passive and redox active thin films, diffusion through thin films and nonideal capacitive effects. A good general rule when using equivalent circuit analysis is only to model what is actually obtained in the spectrum and to keep models as simple as possible. Very complex models may yield better mathematical fits, but may not give physically meaningful information.

Two further points of advice for persons using impedance spectroscopy are: if possible, measure how the impedance changes with time, and use at least one further surface analytical method to monitor the surface.

F.2 A Case Study: Impedance Study of Lipid Vesicle Fusion on a Hydrophilic Self-Assembled Monolayer (SAM)

In this experiment, unilamella egg-phosphatidylcholine lipid vesicles were fused onto a hydrophilic SAM. Impedance spectroscopy was used to follow the change in the film capacitance as the adlayer formed. A schematic of the experiment is shown in Figure F.1, and the results in Figure F.2.

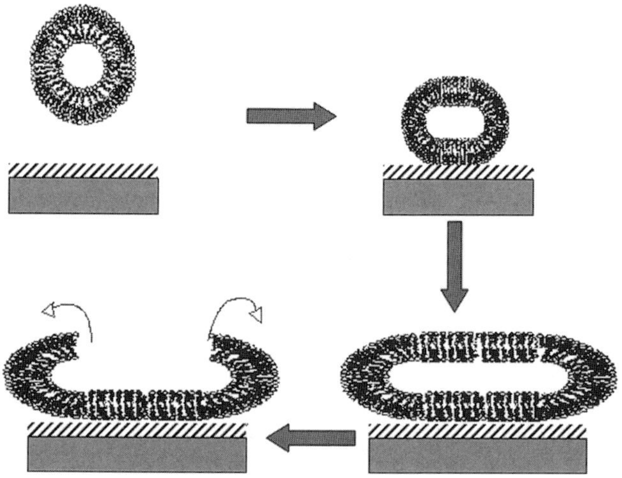

Figure F.1 Schematic of vesicle fusion on a hydrophilic SAM.

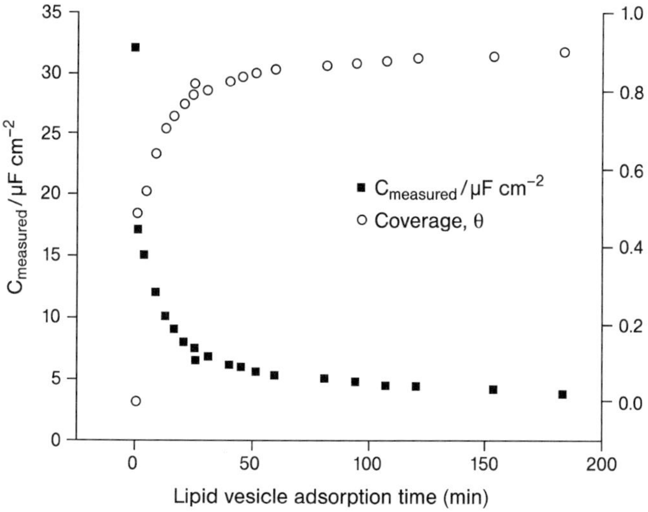

Figure F.2 The change in capacitance of the system as vesicles fused on the SAM, and calculated lipid film coverage.

The surface coverage of the lipid film was estimated by assuming that the measured capacitance was the sum of SAM capacitance, where this was not covered by lipid (C_{SAM}) and the capacitance of the lipid covered SAM (C_{lipid}), which will always be lower than C_{SAM} since the dielectric thickness, d, of the SAM–lipid film is larger than the SAM, and $C\ d^{-1}$. This is discussed in more detail in the references given below.[1,2]

Further Reading

1 A.T.A. Jenkins, J. Hu, Y.Z. Wang, S. Schiller, R. Förch, W. Knoll. Pulse Plasma Deposited Maleic Anhydride Thin Films as Supports for Lipid Bilayers. *Langmuir* **16**, 6381–6384 (2000).

2 A.T.A. Jenkins, H. Schönherr, S.D. Evans, R.E. Miles S.D. Ogier, G.J. Vancso. Retention of Protein Activity in Biomimetic Lipid Bilayers Supported by Microcontact Printed Lipophilic Self-Assembled Monolayers. *J. Am. Chem. Soc.* **121**, 5274–5280 (1999).

3 A.J. Bard, L.R. Faulkner, *Electrochemical Methods – Fundamentals & Applications*, 2nd edn, Wiley, 2001.

Appendix G
Scanning Electron Microscopy

In scanning electron microscopy (SEM) the sample surface is raster-scanned in high vacuum with a focused beam of electrons. Secondary or backscattered electrons are detected to provide an image of the sample surface with up to nanometer-scale resolution. In addition to structural imaging, the elemental composition of the surface can be assessed from characteristic X-rays that are emitted (Figure G.1).

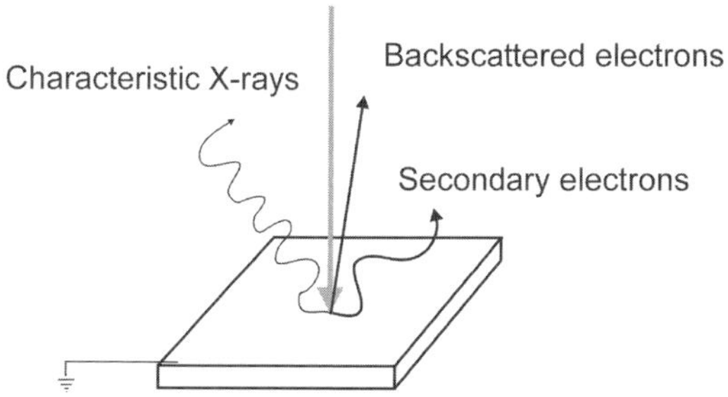

Figure G.1 Rudimentary processes that occur upon impingement of an electron beam onto a sample surface in SEM.

In general, sample surfaces for SEM analysis require pretreatment to render them clean and conductive to prevent the accumulation of charges in the specimens. This pretreatment may consist of a thin coating of metals, such as gold or platinum, or of graphite. In modern environmental SEM the sample chamber may be at higher pressure than the rest of the microscope, thereby allowing for imaging at lower voltages and in some case under more realistic conditions (e.g. hydrated samples).

The advantage of a very large depth of field is compromised in some routine applications by the mentioned need for sample preparation that effectively limits studies of biological processes. However, most materials can very well be analyzed. If e-beam damage becomes a problem, SEM can also be performed at cryogenic temperatures.

Due to its surface sensitivity, the secondary electron imaging mode is widely used. In this mode, low-energy secondary electrons (<50 eV) are detected that

Surface Design: Applications in Bioscience and Nanotechnology
Edited by Renate Förch, Holger Schönherr, and A. Tobias A. Jenkins

ISBN: 978-3-527-40789-7

were ejected from specimen atoms by inelastic scattering interactions with primary electrons. The low energy hence leads to a limited escape depth of only a few nanometers.

Backscattered electrons possess much higher energy compared to secondary electrons and allow one to image with atomic number contrast (relation to Z). High atomic number elements backscatter the primary electrons more effectively compared to lighter elements, hence the contrast is related to atomic mass/number.

Characteristic X-rays are being emitted from atoms at depths of up to a micrometer upon interaction with the incident electrons. The ejection of photoelectrons is followed by subsequent rearrangements of the electronic shells of the energetically excited atoms. One pathway for deactivation of the excited state is the emission of a characteristic X-ray photon. To analyze the X-rays the SEM must be equipped with an energy dispersive X-ray detector (EDX). Since the escaping photon is not charged, like for instance the mentioned secondary electrons, the escape depth is considerable.

Reference

1 L. C. Sawyer, D. T. Grubb, G. F. Meyers, *Polymer Microscopy: Characterization and Evaluation of Materials*, 3rd edn, Springer, Berlin, 2008.

Appendix H
Surface Plasmon Resonance

H.1
Introduction

Surface plasmon resonance (SPR) is an established method for quantifying the attachment of molecules to surfaces, such as in the formation of self-assembled monolayers on metals, protein–protein interaction, DNA hybridization and a range of other interactions, where a mobile phase interacts with a surface-bound phase.

H.2
Methodology

The SPR phenomenon can be understood at various levels of physical complexity. For most users, it is probably sufficient to appreciate that light can be coupled into interacting with the free electron 'gas' in a thin metal film (usually gold or silver). When the light wave vector matches the electron 'gas' or surface plasmon wave vector, resonance occurs. This is illustrated in Figure H.1 and Equations (1) and (2).

Incident light has a wave vector: $K_{\text{in}}{}^{\omega}/_{c}$ where ω is its frequency and c its speed. In general, it is necessary to couple light into the plasmon wave in the thin metal film, either via a prism or a grating. This discussion will focus on the most common approach that utilizes a prism in the so-called Kretschmann configuration (Figure H.1).

Only the incident light that has a component in the plane of the metal interacts with the surface plasmon in the metal, so we can see by simple trigonometry that the incident light wave vector in the plane of the metal is given by:

$$K_{\text{in}}(x) = \frac{\omega}{c}\varepsilon_{\text{p}} \sin\theta \qquad (1)$$

ε_{p} is the dielectric permittivity, (the square of the refractive index) of the prism. The wave vector of the surface plasmons within the metal film depends on the dielectric permittivity of the metal (ε_{m}) and the dielectric itself (ε_{d}).

Surface Design: Applications in Bioscience and Nanotechnology
Edited by Renate Förch, Holger Schönherr, and A. Tobias A. Jenkins

ISBN: 978-3-527-40789-7

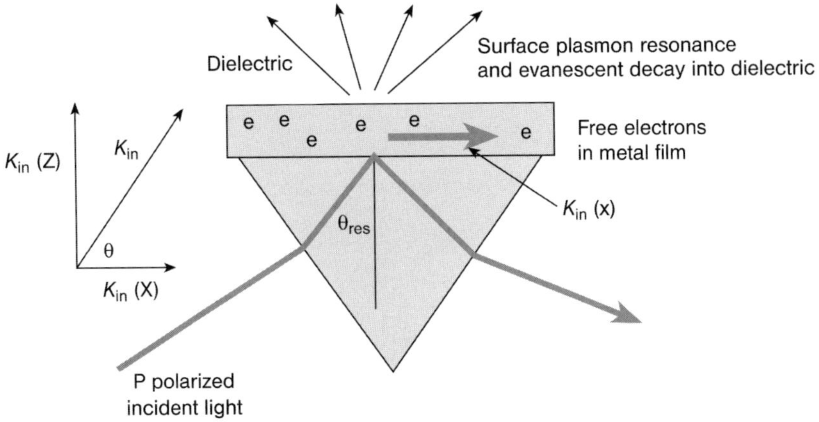

Figure H.1 SPR showing light coupled into thin metal film via a prism.

Hence, we can write:

$$K_{SP} = \frac{\omega}{c}\left(\frac{\varepsilon_m \varepsilon_d}{\varepsilon_m + \varepsilon_d}\right)^{1/2} \tag{2}$$

When $K_{in} = K_{SP}$, surface plasmon resonance takes place. In this situation, the incident light is no longer totally internally reflected by the prism, but instead is dispersed into an evanescently decaying energy field within the dielectric.

When carrying out an experiment on a metal surface (within the evanescent decay field of around 200 nm for 633 nm light on gold), the binding of a molecule or protein increases ε_d. Hence to get the matching condition $K_{in} = K_{SP}$, incident light must be coupled in at a higher angle θ. This is illustrated in Figure H.2, showing the characteristic reflection–incident angle curve obtained in an SPR experiment and the shift in the resonance angle upon binding of a thiol moiety.

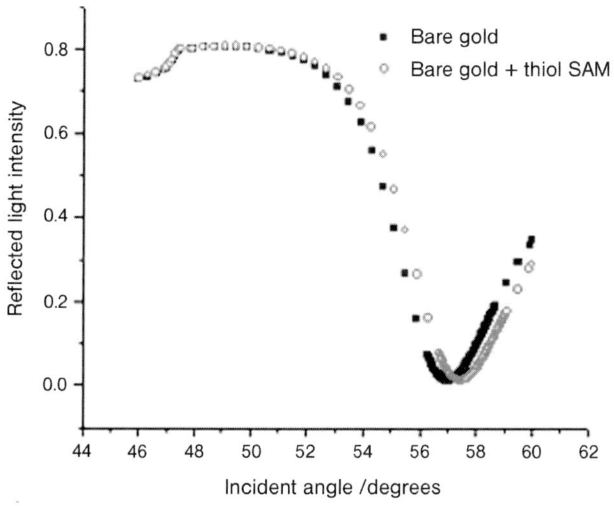

Figure H.2 Shift in the angle of surface plasmon resonance for a surface prior to binding of a thiol moiety on gold (resonance minimum on left) and following (resonance angle shifts to the right).

The binding can be followed in real time and modeled to obtain parameters relating to the thickness and/or mass density of the adsorbed layer. These kinetic measurements can themselves be fitted to surface binding isotherms such as the Langmuir isotherm. From this it is possible to calculate the equilibrium dissociation constant K_D between a surface bound species and a species in solution.

Further Reading

1 Handbook of Surface Plasmon Resonance; Richard B. M. Schasfoort, Anna J. Tudos (eds). RSC Publishing, 2008.

See also references of Chapters 1.2 and 4.5.

Appendix I
Optical Waveguide Spectroscopy (OWS) – μm-Thick Films

Optical waveguide mode spectroscopy takes advantage of the ability of transparent films to guide optical waves through the film. The basic structure of an optical waveguide is a layer of transparent material with a refractive index (n), which is higher than its surroundings. The light is confined and propagates with low attenuation. The optical coupling depends on the dimensions and refractive indices of the different parts in the waveguide system. This makes optical waveguiding suitable for sensor applications. For example, adsorption of a small amount of analyte on the surface of the waveguiding layer can lead to a large change in the optical coupling conditions. Analysis of the coupling conditions (incident angle, and per cent reflected light) gives quantitative information about the adsorbed analyte. Only light with specific momentum and energy may propagate in a waveguide. Depending on the thickness and refractive indices of the waveguide system, more than one "mode" of such guided light may be allowed. Analysis of multiple guided modes can give information on the anisotropic or structural properties of the adsorbed analyte layer.

OWS refers to the application of waveguide spectroscopy for sensing purposes using a 1-dimensional slab waveguide and using prism coupling through a semitransparent metal film in the Kretschmann configuration [1]. Figure I.1 shows the general coupling and detection scheme: The dielectric thin-film waveguide (typically >500 nm) is prepared on top of a semitransparent metal layer (e.g. 50 nm Au), which in turn is deposited on a glass substrate with a refractive index identical to the prism. Optical modes in the dielectric waveguide are confined in the thickness direction by the Au layer and the medium on top of the sample (e.g. air or aqueous buffer). The prism (with refractive index higher than the waveguide layer) is optically joined to the back of the index-matched glass substrate with immersion oil. Laser light is directed through the prism/glass substrate assembly and impinges at an angle on the substrate side of the semitransparent metal layer.

The momentum of the incidence light is defined by the refractive index of the prism, and the component parallel to the surface can be adjusted by changing the incidence angle. At most angles, light is reflected from the back of the metal film and detected by a photodiode. Past the critical angle of total internal reflection (θ_{TIR}), light is coupled into the waveguide at specific combinations of wavelengths

Surface Design: Applications in Bioscience and Nanotechnology
Edited by Renate Förch, Holger Schönherr, and A. Tobias A. Jenkins

ISBN: 978-3-527-40789-7

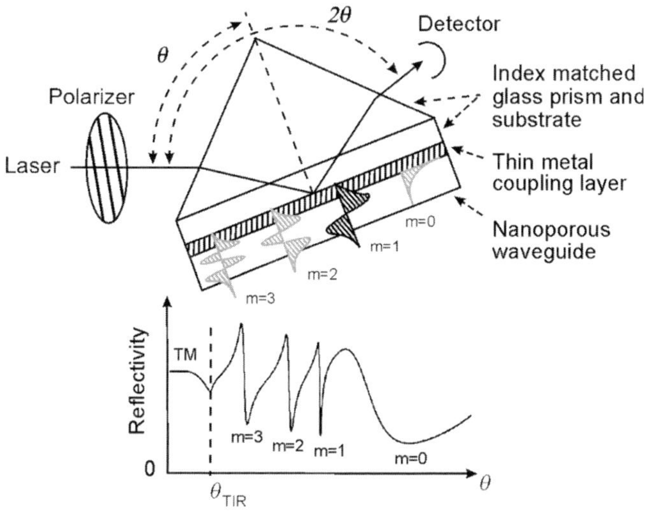

Figure I.1 Schematic of optical coupling in the Kretschmann configuration and the waveguide spectroscopy detection scheme. θ_{TIR} is the critical angle for total internal reflection. The *R* vs. θ trace indicates a measurement of TM modes, for which the electric-field distributions are shown overlaid on the waveguiding thin film.

and incidence angles if there is a match in momentum between the incidence light and a waveguide mode. The different modes are referred to by the number of nodes in the propagating electric field and are excited in turn as the incidence angle (θ is scanned in a θ –2θ configuration. If light is coupled into the waveguide and propagates through the dielectric thin-film layer, optical damping by the metal layer dissipates the energy as heat and little light can be backcoupled by the symmetric prism. Thus, waveguide modes show up as sharp intensity minima in the reflected intensity (*R*) in a *R* vs. θ scan. Different sets of waveguide modes can be excited for light polarized in different orientations. Modes with the magnetic field in the plane of the film are transverse magnetic (TM) modes, while those with the electric field polarized in the plane of the film are transverse electric (TE) modes. Further information on the Kretschmann configuration and OWS, are given in several reviews [1–3]. As a matter of convenience, we typically excite waveguide modes at only one wavelength, using a helium-neon red laser operating at 633 nm.

R vs. θ measurements can be compared to Fresnel calculations [1, 4, 5], which are solutions to the Maxwell equations for the layer system in the Kretschmann configuration. The thickness and refractive index of the waveguide layer can thus be obtained. The optical effect of analyte binding is the addition of a layer of dielectric material with the analyte's refractive index on top of the waveguide and a thickness proportional to the surface coverage. The presence of this additional layer shifts the waveguide mode coupling conditions and shifts the incidence angles for mode coupling (the reflectivity minima). These shifts can be tracked in real time by following the angle minima or the reflected intensity at an angle near a reflectivity minimum (Figure I.2). The shifts are also analyzed with Fresnel calculations and the thickness and refractive index of the analyte layer can be obtained.

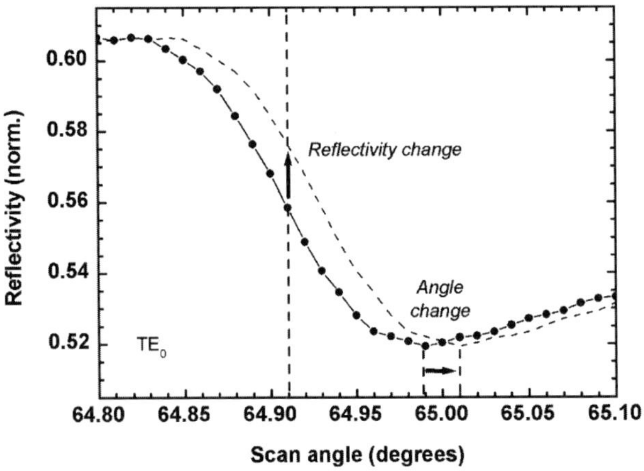

Figure I.2 Symbols and solid line: TE_0 mode of an nanoporous alumina waveguide (see Chapter 4.5). Dashed line: angle shift of the mode in response to surface binding of molecules. The arrows indicate how the shift can be monitored in real time.

Further Reading

1 Knoll, W., Annual Review of Physical Chemistry 1998, 49(1), 569.
2 Hunsperger, R. G., Integrated Optics – Theory and Technology, 5th edn, Springer-Verlag, Berlin, Heidelberg, 2002.
3 Kogelink, H., Theory of dielectric waveguides. In: Integrated Optics, Tamir, T. (ed.). Springer-Verlag, Berlin, Heidelberg, 1975, p. 13.
4 Fowles, G. R., Introduction to Modern Optics, 2nd. edn. Dover Publications, Inc., New York, 1989, p. 328.
5 Raether, H., Surface-Plasmons on Smooth and Rough Surfaces and on Gratings, Vol. 111, 1988.

Appendix J
Waveguide Mode Spectroscopy (WaMs) – nm-Thick Films

Optical waveguide mode spectroscopy can also be used to characterize a very much thinner adlayer on top of the waveguide. In this case very thin planar waveguides of high refractive index with TE0 and TM0 modes are used (Figure J.1). By binding an adlayer to the surface of the waveguide the refractive index of the surrounding medium changes in the vicinity of the evanescent field of the waveguide modes. This in turn induces changes in the effective refractive indices N_{TE0} and N_{TM0} of the guided modes.

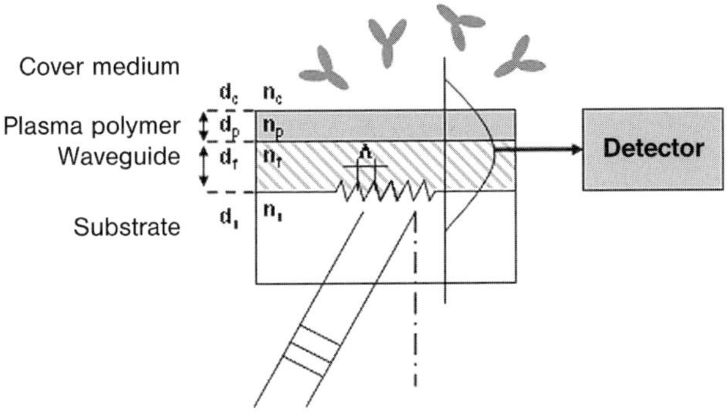

Figure J.1 Schematic of WaMs.

The determination of the propagation constants of waveguide modes is an established procedure for the evaluation of the layer thickness and refractive index of a waveguide. It is based on the selective excitation of waveguide modes by means of a prism or a diffraction grating. The refractive index and the thickness of the layer can be calculated from the measured effective refractive indices of the different modes ($N_{\mathrm{eff}\ i,j..}$ where i, j.. are the mode numbers) using the dispersion relationship, with the assumption that the refractive-index distribution within the layer is in the form of steps. For this procedure the waveguide must be thicker than the wavelength of the laser used in the experiment. The effective refractive index provides the phase velocity of the guided mode ($\varpi = c\ /\ N$, where $c =$ is the velocity of light in a vacuum) and depends on the polarization (TM0 and TE0) and the mode number. Changes in the effective refractive index, ΔN_{eff}, can be mea-

Surface Design: Applications in Bioscience and Nanotechnology
Edited by Renate Förch, Holger Schönherr, and A. Tobias A. Jenkins

ISBN: 978-3-527-40789-7

sured using a grating coupler. A laser beam can be coupled into the planar waveguide at an angle of incidence α if the in-coupling condition:

$$N_{\mathrm{eff}} = n_{\mathrm{C}} \sin \alpha + I \lambda / \Lambda$$

is satisfied. n_{C} is the refractive index of the covering medium (air or solvent), λ the wavelength of the laser used, I the diffraction order and Λ the grating period. Thus, if, for example, a plasma polymer is deposited as an adlayer onto the waveguide, then its optical parameters (thickness d_{a}, and refractive index n_{a}) change the effective refractive indices N_{eff} of the waveguide modes. The effective index shift ΔN_{eff} can be determined from (2) above, by measuring the change $\Delta\alpha$. From the effective index shifts $\Delta N_{\mathrm{eff\ TE0}}$ and $\Delta N_{\mathrm{eff\ TM0}}$ the thickness d_{a} and refractive index n_{a} of the plasma polymer adlayer can be calculated.

Alternatively to the prism coupling described for OWS, a grating with periodicity Λ, incorporated into the waveguide can used to couple a HeNe laser beam (λ = 632.8 nm) into the waveguide. N_{eff} can be determined by scanning the angle of incidence of the incoming laser beam onto the grating, while the in-coupled power is measured by two photodetectors situated at both ends of the waveguide. Such waveguides are designed for optimal sensitivity for the zeroth transversal electric, TE_0, and transversal magnetic, TM_0, modes.

Further Reading

1 M.T. van Os, B. Menges, R. Förch, W. Knoll, G.J. Vancso, Chem Mater, 1999, 11(11), 3252–3257.

2 Z. Zhang, B. Menges, R. B. Timmons, W. Knoll, R. Förch, Langmuir, 2003, 19, 4765–4770.

Appendix K
X-ray Photoelectron Spectroscopy (XPS)

X-ray photoelectron spectroscopy, XPS, or Electron Spectroscopy for Chemical Analysis, ESCA, is probably one of the most widely used surface analytical tools for the analysis of material surfaces and the qualitative and quantitative determination of surface composition. It typically probes a depth of around 5–10 nm depending on the nature of the material and can be used to identify all elements except hydrogen. X-rays are used to probe the substrate and electrons from the core levels leaving the surface are analyzed using an electron analyzer. The technique is not completely *surface specific*, in that whilst *most* of the signal comes from within a few atomic layers of the surface, a small part of the signal comes from much deeper within the solid. The method is probably best described as being a *surface-sensitive* technique.

Photoelectron spectroscopy is based upon a single photon-in/electron-out process.

The energy of the photon is given by the Einstein relation:

$$E = h\,\nu$$

where h is the Planck constant (6.62×10^{-34} J s) and ν is the frequency of the radiation in Hz. The photon is absorbed by an atom in a molecule or solid, leading to ionization of the atoms and the emission of a core (inner shell) electron. Since energy can neither be created nor destroyed, the energy going in must equal the energy going out. Thus:

$$h\,\nu = E_B + E_K$$

or the kinetic energy

$$E_K = h\,\nu - E_B$$

A typical XPS spectrum contains information on both the X-ray-excited photoelectrons as well as the X-ray-excited Auger electrons.

Surface Design: Applications in Bioscience and Nanotechnology
Edited by Renate Förch, Holger Schönherr, and A. Tobias A. Jenkins

ISBN: 978-3-527-40789-7

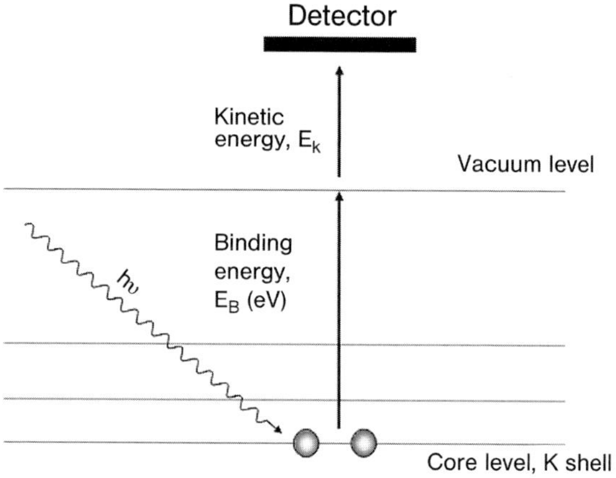

Figure K.1 Schematic of the electronic transitions during XPS.

For each and every element, there is a characteristic binding energy associated with each core atomic orbital, such that each element will give rise to a characteristic set of peaks in the photoelectron spectrum at kinetic energies determined by the photon energy and the respective binding energies. The intensity of the peaks is related to the concentration of the element within the sampled region. Thus, the technique provides a *qualitative and quantitative analysis of the surface composition*. However, when comparing XPS lines from the same element in different chemical environments shifts in the peak positions can be observed. These are a result of a rearrangement of the average charge distribution of the valence electrons. For the interpretation of chemical shifts sophisticated calculations have been carried out that are based on theoretical and semitheoretical models [1–6].

The degree of surface sensitivity of an electron-based technique such as XPS may be varied by collecting photoelectrons emitted at different emission angles to the surface plane. This approach may be used to perform nondestructive analysis of the variation of surface composition with depth. Modern instruments can also rapidly collect X-ray photoelectron images with a lateral resolution of approximately 1 micrometer and sufficient spectral resolution to discern, for example, carbidic and graphitic carbon domains in a few minutes of acquisition time.

The basic requirements for a photoemission experiment (XPS or UPS) are:

1. A source of fixed-energy radiation. Typical sources are AlK_α (1486.6 eV) and MgK_α (1253.6 eV).
2. An electron energy analyzer (which can disperse the emitted electrons according to their kinetic energy, and thereby measure the flux of emitted electrons of a particular energy).
3. A high-vacuum environment (to enable the emitted photoelectrons to be analyzed without interference from gas-phase collisions).

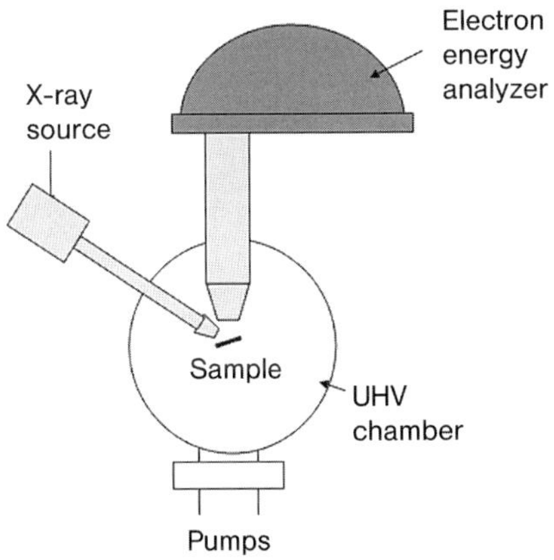

Figure K.2 Schematic of an XPS setup.

XPS can be carried out on virtually all flat samples with a maximum resolution of a few micrometers. It is particularly well suited for metallic and inorganic materials, but can also be used for the analysis of insulating materials such as polymers. In the latter case special care needs to be taken to reduce and control surface charging effects. During the analysis of polymers and other organic materials it is usually the C1s, O1s, N1s and F1s peaks that are of interest. The majority of information about the material is obtained by careful analysis of the C1s peak profile, which provides information about the chemical environment of the different carbon atoms. An example of this can be seen in Figure K.3.

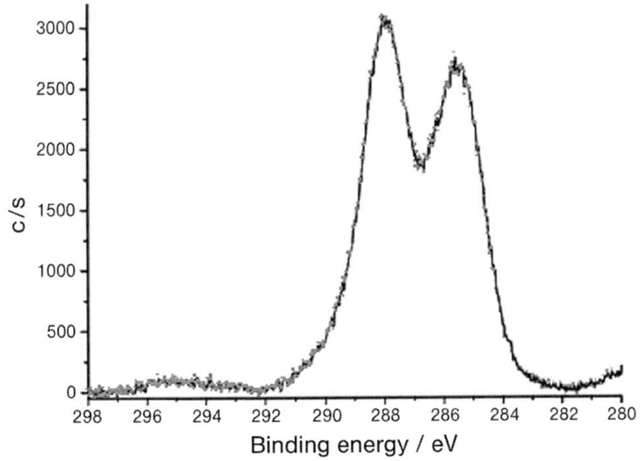

Figure K.3 C1s XPS peak of a fluorinated ester polymer.

Further Reading

1 D. Briggs, M.P. Seah. Practical Surface Analysis: Auger and X-Ray Photoelectron Spectroscopy, 2nd edn, John Wiley & Sons Ltd, 1996, ISBN 978-0-471-95340-1.

2 D. Briggs, J.T. Grant. Surface Analysis by Auger and X-Ray Photoelectron Spectroscopy, IMPublications, ISBN 1-901019-04-7.

3 G. Beamson, D. Briggs. High resolution XPS of organic polymers – the scienta ESCA 300 data base, John Wiley & Sons, Ltd, Chichester, 1992, ISBN 978-0-471-93592-1.

4 N.H. Turner (1992) SIA, **18**, 47–51.

5 N.H. Turner, A.M. Single (1990) SIA, **15**, 215–222.

6 C.I. de Matteis, M.C. Davies, S. Leadlty, D.E. Jackson, G. Beamson, D. Briggs, J. Heller, N.M. Franson (1993) *J. Electron Spectroscopy amd Related Phenomena* **63**(3), 221–238.

Index

Surface Design: Applications in Bioscience and Nanotechnology
Edited by Renate Förch, Holger Schönherr, and A. Tobias A. Jenkins

ISBN: 978-3-527-40789-7

d

q